Kurt Gödel
COLLECTED WORKS
Volume IV

Wien 17./VI. 1932

Sehr geehrter Herr Church!

In einem Gespräch mit Prof. Veblen über Ihre vor kurzem in den Ann. of. Math. 33 erschienene Arbeit, haben sich die beiden folgenden Fragen ergeben, von denen Prof. Veblen meinte, es wäre vielleicht nützlich Ihnen davon Mitteilung zu machen:

1. Wie ist es in Ihrem System möglich absolute Existenz-sätze z.B. das Unendlichkeitsaxiom zu beweisen?

2. Falls das System widerspruchsfrei ist, wird es dann nicht möglich sein, die Grundbegriffe in einem System mit Typentheorie bzw. im Axiomensystem der Mengenlehre zu interpretieren, und kann man überhaupt auf einem anderen Wege als durch eine solche Interpretation die Widerspruchsfreiheit plausibel machen?

Ich übersende Ihnen gleichzeitig einige Separata meiner Arbeiten.

Ihr ergebener   Kurt Gödel

Kurt Gödel

COLLECTED WORKS

Volume IV
Correspondence A–G

EDITED BY
Solomon Feferman
John W. Dawson, Jr.
(Editors-in-chief)
Warren Goldfarb
Charles Parsons
Wilfried Sieg

Prepared under the auspices of the
Association for Symbolic Logic

CLARENDON PRESS • OXFORD

OXFORD
UNIVERSITY PRESS

Great Clarendon Street, Oxford OX2 6DP
Oxford University Press is a department of the University of Oxford.
It furthers the university's objective of excellence in research, scholarship,
and education by publishing worldwide.

First published 2003

First published in paperback 2014

Published in the United States of America by Oxford University Press
198 Madison Avenue, New York, NY 10016, United States of America

ISBN 978-0-19-850073-5(Hbk)
ISBN 978-0-19-968961-3(Pbk)

Preface

This is the fourth volume of a comprehensive edition of the works of Kurt Gödel. Volumes I and II comprised all of his publications, ranging from 1929 to 1936 and from 1937 to 1974, respectively. Volume III consisted of a selection of unpublished papers and texts for individual lectures found in Gödel's *Nachlaß*, together with a survey of the *Nachlaß*. The present volume and its successor are primarily devoted to a selection of Gödel's scientific correspondence and the calendars thereto. In all cases our criterion for inclusion was that letters should either possess intrinsic scientific, philosophical or historical interest or should illuminate Gödel's thoughts or his personal relationships with others. Volume V, being published simultaneously with this one, also contains a full inventory of his *Nachlaß*.

There were several sources for the correspondence from which the selection in this and the following volume were made. The primary one was, of course, Gödel's *Nachlaß*; we have also solicited and obtained from other archives and from individuals (or their estates) copies of correspondence that filled gaps therein. The section on Permissions, below, contains a full list of all those sources, to whom we are, of course, greatly indebted. The total number of items of personal and scientific correspondence in the Gödel *Nachlaß* alone is around 3500, distributed over 219 folders.

In the main body of these two volumes we have selected correspondence with 50 individuals from the indicated sources. The most prominent correspondents among these are Paul Bernays, William Boone, Rudolf Carnap, Paul Cohen, Burton Dreben, Jacques Herbrand, Arend Heyting, Karl Menger, Ernest Nagel, Emil Post, Abraham Robinson, Alfred Tarski, Stanislaw Ulam, John von Neumann, Hao Wang and Ernst Zermelo. In addition, the reader will find in Appendix A to volume V several letters written on behalf of Gödel to others by Felix Kaufmann, Dana Scott and Hao Wang.

There are two major correspondents of Gödel who declined to allow us to publish their side of the exchanges, namely Paul Cohen and Georg Kreisel. In the latter case, except for one item from Gödel which is not included here, the correspondence was entirely a one-way street, but it is revealing of many topics of discussion of mutual interest, and thus we regret that it could not be represented. In the case of Cohen, as the reader will see, the nature of the correspondence could be fully reconstructed from the items found in Gödel's *Nachlaß*.

As in the first three volumes of this edition all the original material was written in German (sometimes in the Gabelsberger shorthand) or English, sometimes in both; those items originally in German are accompanied by facing translations. Credit for the work on the translations is contained in the Information for the reader below. Also as in the first three volumes, a

significant component is played by introductory notes and to a greater extent than previously these notes have been written by the editors. We are additionally indebted in this respect to Michael Beeson, Jens Erik Fenstad, Akihiro Kanamori, Øystein Linnebo, Moshe Machover and David Malament. The purpose of the notes themselves is to provide historical context to the correspondence, explain the contents to a greater or lesser extent, and, where relevant, discuss later developments or provide a critical analysis. Because of these requirements, several of the introductory notes turned out to be quite extensive; in those cases the reader is advised to consult them in tandem with the correspondence itself.

Once more our endeavor has been to make the full body of Gödel's work and thought as accessible and useful to as wide an audience as possible, without compromising the requirements of historical and scientific accuracy. As with the preceding volumes, this one is expected to be of interest and value to professionals and students in the areas of logic, mathematics, computer science, philosophy and even physics, as well as to many non-specialized readers with a broad scientific background. Naturally, even with the assistance of the introductory notes, not all of the material to be found here can be made equally accessible to such a variety of readers; nonetheless, the general reader should be able to gain some appreciation for what is at issue in the various exchanges.

Work on this volume and its successor was supported in its entirety by a grant from the Sloan Foundation, whose generosity and flexibility were indispensable to their successful completion. We are also grateful to the Department of Philosophy at Harvard University for its generous assistance with some last-minute expenses. In addition, by helpful arrangement with William Joyce and Donald C. Skemer of the Princeton University library system, and with Marcia Tucker of the library of the Institute for Advanced Study, the Foundation completely underwrote the preservation microfilming of Gödel's *Nachlaß*, in an effort to prevent further deterioration. That lengthy and delicate task was actually carried out by Preservation Resources Company of Bethlehem, Pennsylvania. As with the *Nachlaß* itself, one copy of the microfilms is housed at the Rare Books and Manuscripts Division of Firestone Library at Princeton University; the inventory of Gödel's *Nachlaß* to be found in volume V includes the finding aid for that. Abridged copies, excluding correspondence, are distributed by IDC Publishers, Inc., 350 Fifth Avenue, Suite 1801, New York, NY 10118 (web address: http://www.idc.nl).

The editorial board for these volumes consists of the undersigned as editors-in-chief, together with Warren Goldfarb, Charles Parsons and Wilfried Sieg. We are especially thankful to Cheryl Dawson, who performed extraordinarily sustained, careful and thorough service as managing editor over a time period that was far longer than any of us anticipated. Under her supervision, the volumes themselves were set in camera-ready copy using

the TEX system, and later $\mathcal{A}\mathcal{M}\mathcal{S}$-TEX, in a form that had been developed for the previous volumes by Yasuko Kitajima; after significant initial TEX work by Kitajima, most of the remaining TEX work in the present volumes was carried out (promptly and with diligence—often under pressure) by Bruce S. Babcock, with the balance done by Cheryl Dawson. We were also ably assisted by Montgomery Link, who joined the effort late in the process to complete the work of researching reference citations; his work under pressure is much appreciated.

From the outset with volume I of these *Works*, the project to produce the volumes has been sponsored by the Association for Symbolic Logic, and the grants under which they were carried out were ably administered for the Association at the hands of its Secretary-Treasurers, C. Ward Henson and, since the beginning of the new millennium, Charles Steinhorn. Clerical support was provided by the Department of Mathematics of Penn State York, with special assistance, especially for the extensive photo-copying required, by Carole V. Wagner. Our editor Elizabeth Johnston at Oxford University Press (in Oxford, England) has been both encouraging and very patient.

We mourn the loss of our dear friend, Stefan Bauer-Mengelberg, who died of a heart attack on 19 October 1996, at the age of 69. Noted for his unusual multi-faceted career as mathematician, symphony conductor and lawyer, Stefan had both given us legal advice and provided us with considerable assistance on the translations in the previous volumes. Following the publication of volume III of these *Works* in 1995, he was much looking forward to further work with us to help bring the entire project to completion.

Solomon Feferman and John W. Dawson, Jr.

Information for the reader

Copy texts. A basic tenet of documentary editing is that published texts of letters should represent what recipients actually saw. In particular, recipients' copies of letters should be used as copy texts whenever they are available, and readers should be made aware of authorial errors and emendations. In the correspondence reproduced herein both of those precepts have been followed. In cases where the recipient's copy of a letter was unobtainable but the author's retained copy has been preserved, we have used the latter as our copy text. Details concerning the copy texts and their sources are provided in the calendar for each correspondent. Errors and emendations are indicated by a variety of devices, chosen with the aim of facilitating proofreading and of distracting as little as possible from readability. They are described in detail below.

Arrangement of letters. Letters are grouped alphabetically according to the names of Gödel's correspondents, and within each group by date.

Dating of letters. The date and author's return address, when included as part of a letter, are placed flush right above the salutation and text, even if not so positioned in the original. Undated letters are identified as such and have been placed in sequence on the basis of annotations on the copy texts, postmarks on retained envelopes or internal references to other correspondence or events of known date. Conjectural dates are enclosed in double square brackets ([[ ]]).

Editorial apparatus. Original pagination of letters, except for the first page of each, is indicated by small numbers in the outer margins of these pages. The symbol | in the text indicates the location of page breaks. Authorial errors and emendations are indicated as follows:

1. Letters or symbols that should have been deleted are backslashed (~~like this~~). Spaces that should be deleted are indicated by a ligature symbol (‿) below the space in question.

2. Letters or symbols that should be replaced by other letters or symbols (including capitalization errors) are backslashed, and the symbols that should replace them are printed in small type in the inner margin of the page on the corresponding line. A "square cup" symbol (⊔) placed in the margin indicates that a symbol is to be replaced by a blank space.

3. Letters or symbols inserted by the editors are enclosed within double square brackets ([[ ]]). Authorial insertions are enclosed within single pointed brackets (⟨ ⟩). (We have not distinguished among insertions made above, below, or on the line.) A caret below the line, together with the "square cup" symbol in the margin, indicates a place where a space should be inserted.

4. Material crossed out by the author is printed with a horizontal over-strike (~~thus~~).

5. Errors or emendations not falling in the above categories, including endorsements (annotations by the recipient or a third party), as well as ancillary details, such as the use of letterhead stationery, are noted in textual notes or editorial footnotes.

Translations. Textual errors in German originals, such as misspellings of names, have been corrected silently in the translations. Authorial deletions have not been translated. Authorial insertions have been translated, but are not indicated as such in the translations, in view of the lack of one-to-one correspondence between the structure of German and English. Likewise, German salutations and closings, which are more varied than those customary in English and would seem affected or obsequious if rendered literally, have been translated by conventional English phrases.

Overall responsibility for the preparation and accuracy of translations was shared by John Dawson and Wilfried Sieg. In general, the editor responsible for preparing the introductory note to each body of correspondence was also responsible for the translation thereof. Where that was not the case, credit for the drafting and/or revision of the translations is indicated at the end of the introductory note.

Introductory notes. To a greater extent than in previous volumes of these *Works*, the introductory notes have been written by the editors. The authorship of each note is given in the Contents and at the end of the note itself. As in the earlier volumes, the notes aim (i) to provide historical context, (ii) to explain the contents of the texts to a greater or lesser extent, (iii) to discuss later developments and, in some cases, (iv) to provide a critical analysis, either of the contents of individual letters or of the correspondence as a whole. The notes also provide biographical information about Gödel's correspondents, cross-references to related correspondence, and, where relevant, indications of the contents of letters not selected for inclusion.

Drafts of each note were circulated among the editors and were subsequently revised by their authors in response to criticisms and suggestions. No attempt was made to impose uniformity of style or point of view. The lengths of the notes vary, depending on the extent and significance of each particular body of correspondence and on how familiar the correspondents were thought likely to be to readers.

Introductory notes are distinguished typographically by a running vertical line along the left- or right-hand margin and are boxed off at their end.[a]

---

[a] A special situation occurs when the note ends in mid-page before facing German and English text. Then the note extends across the top half of the facing pages and is boxed off accordingly.

Footnotes. We use a combination of numbering and lettering, as follows. Authorial footnotes in the letters and their translations are numbered, even in cases where an author (especially Gödel) employed non-numeric (and sometimes idiosyncratic) footnote symbols. Editorial footnotes, on the other hand, used to provide reference citations, to supply ancillary information and to alert the reader to textual issues of various sorts, are lettered and placed below a horizontal line at the bottom of the page. (Where texts are accompanied by facing-page translations such footnotes are divided evenly across the two pages.) When the number of editorial footnotes extends beyond 26, double letters, ordered lexicographically, are employed.

Editorial annotations and textual notes. Editorial annotations within any of the original texts or their translations are signaled by double square brackets: [ ]. Editorial emendations other than those indicated by the editorial apparatus described above are discussed either in editorial footnotes or in the textual notes at the back of the volume. In addition, the following kinds of changes have been made uniformly in the original texts: (i) authorial footnotes have been numbered sequentially within each letter; (ii) spacing, used for emphasis in German texts, as well as underlining, have been rendered by italics; (iii) initial subquotes in German have been raised, e.g., „engeren" becomes "engeren"; (iv) inside addresses to Gödel at the I.A.S. have been omitted.

References. The list of references in volumes IV and V is restricted to items cited therein; however, all citation codes for references listed in earlier volumes remain unchanged. Citation codes consist of the name(s) of the author(s) followed by a date with or without a letter suffix, e.g., "1930", "1930a", "1930b", etc.[b] When the author is clear from the context, that part of the code may be omitted. Where no name is specified or determined by the context, the reference is to Gödel's bibliography, as, e.g., in "Introductory note to *1929, 1930* and *1930a*". For each reference, the date is that of publication, where there is a published copy, or of presentation, for unpublished items such as a speech. (In the case of works by Gödel published posthumously in volume III of these *Works*, the date is that of presentation or composition; the code for such works is preceded by an asterisk.) A suffix is used when there is more than one publication by an author in the same year; however, the ordering of suffixes does not necessarily correspond to the order of publication within that year. A question mark is used when a date, or some part thereof, is uncertain. For works whose composition or publication extended over a range of dates, the starting and ending dates are both given, separated by a slash.

---

[b] "200?" is used for articles whose date of publication is not yet known.

Except in the reference list, citation codes are given in italics. They are employed in the introductory notes, editorial footnotes and textual notes. Where citations occur within letter texts and translations, however, they are reproduced as the author gave them. An accompanying editorial footnote then provides the citation code.

References to page numbers in Gödel's publications are to those of the textual source. References to other items in these *Works* are cited by title, volume number and page number within the volume.

To make the reference list as useful as possible for historical purposes, authors' names there are supplied with first and/or middle names as well as initials, except when the information could not be determined. Russian names are given both in transliterated form and in their original Cyrillic spelling. In some cases, common variant transliterations of the same author's name, attached to different publications, are also noted.

Logical symbols. The logical symbols used by authors of letters are here presented intact, even though these symbols may vary from one letter to another. Authors of introductory notes have in some cases followed the notation of the author discussed and in other cases have preferred to make use of other, more current, notation. Also, logical symbols are sometimes used to abbreviate informal expressions as well as formal operations. No attempt has been made to impose uniformity in this respect. The following is a brief glossary of logical symbols that are used in one way or another in these volumes, where 'A', 'B' are letters for propositions or formulas and '$A(x)$' is a propositional function of x or a formula with free variable 'x'.

Conjunction ("A and B"): $A.B$, $A \wedge B$, $A\&B$
Disjunction ("A or B"): $A \vee B$
Negation ("not A"): $\overline{A}$, $\sim A$, $\neg A$
Conditional, or Implication ("if A then B"): $A \supset B$, $A \to B$
Biconditional ("A if and only if B"): $A \equiv B$, $A \sim B$, $A \leftrightarrow B$
Universal quantification ("for all x, $A(x)$"): $(x)A(x)$, $\Pi x A(x)$, $x\Pi(A(x))$, $(\forall x)A(x)$
Existential quantification ("there exists an x such that $A(x)$"): $(Ex)A(x)$, $\Sigma x A(x)$, $(\exists x)A(x)$
Provability relation ("A is provable in the system S"): $S \vdash A$

Note: (i) The "horseshoe" symbol is also used for set-inclusion, i.e., for sets X, Y one writes $X \subset Y$ (or $Y \supset X$) to express that X is a subset of Y. (ii) Dots are sometimes used in lieu of parentheses, e.g., $A \supset .B \supset A$ is written for $A \supset (B \supset A)$.

Calendars. Separate calendars of Gödel's correspondence with each of seven major correspondents in this volume are included in this volume. Those calendars list all extant letters known to us, whether or not selected for inclusion herein, as well as the source archives for each letter and details concerning the form of each document (whether typed or handwritten,

signed or unsigned, etc.). In addition, each volume contains a general calendar that lists all items of correspondence included in either volume.

Finding aid to the Gödel Nachlaß. In 1998, funds from the Alfred P. Sloan Foundation enabled the preservation microfilming of Gödel's *Nachlaß*, held by the Rare Books and Manuscripts Division of the Firestone Library at Princeton University. At that time a revision was prepared of the finding aid to the *Nachlaß* compiled by John Dawson in 1984. A further revision of that document by Cheryl Dawson, including references to the location of items on the microfilm reels, is included in volume V for the benefit of scholars who may wish to consult the originals or order copies from the microfilm.[c]

Appendices. Rounding out these volumes are two appendices to volume V. The first contains a small number of letters written by others on Gödel's behalf. The second provides an alternate version of Remark 3 of item *1972a*, published in volume II of these *Works*, p. 306.

Typesetting. These volumes have been prepared using the TEX computerized mathematical typesetting system devised by Donald E. Knuth of Stanford University, as described in the preface to volume I. (For the first three volumes, camera-ready copy was delivered directly to the publisher.) The computerized system was employed because: (i) much material, including the introductory notes and translations, needed to undergo several revisions; (ii) proofreading was carried on as the project proceeded; (iii) in the case of previously published letters, texts could be prepared in a uniform format, incorporating our editorial apparatus, instead of being photographed from the original sources. Choices of the various typesetting parameters were made by the editors in consultation with the publisher. After significant initial work by Yasuko Kitajima, primary responsibility for the typesetting in these volumes lay with Cheryl A. Dawson and Bruce S. Babcock.

Photographs. Primary responsibility for securing these lay with John Dawson. Their various individual sources are credited in the Permissions section, which follows directly.

---

[c]A selective subset of the microfilms, *excluding* all correspondence and photographs, is available for purchase from IDC Publishers (Leiden). Further information about that edition is available on line at http://www.idc.nl/catalog/index.php?c=375 .

Copyright permissions

The editors are grateful to the Institute for Advanced Study, Princeton, literary executors for the estate of Kurt Gödel, for permission to reproduce and translate items of correspondence written by him. In addition, we thank the following individuals and institutions for providing us with copies of letters written by Gödel, and for allowing their publication herein:

Dr. Beat Glaus and the ETH-Bibliothek, Zürich, for letters to Paul Bernays;

Dr. Tilo Brandis, then curator of manuscripts, and the Staatsbibliothek zu Berlin, Preussischer Kulturbesitz, for letters to Heinrich Behmann and to Gotthard Günther;

the Carnap Collection Committee, University of Pittsburgh Library, for letters to Rudolf Carnap;

the Wiener Stadt- und Landesbibliothek, for Gödel's letters to his mother.

For permission to reproduce and translate the texts of letters to Gödel included in this volume we thank the following individuals and institutions:

New York University Press, for the letters of Allan Angoff and Wilson Follett;

Mrs. Emmy Behmann and Prof. Dr. Christian Thiel, for the letters of Heinrich Behmann;

Dr. Beat Glaus, on behalf of the ETH-Bibliothek Zürich, and Dr. med. Ludwig Bernays, nephew of Paul Bernays, for the letters of Professor Bernays;

Dr. Kenneth Blackwell, of the Bertrand Russell Archives, McMaster University, Hamilton, Ontario, the late Professor Herbert G. Bohnert of Michigan State University, and Academician Georg Brutian, of the Armenian Academy of Sciences, for their letters;

Mrs. Sylvia Büchi, widow of Professor J. Richard Büchi, for his letters;

Professor J. Richard Creath, on behalf of Mrs. Hanneliese Carnap Thost, for the letters of Rudolf Carnap;

Alonzo Church, Jr., trust administrator for the estate of Alonzo Church, for the letters of his father, Alonzo Church;

the late Professor Burton Dreben, of Harvard University, for his letters;

Frau Regula Lips-Finsler, for the letters of her uncle, Professor Paul Finsler;

Professor Leonard Goddard, Ferry Creek, Victoria, Australia, Professor Burke D. Grandjean of the University of Wyoming, and Professor Emeritus Marvin Jay Greenberg, of the University of California, Santa Cruz, for their letters;

Prof. Dr. Claus Baldus, literary executor for the estate of Gotthard Günther, for Professor Günther's letters.

For supplying portraits of Gödel's correspondents and granting permission for their use as illustrations in this volume we thank:

Dr. Helmut Rohlfing of the Niedersächsische Staats- und Universitätsbibliothek, Göttingen, for the portrait of Heinrich Behmann, from David Hilbert's birthday album (Cod. Ms. D. Hilbert 754: Nr. 115);

Professor Gert Müller of the Mathematisches Institut, Heidelberg, for the portrait of Paul Bernays, taken from the book *Sets and classes* edited by Professor Müller;

the Michigan State University Archives and Historical Collections, for the portrait of Herbert G. Bohnert;

the mathematics department of the University of Illinois at Urbana-Champaign, as well as Theodore Boone and William Boone, Jr., sons of William Boone, for the portrait of him;

the Carnap Collection Committee of the University of Pittsburgh Library, for the portrait of Rudolf Carnap;

Professor Emeritus Paul R. Halmos of Santa Clara University, for the portrait of Alonzo Church, taken from Professor Halmos' book *I have a photographic memory*;

Warren Goldfarb, for the portrait of Burton Dreben;

Professor C. Wade Savage and the Center for Philosophy of Science, University of Minnesota, for the portrait of Herbert Feigl;

the Institute for Advanced Study, Princeton, for the portrait of Gödel's mother Marianne, from Gödel's *Nachlaß*;

the University of Illinois Archives, for the portrait of Gotthard Günther (Record Series 39/2/26).

Contents

Volume IV

List of illustrations

Kurt Gödel
COLLECTED WORKS
Volume IV

Allan Angoff

Allan Angoff was one of the editors at New York University Press during the negotiations between Gödel, Ernest Nagel and NYU Press concerning the publication of one or both of *1934* and a translation of *1931* as an appendix to the planned *Nagel and Newman 1958*. The negotiations came to nothing, in large part because of demands made by Gödel that Nagel found unacceptable. For a full account of the background, see the introductory note to the correspondence with Ernest Nagel, vol. V, p. 135.

1. Gödel to Angoff

Princeton, April 9, 1957.

Dear Mr. Angoff,

After having discussed with Professor Nagel the plan of reprinting some of ⟦my⟧ papers as an appendix to the proposed book on my results, I feel that, if this idea is carried out, the following requirements would have to be met:

1. In view of the fact that giving my consent to this plan implies, in some sense, an approval of the book on my part, I would have to see the manuscript and the proof sheets of the book, including the appendix.

2. For the same reason I would have to write an introduction to the appendix, on the one hand in order to mention advances that have been made after the publication of my papers,[a] on the other hand in order to supplement the considerations, given in the book, about the philosophical implications of my results. I am not very well pleased with the treatment of these questions in the articles that came out in the Scientific American and in the "World of Mathematics".[b] This, for the most part, is not the fault of the authors, because almost nothing has been published on this subject, while I have been thinking about it in the past few years.

u

---

[a] *Gödel 1931* and *1934.*

[b] *Nagel and Newman 1956* and *1956a.*

3. If a sufficient number of copies of the book are sold, I believe I would be entitled to share in the proceeds the book yields. I am, therefore, asking you what you would suggest in this respect.

In support of item 1. above I would like to mention that the article on my work published in the "World of Mathematics" contains some very troublesome mistakes.

2 | I would like to add two more suggestions:

1. To reprint both my ~~book~~ paper of 1931 and my lectures of 1934[c] would involve an undue amount of duplication. I would suggest to reprint only my lectures and, perhaps, the first chapter of my paper. Some of the space saved in this way could be used for my introduction and, perhaps, for reprinting a few pages of the work mentioned in it.

2. The proposed title of the book does not seem very good to me. "Gödel's Theorems on the Principles of Mathematics" would give more information about the content of the book.

Very truly yours

Kurt Gödel

---

[c] *Gödel 1931* and *1934*.

2. Angoff to Gödel[a]

April 22, 1957

Dear Professor Goedel:

This is a point by point reply to your letter of April 9, regarding the volume on Goedel's theorem by Ernest Nagel and James R. Newman, which we plan to bring out:

1. We are glad to accede to your request that you examine the manuscript and galley proofs of the book prior to publication.

2. We accept your suggestion that the appendix should carry an introduction by you.

---

[a]On letterhead of New York University Press, Washington Square, New York 3.

3. You ask about any possible share for yourself of any income derived from the book. I have discussed this with Messrs. Nagel and Newman and they have agreed to forego a portion of the royalties they would have received under their present contract and to allocate this portion to you. At the present time, their contract calls for royalty payments of 10% of the list price on the first 3,000 copies sold, $12\frac{1}{2}$% of the list price on the second 3,000 copies sold, and 15% of the list price of all copies sold in excess of 6,000 copies. I therefore suggest that we pay you 2% of the retail price on all copies of the book sold. That would leave a royalty of 8% for Messrs. Nagel and Newman to divide between them, or 4% each on the maximum number of copies we are likely to sell.

If this is satisfactory to you, I shall have our business manager, Mr. Saverio Procario, send you a contract or an agreement binding New York University Press to this arrangement. If for any reason you wish to modify this, please write me accordingly.

4. We understand your desire to save space and your suggestion that only your lectures[b] and perhaps the first chapter of your paper[c] be reprinted. But we are willing to make the book somewhat longer if we | 2 are permitted to reprint both the paper of 1931 and the lectures of 1934.

5. We are glad to accept your suggestion that the title of the book be changed to the following: "Goedel's Theorems on the Principles of Mathematics".

I look forward to hearing from you. As always, please feel free to make any other suggestions. I am sure they all will make for a more effective book.

Sincerely yours,

Allan Angoff

Editor

---

[b] *Gödel 1934.*
[c] *Gödel 1931.*

3. Gödel to Angoff

Princeton, May 6, 1957

Dear Mr. Angoff,

In reply to your letter of April 22 I wish to say that the royalty agreement you suggest is not acceptable to me. I'll I don't want to lessen the

royalties of Messrs. Nagel and Newman in case the number of copies
sold is not large. I'll be glad to resign my share entirely if fewer than
6000 copes are sold. On the other hand if, contrary to expectation, more
copies should be sold, I feel I would be entitled to a larger share than
you suggest, because my papers and my introduction probably will contribute more toward a larger sale than would be indicated by the number
of pages. I therefore suggest that I receive half the royalties for all copies
sold in excess of 6000, i.e., $7\frac{1}{2}$ .5- $7\frac{1}{2}\%$ of their list price.

Of course I shall have to see the manuscript of the book before I sign
the contract, so that I can make sure that I am in agreement with its
content, or that passages with which I don't agree may be eliminated, or
that I may express my view about the questions concerned in the introduction. Furthermore I wish to add that I can't permit any "editing"
(without my consent) of the parts of which I am the author and that I
must insist on receiving galley and page proofs of these parts. You will
understand these requirements in the light of what I said in my preceding
letter.

Very truly yours

Kurt Gödel

4. Angoff to Gödel[a]

May 28, 1957

Dear Professor Gödel:

I am glad to inform you that your request for a royalty of $7\frac{1}{2}\%$ for all
copies sold above 6,000 is agreeable to all of us—to Messrs. Nagel and
Newman and to the Press. Our business manager will be glad to send
you the appropriate contract upon your acceptance of this arrangement.

It would be extremely difficult to send you the manuscript prior to
printing because it might well result in a delay we would like to avoid,
particularly at this time when so many printers have tight schedules because of summer vacations. But I am glad to give you every assurance
you will be sent the galleys for the entire book as soon as we receive

---

[a]On letterhead of New York University Press, Washington Square, New York 3,
erroneously addressed to Gödel at the "Institute for Advanced Education".

them from the printer, and we shall be grateful for any suggestions you would like to make regarding them. Messrs. Nagel and Newman have in turn assured me they would be delighted to consider all your suggestions.

We plan to reproduce your original German paper[b] and the English translation of it made by Professor van Heijenoort, and indeed plans for this are already far advanced. If you want to comment on any of the material in your original paper, then you are free of course to do so in a footnote or in an appendix. I hope you won't mind my urging you to keep these comments down to a minimum consistent with accuracy, for, as you can understand, this will be an expensive book to produce and we are making every effort to keep costs down. I am hopeful you will be able to make these changes within three weeks of receiving the galleys and we are taking the liberty of making our production and publishing plans accordingly. It would be most helpful to us if long before then, perhaps within the next two weeks, you can send us the introduction you spoke of in your letter to me of May 6.

I wonder if you could reply to this letter at your earliest convenience if at all possible. This would help us tremendously in meeting various schedules with printers and binders.

Sincerely yours,

Allan Angoff

---

[b] *Gödel 1931.*

5. Gödel to Angoff

Princeton, June 3, 1957

Dear Mr. Angoff,

I have received to/day your letter of May 28. I would like to remind you that I have not yet given my consent to my papers being reprinted, but rather, in my letter of April 9, my consent was conditioned *on my seeing the manuscript of the book and approving of its content.* In view of the fact that I may be in substantial disagreement with some parts of it it does not seem advisable to set the book in type before I had a chance to see it. Moreover I cannot write the introduction before I have ~~seen~~ read the book

and I shall need at least six weeks for writing the introduction. I wish to emphasize that I will not be responsible for any expenses you may incur by doing something rash.

I think it is quite unnecessary to reprint both the original and the translation,[a] if the author revises the translation. Such duplication may be advisable in case of a philosophical article, but certainly not in case of a mathematical one. As you know I am even doubtful whether both my lectures and my article in the Monatshefte[b] should be reprinted.

Sincerely yours

Kurt Gödel

P.S. I think my introduction will be about 10 typewritten pages, but I cannot fix the number of pages too precisely now. It will[c] the comments on my original papers which I want to make.

---

[a]Of *Gödel 1931*.
[b]I.e., *Gödel 1934* and *1931*.
[c]A word is evidently missing here; "contain" would make sense.

6. Angoff to Gödel[a]

June 6, 1957

Dear Professor Gödel:

After discussing your letter of June 3 with Messrs. Nagel and Newman, I have come to the conclusion that it will be impossible for us to send you the manuscript of their book. Publication has already been delayed considerably and for that reason we are forced to proceed with the composition of the manuscript. I shall be glad to send you the galleys, as I have already written you, as soon as we receive them from the printer. At that time, you may of course make suggestions or add footnote material which you feel is necessary and which Messrs. Nagel and Newman would be delighted to consider for inclusion in their book. Needless to

---

[a]On letterhead of New York University Press, Washington Square, New York 3.

say, you are free to consult the manuscript at the office of our copyeditor
before it is sent to the printer a week or two from now. We shall be glad
to make a desk available to you, should you wish to come in during this
period.

I am hopeful that after you receive the galleys you will be able to complete your introduction in less than six weeks, since that period would further delay the publication of the book and might well cause very considerable difficulty in our production and sales departments. But perhaps we
can go into that after you receive the galleys. I mention it now in the
hope that you may be able to give some thought to the possibility of completing it in less time. That would help us tremendously.

Sincerely yours,

Allan Angoff

7. Gödel to Angoff[a]

June 25, 1957

Mr. Allan Angoff
New York University Press
Washington Square
New York 3, N.Y.

Dear Mr. Angoff:

If you wish to send me the galleys of the text of the book instead of
a copy of the manuscript, this is all right with me, if you are willing to
bear the costs in case changes should be necessary. Of course it is clear
that, if the book should contain serious mistakes, as the article in the
"World of Mathematics"[b] does, I cannot give my permission to reprinting
my papers as an appendix, until I have made sure that these mistakes
have been corrected. To look up the manuscript in your office would not
give me sufficient time. I have to consider it carefully for a few days.

---

[a]It is likely that Angoff personally did not receive this letter. June 6, the date of the
previous letter, was his last day working for NYU Press. See note g of the introductory
note to the correspondence with Ernest Nagel, these *Works*, vol V.

[b] *Nagel and Newman 1956a.*

Of course I assume that you won't set the appendix in type before the matters mentioned have been settled.

Since you seem to be very intent on speeding up matters, I hope to receive the galleys or a copy of the manuscript soon. Of course I will need a few weeks for writing the introduction. I don't understand why you didn't send me a copy of the manuscript sooner. I wrote you already in my letter of April 9 that I have to see it.

Sincerely yours,

Kurt Gödel

P.S. For the reasons mentioned in my letter of April 9 I cannot permit the appendix to be published without an introduction.

K. G.

Yossef Balas

From time to time Gödel received inquiries about his work from students or mathematical amateurs. Most were routine, but a few prompted him to expound on the background or implications of his discoveries.

One such query came from Yossef Balas, a master's degree student at the University of Northern Iowa. On 27 May 1970 Balas wrote the American Mathematical Society seeking information concerning the genesis of Gödel's first incompleteness theorem. The letter was subsequently forwarded to Gödel himself, who took the time to draft (but not to send) a detailed reply. Found in his *Nachlaß* after his death, the draft describes the circumstances that caused Gödel to compare the notions of truth and provability, namely, his attempt to further Hilbert's program by giving a consistency proof of analysis relative to arithmetic. Gödel also concisely and conclusively criticizes Finsler's earlier attempt to construct a formally undecidable proposition.

For related commentary on the incompleteness results, see Gödel's direct response to Finsler as well as his letters to Georg Brutian, David Plummer and Leon Rappaport.

John W. Dawson, Jr.

---

1. Gödel to Yossef Balas

Dear Mr. Balas:

I have explained the heuristic principle for the construction of propositions undecidable in a given formal system in the lectures I gave in Princeton in 1934.[a] They came out in print in the book "The Undecidable" edited by M. Davis in 1965.[b] See Sect. 7, p[[p]]. 63–65, ~~of this edition,~~

---

[a] *Gödel 1934*.

[b] *Davis 1965*.

in particular footnote 26. The occasion for comparing truth and demonstrability was an attempt to give a relative model-theoretic consistency proof of analysis in arithmetic. This leads almost by necessity to such
2 a comparison.[c] | ⟨For, an arithmetical model of analysis is nothing else but an arithme⟨tical ε-relation satisfying the comprehension axiom:⟩ $(\exists n)(x)[x \epsilon n \equiv \varphi(x)]$. ~~by an arithmetical~~ Now, if in the latter "$\varphi(x)$" ⟨is replaced⟩ by "$\varphi(x)$ is provable", such an ⟨ε-⟩relation can easily be defined. Hence, if truth were equivalent to provability, we would have reached our goal. However (and this is the decisive point) it follows from the[d] <u>correct</u> solution of the semantic paradoxes ~~i.e., the fact~~ that ~~the concept of~~ "truth" of the propositions of a language *cannot be expressed* in the same language, while provability (being an arithmetical relation) *can*. Hence true $\neq$ provable.⟩

As for work done earlier about the question of ⟨formal⟩ decidability of mathematical propositions, I know only a paper by Finsler ~~published a few years before mine (I believe~~ in Math. Zeits. 25 (1926),[e] p. 676.) ~~He also applies a diagonal procedure in order to construct undecidable propositions.~~ However, he ⟨Finsler⟩ omits exactly the main point which makes a proof possible, namely restriction to ~~a definite~~ ⟨some well∧defined⟩ –
4 formal system[f] in which the proposition is undecidable. For, | he had the nonsensical aim of proving formal undecidability in an absolute sense. This leads to the nonsensical definition given in the first two parags. of sect. 5, p. 678, ⟨supplemented⟩ by the 2nd paragr. of sect. 9, p. 680; and

---

[c] A paragraph crossed out at this point is of independent interest. It reads:

> However in consequence of the philosophical prejudices of our times 1. nobody was looking for a relative consistency proof because i̸s̸ was considered axiomatic that a consistency proof must be finitary in order to make sense
3 2. a concept of objective mathematical | truth as opposed to demonstrability was viewed with greatest suspicion and widely rejected as meaningless.

[d] A struck-out passage reads:

> in this model it would have to be shown that truth and demonstrability are equivalent [i.e. Dem(p) $\equiv p$]. But the decisive point is now to apply the

There is also evidence of erasures in this area as well as others.

[e] *Finsler 1926.*

[f] At this point the following passage, which originally closed the letter, was crossed out and the remainder of the text given above was marked to be inserted. The words in editorial brackets are not completely clear.

> ~~The distinction between [truth and] demonstrability is~~ This "proof" (which incidentally was not known to me when I wrote my paper) is therefore worthless and the result claimed, namely the existence of "absolutely" undecidable propositions is very likely wrong.

~~this leads~~ to the flagrant inconsistency that he decides the ⟨"formally⟩ undecidable⟨"⟩ proposition by an argument ⟨(sect. 11, p. 681)⟩ which, *according to his own ⟨definition* in⟩ the two passages just cited, is *a formal proof* ~~while on the other hand he asserts on p. 681 it/s formal undecidability~~. If Finsler had confined himself to some well∧defined formal system S, his proof (by ~~changing~~ replacing the nonsensical section 11 with a proof that the ⟨proposition in question⟩ is expressible in S) could be made correct and applicable to any formal system. I myself did not know his paper when I wrote mine, and other mathematicians or logicians probably disregarded it because it contains the obvious nonsense just mentioned.

Sincerely   5

[[No signature]]

Heinrich Behmann

Heinrich Behmann

Heinrich Behmann (1891–1970) was a logician of some prominence in inter-war Germany, probably best known today for his solution to the decision problem for monadic predicate logic (*Behmann 1922*).[a] Indeed, Behmann contributed importantly to the formulation of the general decision problem for first-order logic and probably coined the term *Entscheidungsproblem*.[b] From 1925 to 1945, a period that includes that of his correspondence with Gödel, he taught mathematics at the University of Halle-Wittenberg.

Behmann also pressed the requirement, originating with Pascal, that definitions should allow the complete elimination of defined terms in favor of their definientia. This was a central idea of an approach to logical paradoxes that he presented in a lecture that he gave at a conference in September 1929 in Prague.[c] The proposal was in effect a suggestion for a type-free logic. One can see the idea in a simple case. Suppose one thinks of class abstracts as defined symbols and of the comprehension schema as a rule for the introduction of abstracts, e.g. by putting it in the form

(A) $$\forall y[y \in \{x \colon Fx\} \leftrightarrow Fy].$$

Then one sees immediately that in the case of Russell's paradox, the rule does not allow the elimination of the abstract $\{x \colon x \notin x\}$, since the rule reduces $\{x \colon x \notin x\} \in \{x \colon x \notin x\}$ to $\{x \colon x \notin x\} \notin \{x \colon x \notin x\}$ and vice versa. Behmann accordingly states as a first rule:

(1) Expressions containing abbreviations are only admissible insofar as the complete replacement of the abbreviations by their meanings can be carried out symbolically.[d]

---

[a]For a very brief sketch of his contributions to logic, see *Thiel 1980*. *Haas and Haas 1982* is an account, still quite brief, of Behmann's life and work; it contains a list of his publications. *Mancosu 1999* discusses his dissertation work under Hilbert and its importance for the early development of the work of the Hilbert school, as well as other contributions of Behmann to the work of the Hilbert school in the period 1914–1922.

[b]*Mancosu 1999*, pp. 320–321 and note 36; cf. *Zach 1999*, p. 355.

[c]The conference was the 5. Deutsche Physiker- und Mathematikertagung. The lecture was published as *Behmann 1931*.

[d]*Behmann 1931*, pp. 42–43.

Behmann also states a second rule to deal with the case of expressions containing quantifiers:

(2) A variable may not automatically be considered as ranging over the whole domain that would be in question according to the immediate symbolic context—i.e. in the case of logical variables all things, all statements, or all properties, etc. as the case may be—but rather only insofar as the result of substitution can be written without abbreviations.[e]

(2) apparently implies that the instantiation of $\{x : x \notin x\}$ for y in (A) is impermissible because the resulting instance violates rule (1).

The correspondence of Behmann with Gödel published here originates with a misunderstanding by the Berlin philosopher Walter Dubislav of a comment made by Gödel about Behmann's proposal while they were returning by train from the Conference on Epistemology of the Exact Sciences in Königsberg in September 1930.[f] From the correspondence and from the note *Behmann 1931a* that resulted from it, it is evident that Gödel's observation was that rule (1) was not sufficient to avoid paradoxes, because one could derive them without any use of abbreviations. Dubislav, however, inserted into the third edition *1931* of his book on definition, which must have been about ready to go to press, the assertion that Behmann's proposal had been refuted by Gödel and that Gödel intended to set forth his argument in print.[g]

Neither Dubislav nor Gödel was adequately informed about Behmann's proposal, although Gödel was quite aware of this; in fact *Dubislav 1931* appeared before Behmann's own paper. When he saw the book, Behmann wrote immediately (3 February 1931) to Felix Kaufmann in Vienna, with whom he had corresponded about another matter in which Gödel had also been involved (see below).[h] On 10 February he wrote to Gödel (letter 1). His concern was to have the matter corrected as soon as possible in the journal *Erkenntnis*, which he no doubt chose because *Dubislav 1931* appeared as a supplementary volume of that journal. There followed an exchange between Gödel, Behmann and Dubislav, in

[e] *Ibid.*, pp. 43–44.

[f] It was at this conference that Gödel delivered the lecture **1930c* on his completeness proof and also announced his incompleteness theorem. On the latter see **1930c*, p. 28, and *1931a*, pp. 147–148.

[g] *Dubislav 1931*, §48, pp. 94–96.

[h] Felix Kaufmann (1895–1949) was a member of the Vienna Circle whose philosophical views were also influenced by the work of Husserl. In addition to *Kaufmann 1930* on the philosophy of mathematics, he also wrote on the philosophy of law and the methodology of the social sciences. In 1938 he emigrated to the United States.

which the correcting note was in effect negotiated between Behmann and Gödel. All the letters between Behmann and Gödel in this exchange are included here. Although it might have been up to Dubislav to publish a correction, after the early stages he took only a passive role. As the reader of the letters below will see, Gödel quickly drafted a version of the correction, but in the end the correction was published in Behmann's name, although language drafted by Gödel entered into it.

As a narrative of the episode, which went on from February to May 1931, the correspondence speaks pretty well for itself.[i] This was an early case in which Gödel's well-known meticulousness about details of formulation was manifested, and Behmann also had concerns, not only to set matters straight from his point of view but also, for example, to use his own logical symbolism. But matters were settled to the apparent satisfaction of both parties, and with reasonable dispatch.

In response to Behmann's letter Gödel drafted a correction in the form of a letter to Dubislav to be published in *Erkenntnis* (letter 2). He confirmed that the point was the same as one he had made to Behmann himself on a visit of the latter to Vienna.[j] Behmann evidently thought more detail was called for in order to make the point completely clear. Gödel's concern throughout the exchange was to make clear what he had proved and not to be taken to endorse Behmann's approach. After Behmann had concluded (see letter 5) that the correction should appear in his name and had written his own draft of a note (document 010015.426 in Gödel's papers), Gödel in letter 6 pressed for revisions, to clarify his own "formal and mathematical" claim and not to allow Behmann to say that his reply had been "recognized as indisputable" by Gödel. About this he makes the interesting remark:

> For me an indisputable solution of the problem of the antinomies lies only in the presentation of a formalism with precisely formulated axioms and rules of inference that avoids the contradictions, and such a [formalism] has up to now only been indicated by you (you have by the way not carried out at all the approach communicated to me in Vienna in the form of a restriction on the rule of implication).

(Gödel had by now seen *Behmann 1931*.)

[i]A fuller account of the episode is given in *Thiel 2002*, to which I am much indebted.

[j]Behmann visited Kaufmann in Vienna from 9 to 20 September 1930, after the Königsberg conference. (See *Mancosu 2002*, §1.) On 16 September he gave a lecture "Widersprüche in Mathematik und Logik" (*Thiel 2002*, p. 389). It seems likely that Gödel made his comment on that occasion. As one can see from letter 6 and from note 1 of *Behmann 1931a*, much the same argument as Gödel's had been communicated to Behmann earlier by Bernays and Ackermann.

Happily, Behmann was conciliatory about the points Gödel raised, and
the version he sent with his reply was, with a small omission that he
invited Gödel to make if he wished, what was published in *Erkenntnis*.[k]
That ended the episode.[l]

The above-quoted remark is the only direct indication we have of
what Gödel thought of Behmann's approach to the paradoxes. Gödel
may, however, have had Behmann's ideas in mind in a remark in *1933o*.
He says that the theory of types (generalized in a way he has just explained) is "until now the only solution" to the problem of restricting
"the so-called naive logic" so as to avoid paradoxes and retain all of
mathematics. He goes on to say:

> All other solutions that have been presented up to now either remained
> vague promises, i.e., have not been followed up to the point of setting
> up a formal system, or led to contradictions.[m]

Still later, when he discussed the "logical paradoxes" as a general problem somewhat independently of the foundations of mathematics, Gödel
showed a friendlier attitude toward attempts toward a type-free logic.
In *1944*, p. 150, he makes the tantalizing suggestion:

> It might even turn out that it is possible to assume every concept to be
> significant everywhere except for certain "singular points" or "limiting
> points", so that the paradoxes would appear as something analogous to
> dividing by zero.

This might be compared with Behmann's:

> It will turn out that they [the paradoxes] rest on leaving out of account
> a requirement for working with any symbolism (a trivial one once one
> has noticed it), that from the standpoint of logical symbolism it's a

[k]*Behmann 1931a*. What was sent with letter 7 was clearly document no. 01005.
424 in Gödel's papers. Footnote 1 of *Behmann 1931a* differs from the draft offered
by Gödel in letter 6 by the omission of the phrase "die mir Sept 1930 von Herrn
Gödel vorgelegt wurde". Behmann had expanded this to "die... vorgelegt und von
mir in dem hier angegebenen Sinne aufgeklärt wurde". In letter 7 he says that if
Gödel finds the addition unsuitable, the whole phrase can be omitted. Evidently
Gödel did find it unsuitable.

[l]The only known further correspondence between Gödel and Behmann is a card
sent by Behmann from a Baltic resort in September 1932. He thanks Gödel for
sending "Semantik II". This is evidently the manuscript by Carnap that Gödel
reports to Carnap having sent to Behmann in August; see the letter to Carnap of 11
September 1932 in this volume. Gödel's papers also contain a three-part typescript
by Behmann dating from 1935 (document 010015.429), but there is no accompanying
correspondence.

[m]*1933o*, p. 49.

matter of a mere mistake in calculation, for example of the type of the impermissible dividing by 0 or careless interchanging of limit processes.[n]

Gödel says, "Unfortunately, the attempts made in this direction have failed so far," and cites *Church 1932* and *1933*. Possibly Behmann's failure to offer a definite formal system made Gödel think his effort not even worthy of mention at this point. It does not seem that Behmann made up this deficiency in his later work, not even in the extensive *Behmann 1959*.

The correspondence between Gödel and Behmann was preceded by an indirect exchange a few months earlier. After his visit to Vienna in September 1930 (see note j), Behmann wrote and sent to a number of people a paper "Zur Frage der Konstruktivität von Beweisen". It was prompted by a not very precise conjecture of Kaufmann, that in the case of proofs where the uncountable infinite is not involved, it will be possible to extract even from an indirect proof of an existential statement $\exists x A(x)$ an instance, i.e. a term t for which $A(t)$ is provable.[o] This would vindicate at least part of classical mathematics against Brouwer's criticism, since proofs in the relevant (not very precisely delimited) parts of mathematics could be made constructive. Behmann's paper claimed to prove Kaufmann's conjecture. The proof was very sketchy and based on ideas about how one might transform proofs by contradiction into direct proofs.[p] Neither Kaufmann nor Behmann seems to have tackled the conflicts between the demand for constructivity and the law of excluded middle that had been brought forth in a number of publications of Brouwer.[q]

By the mediation of Carnap, Gödel saw Behmann's paper and came up with a counterexample to the claim that classical proofs of existential claims can always be made constructive. The example was communicated to Behmann by Kaufmann in a letter of 19 October 1930 (letter 1, Appendix A, vol. V).[r] The example is of a kind familiar to readers

[n] *1931*, p. 39, my translation. The comparison of the reasoning leading to paradoxes with dividing by 0 also occurs in the first part of the typescript referred to in note l.

[o] *Kaufmann 1930*, pp. 66–67.

[p] For discussion of Behmann's argument, as well as a fuller treatment of the history surrounding Behmann's paper, see *Mancosu 2002*.

[q] Especially relevant is perhaps *Brouwer 1925*.

[r] No copy of this letter is in Gödel's papers, although they do contain copies of Behmann's reply and Kaufmann's reply to that (21 and 23 October). The counterexample is also documented by a note of Carnap on a conversation with Gödel; see *Dawson 1997*, p. 73 n.

of Brouwer: Gödel describes a bounded sequence of rationals that has (classically) an accumulation point 0 or 1, but which it is depends on the truth of Goldbach's conjecture. Behmann was not immediately moved by the objection. In his reply of 21 October to Kaufmann's letter, he says that the example affects "only the *presupposition* of my proof, that every *direct* existence proof rests on an exhibition [of an instance]". Moreover, he is not bothered by the possibility that a proof might branch and offer different "exhibitions" depending on the truth of some statement; he even remarks, as Gödel did, that the branching might be infinite. This response perhaps reveals the imprecision of Behmann's claim. It is hard to escape the conclusion that neither he nor Kaufmann understood the force of Brouwer's examples, as Gödel clearly did. The latter's intervention in the response to Behmann's manuscript ended with Kaufmann's letter of 19 October. Eventually, however, Behmann was persuaded by Paul Bernays, to whom he had sent the paper with an eye to publication

1. Behmann to Gödel[a]

Halle, 10. Februar 1931.
Moltkestr. 5.

Sehr geehrter Herr Gödel!

Da ich mich nicht erinnerte, Ihre Anschrift zu besitzen, hatte ich vor einigen Tagen an Herrn Kaufmann die Bitte gerichtet, sich bezüglich der folgenden Angelegenheit mit Ihnen in Verbindung zu setzen. Herr Kaufmann ist allerdings in diesen Tagen in Breitenstein; er hat mir bereits zugesagt, gleich nach seiner Rückkehr bei Ihnen anzufragen. Da ich nun inzwischen zufällig Ihren Vermerk auf dem Umschlag Ihres damals freu[n]dlicherweise übersandten Aufsatzes wiedergefunden habe, möchte ich gleich selbst bei Ihnen anfragen.

Im Beiheft der "Erkenntnis" (Dubislav, Die Definition[b] lese ich zu meinem Erstaunen den §48 "Der Behmannsche Auflösungsversuch und

---

[a]Page 2 of this letter contains a shorthand comment evidently by Gödel. It clearly refers, however, to a draft of *Behmann 1931a* sent to Gödel with a later letter, most likely to document no. 010015.426 sent with letter 5 (see note a to that letter). It

in *Mathematische Annalen*, that his claim had not been established and that the paper was not ready for publication.[s] Neither it nor a descendant of it was ever published.[t]

Charles Parsons

The translation is by Charles Parsons, revised using suggestions from John Dawson, Dirk Schlimm, and Wilfried Sieg. A complete calendar of the correspondence with Behmann appears on p. 551 of this volume.

[s]See §6 of *Mancosu 2002*.

[t]I am greatly indebted to Christian Thiel for information concerning Behmann and this correspondence and for supplying copies of documents. In the matter of Behmann's paper on the constructivity of proofs, I am similarly indebted to Paolo Mancosu. I wish to thank John Dawson and Warren Goldfarb for comments.

1. Behmann to Gödel[a]

Halle, 10 February 1931
Moltkestr. 5

Dear Mr. Gödel,

Since I did not recall having your address, I had some days ago directed to Mr. Kaufmann the request that he contact you about the following matter. Kaufmann is, to be sure, in Breitenstein these days; he has already promised to inquire with you just after his return. Since in the meantime I happened to find again your return address on the envelope of the paper you had kindly sent me, I would like to inquire of you myself.

In the supplement to *Erkenntnis* (Dubislav, Die Definition[b]) I read to my astonishment §48, "Behmann's attempted solution and its refuta-

would thus be a draft of the part of Gödel's reply (letter 6) beginning with the second paragraph. For a full transcription of the shorthand, see the Textual Notes.

[b]*Dubislav 1931*.

seine Widerlegung". Ich finde es schon den wissenschaftlichen Gepflogen-
heiten wenig entsprechend, wenn Herr Dubislav, lediglich indem er über
einen gegen sie erhobenen Einwand *berichtet*, ohnen ihnen zu nennen,
sich bereits das Recht nimmt, meine Auflösung für widerlegt zu erklären.
Ich hatte anfangs geglaubt, daß Sie seit dem Herbst einen neuen Einwand
gemacht hätten, von dem ich unbegreiflicherweise nichts erfahren hätte;
indessen scheint es mir wahrscheinlicher, daß Sie den damals gemachten
Einwand, aber nicht später seine Entkräftung, Herrn D. mitgeteilt haben, worauf dann Herr D. unvorsichtigerweise ohne Rückfrage bei mir
die Sache veröffentlicht hat. Würden Sie die Freundlichkeit habe⟨n⟩, mir
möglichst umgehend zu schreiben bzw. mich durch Herrn Kaufmann wissen zu lassen, wie die Angelegenheit sich verhält. Selbstverständlich liegt
es mir, falls ich mit meiner eben geäußerten Vermutung Recht habe, völlig fern, Ihnen irgend einen Vorwurf zu machen, da es natürlich Herrn
D.s Sache war, sich in diesem Punkte ausreichend zu vergewissern.

Auf alle Fälle muß natürlich so rasch wie möglich etwas geschehen, um
dem leider entstandenen ungünstigen Eindruck entgegenzuwirken. Ich
würde es—falls es sich tatsächlich noch um den damaligen Ei⟨n⟩wand
handelt—für das richtigste halten, wenn Sie im nächsten Heft der "Erkenntnis" bzw. so rasch, wie es technisch möglich ist, eine kurze Berichtigung bringen würden. Sie könnten ja mit Herrn Carnap darüber sprechen. Sollte Herr Carnap es für besser halten, daß ich die Berichtigung
schreibe, so würde ich natürlich auch dazu bereit sein.

2   Eben ist mein Prager Vortrag^c erschienen; sobald ich die Abdrucke |habe,
werde ich natürlich auch Ihnen einen zusenden.

Mit den besten Grüßen, auch an Herrn Carnap,

Ihr

Heinrich Behmann

---

^c *Behmann 1931.*

2. Gödel to Behmann

Wien 22./II. 1931

Sehr geehrter Herr Behmann!

Ich komme erst heute dazu Ihr Schreiben vom 10./II. zu beantwor-

tion." I find it little in accord with scientific customs when Mr. Dubislav already assumes the right to declare my solution refuted, while he merely *reports* on an objection made against it without naming it. I had initially believed that since last fall you had made a new objection, of which I had incomprehensibly heard nothing; however, it seems to me more probable that you communicated to D. the earlier objection but did not later communicate its disabling, after which D. carelessly published the matter without checking with me. Would you have the kindness to write to me as soon as possible or let me know via Kaufmann how the matter stands? Obviously, in case I am right in the conjecture I have just expressed, the last thing I would do is to make any reproach to you, since it was of course D.'s responsibility to assure himself sufficiently on this point.

In any case, something must happen as quickly as possible in order to counteract the unfavorable impression that has unfortunately arisen. If it is really still the previous objection that's involved, I would consider it the most correct course if you would bring out a short correction in the next issue of *Erkenntnis* or as quickly as is technically possible. You might talk about it with Carnap. If Carnap considers it better for me to write the correction, I would naturally be prepared to do it.

My Prague lecture[c] has just appeared; as soon as I have offprints, of course I will send you one.

With cordial greetings, also to Carnap,

> Yours,
>
> Heinrich Behmann

---

2. Gödel to Behmann

Vienna, 22 February 1931

Dear Mr. Behmann,

I am only today getting to replying to your letter of 10 February, be-

ten, weil ich mir zunächst den Aufsatz von Herrn Dubislav[a] verschaffen
wollte, was ~~durch~~ sich durch einen Zufall verzögert hat. Wie Ihnen Herr
Kaufmann ja schon mitgeteilt hat, werde ich (auf dessen Rat) die Angelegenheit in der Weise erledigen, daß ich an Herrn D. ein Schreiben richte,
mit dem Ersuchen, es so bald als möglich in der "Erkenntnis" zu pub-

2   lizieren. Der Einwand, um den es sich handelt, ist | tatsächlich derselbe,
den ich Ihnen seinerzeit in Wien mitgeteilt habe. Ich erzählte ihn Herrn
D. auf der Rückreise von Königsberg, ohne im geringsten an eine Publikation von meiner oder seiner Seite zu denken. Eine solche lag mir umso
ferner, als mir damals Ihr Auflösungsversuch gar nicht hinreichend genau
bekannt war. Der ganze Wortlaut der zu publizierenden Mitteilung, die
noch heute an Herrn D. abgeht, folgt auf der nächsten Seite.

Mit den besten Grüßen

Ihr Kurt Gödel

3 | In Ihrer Arbeit[b] .... §48, "Der Behmannsche Auflösungsversuch..."
S 96 scheint ein Mißverständnis vorzuliegen. Ich habe niemals behauptet,
den Behmannschen Versuch widerlegt zu haben. Die von mir gelegentlich
gemachte Bemerkung (deren Publikation übrigens von meiner Seite niemals beabsichtigt war) bezog sich vielmehr lediglich auf das folgende:
Man kann den Russellschen Widerspruch jedenfalls nicht allein durch
Ausmerzung eines Kur[[z]]zeichens vermeiden, sondern man muß notwendig auch irgendwelche Einschränkungen im Operieren mit *kurzzeichen-*

4   *freien* Ausdrücken machen, oder anders aus|gedrückt: Wenn man die
Regeln der formalen Logik *völlig uneingeschränkt* anwendet, (d. h. insbes. für Funktionsvariable *beliebige* Ausdrücke mit der entsprechenden
Anzahl freier Variablen einsetzen darf) dann kommt man, *auch ohne*
überhaupt ein Kurzzeichen einzuführen, zu einem Widerspruch. (Womit
~~aber~~ ⟨übrigens⟩ noch nicht gesagt ist, daß die erforderlichen Einschränkungen gerade in der Richtung der Russellschen Typentheorie liegen
müßten[[.]]) Als ich Ihnen gegenüber diese Bemerkung machte, war mir
der Inhalt des Behmannschen Vortrages nicht hinreichend genau bekannt,
um beurteilen zu können, ob und wieweit dieser dadurch betroffen wird.

5   Aus einer Mitteilung Herrn Behmanns, | mit dem ich gelegentlich über
diese Sache gesprochen habe, entnehme ich, daß dies *nicht* der Fall ist,
weil aus seiner Auffassung sich auch gewisse Einschränkungen im Operieren mit *kurzzeichenfreien* Ausdrücken ergeben u. zw. Einschränkungen

---

[a]Evidently *Dubislav 1931*.

cause I wanted first to obtain Dubislav's essay,[a] which was delayed by chance. As Kaufmann has already informed you, I will, on his advice, take care of the matter by writing a letter to D. with the request that he publish it as soon as possible in *Erkenntnis*. The objection that is at issue is in fact the same as the one I previously communicated to you in Vienna. I told D. about it on the trip back from Königsberg, without in the least thinking of a publication either on my part or on his. Such was even further from my mind, since at the time your attempted solution was not known to me in sufficient detail. The entire wording of the communication to be published, which is going off to D. today, follows on the next page.

With cordial greetings,

Yours, Kurt Gödel

In your monograph[b]...§48, "Behmann's attempted solution..." p. 96, there seems to be a misunderstanding. I never claimed to have refuted Behmann's attempt. The remark made on occasion by me (whose publication, moreover, I never intended) rather concerned only the following: One can in any case not avoid Russell's contradiction just by elimination of an abbreviation, but one must necessarily also make some restrictions on operating with *unabbreviated* expressions, or, in other words: If one applies the rules of formal logic *without any restriction* (that is, in particular, one may substitute for functional variables *arbitrary* expressions with the corresponding number of free variables), then one still arrives at a contradiction *without introducing any abbreviation at all*. (This would, by the way, not yet say that the required restrictions would have to be just in the direction of Russellian type theory.) When I made this observation to you, the content of Behmann's lecture was not known to me in sufficient detail for me to judge whether and to what extent it was affected. From a communication of Mr. Behmann, with whom I have occasionally spoken about this matter, I gather that this is *not* the case, because from his approach certain limitations in operating with expressions

[b] *Dubislav 1931.*

die nicht in der Richtung der Typentheorie liegen, sondern die Implikationsregel betreffen.

3. Behmann to Gödel

Halle (Saale), 25. Februar 1931.
Moltkestr. 5.

Sehr geehrter Herr Gödel!

Ihren Brief vom 22. erhielt ich gestern Abend. Ich bin natürlich mit dem ei⟨n⟩geschlagenen Weg einverstanden. Doch erscheint es mir notwendig, daß der springende Punkt nicht auch hier wieder dem Leser vorenthalten wird. Ich möchte daher vorschlagen, die Argumentation und ihre Kritik gemäß dem beiliegenden Entwurf[a] in die Veröffentlichung aufzunehmen. Zugleich habe ich mir erlaubt, den entscheidenden Kern Ihrer Erklärung bei möglichster Wahrung des Sinnes noch etwas schärfer herauszuarbeiten, damit der uneingeweihte Leser es leichter hat, zu verstehen, worauf es ankommt. Allerdings wäre nunmehr zu überlegen, ob, da Sie den Brief an Herrn Dubislav ja bereits in der ursprünglichen Fassung abgeschickt haben, es noch tunlich erscheint, die Briefform beizubehalten; doch würde das ja gegebenenfalls durch Änderung weniger Worte erreicht werden. Ein anderer Punkt ist der, ob es zweckmäßig erscheint, meine Symbolik zu verwenden. Man wird ja verstehen, daß mir daran liegt, die Kenntnis und Anwendung meiner Symbolik, für deren Aufstellung ich gewichtige Gründe hatte, verbreiten zu helfen. Sollte indessen namentlich Herr Carnap triftige Gründe für die Verwendung der Russellschen Symbolik in dem vorliegenden Fall sehen, so wäre ja auch dies leicht zu ändern. Den Hinweis auf mein Buch "Mathematik und Logik"[b] möchte ich als einen kleinen Ersatz für die mir nicht recht verständliche, gewiß versehentliche Übergehung meiner Schriften in der Bibliographie des Berichtes der Prager Tagung betrachten. Für den Fall, daß jene Bibliographie gelegentlich ergänzt wird, würde ich natürlich gern eine Aufzählung meiner Schriften einsenden.

Einen Durchschlag des beiliegenden En⟦t⟧wurfes sende ich gleichzeitig Herrn Dubislav.

---

[a]This is very probably document no. 010015.423 in Gödel's papers, a typescript consisting of a revision of Gödel's proposal attached to the preceding letter, and a further section entitled "Zusatz von H. Behmann", giving the argument in more formal detail.

without abbreviations result, namely limitations that don't lie in the direction of type theory but concern the rule for implication.

3. Behmann to Gödel

Halle (Saale), 25 February 1931
Moltkestr. 5

Dear Mr. Gödel,

I received your letter of the 22nd yesterday evening. Naturally I am in agreement with the course that has been taken. But it still seems to me necessary that the salient point should not here again be withheld from the reader. Thus I would like to propose that the argumentation and the criticism of it be taken up into the publication according to the enclosed draft.[a] At the same time I allowed myself to work out somewhat more sharply the decisive core of your explanation, preserving the sense as much as possible, so that it is easier for the uninitiated reader to understand what the matter turns on. Since you have already sent off the letter to Dubislav in the original version, it should certainly be considered whether it is still wise to stick to the letter form; still, in that case that could be achieved by changing a few words. Another point is whether it seems appropriate to use my symbolism. One will understand that I have an interest in helping to spread knowledge and application of my symbolism, for whose construction I had weighty reasons. Should, however, Carnap in particular see convincing reasons for the application of the Russellian symbolism in the present case, then this too could be easily changed. I would like to consider the reference to my book "Mathematik und Logik"[b] as a small compensation for the no doubt unintentional but to me not quite understandable passing over of my writings in the bibliography of the report of the Prague conference. For the case that that bibliography is enlarged on some occasion, I would of course gladly send in a list of my writings.

I am sending a carbon copy of the enclosed draft simultaneously to Dubislav.

---

[b] *Behmann 1927.*

Mit den besten Grüßen, auch an Herrn Carnap,

Ihr

Heinrich Behmann

4. Gödel to Behmann

Wien, 18./III. 1931

Sehr geehrter Herr Behmann!

Ihr Schreiben vom 25./II sowie Ihr Separatum,[a] für das ich bestens danke, habe ich erhalten. Ihrem Abänderungsvorschlag für meine Bemerkung konnte ich mich nicht in allen Punkten anschließen, vor allem weil mir dann nicht hinreichend deutlich formuliert schien, was ich bewiesen habe. Dagegen habe ich Ihrem Wunsche entsprechend das Verhältnis zu Ihrer Theorie genauer auseinandergesetzt (Vgl. das beiliegende Manuskript, das auch Herrn Carnaps Zustimmung gefunden hat). Ferner habe ich die Grundzüge meines Arguments in Formeln angeführt, damit Sie sich in Ihrem Zusatz direkt darauf beziehen können. Übrigens glaube ich (auch Herr Carnap ist dieser Meinung), daß es zur Klarstellung des Sachverhalts beitragen würde, wenn Sie in Ihrem Zusatz deutlich zum Ausdruck brächten, daß Sie gegen meine die Kurzzeichenregel betreffende
2 | Behauptung nichts einzuwenden hätten, daß aber der von mir konstruierte Widerspruch durch andere in Ihrem Prager Vortrag formulierte Regeln ausgeschlossen wird.

Eine Kopie des beiliegenden Manuskripts sende ich gleichzeitig an Herrn Dubislav und hoffe damit die Angelegenheit erledigt zu haben. Für die Zusendung der endgültigen Formulierung Ihres Zusatzes wäre ich Ihnen sehr verbunden.

Mit den besten Grüßen

Ihr Kurt Gödel

P.S. Herr Carnap, der Ihren Brief ebenfalls gelesen hat, läßt Ihnen mitteilen, daß anläßlich des Berichts über die Königsberger Tagung in der "Erkenntnis" eine Bibliographie der Grundlagen der Mathematik in

---

[a]Evidently *Behmann 1931*.

With cordial greetings, also to Carnap,

Yours,

Heinrich Behmann

4. Gödel to Behmann

Vienna, 18 March 1931

Dear Mr. Behmann,

I have received your letter of 25 February as well as your reprint,[a] for which I thank you. I could not endorse your proposed revision of my remark in all points, above all because then what I proved seemed not to be formulated distinctly enough. On the other hand, accommodating your wish, I have set forth the relation to your theory more exactly (cf. the enclosed manuscript, which also found Carnap's assent). Furthermore, I have given the outline of my argument in formulae, so that you can refer to it directly in your addition. Moreover I believe (and Carnap is also of this opinion) that it would help to clarify the matter if you were to bring out clearly in your addition that you had no objection to my claim concerning the rule for abbreviations, but that the contradiction I constructed is excluded by other rules formulated in your Prague lecture.

I am sending a copy of the enclosed manuscript right away to Dubislav and hope that with it the matter has been settled. I would be much obliged to you for sending the final formulation of your addition.

With cordial greetings,

Yours, Kurt Gödel

P.S. Carnap, who has also read your letter, sends word to you that on the occasion of the report in *Erkenntnis* of the Königsberg conference, a

Aussicht genommen sei und daß, im Falle dieses Projekt zur Ausführung gelangt, selbstverständlich auch Ihre Schriften angeführt sein werden.[b]

---

[b] *Hahn et alii 1931*, p. 151, gives a bibliography on foundations of mathematics, which lists *Behmann 1922, 1927* and *1931*.

5. Behmann to Walter Dubislav and Gödel

Halle (Saale), 25. März 1931.
Moltkestr. 5.

Sehr geehrter Herr Dubislav!
Sehr geehrter Herr Gödel!

Gestatten Sie mir, daß ich mich der Einfachheit halber mit diesem Brief an Sie beide zugleich wende.

Aus Ihren Sch[[r]]eiben ersehe ich, daß es doch große Schwierigkeiten macht, die Berichtigung in eine Form zu bringen, mit der alle Teile gleichmäßig einverstanden sind. Ich hatte es natürlich als am wirkungsvollsten betrachtet, wenn die Berichtigung von einem von Ihnen ausginge. Doch hat die Sache den Haken, daß die Aufklärung der in Rede stehenden Argumentation doch auf jeden Fall aus technischen Gründen meine Aufgabe bleibt und dadurch bereits der Hauptteil der Berichtigung, indem er jenen Punkt beseite läßt, notgedrungen bis zu einem gewissen Grade der Geschlossenheit und damit vielleicht auch der Klarheit entbehren muß, womit zugleich die Gefahr entsteht, daß manche Gesichtspunkte, die sachlich nicht von wesentlicher Bedeutung sind, sich unnötig in den Vordergrund drängen und damit der eigentlichen Absicht der Berichtigung, die doch der rein sachlichen Aufklärung des Lesers und nur dieser dienen soll, entgegenwirken.

Um diesen Übelstand zu vermeiden, schlage ich die beiliegende Fassung der Berichtigung vor, die diese als von mir ausgehend erscheinen läßt.[a]

---

[a] This is in all probability document no. 010015.426 in Gödel's papers, a typescript entitled "Zur Richtigstellung einer Kritik meiner Auflösung der logisch-mengentheoretischen

bibliography of the foundations of mathematics is envisaged, and in case this project is carried out, of course your writings will be listed.[b]

---

5. Behmann to Walter Dubislav and Gödel

Halle (Saale), 25 March 1931.
Moltkestr. 5.

Dear Mr. Dubislav,
Dear Mr. Gödel,

Permit me, for the sake of simplicity, to address both of you simultaneously with this letter. From what you have written I see that bringing the correction into a form with which all parties are equally in agreement does have great difficulties. Of course I had thought it most effective if the correction were to come from one of you. But there's a catch in the matter, that in any case, for technical reasons, the explanation of the relevant argumentation remains my task, and as a result the principal part of the correction, in that it leaves that point aside, necessarily must to a certain degree lack solidity and therewith possibly also clarity, whereby at the same time the danger arises that several aspects that are not objectively of essential significance are pressed unnecessarily into the foreground and work against the true intention of the correction, which is the purely objective enlightenment of the reader and only that.

To avoid this evil, I propose the enclosed version of the correction, which lets it appear as coming from me.[a]

Widersprüche", i.e., the title of *Behmann 1931a*. From Gödel's following letter it can be determined that this is the draft on which he is commenting.

Zur Beurteilung möchte ich Sie bitten, die folgenden Gesichtspunkte zu bedenken:

Der entscheidende Umstand ist, daß Herr D. in seinem Buch eine Beurteilung ausgesprochen hat, die er gegenwärtig nicht mehr vertritt, und daß eine diesbezügliche Aufklärung der Leser des Buches innerhalb der Grenzen der technischen Möglichkeit hierfür nach unserer übereinstimmenden Ansicht dringend erwünscht erscheint. Darum und nur darum handelt es sich, dies zu bewirken. Damit ist gegeben, daß alles Persönliche tunlichst aus dem Spiel bleibt. Es soll sich also weder um eine Entschuldigung (d. h. eine Bitte um E.) noch um eine Rechtfertigung in persönlicher Hinsicht, um eine Feststellung darüber, daß man selber aus diesen oder jenen Gründen für das Mißverständnis nichts gekonnt habe, handeln. Da wir unter uns darüber vollkommen einig sind, daß keinerlei böse Absicht im Spiel gewesen ist, besteht m. E. kein Anlaß, diese vorhandene Basis der Ver⟨s⟩tändigung und ebenso, daß wir in der Sache selbst, wie ich doch glaube, völlig einig gehen, vor dem Leser | dem Anschein nach in Frage zu stellen.

2

So erscheint es mir durchaus nicht nötig, auf den Inhalt Ihres damaligen Gespräches in der Berichtigung einzugehen—vielleicht wäre es nicht einmal unbedingt erforderlich, zu sagen, daß er stattgefunden hat—; denn es wird ja z. B. kein vernünftiger Mensch verlangen, daß das damals von Herrn G. in durchaus vorläufiger und unverbindlicher Form Ausgesprochene in vollkommener Übereinstimmung mit seiner jetzigen Ansicht stehen solle, und es wird den Leser daher kaum interessieren, sondern eher ablenken, zu erfahren, ob oder inwieweit dies doch der Fall ist. Allenfalls würde Herrn G.s Bemerkung wegen der Fußnote eine gewisse Berechtigung haben.

Ebensowenig scheint es mir der ausdrücklichen Hervorhebung zu bedürfen, daß Herr G. den *Beweis* geliefert hat, daß mein erstes Prinzip zu[[r]] Auflösung der Widersprüche nicht ausreicht. Denn einmal ist das zweite Prinzip eine so natürliche, zwang[[s]]läufige F̶polge des ersten, daß die Beschränkung auf das erste nur eine sklavisch-formale, aber keine vernünftig-sachliche Durchführung eben dieses Printzips bedeuten würde (ich glaube, daß dieser Umstand aus meiner Darstellung auf S. 43 hinreichend hervorgeht; man könnte ja auch unschwer beide Prinzipien in eines zusammenfassen.) Und andererseits war dieser Beweis ja schon früher von anderen Seiten geliefert worden; ich glaubte den diesbezüglichen Vermerk aus Gründen der Gerechtigkeit hinzufügen zu sollen. Natürlich besagt dies nichts gegen die Tatsache, daß ich selbst Herrn G. für die Mitteilung seines Beispiels sehr dankbar bin und dieses als eine W̶i̶ willkommene Gelegenheit zur Erprobung meiner Theorie betrachtet habe. (Herrn G.s Wunsch glaube ich im Schlußabsatz erfüllt zu haben.)

Die Argumentation selbst habe ich in der einfacheren Form belassen. Ich glaube, daß man auf die formale Herleitung der richtigen Alterna-

In assessing it, may I ask that you keep in mind the following points:

The decisive circumstance is, that D. has pronounced a judgment in his book that at present he no longer stands by, and that in our shared opinion an informing of the reader of the book about this matter seems, within the limits of technical possibility, urgently to be desired. To bring this about, and only this, is what's at issue. Therewith it is a given that everything personal stays as much as possible out of things. Neither an excuse (that is, an apology) nor a justification from a personal viewpoint, a statement that one for this or that reason could do nothing about the misunderstanding, is at issue. Since we are completely in agreement among ourselves about the fact that no bad intention of any kind was involved, there is in my opinion no occasion for calling in question before the reader, even in appearance, either this present basis of understanding or that we are proceeding in complete agreement about the thing itself, as I do believe.

So it appears to me not at all necessary in the correction to go into the content of your conversation of that time—perhaps it wouldn't even be absolutely required to say that it took place—; after all, no reasonable man would demand, for example, that what G. said at that time in a completely provisional and non-binding form should be in complete agreement with his present view, and it will scarcely interest the reader, but rather distract him, to learn whether or to what extent this is the case. At best G.'s remark would have a certain justification, on account of the footnote.

Just as little does it appear to me to need explicit emphasis that G. provided the proof that my first principle does not suffice to resolve the contradictions. For, for one thing, the second principle is such a natural and inevitable consequence of the first that the limitation to the first would mean only a slavishly formal and not an objectively reasonable execution of just this principle. (I believe that this condition follows sufficiently from my exposition on p. 43; indeed one could without difficulty combine both principles into one.) And on the other hand this proof had already been given by others; I thought I should add the notation to this effect on grounds of justice. Of course this says nothing contrary to the fact that I myself am very grateful to G. for communicating his example and have considered this a welcome opportunity for testing my theory. (I think I have fulfilled G.'s wish in the closing paragraph.)

I have left the argumentation itself in the simpler form. I think one can by all means do without the formal derivation of the correct alter-

tive ruhig verzichten kann—auf diese Weise wird der Leser nur mit dem unumgänglichen Minimum von Symbolik behelligt.

Daß Herr G. die Ergänzugng meines ersten Prinzips nicht in der Richtung der Typentheorie hat suchen wollen, bedarf wohl nicht der Bekräftigung, da Herr D. sich in diesem Punkt ja nicht auf ihn berufen hat. Aus dem gleichen Grunde darf die Feststellung fehlen, ob oder inwieweit Herr G. meine Theorie damals überhaupt als unzureichend betrachtet hat.

Natürlich würde die Anfügung einer kurzen zustimmenden Äußerung Herrn D.s sehr erwünscht sein. Wir sind uns wohl so weit einig oder werden uns noch einigen, daß diese nicht auf die ⟨eine⟩ "Berichtigung der Berichtigung" hinauszulaufen braucht.

Es würde mich herzlich freuen, durch meinen vermittelnden Vorschlag die deutsch-österreichische Einigung auch in unserm Fall erreicht zu haben.

Mit den besten Grüßen

    Ihr

      Heinrich Behmann

6. Gödel to Behmann

Wien 22./IV. 1931

    Sehr geehrter Herr Behmann!

Mit Ihrem Vorschlag vom 25./III. bin ich im Prinzip einverstanden; ich bin auch ganz Ihrer Meinung, daß alles Persönliche möglichst in den Hintergrund treten soll. Doch muß ich gerade mit Rücksicht auf die sachliche Klarstellung darauf dringen, daß meine Argumentation in ausführlicher Form von Ihnen dargestellt wird, falls Sie allein die Berichtigung schreiben. Meine Behauptung ist ja eine rein formal-mathematische und kann daher ohne ausreichenden Gebrauch des Formalismus nicht adäquat dargestellt werden. Sonst entsteht die Gefahr, ⟨daß der Leser glaubt,⟩ es handle sich um irgendwelche inhaltliche Überlegungen mittels der Formeln, was durchaus nicht der Fall ist. Ich schlage daher vor, an Stelle des 2. und 3. Absatzes (Zeile 6–14) Ihres Manuskripts folgendes zu setzen:

2 | Die fragliche Argumentation bezieht sich nur darauf, daß man bei uneingeschränkter Anwendung der formalen Regeln des typenlosen Logikkalküls Widersprüche auch *ohne Verwendung von Kurzzeichen* konstruieren

native—in this way the reader is bothered only with the unavoidable minimum of symbolism.

That G. did not want to look for amplification of my first principle in the direction of type theory does not need confirmation, since D. did not appeal to him on this point. For the same reason no statement needs to be made whether or how far G. at the time considered my theory generally inadequate.

Naturally the addition of a short statement of D.'s agreement would be very desirable. We are evidently so far in agreement or will come to agreement, that this doesn't need to lead to a "correction of the correction."

It would please me heartily to have achieved through my mediating proposal German–Austrian unity also in our case.

With cordial greetings,

Yours,

Heinrich Behmann

6. Gödel to Behmann

Vienna, 22 April 1931

Dear Mr. Behmann,

I am in principle in agreement with your proposal of 25 March; I am also entirely of your opinion that everything personal should as much as possible step into the background. Still, I must insist precisely out of concern for objective clarification that my argument be presented by you in complete form, in case you alone write the correction. My claim is evidently a purely formal and mathematical one and cannot be adequately presented without sufficient use of the formalism. Otherwise the danger arises that the reader will think the issue is about some considerations or other about the content expressed by the formulae, which is not the case at all. Thus I propose to put in place of the second and third paragraphs of your manuscript (lines 6–14) the following:

The argumentation in question relates only to the fact that with unlimited application of the formal rules of the type-free logical calculus contradictions can be constructed also *without use of abbreviations*, in the

kann u. zw. auf folgende Weise: (Fortsetzung siehe beiliegende Maschinenschreibseite!ᵃ)

Ferner möchte ich Sie ersuchen, die historischen Details über Zeitpunkt u. Inhalt des Privatgesprächs zwischen Herrn D. und mir bzw. Ihnen und mir—ganz im Sinne Ihres letzten Schreibens—wegzulassen, weil dadurch wieder der falsche Anschein Erweckt wird, als hätte ich irgendeine Herrn D. gegenüber ausgesprochene Behauptung später zurückgenommen. Wenn schon derartige historische Details (welche auch nach meiner Meinung den Leser nicht interessieren werden) gebracht werden sollten, so müßte auch festgestellt werden, *was* ich Herrn D. gegenüber geäußert habe, und das soll doch gerade vermieden werden, um unnötige

3 Auseinandersetzungen aus dem Wege zu gehen. | Die Feststellung, daß der Widerspruch im Sept 1930 von Ihnen aufgeklärt wurde, kann doch für Sie umsoweniger in's Gewicht fallen, als Sie ja dieselbe Aufklärung schon früher Herrn Bernays u. Ackermann gegeben haben müssen, und die Worte "auch von Herrn G. als einwandfrei anerkannt" würde ich nicht bedingungslos unterschreiben, weil für mich eine einwandfreie Lösung des Antinomienproblems nur in der Angabe eines die Widersprüche vermeidenden Formalismus mit präzis formulierten Axiomen u. Schlußregeln liegt, und ein solcher von Ihnen bisher nur andeutungsweise gegeben wurde (den in Wien mir mitgeteilten Ansatz in Form einer Beschränkung der Implikationsregel haben Sie ja übrigens gar nicht ausgeführt)—Ich möchte Sie ferner bitten anzuführen, daß eine Publikation von meiner Seite nicht beabsichtigt war u schlage daher folgende abgeänderte Form Ihrer Fußnote ¹ vor:

4 Die obige Argumentation, die mir im Sept 1930 | von Herrn Gödel vorgelegt wurde, war mir übrigens in nahezu dergleichen Gestalt im Jun. 1930 von Herrn B. und in etwas abweichender bereits im Nov. 1928 von Herrn A. mitgeteilt worden. Herr Gödel ersucht mich noch, festzustellen, daß eine Publikation von seiner Seite nicht in Aussicht genommen war u. die diesbezügliche Fußnote des Herrn Dubislav auf einem Mißverständnis beruht.

Ich möchte Sie weiter noch ersuchen, nicht von der Gödelschen sondern von der "obigen" Argumentation zu sprechen, da sie ja bereits früher von anderen angegeben wurde u. schließlich möchte ich bemerken, daß mir der letzte Absatz Ihres Manuskripts— ~~den~~ ⟨falls⟩ Sie ~~, wenn ich~~

ᵃThis text is what is printed at the end of the letter; concerning its identification see the textual notes.

following way: (for continuation see the enclosed typed page[a])

Further I would like to ask you to leave out the historical details about the time and content of the private conversations between D. and me and between you and me—entirely in the sense of your last letter, because thereby the false appearance is aroused as if I had later taken back some assertion or other made to D. If historical details of that kind (which in my opinion would not interest the reader) should be brought out, it would also have to be stated *what* I said to D., and that is just what should be avoided in order to circumvent unnecessary disputes. The statement that the contradiction was cleared up by you in September 1930 can be of even less consequence to you since you must have already given the same solution earlier to Bernays and Ackermann, and I wouldn't subscribe without conditions to the words "also recognized as indisputable by G.", since for me an indisputable solution of the problem of the antinomies lies only in the presentation of a formalism with precisely formulated axioms and rules of inference that avoids the contradictions, and such a ⟦formalism⟧ has up to now only been indicated by you (you have by the way not carried out at all the approach communicated to me in Vienna in the form of a restriction on the rule of implication). I would like to ask you further to mention the fact that a publication on my part was not intended, and thus I propose the following revised form of your footnote 1:

The argumentation above, which was presented to me in September 1930 by Gödel, had been communicated to me in almost the same form in June 1930 by B⟦ernays⟧ and in a form somewhat deviating from that already in November 1928 by A⟦ckermann⟧. Gödel requests that I further state that a publication on his part had not been envisaged and that the footnote of Dubislav to this effect rests on a misunderstanding.

I would like to ask you further not to speak of "Gödel's" but of the "above" argument, since it was already given earlier by others, and finally I would like to remark that the last paragraph of your manuscript—in case you inserted it only on my wish—seems superfluous in the light of what was said previously, since you have already clearly

nicht verstanden habe, ⟨denselben nur⟩ auf meinen Wunsch hineinfüg-
ten—mit Rücksicht auf das vorher Gesagte überflüssig erscheint, weil Sie
ja bereits ⟨im⟩ 4. Absatz deutlich erklärt haben, daß Ihr erstes Prinzip
allein nicht hinreicht.

　　Ich hoffe, daß Sie mit meinen Abänderungsvorschlägen einverstanden
5　sind, da diese ja in | keiner Weise Ihre sachliche Stellungnahme berühren,
und daß auf diesem Wege Ihr Vermittlungsvorschlag doch zu dem wün-
schenswerten Ziele führen wird.

　　　　Mit den besten Grüßen

　　　　Ihr Kurt Gödel

　　Anbei ein Separatum meiner jüngst erschienenen Abhandlung.[b]

6　| Man setze in die beweisbare Formel $F(\psi) \supset (\exists\phi)F(\phi)$ an Stelle von
$F(\psi)$ den Ausdruck $(\chi)[\psi(\chi) \equiv \sim \chi(\chi)]$ und in in der so entstehenden
Formel für $\psi(\chi)$ den Ausdruck $\sim \chi(\chi)$ ein. Dann ergibt sich durch Anwendung der Implikationsregel

$$(\exists\phi)(\chi)[\phi(\chi) \equiv \sim\chi(\chi)] \qquad (1)$$

Ferner kann man bei den oben genannten Voraussetzungen ⟨uneinge-
schränkter Anwendung der logischen Regeln⟩ in gewöhnlicher ⟨bekannter⟩
Weise die Formel $(\exists\phi)(\chi)G(\phi,\chi) \supset (\exists\phi)G(\phi,\phi)$ ableiten. Setzt man in
dieser für $G(\phi,\chi)$　$\phi(\chi) \equiv \sim \chi(\chi)$ ein, so folgt unter Berücksichtigung von
(1):

$$(\exists\phi)[\phi(\phi) \equiv \sim \phi(\phi)] \qquad (2)$$

während andererseits natürlich auch die Negation von (2) beweisbar ist.

---

[b]Presumably *Gödel 1931*.

7. Behmann to Gödel

　　　　　　　　　　　　　　Halle (Saale), Moltkestr. 5.
　　　　　　　　　　　　　　18. Mai 1931.

　　Sehr geehrter Herr Gödel!

　　Da ich inzwischen längere Zeit gebraucht hatte, um mich von dem
langen Winter zu erholen, komme ich erst jetzt auf Ihren Brief vom 22.
April und auf die Angelegenheit der Berichtigung zurück.

explained in the fourth paragraph that your first principle alone is not sufficient.

I hope that you are in agreement with my proposals for revision, since they do not touch your position on the matter at issue, and that in this way your proposal for mediation will lead to the desirable goal.

With cordial greetings

Yours, Kurt Gödel

Enclosed is a reprint of my recently published paper.[b]

In the provable formula $F(\psi) \supset (\exists\phi)F(\phi)$, substitute for $F(\psi)$ the expression $(\chi)[\psi(\chi) \equiv \sim \chi(\chi)]$ and in the resulting formula for $\psi(\chi)$ the expression $\sim\chi(\chi)$. Then by applying the rule of implication we obtain

$$(\exists\phi)(\chi)[\phi(\chi) \equiv \sim\chi(\chi)] \qquad (1)$$

Further we can by unrestricted application of the logical rules derive the formula $(\exists\phi)(\chi)G(\phi,\chi) \supset (\exists\phi)G(\phi,\phi)$ in a well-known way. If in this one substitutes $\phi(\chi) \equiv \sim\chi(\chi)$ for $G(\phi,\chi)$, then, taking account of (1),

$$(\exists\phi)[\phi(\phi) \equiv \sim \phi(\phi)] \qquad (2)$$

follows, while on the other hand of course the negation of (2) is also provable.

---

7. Behmann to Gödel

Halle (Saale), Moltkestr. 5
18 May 1931

Dear Mr. Gödel,

I am only now getting back to your letter of 22 April and the matter of the correction, since in the meantime it took a long time for me to recover from the long winter.

Zunächst meinen besten Dank für die Übersendung Ihrer Arbeit! Ich hatte gerade kurz vorher durch Herrn Herbrand von ihr gehört. Er hält sie für sehr wichtig und für sachlich durchaus in Ordnung. Ich selbst habe noch keinen gründlichen Einblick in sie nehmen können, werde das aber, sobald es mir möglich ist, nachholen.

Beifolgend sende ich Ihnen den gemäß Ihren Wünschen geänderten Artikel im Durchschlag.[a] Die Urschrift geht gleichzeitig an Herrn Carnap zur Aufnahme in die "Erkenntnis", ein weiterer Durchschlag an Herrn Dubislav.

Ich habe nur noch in der Fußnote 1 den Zusatz "und von mir in dem hier angegebenen Sinne aufgeklärt" eingefügt, der ja über Ihre Anerkennung meiner Aufklärung nichts aussagt. Sollte er Ihnen gleichwohl untunlich erscheinen, so bitte ich Sie, die Streichung des ganzen Nebensatzes "die...wurde" zu veranlassen.

Es grüßt Sie bestens

Ihr

Heinrich Behmann

---

[a]This copy is in all probability document no. 010015.424 in Gödel's papers, which differs from *Behmann 1931a* only in the wording of footnote 1; see the introductory note.

First, thank you very much for sending your paper! I had heard about it from Herbrand only just before. He considers it very important and completely correct as to content. I myself have not yet been able to take a thorough look at it but will make up for that as soon as possible.

I am sending along a carbon copy of the article, altered corresponding to your wishes.[a] The original is going simultaneously to Carnap for inclusion in *Erkenntnis*; a further carbon is going to Dubislav.

I have only introduced into footnote 1 the addition "und von mir in dem hier angegebenen Sinne aufgeklärt" [["and resolved by me in the sense given here"]], which says nothing about your acceptance of my solution. If it should nevertheless seem to you unwise, then I ask you to bring about the deletion of the whole clause "die...wurde".[b]

With most cordial greetings,

Yours,

Heinrich Behmann

---

[b]The whole clause reads "die mir im September 1930 von Herrn Gödel vorgelegt und von mir in dem hier angegebenen Sinne aufgeklärt wurde" [["which was presented to me by Gödel in September 1930 and resolved by me in the manner given here"]]. This clause is indeed omitted in *Behmann 1931a*, which otherwise agrees with document no. 010015.424, mentioned in the previous note.

Paul Bernays

Paul Bernays

The correspondence between Paul Bernays and Kurt Gödel included below is one of the most extensive in this collection; it comprises 85 items, ranging in dates from 1930 to 1975, drawn in part from the Bernays archives at the ETH (Eidgenössische Technische Hochschule) in Zürich and in part from the Gödel archives at the IAS (Institute for Advanced Study) in Princeton. Besides being a reflection of a long, warm and deep personal and intellectual relationship, the correspondence deals with a rich body of logical and philosophical issues. This introductory note is divided into three parts, according to three natural divisions between the general range of subjects of the exchanges, covering, respectively, the periods 1930–1942, 1956–1964 and 1965–1975.

Bernays was senior to Gödel by eighteen years: he was born in 1888 of Swiss parents in London, from where the family soon moved to Paris and then in 1895 to Berlin.[a] He began his studies of mathematics at the University of Berlin with Landau and Schur and continued them at the University of Göttingen with Hilbert, Landau, Weyl and Klein, as well as of physics with Voigt and Born. During that same period in Göttingen he studied philosophy with Leonard Nelson; the latter was identified as the leading neo-Friesian, having renewed the Kantian school of Jakob Friedrich Fries that flourished in the first part of the nineteenth century. Bernays' first philosophical papers were heavily influenced by Nelson.[b] Bernays completed a doctorate on the analytic number theory of binary quadratic forms under Landau's direction in 1912. Later that same year he wrote his Habilitationschrift on analytic function theory at the University of Zürich, and held a position there as Privatdozent until 1917. In that same year Hilbert came to Zürich to deliver his famous lecture, "Axiomatisches Denken", published as *Hilbert 1918*. Having resumed his interest in foundational problems, Hilbert invited Bernays to become his assistant in Göttingen to work with him on those questions. In 1918 Bernays did a second habilitation

---

[a]The biographical information here is drawn from Bernays' short autobiography *1976b*.

[b]Bernays' list of publications in *Müller 1976*, pp. xvii–xxiii, is revealing in this respect. He published five papers in the *Abhandlungen der Fries'schen Schule* between 1910 and 1937, a 1928 article on Nelson's philosophy of mathematics, a 1953 piece on the Friesian viewpoint for a memorial volume for Nelson, and in 1970 and 1974 some contributions to Nelson's collected works. Some of these items were the subject of the correspondence with Gödel, and references to them will be found at the corresponding points below.

thesis, this time on the completeness of the propositional calculus, eventually published, but only in part, as *Bernays 1926*.[c]

As Hilbert's assistant in Göttingen from 1917 to 1922, Bernays was significantly involved in helping Hilbert elaborate his developing ideas about mathematical logic and the foundations of mathematics; by the end of that period these explorations had evolved into the program for finitist consistency proofs of axiomatic systems of mathematical significance.[d] At Hilbert's urging, Bernays was promoted to the (non-tenured) position of Professor Extraordinarius at the University of Göttingen in 1922; he held that until 1933, when he was forced to give up his post by the Nazis as a "non-Aryan" because of his Jewish origins. Hilbert kept Bernays on as his personal assistant for an additional year at his own expense.[e] The following year, Bernays returned to Zürich, where he held a temporary position at the ETH from 1934 on; that finally turned into a regular position as an extraordinary professor in 1945. Despite the move from Göttingen, all through the 1930s he continued his work with Hilbert, whose major output was the publication of the two volumes, *Hilbert and Bernays 1934* and *1939*. Bernays was completely responsible for its preparation; it was the first full exposition of Hilbert's finitist consistency program—and beyond[f]—in a masterful, calm and unhurried presentation. Alongside that, beginning by 1930, Bernays was developing his axiomatic system of sets and classes as a considerable improvement of the axiomatization due to von Neumann (*1925, 1928a*); this was eventually published in *The journal of symbolic logic* in seven parts, stretching from 1937 to 1954.[g]

[c]Surprisingly, that did not include the proof of completeness; only the independence results were presented.

[d]Cf. *Mancosu 1998*, pp. 149–188, and *Sieg 1999* for full accounts of this development.

[e]Cf. *Reid 1970*, p. 205.

[f]Section 5 of *Hilbert and Bernays 1939* is entitled, "Der Anlass zur Erweiterung des methodischen Rahmens der Beweistheorie" (The reason[s] for extending the methodological framework of proof theory). Its subsections cover Gödel's incompleteness theorems and Gentzen's consistency proof for arithmetic, the last in a subsection entitled "Überschreitung des bisherigen methodischen Standpunktes der Beweistheorie" (Going beyond the methodological standpoint of proof theory [adopted] hitherto).

[g]All seven parts are brought together in *Müller 1976*, pp. 1–121, as "A system of axiomatic set theory". The same volume, pp. 121–172, also contains a further significant contribution of Bernays to that subject: "On the problem of schemata of infinity in axiomatic set theory" (*Bernays 1961*, in translation). That makes use of strong reflection principles in a theory of sets and classes to derive the existence of "large" cardinals in the Mahlo hierarchies.

Bernays' distinctive voice in the philosophy of mathematics began to emerge early in the 1920s with his article *1922* on the significance of Hilbert's program, and then more clearly on the same subject in his *1930*.[h] Distancing himself from a strictly finitist position, in a number of articles from then on Bernays expressed a more liberal, yet nuanced, receptiveness to alternative views.[i] This was reinforced through his contact at the ETH in Zürich with Ferdinand Gonseth, who held an "open" philosophy that rejected the possibility of absolute foundations of mathematics or science.[j] He participated in several of the "Entretiens de Zürich" organized by Gonseth beginning in 1938, and with the latter and Gaston Bachelard founded the journal *Dialectica* in 1945 and joined its editorial board.

Bernays visited the Institute for Advanced Study in Princeton in 1935–1936 and again in 1959–1960, at which time he had extensive contact with Gödel. He also paid visits to Gödel during three stays that he had as visiting professor at the University of Pennsylvania between 1956 and 1965. He died in Zürich at the age of 89 in September 1977, just four months before Gödel's death early in 1978.

1. Correspondence 1930–1942

1.1 *The incompleteness results and Hilbert's program.* In letter 1, dated 24 December 1930, Bernays thanked Gödel for an offprint of his paper *1930* on the completeness of the first order functional (or predicate) calculus, comparing it favorably with the "quite difficult [related] considerations" in *Herbrand 1930*. Bernays went on to remark that the non-finitary part of Gödel's proof (implicitly, by use of König's Lemma) can be avoided in "a more modest version" by formalizing it in classical arithmetic (Z); the argument was eventually spelled out in full in *Hilbert and Bernays 1939*, pp. 234–253.[k] For finitary mathematics, this would

---

[h]See *Mancosu 1998*, p. 168ff.

[i]See the collection *Bernays 1976*.

[j]The mathematician and philosopher of science Ferdinand Gonseth (1890–1975) was appointed professor of mathematics at the ETH in 1929. For information about his life and work see *Heinzmann 1982* and *Pilet 1977*; the latter appears in volume 31 (1977) of the journal *Dialectica*, the first two numbers of which consist of articles dedicated to the memory of Gonseth.

[k]It is not clear whether Bernays understood what would have to be done for such a result when he wrote in 1930, at which point the method of arithmetization of syntax was not yet available to him.

constitute progress only if the proofs in the system Z were shown finitarily to be sound. At the time, it was believed mistakenly that Ackermann (*1924*) and von Neumann (*1927*) had accomplished that; in fact it was subsequently realized that they had achieved much less, and the question whether a finitary consistency proof could be given for Z at that point became live again.

This first letter concluded with Bernays' desire to see Gödel's "significant and surprising results" in the foundations of mathematics that he had heard about from Courant and Schur. Gödel sent Bernays the galleys of his *1931* forthwith, and the four items of correspondence between them in 1931 are largely devoted to Bernays' reflections on the incompleteness theorems—especially in relation to a kind of finitary ω-rule applied to (primitive) recursive formulas proposed by Hilbert as a means of extending Z to a system Z*. The matter has been discussed at length in the introductory note to *Gödel 1931c* (a brief review of *Hilbert 1931*) in these *Works*, vol. I, pp. 208–213, and only some additional points will be remarked upon here.[1] Bernays wrote that Z* was shown (finitarily) by Hilbert to be consistent, again relying mistakenly on the work of Ackermann and von Neumann cited above. For "aesthetic" reasons he went on to suggest an extension Z** by the ω-rule applied in a finitary way to arbitrary formulas, and said that he has thought about a consistency proof for Z**. After an observation about Theorem VII of *Gödel 1931* and a puzzlement about the reach of the incompleteness theorems when higher type objects are added, Bernays concluded with mention of his improvement of von Neumann's axioms for set theory, which he had lectured on in Göttingen during the summer of 1930.

Gödel's response in letter 3 (of 2 April 1931) was regrettably not available when the introductory note to *1931c*, cited above, was written. He began by sketching an argument why the systems Z* and Z**, too, are incomplete; this prefigures the incompleteness results of *Rosser 1937* for systems with the ω-rule. Gödel then related this to the use of higher types to decide previously undecidable propositions, for example in the case of Z by use of a truth predicate W for the sentences of Z; W can

---

[1]Though Bernays writes in letter 2 that Gödel's results have a "topical" interest to him beyond their general significance, because "they cast light on an extension [namely to Z*] of the usual framework for number theory recently undertaken by Hilbert", this does not answer the question raised in the introductory note to *1931c* whether Hilbert undertook that extension in response to some awareness of Gödel's incompleteness theorems, or came to it quite independently, as would be suggested on the face of it. Since Hilbert had previously conjectured the completeness of Z, he would have had to have a reason to propose such an extension, and the only obvious one is the incompleteness of Z.

be defined by a recursion of higher type.[m] Of particular note is Gödel's statement that he does not think "that one can rest content with the systems Z*, Z** as a satisfactory foundation of number theory (even apart from their lack of deductive closure), and indeed, above all because in them the very complicated and problematical concept 'finitary proof' is assumed (in the statement of the rule for axioms) without having been made mathematically precise." This resonates with Gödel's statement to Carnap in May of 1931 that he viewed the step to Z* as a step compromising Hilbert's program;[n] However, he did not make this fundamental criticism in the review of *Hilbert 1931* itself, as he might well have.

In letter 4 to Gödel later in April, Bernays was puzzled by what he took to be the recursive definition of W but cleared that up for himself in the subsequent letter 5, the last one dated 1931. But then he was worried there about Ackermann's supposed consistency proof for Z (*1924*) and didn't see at what point its formalization in Z should become impossible. We don't know when the inadequacy of Ackermann's proof of the consistency of Z became generally realized nor that what it *did* prove could indeed be formalized in Z; the nested double recursion which Ackermann used and which Bernays conjectured could not be formalized in Z, is an example of such. At any rate, there is no answer from Gödel to this question in the further (available) correspondence. Letter 5 is mainly devoted to a description at length of an early version of Bernays' theory of sets and classes, which Gödel would later modify and use in *1940* as the framework for his proof of the consistency of AC and GCH with the axioms of set theory.[o]

1.2 *The metamathematics of set theory.* Apart from a brief personal exchange in 1935, the correspondence of record lapsed at that point and

[m]In letter 2, Gödel states that Tarski had arrived at this way of defining truth simultaneously and independently of him. However, the specific definition of W that he describes depends on the fact that every element of the standard model of Z is denoted by a numeral; the more general definition of truth for languages of other structures given by Tarski in terms of satisfaction is not noted by Gödel. For a discussion of the relationship between the two on the concept of truth, see *Feferman 1984a*.

[n]See p. 212 of the introductory note to *Gödel 1931c* cited above.

[o]Amusingly, Bernays ends this letter with a footnote saying that he intends to publish his system "sometime soon". In his short autobiography *1976b* he says that the reason he actually took so long to publish it is that he felt that axiomatization was somewhat artificial. When he expressed this feeling to Alonzo Church, the consoling but smiling response was: "That cannot be otherwise." According to this story, that is what finally encouraged Bernays to begin the publication of *1937* and its sequels.

only resumed more substantively in June 1939 with a letter to Bernays from Gödel (letter 6). Written from Bremen on his return from the visits to the IAS and Notre Dame in 1938–1939, Gödel reported that in his lectures on the consistency of CH in Princeton he made use of the system of axiomatic set theory that Bernays had communicated to him by letter "a few years ago" (presumably the one of May 1931 just described). Of personal note is the change in salutations: where, previously, Bernays was addressed as "Sehr geehrter Herr Professor!" and Gödel as "Sehr geehrter Herr Dr. Gödel!", these now became "Lieber Herr Bernays!" and "Lieber Herr Gödel!", respectively, and so remained throughout their correspondence thenceforth.

At the outset, the 1939 correspondence concerns the relationship between Bernays' axioms for sets and classes as he had communicated them to Gödel in 1931, the published version in *Bernays 1937*, and Gödel's version in his Princeton lectures, to be published in *1940*. In Bernays' letter of response (21 June), he went on to ask about the rumor, via Mostowski, that Gödel had done "something" about the independence of CH. Gödel replied that he had indeed made a few starts on that, but had not yet succeeded in obtaining a final result, saying that Mostowski's report must be based on a misunderstanding.[p] In turn, he asked Bernays about something that Mostowski had told *him*, namely that Bernays had claimed to prove that there are uncountably many constructible sets of natural numbers. This Gödel found surprising, since he conjectured, on the contrary, that it is consistent (with ZF) that there are only countably many such sets. (That is indeed the case, as was established 25 years later by Cohen's method of forcing.[q]) In the latter part of letter 9 (28 September 1939), Bernays answered that he doesn't remember what he said to Mostowski in that respect, but suggested that this would hold from the "semi-intuitionistic standpoint" (presumably of the French school of Borel, Lebesgue and Lusin and/or the work of *Weyl 1918*). Earlier in that letter, he said of Gödel's constructible sets model (as presented in *1939a*) that it is "a very penetrating and efficient approach" underlaid by a "considerable mathematical tour de force". He then went on to remark that there is a certain analogy between it and Hilbert's (unsuccessful) *1926* approach to a "proof" of the continuum hypothesis. Other matters that were raised are the question

---

[p]See the introductory notes to the Church and Rautenberg correspondence on this matter in this volume and volume V of these *Works*, respectively.

[q]This result was conjectured by Cohen and established by Azriel Lévy (*1963, 1970*) using the technique of "collapsing a cardinal"; cf. *Moore 1988*, p. 160 and the introductory note to the correspondence of Gödel with Cohen in this volume.

how the constructible sets model can be handled in the absence of the replacement axiom scheme (using the constructibles up to the ωth initial ordinal ω_ω), and a suggestion of Gentzen that the Gödel "negative" interpretation of classical logic and arithmetic into the corresponding intuitionistic systems (*1933e*) can be extended to analysis. In his reply near the end of December 1939, practically on the eve of his departure with Adele for the United States via Russia and Japan, Gödel explained how to deal with the constructibles up to ω_ω without replacement; of the proposed extension of the negative interpretation to analysis he said that it can of course be done syntactically, but that doing so has no particular interest, since the constructivity of the system into which the interpretation is made "is very problematical".[r]

1.3 *War-time exchanges.* The correspondence lapsed again, this time until 1942, where we have three letters, the first (of 16 January) a brief one from Gödel to Bernays, in which he returned to the relation of his constructible sets model to the ideas sketched in *Hilbert 1926*. He said that in the past year he had brought his work on the consistency of the continuum problem to a form close to Hilbert's approach.[s] In the same letter, Gödel expressed surprise at a statement in Bernays' article *1941a* on p. 152, lines 8–11, in the 1938 Entretiens de Zürich which he found hard to comprehend, since it seemed to him to be "tantamount to giving up the formalist standpoint".[t] The letter made its way to Switzerland

[r]Cf. the related discussion in *Kreisel 1968a*, section 5.

[s]In the correspondence with Bernays, this matter came up again in Letters 68b and 70; see the discussion below. Actually, Gödel had stressed the analogy with Hilbert's plan in his lectures of 15 December 1939 at Göttingen and of 15 November 1940 at Brown University, the unpublished manuscripts for which were published in volume III of these *Works* as *1939b* and *1940a*, respectively. However, the relation between his approach in these lectures and *Hilbert 1926* is, as explained by Robert Solovay in the introductory note to *1939b* and *1940a*, op. cit., p. 120ff., much more tenuous (if not "downright misleading") than this would suggest. Cf. also, in this respect, *Dawson 1997*, p. 148.

[t]The passage in question is the last sentence in the following paragraph, here translated from the French:

Now the question arises as to what the character is of the methodological limitation of the theory of proofs, if one does not require of it the kind of elementary evidence which distinguishes the finitist point of view. The response is as follows: The tendency to methodological limitation is at bottom the same [as it is for the latter]; however, one should not conceive of evidence and certainty in a too absolute fashion, if one wishes to leave open the possibility of conserving the methodological framework. Moreover, by proceeding thusly, one is all the more assured of not being obliged to declare illegitimate or doubtful the traditional methods of analysis.

two months later, in March, but Bernays did not respond until 7 September 1942 (letter 13). Of Gödel's astonishment, he said that it is understandable given the brevity of the passage in question, and then he elaborated the view expressed in an interesting way—commended to the reader's attention—that will not be repeated here. Bernays added that he never held strictly to the "formalist" standpoint, and in fact had conspicuously distanced himself from it as far back as in his article *1930* ("Die Philosophie der Mathematik und die Hilbertsche Beweistheorie") and, further, in the article *1935* ("Sur le platonisme dans les mathématiques"). The body of the September 1942 letter is concerned with the question of how the notion of constructibility might be restricted to much weaker systems than even Zermelo's, by omitting the power set axiom.[u] After remarking on the mild effects of the war (on the Swiss) thus far, Bernays—a bachelor—charmingly concluded the letter as follows:

> Hopefully you are now well settled in Princeton and married life and the domesticity associated with it is quite salubrious for your physical and emotional health, and thereby also for your scientific work.
>
> Would you please convey my respects to your wife, even though I am not personally acquainted with her.
>
> Friendly greetings to you yourself.

In his undated response (letter 14), mainly dealing with the questions about extending the method of constructible sets, Gödel added as a postscript, "Though she is not personally acquainted with you, my wife also sends her regards."

2. Correspondence 1956–1964

After 1942 the correspondence between Bernays and Gödel lapsed for fourteen years; no doubt the difficulties of transmission during the war was the initial reason. The gap of over ten years in the immediate post-war period is surprisingly long, but once resumed in 1956, it was to continue in a steady stream until their final exchange in 1975. Also, the nature of the correspondence that is available from 1956 on is somewhat different from what it had been prior to the war. It ranges over a variety of topics—both logical and philosophical—and, somewhat as in

---

[u]The treatment of constructible sets in Kripke–Platek set theory, due to *Jensen and Karp 1971*, is one development of this idea. See also *Barwise 1975*, pp. 57–69.

a conversation, many of the letters shift in content from one to another of these, but with certain recurrent threads.[v] The first letter (28 December 1956) refers to several conversations that Bernays had with Gödel "last year". We know from his short autobiography *1976b* that Bernays held visiting positions at the University of Pennsylvania three times between 1956 and 1965, so he was no doubt referring to the opportunities he had to meet personally with Gödel during the first such stay. In subsequent years the correspondence often refers to conversations made during these and other kinds of visits (plans for which are a frequent subject)—conversations whose content can only be surmised from the letters themselves.

The second of Bernays' visits, in January 1958, was brief, on return from his participation in a symposium on the axiomatic method held in Berkeley, California. Over a year later, during the academic year 1959–1960, Bernays enjoyed an extended stay at the IAS as a visiting fellow.[w] In later years, when he held his second and third appointments at the University of Pennsylvania, Bernays again paid relatively brief visits to Gödel in Princeton, and the letters continue where their conversations left off.

Rather than try to trace chronologically the substantive issues discussed in the correspondence, we here separate out the more prominent threads, to give some background as to the questions that were dealt with, beginning first with matters concerning logic and logicians (2.1–2.7), followed later by those concerning philosophy and philosophers (2.8–2.9).

2.1 *Gentzen's last days; proof theory.* The circumstances of Gerhard Gentzen's death, at the age of 36, in a Prague prison in August, 1945, are one of the subjects of the first three letters in this period. Gödel had evidently asked Bernays during the 1956 visit about the fate of Gentzen

---

[v]There are a total of 38 items reproduced here for the period 1956–1964, ending with only one letter from 1964; thus the exchange averaged between four and five letters per year, though with considerable variation in frequency.

[w]Beginning in the latter part of the 1950s, many logicians came to the IAS as visiting fellows, attracted especially by Gödel's presence. Other fellows in logic at the IAS during the year of Bernays' visit were Roger Lyndon, Anne Davis Morel, Kurt Schütte, Gaisi Takeuti and the undersigned. Paul Cohen, whose contributions to logic were not to begin until 1962, was a fellow there as well in 1959–1960 and for the following year, and Clifford Spector visited in 1960–1961. Earlier, Georg Kreisel had been a fellow for two years, 1955–1957, Willard Van Orman Quine visited in 1956–1957 and John Myhill was there from 1957 to 1959. Gödel's role in deciding who was to be invited in logic to the IAS has been described in these *Works*, vol. I, p. 14.

and of his *Nachlaß*. When Bernays returned to Zürich, he sent Gödel
copies of letters in 1946 from a Dr. Franz Krammer to a Dr. M. Pinl
and in 1948 from a Dr. Fritz Kraus to Bernays.[x] Gentzen had been lecturing at the Mathematical Institute of the Deutsche Karls-Universität
in Prague when Germany surrendered. He refused to leave Prague and
was taken prisoner, along with other academics, including Krammer and
Kraus, by the Czechs who had come to power. The prisoners were employed on a work gang, but Gentzen became severely weakened, had to
be relieved from work, and died of malnutrition in prison. As to the
question concerning Gentzen's *Nachlaß*, according to *Vihan 1995* and
Menzler-Trott 2001 an intensive search of the archives in Prague following the war failed to uncover any of Gentzen's papers, and none have
surfaced subsequently.

Another topic in the early correspondence is Gödel's functional interpretation of intuitionistic arithmetic (eventually published as *Gödel
1958*). Bernays asked whether there was an extant presentation, and
Gödel answered (letter 17) that he had lectured on it at the IAS in
1941 but that no transcript of that talk remained. (In fact, he had also
lectured on it that year at Yale, and a lecture text was eventually discovered in his *Nachlaß*; it was published in volume III of these *Works*
as **1941*.) In that same letter, Gödel asked Bernays whether it might
be possible to obtain an interpretation of analysis in an extension of
his functional interpretation of arithmetic by functionals defined over
sufficiently large constructive ordinals, using the translation of *1933e*
to first reduce classical to intuitionistic logic. (The related suggestion
to use an extension by functionals of transfinite type to interpret some
unspecified systems stronger than arithmetic was later included at the
end of *1958*.)

In a postscript to letter 18 Gödel asked Bernays about the situation of
Kurt Schütte, saying that he was very impressed with the latter's treatment of infinite induction for proof-theoretic purposes (*Schütte 1951*)
and wanted to propose that he be invited to the IAS for 1958–1959.
After further correspondence, a visit was eventually arranged for 1959–
1960, the same year that Bernays was to visit there. It was then that
Schütte completed for publication his major work *Beweistheorie (1960)*,
which brought together various applications of proof-theoretical methods via systems with infinitary rules of inference.

---

[x]According to *Vihan 1995*, which is a succinct source of information on Gentzen's
last days, these two items along with two others on this topic are in the Bernays
archives at the ETH (numbered 1666–1669). A full account is to be found in the
biography of Gentzen, *Menzler-Trott 2001*.

2.2 *Set theory.* The exchange between Bernays and Gödel turned to set theory in the letter of 5 March 1958 (number 20) and follow-up correspondence. At first this focussed on the use of relative constructibility in *Hajnal 1956* and *Lévy 1957* (cf. also *Lévy 1960b*). More novel was the work by Lévy in his 1958 Ph. D. dissertation, published in *Lévy 1959*, on an unusual system A of set theory that had been proposed by Ackermann (*1956*). The system A has very few axioms compared to usual systems, but from them Ackermann had been able to derive all of Zermelo set theory. He attempted to prove Fraenkel's replacement scheme as well, but Lévy (op. cit.) found an error in that argument and showed that A is in fact weaker than ZF, though it does prove "very strong" theorems whose proofs in ZF make essential use of replacement. In his letter to Bernays of September 1958 (number 23) Gödel said he found Lévy's results very surprising, and he said he had heard from Kreisel about some lectures that Bernays had given in England in which he "discussed the combinatorial concept of set in detail." He went on to say that he very much regretted "that nothing about that has appeared in print," since "[c]onceptual investigations of that sort are extremely rare today."

Set theory again became a subject of correspondence in letter 29 from Gödel. There he thanked Bernays for his new book on set theory (with Abraham Fraenkel, *1958*) and said he thought it "very desirable, given the contemporary attitude of many mathematicians, to cultivate the *infinitary* methods," about which he had heard that Bernays had "said interesting things at the Congress in Warsaw that was devoted to that theme." The reference is to the Symposium on Infinitistic Methods held in Warsaw in September 1959; Bernays' lecture there, *1961b*, concerned the derivation of large cardinal statements of Mahlo type in axiomatic set theory from a reflection principle that had been suggested in *Lévy 1960a*. Interestingly, in his reply to Gödel, Bernays remarked that there was hardly any methodological difference between the Warsaw meeting and that on Constructivity in Mathematics that was held in Amsterdam two years earlier, despite the differences suggested by their titles. Perhaps that was because at both meetings there was considerable use of metamathematical methods without regard to doctrinal positions.

The set-theoretical thread was continued in letter 32 of December 1960 from Bernays, which again concerned the use of strengthened reflection principles to establish strong axioms of infinity, in *Lévy 1960* and *1960a* and in Bernays' then forthcoming contribution *1961* to a Fraenkel festschrift; the novelty of Bernays' method in the latter was to introduce a second-order reflection principle to derive the existence of the cardinals in the transfinite Mahlo hierarchies. Eight months then

passed before Gödel asked further about the publications in question (letter 40). In that same letter, he also mentioned results that had been obtained by Hanf, Keisler and Tarski—work that eventually appeared as *Hanf 1964* and *Keisler and Tarski 1964*—concerning the "largeness" of the first measurable cardinal (which exceeds all the cardinals produced in the Mahlo hierarchies). No one else, he supposed, had published on those kinds of matters.

Earlier, in his long letter of 20 April/5 May 1961 (number 36), Bernays brought up Takeuti's paper on "Cantor's absolute" (eventually published as *Takeuti 1961*). In this paper Takeuti considered well-founded models of set theory that satisfy $V = L$ and enjoy a kind of maximality condition. Bernays reformulated Takeuti's notions and constructions, but concluded that "it seems in no way appropriate that Cantor's absolute be identified with a set theory formalized in standard logic, which is considered from a more comprehensive model theory." In picking up that thread much later (30 July 1962), Gödel replied that he agreed with Bernays that "[Takeuti's] result is insignificant for the absolute", but said that he found "many of his definitions, lemmas and problems not without interest. With certain modifications they would even be significant for the absolute."[y]

2.3 *Proof theory and proof-theorists: the work of Tait and Spector.* Proof theory and related schemes of recursion also surfaced as a topic in Bernays' letter of 20 April/5 May 1961, where he mentioned four preprints by Tait that Gödel had given him to look over, including two on number theory, one on analysis and one on nested recursion, only the last of which clearly refers to work that was eventually published (*Tait 1961*).[z] In response, in letters 37 and 40, Gödel spoke of Tait's

---

[y] According to p. 149 of the introductory note to *1946* in volume II of these *Works*, the idea of ordinal definability is implicit in *Takeuti 1961*. One wonders whether Gödel noted that.

[z] In an e-mail communication (12 March 2001) Tait clarified the nature of the mentioned preprints, as follows. The first of those on arithmetic was an early form of his paper on the substitution method, *Tait 1965b*. The second provided a direct construction of the no-counterexample interpretation for arithmetic by induction on derivations in a Hilbert-style formalization of the system; it was never published. (Incidentally, in his unpublished lecture *1938a* at Zilsel's seminar, Gödel had sketched such a direct construction, as pointed out in the introductory note to that lecture, pp. 83–84. The matter is elaborated by Tait on pp. 120–123 of his review *Tait 2001* of volume III of these *Works*.) Tait's work on analysis referred to in the correspondence was not so much a preprint as an outline of an idea to obtain the consistency of analysis by the substitution method; the idea broke down during his stay at the Institute during the year 1961–1962, and no form of it was ever published. The four preprints had been sent to Gödel the year before.

continuing work on the consistency of analysis. In the second of these he said that "So far Tait has apparently proved the consistency of analysis only with non-constructive means."[aa]

Interjected in letter 40, dated 11 August 1961, is Gödel's communication to Bernays of the sudden death from leukemia of Clifford Spector on 29 July. The report is surprisingly terse and unemotional, except for Gödel's saying, "It is really astonishing how many highly gifted logicians died young (Nicod, Herbrand, Gentzen, Spector, Ramsey, etc.)." Spector, who was thirty at the time of his death, had been a visiting fellow at the IAS during the preceding academic year 1960–1961. A student of Kleene, he had made important contributions to the theory of hyperarithmetic sets and constructive ordinals among other parts of higher recursion theory, but then became interested in the functional interpretation of analysis through his contact with Georg Kreisel. Extensions to analysis of Gödel's functional interpretation of arithmetic using recursively continuous functionals of finite type had been considered in *Kreisel 1959*. During Spector's year at the Institute, he arrived at an interpretation of (the negative translation of) classical analysis in a quantifier-free theory of "bar recursive" functionals (formally related to Brouwer's bar theorem), which are generated by a scheme of transfinite recursion on the unsecured sequences of given (continuous) functionals of finite type.[ab] Spector was preparing a draft paper detailing this work when he died; the paper itself was brought to final form by Kreisel as *Spector 1962*. On p. 2 of that paper, Spector referred to conversations he had had with Gödel and Bernays concerning whether bar recursion is intuitionistically acceptable. Both, he reported, thought that it is—*if* the bar theorem is acceptable; however, Spector believed that the bar theorem still required a "suitable foundation."

Clearly, Gödel was interested in the progress and possible significance of Spector's work at the Institute, related as it was to his suggestion that there might be a functional interpretation of analysis somehow using recursion on constructive ordinals (letter 17). In a postscript to *Spector 1962* (p. 27) he wrote that "This important paper was written by Clifford Spector during his stay at the Institute for Advanced Study in 1960–61... The discussions P. Bernays and I had with Spector... took

[aa]Tait found this remark puzzling, even given that Gödel had only seen an outline of the proposed proof (previous footnote). The defect in Tait's attempt lay in the assumption that all functionals of type 3 involved would have a certain continuity property, which turned out to be false.

[ab]For more information about this and related interpretations see pp. 236–237 of the introductory note to *1958* and *1972* in volume II of these *Works*.

place after the main result. . . had been established already." Gödel ended
the postscript by saying that at the time of his death Spector was working on an extension of his result "in the direction of greater constructivity." And, as explained in footnote 1 to *Spector 1962*, it was Gödel
who suggested its title, "Provably recursive functionals of analysis: A
consistency proof of analysis by an extension of principles formulated in
current intuitionistic mathematics."

Movingly, in his response to the news (letter 41), Bernays wrote, "I
received the report of the sudden death of Clifford Spector simultaneously from you and from Mr. Kreisel. This loss is really very much to
be lamented. That I became more closely acquainted with Mr. Spector
in the spring I now value all the more."

2.4 *The limits of finitist reasoning; Kreisel's work.* A subject that should
have aroused more sustained discussion than it did in the correspondence
was Kreisel's proposed characterization *1960* of the notion of finitist
proof, since the general issue of the limits (if any) of Hilbert's program
was of great mutual interest to Bernays and Gödel. Kreisel's work was
referred to indirectly in Gödel's letters 23 and 40, but there is practically no substantive attention to it in the available correspondence;
perhaps it was explored more in personal conversations. We have only
Gödel's statement from letter 40 that he had had interesting discussions with Kreisel about his work and that "[h]e now really seems to
have shown in a mathematically satisfying way that the first ϵ-number
is the precise limit of what is finitary. I find this result very beautiful,
even if it will perhaps require a phenomenological substructure in order
to be completely satisfying." What Kreisel provides (op. cit.) is a characterization of finitist proof in terms of a transfinite sequence of proof
predicates for formal systems Σ_ν, or ordinal logics in the sense of *Turing
1939*, under the restriction that the ordinal stages ν to which one may
ascend are controlled autonomously—i.e., there must be for each such
ν a recognition at an earlier stage μ that the iteration of the process ν
times is (finitarily) justified. It is claimed in *Kreisel 1960* that the least
non-autonomous ordinal in this sense is ϵ_0, the first (Cantor) ϵ-number,
and that the finitarily provably recursive functions are exactly the same
as those of the system PA of Peano Arithmetic. Both the description
of the ordinal logic for the proposed characterization and the proofs of
the main results were very sketchy (as acknowledged by Kreisel, due
to limitations of space); regrettably, full details, though promised, were
never subsequently published.[ac] In addition, Kreisel offered little in the

---

[ac] A variant formulation of the proposed ordinal logic was presented in *Kreisel
1965*, section 3.4 (p. 168ff.); it is a little more detailed, but still far from fulfilling
the promise.

way of convincing arguments to motivate his proposed explication of the informal concept of finitist proof; this was perhaps the reason for Gödel's statement that more would be needed to make it "completely satisfying."[ad]

This is an appropriate point to remind ourselves of Gödel's unsettled views over the years as to the exact upper bound of finitary reasoning; the matter arises in another guise below during the third period of his correspondence with Bernays. One thing is clear: except for the cautious statement near the end of *1931* where he discussed the possible significance of the second incompleteness theorem for Hilbert's program, in all his subsequent writings—both published and unpublished—Gödel accepted PA as *an* upper bound for finitist reasoning.[ae] Most immediate in this respect are the remarks in *1933o*, p. 26, then *1938a*, section III (p. 93 in volume III of these *Works*), and, further, the opening paragraph of *1958*, where finitary mathematics is defined as "the mathematics in which evidence rests on what is *intuitive*," it is asserted that "certain *abstract* notions are required for the proof of the consistency of number theory," and abstract notions are equated with "non-intuitive" ones, such as objects of second or higher order. A closer look gives indications that Gödel would put the upper bound of finitary reasoning much lower down than number theory, namely to the system PRA of Primitive Recursive Arithmetic.[af] That is one interpretation of the system A sketched in *1933o*, pp. 23–25.[ag] Another point indicating that he would put the upper bound below that of PA, though not specifically down to PRA, occurs in footnote 4 to *1958*, where Gödel says that a possible extension of the "original finitary standpoint... consists in adjoining to finitary mathematics abstract notions that relate, in a combinatorially

---

[ad]In *Kreisel 1965*, p. 169, it is said that the concept being elucidated is "of proofs that one can *see* or *visualize*. ...our primary subject is a *theoretical* notion for the actual visualizing, not that experience itself." The matter was revisited in *Kreisel 1970a*, where the project is to determine "[w]hat principles of proof...we recognize as valid once we have understood...certain given concepts" (p. 489), these being in the case of finitism, "[t]he concepts of ω-sequence and ω-iteration." (p. 490). A competing view of the nature of finitism has been offered by *Tait 1981* and *2002*; the papers *Parsons 1979* and *1998a* are relevant in their discussion of the relation of finitism to intuitive knowledge. Cf. *Zach 2001*, chapter 4 for an informative comparative discussion.

[ae]But even this is subject to a possible qualification, as we shall see when we take up the issue again in the third period of correspondence.

[af]Which then would put him in accord with the thesis of *Tait 1981*.

[ag]Another interpretation of the system A is possible that puts it beyond PRA in strength but still well below that of PA; cf. the introductory note to the correspondence with Jacques Herbrand in volume V of these *Works*. On the other hand, the system A indicated in *1938a*, section III seems to be exactly PRA.

finitary way, only to finitary notions and objects, and then iterating this procedure. Among such notions are, for example, those that are involved when we reflect on the content of finitary formalisms that have already been constructed. A formalism embodying this idea was set up by G. Kreisel." This is followed by a reference to Kreisel's 1958 Edinburgh lecture published in his *1960*, i.e. the subject of discussion in the above correspondence. Since Kreisel's result was that the result of this iteration is exactly PA (in proof-theoretic strength), the "original finitary standpoint" in the given quote would have to be somewhere below that. Certain modifications of these statements, but none in any essential respect, appear in the revised translation of *1958* that was found in Gödel's *Nachlaß* and that was published in volume II of these *Works* as *1972*; we shall take those up again in section 3.4 below.

2.5 *Type-free systems.* The exchange in letters 40 and 41 touches on a number of other topics; those in set theory have already been described, while those concerning the philosophy of the Friesian school are taken up separately in section 2.9. Another matter of interest, raised by Gödel in letter 40, is a question concerning type-free systems that Ackermann developed for the membership relation, in which the usual set-theoretical paradoxes are avoided. The systems in question were formulated in a number of variants over the years, both prior to and subsequent to this correspondence; at the time of the exchange, the most recent publications had been *Ackermann 1952* and *1953*. In these systems, the law of excluded middle holds only for formulas satisfying certain criteria for being "meaningful". Gödel asked Bernays whether Ackermann had come up with formulas which are not meaningful for any substitution instance. He thought that such examples would refute his own proposed "solution" of the paradoxes, suggested in *1944*, p. 150, according to which "it is possible to assume every concept to be significant everywhere except for certain 'singular points' or 'limiting points', so that the paradoxes would appear as something analogous to dividing by zero." Now he suggested that Ackermann's supposed examples (if there are such) "would necessitate only a minor modification of the resolution method if *1.* only *a few* entirely meaningless functions exist [and] *2.* a propositional function that has a meaning for at least one argument is meaningful almost everywhere." In his reply (letter 41) Bernays said he had checked many of Ackermann's (relevant) papers but had failed to find examples of formulas meaningful for no substitution instance. He then raised problems with Gödel's proposed modified

solution of the paradoxes, pointing also to the work of *Behmann 1959*.[ah] Regrettably, Gödel did not elaborate his ideas in this direction and the matter was not discussed further in their correspondence. We do know that there are notes in the *Arbeitshefte* of his *Nachlaß* headed "Attempts to solve the paradoxes," and these may reward further study; the fact that they are written in the Gabelsberger shorthand script is a partial obstacle (cf. these *Works*, vol. I, p. 28).

Apropos of the article *1944* on Russell's mathematical logic, Gödel also recalled in letter 40 its review fifteen years earlier by Bernays (*1946*) in which he had criticized footnote 47. The issue was that Gödel had said in the latter that under one interpretation of analyticity, all the axioms of *Principia mathematica*, except for the axiom of infinity, are analytic, adding that "this view about analyticity makes it... possible that every mathematical proposition could perhaps be reduced to a special case of $a = a$, namely if the reduction is effected not in virtue of the definitions of the terms occurring, but in virtue of their meaning ... " (*1944*, p. 151). In his review, Bernays objected that this reduction would not preserve meaning considered intensionally (its "sense"), but only extensionally. In returning to the objection in letter 40, Gödel tried to clarify matters by saying that what he meant was that "mathematical propositions are perhaps interpretable as identities of signification (truth values, classes, properties, etc.) *on the basis of the sense*, but not as sense-identities."[ai] In his response (letter 41) Bernays tried to give that a charitable reading, but added: "With regard to the philosophers, it is perhaps important to emphasize that the possibility of that reduction [to the form $a = a$] has a purely technical character and gives us no real epistemological information.... [It is] of course... not really a reduction, but rather an additional construction." There the matter rested.

2.6 *The work of Esenin-Vol'pin.* Another thread of logical interest in this part of the correspondence concerns the controversial work of Alexander Sergeievich Esenin-Vol'pin,[aj] beginning with letter 36 and followed

---

[ah]Behmann's ideas in this respect go back at least to his *1931*; cf. the introductory note to the correspondence between Behmann and Gödel in this volume.

[ai]Cf. also the (essentially the same, but independent) objection to Bernays' argument on p. 117, note v, of the introductory note to *1944* in volume II of these *Works*.

[aj]Alexander S. Esenin-Vol'pin was a noted dissident in the former Soviet Union. He is described in the web site of the Andrei Sakharov Archives and Human Rights Center at Brandeis University <http://www.brandeis.edu/departments/sakharov/Exhibit/volpin.html> as follows: "Mathematician; poet; philosopher. First arrested in 1949, charged with anti-Soviet agitation in some of his poems and held in a psychiatric hospital until 1953. One of the founders of the human rights movement; member of the Human Rights Committee; samizdat author. Emigrated [to the U.S.] in 1972, lives in Massachusetts."

through in letters 44, 45, 47 and 48; he is also simply referred to as Vol'pin in some of these letters. The name has been transliterated in the literature in various forms, including Essenin-Volpin, Yessenin-Volpin and Ésénine-Volpine; this last is the way it appears in his paper *1961*, written in French. Entitled "Le programme ultra-intuitionniste des fondements des mathématiques", it was presented at the conference on Infinitistic Methods held in Warsaw in 1959, at which Bernays was also one of the invited lecturers. In the second part of letter 36, dated 20 April/5 May 1961, Bernays called attention to this publication, since it superseded an earlier draft that had been sent to Gödel. In his paper, Esenin-Vol'pin claimed to prove the consistency of Zermelo-Fraenkel set theory on the basis of an extension of ordinary mathematics by some principles concerning "realizable" or "executable" natural numbers, both in an absolute and relative sense.[ak] The proofs are very sketchy, and the plausibility of these new principles would naturally have to be called into question. Indeed, Gödel was skeptical concerning the project, when he wrote in letter 44 (30 July 1962), that "As to Vol'pin's idea, I would very much like to see *some*, even just halfway plausible axioms about the concept of 'accessible number' which imply the consistency at least of analysis.... It would also be really surprising if one could base mathematics (including number theory) on the insight that the concept of natural number is nonsensical." In his response (letter 45), Bernays asked how one could have number sequences of the kind postulated by Esenin-Vol'pin—not infinite, yet closed under the successor and even stronger operations—unless one admits an essentially "unsharp" notion. Gödel continued the discussion in letter 47, pointing out (in a postscript) that "If one somehow weakens the meaning of his axioms (e.g., also in the way you indicated) so that they become compatible with classical mathematics, then the existence of such a concept of accessibility (or its consistency with number theory) becomes provable in classical mathematics, which makes a consistency proof for it impossible." Bernays seemed to put the finger on the essentials, and brought the exchange on this topic to a close as follows in letter 48: "With respect to the investigations of Esenin-Vol'pin, you speak of the axioms that he assumes about the concept 'accessible' and that are false in the sense of classical mathematics. But in the version of his deliberations that is published in the Warsaw Congress volume I find no axioms

---

[ak]The word "feasible" is the current terminology for the idea involved; Gödel in the correspondence about this work used the word "accessible". In the review *Ehrenfeucht and Kreisel 1967* of *Esenin-Vol'pin 1961* it is suggested that the author's program is much better described as being "ultra-finitist" rather than "ultra-intuitionist", and that too is the subsequent preferred terminology.

at all that directly contradict classical mathematics; rather, only the rejection of many familiar assumptions of classical mathematics. Furthermore, there is a distinction from the usual proof-theoretic consideration, that it is not *consistency* per se that is to be proved, but only this: that a contradiction can only arise with a proof of a certain minimal length, which then no longer is viewed as a concrete one."[al] Subsequent proof-theoretical work has at least given substance to this attempt to understand what Esenin-Vol'pin might have accomplished.[am]

2.7 *Foundations of category theory.* The foundational aspects of the subject of category theory came up for discussion in Gödel's letter of 9 January 1963 (number 47); his point of departure was a remark in *Bernays 1961a* to the effect that "the 'newer' abstract disciplines of mathematics' [are] something lying outside of set theory." Gödel assumed (mistakenly as it turned out) that Bernays was "thereby alluding to the concept of category and to the self-applicability of categories". With reference to self-applicability, what Gödel presumably had in mind are examples such as the category of all categories, for which there is no straightforward set-theoretical interpretation. He went on, interestingly, to suggest that such cases of self-reference could be handled through "typical ambiguity" applied to an extension of set theory by classes of finite type. Actually, something like that had been pursued in the Grothendieck school of homological algebra (e.g., in *Gabriel 1962*), which assumes (instead of higher types) the existence of arbitrarily many "universes", i.e., collections of sets closed under various standard operations, such as the stages V_α in the cumulative hierarchy for α strongly inaccessible. But for typical ambiguity, one would need that the properties (in the language of set theory, or higher type theory as suggested by Gödel) of any universe used are the same as in any other universe. Kreisel

---

[al]Esenin-Vol'pin's claimed proof of the consistency of ZF has received little if any credence; a further sketch of "a very small part of the consistency proof" in *Esenin-Vol'pin 1970* did not increase conviction. However, his ideas about feasible numbers and constructions have been influential, to begin with on *Parikh 1971* and *Geiser 1974*. The former interprets such notions in proof-theoretical terms, while the latter gives an interpretation on the basis of non-standard arithmetic. Computational interpretations of feasibility notions (e.g., in terms of polynomial time computability, etc.) and related formal systems have dominated in recent years; cf. *Clote 1999* and *Buss 1986*, respectively.

[am]Parikh (*1971*) gave an example of an inconsistent theory of arithmetic for which there is no "short" proof of its inconsistency. P. Pudlák (*1987*) showed for a wide class of consistent theories T (including arithmetic, analysis and theories of sets) that the length of the shortest proof of the statement $\text{Con}_T(n)$—that there is no proof of a contradiction of length less than or equal to n—grows as $O(n)$; this improves an earlier unpublished result (1979) of H. Friedman. See *Pudlák 1996* for an informative survey of work on lengths of consistency proofs.

suggested in *1965*, pp. 117–118, that the required applications could just as well be taken care of by use of the reflection principle in set theory without the assumption of inaccessible cardinals; the idea was spelled out and verified to a considerable extent in *Feferman 1969*, though not all the problems were dealt with thereby.

Also in letter 47, Gödel said that he had heard "that someone has formulated the axioms of set theory with the aid of the concept of category and that this has perhaps even been published," and he asked Bernays if he knew anything about it. The first publication of an axiomatic theory of the category of sets was by F. William Lawvere in his *1964*; since that appeared in the *Proceedings of the National Academy of Sciences*, it is possible that Gödel had heard of its submission through one of the other members of the academy. In any case, Bernays' response in letter 48 was not really helpful in dealing further with either question concerning category theory, and the subject was not pursued further in the correspondence.

The next two items of correspondence are both from Bernays. In addition to matters of logical interest already discussed and philosophical interest taken up below, in letters 48 and 49 Bernays urged Gödel to help invite Gert Müller and John Myhill, respectively, as visitors to the IAS.[an] We don't have any specific responses to either of these recommendations. In the spring of 1963, Gödel learned of Paul Cohen's proofs of the independence of the axiom of choice and the continuum hypothesis, and his involvement thereafter in the publication of those results[ao] drew a great deal of his attention. To say the least, this capped Gödel's interest in logical matters during the period 1956–1964. We now turn to recurrent philosophical topics in this period.

2.8 *Philosophy of mathematics: Wittgenstein, Wittenberg and Speiser.*
Roughly speaking, the areas of philosophical discussion between Bernays and Gödel in the period 1956–1964 were the philosophy of mathematics and the philosophy of the Friesian and neo-Friesian school. We begin with the former, which, surprisingly, received much less attention than the latter.

First, there is a brief exchange in 1958 concerning Wittgenstein's *Remarks on the foundations of mathematics (1956)*. At the end of letter 24, Bernays wrote that he was busy preparing a review of that

---

[an]Müller had been Bernays' assistant at the ETH from 1952 to 1959, from where he moved on to a position in Heidelberg. Myhill was much more advanced in his career, holding a professorship at Stanford University at the time of Bernays' recommendation.

[ao]For which, see the correspondence with Cohen in this volume.

book (*Bernays 1959*), in which Gödel's incompleteness theorem is mentioned several times: "To be sure, in that regard Wittgenstein does not view it as his task to speak about your proof, but rather, as he himself declares, 'to talk past it'. The book is not entirely uninteresting, but is unsatisfying in many respects." Gödel's direct response is unusual for its snide dismissal: "As to Wittgenstein's book on the foundations of mathematics, I also read parts of it. It seemed to me at the time that the benefit created by it may be mainly that it shows the falsity of the assertions set forth in it. Ftn: and in the *Tractatus* (the book itself really contains very few assertions.)"[ap]

A more substantial discussion between Bernays and Gödel (in letters 23 and 24 of 1958 and 27 and 29 of 1959) concerned the ideas that had been developed by Bernays' student Alexander Wittenberg, presented first in an article *1953* for *Dialectica*, and then at length in his book *Wittenberg 1957*, based on his doctoral dissertation at the ETH under Bernays' direction.[aq] The article *1953* brought forth critical comments in *Bernays 1954a*, with a response (also to other comments) in *Wittenberg 1954*. The thesis of *Wittenberg 1953*, as summarized in English by its author, is that "the usual methods in the foundations of mathematics (platonist, formalist, Hilbertian) do not enable us to reach a purely objective standpoint devoid of dogmatic components. ... This entails the necessity of taking into account quite a different type of theory for the foundations of mathematics. Mathematical reflection about abstract objects appears as an epistemological *datum* that ought to be studied on its own account—previous to any interpretation, platonist or not. The starting-point of such a study will be the fact that merely by disposing of certain concepts we dispose already of a certain knowledge. Its object will be an epistemological study of this knowledge. This way of posing the problem necessarily takes us beyond the limits of a purely mathematical-logical investigation; it refers us to an epistemological investigation of human language." The book *Wittenberg 1957* is an attempted elaboration of that program.

[ap] Bernays' and Gödel's critical comments in this respect are supported at length in the review *Kreisel 1958a*. Of course, Wittgenstein has his defenders of this work; cf., for example, *Shanker 1987* and *1988*.

[aq] Bernays also wrote a short *Geleitwort* to the book. There is little information available as to Wittenberg's life and career. According to the preface to the book *1957*, completed in Québec, Canada, he had been Gonseth's assistant at the ETH for six years. In Bernays' assessment (*1966*) of a later book, *Wittenberg 1963*, on how best to teach mathematics at the gymnasium level, he wrote that a second part to that work was planned, but that Wittenberg died prematurely of a "malignant disease". It is the recollection of Gert Müller, who knew Wittenberg personally during his period as assistant to Bernays in the 1950s, that the cause of death was a brain tumor. (Personal communication.)

In his first comment on Wittenberg's book (letter 23 of September 1958), Gödel wrote that "The arguments against platonism he gave in *Dialectica* [i.e., *Wittenberg 1953*] seem to me not at all convincing, but I like his attitude toward the problems very much."[ar] Bernays, in his direct response, was pleased at Gödel's "lively interest" in the book, and added that he is discussing various fundamental points from it in his correspondence with Wittenberg. But in letter 27 of January 1959 Gödel's view of *Wittenberg 1957* turned negative: "...it seems to me in general to take quite a pessimistic view of the philosophical problems of the foundations. But I've still not read the treatment of the individual questions." Finally, in letter 29 of October 1959, after having read the book to some extent, he concluded: "Unfortunately, I find in it almost no objective position as to the content of mathematics, which one would certainly have expected on the basis of his paper in *Dialectica*." Gödel added that "I am very pleased to be able to speak with you soon about foundational questions", so his criticisms may have been adumbrated in their subsequent conversations. At any rate, there is no further discussion of Wittenberg's ideas in the available correspondence; one gathers that Bernays was sympathetic to Gödel's views in this respect.

The book by Andreas Speiser, *Elemente der Philosophie und der Mathematik (1952)*, became the subject of a brief exchange in letters 41–45. Speiser, a professor of mathematics at the University of Basel, was noted for his work in group theory as well as in the history of mathematics, and he was also broadly cultured, with strong interests in music, the arts and philosophy.[as] The work in question is an odd attempt to derive mathematical and physical notions from Hegelian idealistic philosophy. It is laid out in three parts: (i) a prelude ("Präludium"), (ii) comments on the prelude and preparations for the fugue, and (iii) the fugue. The prelude is the text of a lecture by Speiser given for a general audience in the old Aula in Basel in 1949. It ties together in a literary manner ideas and works of Plato, Plotinus, Goethe, Bach, Beethoven, Mozart, Newton, Euler, Riemann, Fichte and Hegel, among others. The fugue is divided into a series of headings ("Das Sein", "Das Wesen", "Der

---

[ar] In a comment to the undersigned concerning this, Charles Parsons has suggested that "[T]here is something phenomenological about Wittenberg's statement of his starting point. Even though it was a little later that Gödel started studying Husserl, perhaps this appealed to him."

[as] Andreas Speiser completed a dissertation on quadratic forms under the guidance of H. Minkowski at the University of Göttingen in 1908–1909. Minkowski died before the oral examination, which was then guided by Hilbert. Speiser held a professorship of mathematics at the University of Zürich until 1944, when he moved to Basel; he died in 1970. For further information, see *Burckhardt 1980*.

Begriff"), with subheadings and sub-subheadings, some of which are elaborated but many of which are not. Bernays introduced the discussion of the book (which he had sent to Gödel) in letter number 41 of October 1961, saying that Speiser "offers a novel account of Hegel's philosophy. ...I surely needn't tell you that I do not feel tempted to become converted to this sort of philosophizing. But many remarks in the book are interesting." Gödel responded, saying initially that he found the book interesting; but then, in letter 44 of July 1962, he assessed it more specifically as follows: "As to the book by Speiser, I find the introduction [i.e., the Präludium] very beautiful, particularly because a precise formalism is given. But the detailed execution, as far as I've read it, I find not a bit clearer than Hegel himself." He went on to say that he found Speiser's proposed clarification of the distinction between arithmetic and geometry (op. cit., pp. 20–21) incomprehensible. In response, in letter 45, Bernays wrote that "the introduction ('Das Präludium') contains much that is stimulating. That, however, the trains of thought here are nonetheless in places very unsatisfying you too have noted..." Concerning Speiser on the subject-object relation following Fichte, he continues: "This dependence on a mystical attitude, according to which actual knowledge is supposed to be oneness with the object, seems to me not to be conducive for philosophy at all. ...Through such a conception the specific function of thought...is not recognized ... (That, in thinking, we are one with our thinking in no way means that we thereby recognize the real nature of the thinking process.)" Their discussion of Speiser's book ceased with that.

2.9 *Philosophy of the Friesian and neo-Friesian schools (Nelson).* We turn finally to the exchanges concerning the philosophy of the Friesian and neo-Friesian schools. Of the latter, one recurrent matter of discussion was the philosophical views of Leonard Nelson (1882–1927), especially as they related to ethics and the philosophy of law. Nelson had begun as a docent in the natural science division of the philosophical faculty at Göttingen in 1909 and was appointed extraordinary professor there in 1919.[at] As explained at the beginning of this introductory note, Bernays was strongly influenced by Nelson in his early philosophical work, which was no doubt the reason that Gödel engaged him in that respect. Considered a neo-Kantian, Nelson was mainly concerned with Kant's critical method in the way that it had been (to some extent)

[at] For general information about Nelson and his philosophical work, see the entry *Nelson, Leonard*, in *Edwards 1967*, vol. 5, pp. 463–467. For his "critical" philosophy of mathematics and relations with the Hilbert school in Göttingen, see *Peckhaus 1990*, especially chapters 5 and 6, and (more briefly), *Mancosu 1998*, pp. 170–175.

clarified and extended by Jakob Friedrich Fries (1773–1843).[au] Nelson also gave considerable attention to ethical principles examined from a critical point of view. Together with a few friends, he began to publish a *Neue Folge* of the *Abhandlungen der Fries'schen Schule* in 1904, which appeared in six volumes between then and 1937, when the journal was discontinued by the Nazis. In 1958, Julius Kraft, a former student of Nelson's, founded the journal *Ratio* as a continuation of the *Abhandlungen*.

Bernays' publications concerning the Friesian school and Nelson have been indicated in footnote b. Apparently, these were brought up in conversations during the visit that Bernays paid to Gödel on his way back from Berkeley in January 1958; in his first letter of that year (number 20) Bernays refers to "the promised Fries offprint" (presumably, *Bernays 1953*). Fries and Nelson were no doubt discussed further during Bernays' extended visit to Princeton in 1959–1960. In the correspondence these names surface again in letter 31, written by Bernays two weeks after he left the Institute; with that he sent some further older reprints of his on a variety of relevant subjects. One, *Bernays 1913*, treated certain "antinomies" from the point of view of transcendental idealism, beginning with some familiar epistemological problems and then four antinomies explicitly modeled on the well-known ones of Kant's *Critique of pure reason*, though with significant differences. Another, *Bernays 1928*, provided a critique of Nelson's views in the philosophy of mathematics from the point of view of the Hilbert program.[av] Concerning the first of these, Gödel wrote on 21 December 1960 (letter 33) that he found Bernays' "treatment of the antinomies of the continuum side by side with those of Zeno to be very beautiful and justified," and added some pregnant remarks about other antinomies. In the same letter, with respect to another of these early reprints (presumably *Bernays 1910*), he said that he found Bernays' "treatment of moral law very original and interesting. But that probably belongs only to you and not to the Friesian school." He also asked, then and on a number of subsequent occasions, whether Nelson's history of philosophy had yet appeared. In the postscript to his letter three months later, Gödel wrote, "To my great sorrow, I heard that Julius Kraft died. How do things stand now with the editing of Nelson's history of philosophy?" Bernays responded directly, "I also learned of the too early death of Mr. Julius Kraft. . . . I think he had gotten so far with the editing of Nelson's lectures on the

[au] For general information about Fries's career and views, see the entry, *Fries, Jakob Friedrich*, in *Edwards 1967*, vol. 3, pp. 253–255.

[av] Cf. *Mancosu 1998*, pp. 173–174.

history of philosophy that ⟦the project⟧ is not at risk." Indeed, the volume, edited by Kraft and published posthumously, appeared as *Nelson 1962*, with a preface by Bernays.

Substantive discussion—in the correspondence itself—of Fries' and Nelson's ethical views only began with Bernays' letter 39 of 10 July 1961. There he clarified and criticized Fries' conception of punishment for violation of the law; in his response, Gödel agreed strongly with Bernays, elaborating his views in an interesting, measured, way: "punishment is also a rendering of compensation, namely to the organized society for the disruption of the order in which it...consists. Since the purpose of this order consists in the protection of the individual from damages, the damages caused to the individual are obviously the proper measure for the magnitude of the disruption." In their subsequent correspondence on these matters, the two generally continued to be in accord with each other. But Gödel seemed to want to bring that aspect of their discussions to a halt when he wrote (in letter 44 of July 1962) concerning a volume of papers from the Friesian school that Bernays had sent him: "Since my work on non-ethical philosophical problems, which are at present my principal interest, takes up my time completely, I was unfortunately able to read only a small fraction of the book." But he took issue there with Nelson's rejection of "the Kantian interpretation of the *Sollen*" and with Nelson's "restriction to psychological deduction, whereas it is precisely the transcendental and metaphysical deductions that are philosophically interesting." In response, Bernays agreed about the first but took qualified exception to the second.

A clear divergence of views emerged only after Gödel finally had *Nelson 1962* in hand, of which he wrote (letter 47, January 1963) that he found the book on the history of philosophy since Kant "*very* interesting and stimulating, ... ⟦but felt that⟧ he does not at all do justice to idealistic philosophy," and especially to Hegel. In his long response (letter 48, of February 1963), Bernays returned to the questions of Nelson's derivation of the requirements of ethics and law, saying that Gödel had mischaracterized it; he also took issue with Gödel's distinction between Nelson's deduction of the moral principle and Kant's transcendental deduction. Finally, as to the supposed defect of Nelson's book with respect to idealistic philosophy, he said that Nelson's intention in not discussing it was perfectly justified, especially as concerned Hegel, with whom Bernays could not relate at all. Given Gödel's sympathies for idealistic philosophy,[aw] he may well have been put off by this largely negative response. It was not until 18 December 1963 that he acknowl-

---

[aw] See the introductory note to *Gödel *1961/?*.

edged it, saying: "I thank you very much for the interesting letter that you wrote me a long time ago. Unfortunately, up to now it was not possible for me to answer it in detail, since I was so busy with the results of Cohen, which are of the highest importance, as well as with a new edition of some of my papers and with Institute affairs, that I had to defer my interest in philosophy for a long time." Bernays wrote back early in 1964, saying that it was very understandable that Gödel had to turn aside from his philosophical deliberations. The correspondence between the two then lapsed for over a year, and when it resumed there was no further discussion of the philosophical views of the Friesian school.[ax] One may speculate that Gödel did not tolerate such differences of opinion well, even granting the reasons offered for turning his attention to other matters—and compelling as those might be in the case of Cohen's discoveries.

3. Correspondence 1965–1975

Although there are about as many letters in this period as in the preceding, there are far fewer topics of discussion. The predominant one concerns the planned but eventually aborted publication in *Dialectica* of a translation and revision of *Gödel 1958*. There are also interesting ancillary topics which will be taken up in that connection; again we follow the exchanges via threads rather than chronologically.

The correspondence in this period began in February 1965 (letter 52) with arrangements for visits that spring to the IAS by Bernays (while at the University of Pennsylvania) on the order of once weekly for a couple of months; Gödel mentioned as part of the attractions for Bernays the presence that year of several logicians as visiting fellows, namely Gisbert Hasenjäger, William Boone, Robert Solovay, C. E. M. (Michael) Yates and Erik Ellentuck. The following September, Bernays wrote apologizing for not having had a chance to write since their time together earlier in the year, saying, "I think back with pleasure on our conversations during the months of the spring term. I needn't say to you how enjoyable it was for me that at that time I could often be together

---

[ax] It is noteworthy that through this period in which there was the most philosophical discussion between Gödel and Bernays, there is no mention of Husserl, despite the fact that Gödel began to study him seriously in 1959 (cf. the introductory note to *1961/?*). Bernays was surely aware—and perhaps Gödel was too—of the considerable disagreement between Hilbert and Husserl over the *Habilitation* and appointment of Nelson in Göttingen (Husserl having opposed these); cf. *Peckhaus 1990*.

with you and discuss the philosophical questions of mathematics." No
mention was made of other philosophical topics except in connection
with a book by Friedrich Schelling that Bernays was sending Gödel,
and information about a monograph by Albert Lautman (*1938*), who
applied "a sort of Hegelian perspective to mathematics."

3.1 *The proposed translation and revision of Gödel 1958.* It is at the end
of letter 52 from Bernays that the proposed publication in *Dialectica* of
the English translation of *Gödel 1958* made by Leo F. Boron (with the
assistance of William Howard) was first mentioned. The story of the
subsequent work by Gödel on his intended revised translation of *1958*
is told by A. S. Troelstra in the introductory note to *1958* and *1972*
in volume II of these *Works*; the reader is referred to that especially
for the substance of the modifications. What we concentrate on here
are the stages through which this effort evolved and how they relate
to other topics in the correspondence. Gödel had previously seen a
draft translation by Boron alone, and expected that it would have been
improved through Howard's assistance, in view of the latter's work on
applications of his functional interpretation. But when he received a
copy of the translation proposed for publication, he wrote (letter 56) that
"[t]he text differs hardly at all from the first (uncorrected) version...
I think, therefore, that a few changes would be absolutely necessary
and a series of others very desirable." Bernays was of course agreeable
and even went so far as to prepare a draft editorial note explaining the
nature of the anticipated publication. The first sign that things would
not be so simple came in Gödel's letter 58 of January 1966 when he
wrote that "in looking [the revised translation] over in greater detail I
found that substantially more needs to be corrected in it than appeared
at first." But, he added, by way of reassurance, "I hope to be able to
send you the print-ready manuscript in a few weeks at the latest."

Gödel's 60th birthday was coming up on 28 April of 1966, and Bernays
wrote that very month (letter 59) apologizing for a serious dereliction,
namely that he "had not prepared anything suitable for the *Festschrift*
in honor of your 60th birthday—all the more so, as you made such a
significant contribution to the *Dialectica* issues for my 70th birthday."
The *Festschrift* volume in question is presumably *Bulloff, Holyoke and
Hahn 1969* for a symposium commemorating the birthday occasion in
1966. It is indeed surprising, considering Bernays' continued high rate
of publication and special connection with Gödel, that he did not submit anything. Perhaps he had too many other commitments at that
time, or was wary of contributing to a publication organized out of the
blue by unknowns on the logical scene. By way of personal recompense,
Bernays' letter contains something of a formal tribute: "In view of the

situation in foundational investigations, you can certainly ascertain with much satisfaction that the discoveries and methods that you brought to metamathematics are dominant and leading the way in the research of today.... Yet the foundations of mathematics are of course only one of the concerns of your research; and I would also like to wish that your philosophical reflections may turn into such results that you are induced to publish them." In his response, Gödel said "I was very pleased by your good wishes.... Most especially, I thank you for the wishes concerning my philosophical investigations. For they have been my principal interests for a long time."

In the next letter (61) from Gödel, of January 1967, substantive issues connected with the revised translation of *1958* made their explicit appearance. In the very opening of *1958*, Gödel had written: "P. Bernays has pointed out on several occasions that, since the consistency of a system cannot be proved using means of proof weaker than those of the system itself, it is necessary to go beyond the framework of what is, in Hilbert's sense, finitary mathematics if one wants to prove the consistency of classical mathematics, or even that of classical number theory. Consequently, since finitary mathematics is defined as the mathematics in which evidence rests on what is *intuitive*, certain *abstract* notions are required for the proof of the consistency of number theory (as was also explicitly formulated by Bernays in his *1935...*)." Now Gödel said that he had formulated "the epistemological considerations on the first two pages somewhat more clearly, but without changing any of the content." Here he added a footnote to the letter, saying that "[i]t seems to me, too, that I ascribed to you somewhat too strong an assertion. You never asserted the *necessity* of abstract concepts, but posited the use of them merely as a *possible way* of extending finitism." As to the proposed changes, he said that he had made hardly any but merely added a few footnotes, now hoping to "bring this matter to an end." He added that "My views have hardly changed since [the original publication], except that I am now convinced that ϵ_0 is a bound on Hilbert's finitism, not merely in practice [but] in principle, and that it will also be possible to prove that convincingly." In October 1967, when Bernays next wrote Gödel, no text of the revised translation had as yet been sent, and Bernays wondered what had been holding him back. He also added in a postscript that Gödel had been right to correct his characterization of Bernays' views—referenced by "Sur la platonisme dans les mathématiques"—as to the necessary use of abstract concepts.

3.2 *Bernays' proof of transfinite induction up to ϵ_0.* During the same period of this correspondence, Bernays had been working on a second

edition of his major opus with Hilbert, *Grundlagen der Mathematik*. Its first volume eventually appeared in 1968 and the second in 1970. As it happens, the latter was to contain a new supplement (V) with an exposition of proofs due to Kalmár and Ackermann—subsequent to Gentzen's—of the consistency of the system Z of (Peano) arithmetic by means of transfinite induction on the natural ordering $\leq$ of 0-ω-figures[ay] of order-type ϵ_0. Bernays also mentioned this plan in his letter (62) of October 1967. In letter 64, Gödel asked whether Bernays planned for that to take into account the proof of transfinite induction up to ϵ_0 (i.e., up to each initial segment of the ordering $\leq$) given in *Kreisel 1965*, p. 172.[az] Bernays thanked him for the reference in his follow-up letter (20 July 1968), saying that he had his own version of a proof of that to be included, and sent him a copy with his letter of January 1969.[ba] As he puts it there, what he establishes with this proof is "the weak form of induction, which says that every decreasing sequence comes to an end after finitely many steps", i.e., that the relation is well-founded. This is to be contrasted implicitly with the "strong" or "accessibility" form of induction, which says that every $<$-progressive predicate A holds of all numbers. Symbolically, well-foundedness takes the form $(\forall \alpha)(\exists x) \neg (\alpha(x+1) < \alpha(x))$, where α ranges over sequences of natural numbers, while accessibility takes the form, for each predicate A, $(\forall x)[(\forall y < x)A(y) \to A(x)] \to (\forall x)A(x)$. Gödel became quite excited about Bernays' proof and in July 1969 prepared a draft of a letter (68a) in which he wrote "you undoubtedly have given the most convincing proof to date of the ordinal-number character of ϵ_0. You employ only constant functions of the second level and variables of the first level [i.e., variables 'α'] and a recursion scheme that is undoubtedly finitary. Since functions of the first level can be interpreted as free choice sequences and that concept is obviously decidable, a statement of the form 'For all free choice sequences...' contains no intuitionistic implication, and you have consequently completely eliminated the intuitionistic logic. If one reckons choice sequences to be finitary mathematics, your proof is even finitary." Then Gödel included a draft remark to footnote 2 of the revised version of *1958* to this effect, adding: "Hilbert did not regard

[ay] This is Bernays' terminology for the terms generated from the constant symbols 0 and ω as sums of exponents to base ω of these kinds of terms in decreasing order, i.e., $a = \omega^{a_1} + \omega^{a_2} + \cdots + \omega^{a_n}$ with $a_1 \geq a_2 \geq \cdots \geq a_n$.

[az] Actually, what is found (loc. cit.) is a statement, without proof, that induction up to ϵ_0 can be established in an ordinal logic that Kreisel had proposed for the characterization of finitist reasoning. One finds on p. 173 of *Kreisel 1965* a brief indication how induction up to ϵ_0 can be established in yet another formalism for finitism by use of the results on nested recursion due to *Tait 1961*.

[ba] The proof itself eventually appeared in *Hilbert and Bernays 1970*, pp. 533–535.

choice sequences. . . as finitary, but this position may be challenged on
the basis of Hilbert's own point of view." The terminology "[free] choice
sequences" used here is that due to Brouwer,[bb] and the assertion that
the concept of them is "obviously decidable" makes sense if they are
considered as given intensionally, i.e., that a presentation of an object
either is or is not, for one, a presentation of a choice sequence, and that,
given such a presentation, one can effectively determine of any given
term what its value is.

In the letter of 25 July 1969 actually sent (68b), Gödel changed his
mind about the significance of Bernays' proof, which he still regarded as
"extraordinarily elegant and simple. At first one also has the impression
that it comes closer to finitism than the other proofs. But on closer
reflection that seems very doubtful to me. The property of being 'well-
founded' contains two quantifiers after all, and one of them refers to
all number sequences (which probably are to be interpreted as choice
sequences). In order to eliminate the quantifiers. . . one would use a
nested recursion . . . But nested recursions are not finitary in Hilbert's
sense (i.e. not intuitive). . . Or don't you believe that?" Referring next
to the work of *Tait 1961* on nested recursion which takes one up to ϵ_0
without the use of choice sequences, Gödel added *"Hilbert, I presume,
didn't want to permit choice sequences?* To me they seem to be quite
concrete, but not to extend finitism in an essential way."

3.3 *Further problems with the proposed revision of Gödel 1958.* A second
issue raised in letter 68b concerns footnote 1, p. 283 of *1958.*[bc] The matter is explained on p. 232ff. of the introductory note to *1958* and *1972* in
volume II of these *Works.* Some repetition is necessary here, since it has
to do with the further correspondence. Briefly, the footnote in question
relates to the informal inductive explanation on p. 283 of *1958* of the notion "computable function of type *t*" by: "(1) the computable functions
of type 0 are the natural numbers; (2). . . a computable function of type
$(t_0, t_1, \ldots, t_k)$ is defined as an operation, always performable (and constructively recognizable as such), that to every k-tuple of computable
functions of types $t_1, \ldots t_k$ assigns a computable function of type t_0.
This notion is to be regarded as immediately intelligible . . . ". Footnote
1 (loc. cit.) pointed out that it results from Gödel's functional interpretation in *1958* that the intuitionistic notion of proof is interchangeable
"within certain limits" with the notion of computable function of finite
type. But then the question would be what is more

[bb]Cf., e.g., *Troelstra and van Dalen 1988*, vol. II, chapter 12.

[bc]This appears as footnote 5 on p. 244 of volume II of these *Works.*

special about the process of constructive recognition in (2) than the general notion of intuitionistic proof. This was addressed at the end of the footnote as follows: "If the notion of computable function is not to implicitly contain the notion of proof, we must see that it is immediately apparent from the chain of definitions that the operations can be performed, as is the case for all functions in the system T ..." In the letter 68b, Gödel said "I've now at last found a satisfying way of making precise the second half of footnote 1, p. 283, of my *Dialectica* paper. One can actually avoid every circularity in the definition of the logical constants. I think I will now change hardly anything else in the revised translation... and in the added remarks."

A revised translation of *1958* did indeed reach Bernays' hands,[bd] but only a year later, in July 1970. Very pleased to have finally received the manuscript, in his acknowledgment (letter 70), Bernays said he had gone over the verifications in detail and "saw in them how beneficial your added notes are for these verifications". But he was troubled about the reworking of the passage to which the issue of the footnote 1 above was relevant; in place of "and constructively recognizable as such" in part (2) of the definition of computable function of type t, the manuscript now read "for which, moreover, this general fact is intuitionistically demonstrable." Bernays objected that "the reader could well be taken aback, since your procedure is surely intended to avoid the concept of intuitionistic proof. It seems to me, however, that in fact you do not need that concept here at all, and that only a suitable reformulation is needed in order to make that clear." Gödel's letter (71) of 14 July 1970 crossed with this one of Bernays'. In it he mentioned a technical problem that had been raised by J. Diller concerning the interpretation of the axiom $p \supset p \wedge p$. On the face of it, this seemed to require the decidability of equations between terms of any given type. Gödel pointed out that this would be perfectly justified on an intensional interpretation of the notion of computable function of finite type: "The mathematicians will probably raise objections against that, because contemporary mathematics is thoroughly extensional and hence no clear notions of intensions have been developed. But it is nevertheless certain that, at least within the framework of a particular language, completely precise concepts of this kind could be defined." He did not elaborate; one corroboration of this idea in current terms is that there is a model of T in its closed terms in normal form under a suitable reduction procedure.[be]

[bd] Thanks to the assistance of Dana Scott, then at Princeton University.

[be] Cf. these *Works*, vol. II, pp. 234–235.

Bernays received the proof sheets of the revised translation by September 1970 and, as he wrote in letter 72, instead of sending them on to Gödel went over them himself to correct the "rather many printing errors." He had a few other corrigenda to suggest and raised anew the issue of intuitionistic demonstrability in the definition of computable function of finite type. At the end of the letter Bernays added a postscript, evidently penned some time after the body, in which he said he now had a second set of proofs incorporating his corrections and was sending them on to Gödel. Barely three weeks later, the response came by telegram (73): "Please wait with the page layout. Have reformulated notes c and k. Am sending corrections next week." But Gödel never did return the proof sheets, despite Bernays' further encouragement. They were found in his *Nachlaß*, with drafts of the new notes (c) and (k), and published in these *Works*, vol. II, as *1972*, since that is the last date in the correspondence in which Gödel indicated he was still working on the revisions. It is with reference to that publication that we can explain the remaining exchanges concerning these lengthy notes.[bf] Note (c) contains a more precise description of the quantifier-free system T for computable functions of finite type; in particular it gets around the problem raised by Diller, without having to invoke the intensional interpretation, by the observation that for the particular application of the interpretation of HA in T only equations between terms of type 0 need be considered, and those are decidable. Note (k) on the other hand aims to explain how the notion of computable function of finite type replaces, through the functional interpretation, the general notion of intuitionistic proof, without invoking the latter in its definition. This is to be accomplished by a new special notion of "reductive proof". Here the reader must be referred to the full text as it appears in *1972*, pp. 275–276, and Troelstra's discussion of that in his introductory note thereto, pp. 232–236.

As explanation for the new delay, in his letter 74 of December 1970, Gödel wrote that "completing the new note (k) has taken substantially more time than I projected. [It] has now become four times as long. I think that the precise formalization of what is said in it will create no

---

[bf]The galley proofs and added typescripts contained two series of footnotes, the first numbered, as in *1958*, and the latter lettered (a)–(m). The lettered notes served as new footnotes to the text, to numbered notes and to other lettered notes. Their order was somewhat haphazard and confusing, and in order to have them appear in a more sequential manner when *1972* was prepared for publication in these *Works*, the lettered notes were re-ordered, so that note (c) became (i) and (k) became (h). Here we retain Gödel's reference to them as (c) and (k), respectively; their content is taken to be as found in typescripts in the *Nachlaß* attached to the galley proofs and reproduced in *1972*.

further difficulty, but the matter is not entirely simple, on account of the unavoidable 'self reflexivities'." By the latter he meant that in the inductive definition of computable function of finite type there is a certain impredicativity, since the notion is defined in terms of itself. And where, earlier, he had seemed eager to see the revised translation into print, he now wrote that "[t]he time of publication seems to me to be less important than the improvements to the text." In his New Year's greeting at the end of December 1970, Bernays had no choice but to say that he looked forward with great interest to receiving the corrected version of the paper. A year passed before there was another exchange, again at the New Year, and finally Bernays wrote Gödel in March of 1972 (letter 76) that he thought he was to hear again in detail on the matter but "[n]ow I heard recently from Mr. Kreisel that...you still have certain doubts. Those are presumably related to the definition of the concept of 'computable function of finite type'... and the remark following shortly thereafter: 'and for which, moreover, this general fact is intuitionistically demonstrable'." Of course, Bernays had not seen the revised footnote (k), and suggested some more ways of looking at it, through the idea of some sort of "reserved intuitionism", though some impredicativity was unavoidable. In his response to *that* the following December 1972 (letter 79), Gödel wrote saying that he agreed that the general intuitionistic notion of proof is not necessary for the interpretation of the system T, which otherwise "would make my interpretation of the logical operators epistemologically worthless. ... rather... a *much* narrower and in principle *decidable* concept of proof suffices, which I introduced in note k... and called 'reductive provability'. But to carry that through satisfactorily in detail is not all that easy, mainly on account of the *non-eliminable* impredicativity also of this narrower concept of proof... It is doubtful whether carrying it through would be worth the trouble. Up to now, therefore, I have not been able to make up my mind to do it, although the further pursuit of the question could perhaps contribute in an essential way to the clarification of the foundations of intuitionism." In his follow-up letter of February 1973, Bernays made one more effort to deal with this impasse, by suggesting that the interpretation of T could be given in the class of functions generated by its schemata (in other words, in some sort of term model, as it is currently explained). There is no further reference to the proposed revised translation in the remaining two years of the available correspondence. As Troelstra has aptly put it at the beginning of his introductory note to *1958* and *1972*, "Gödel never managed to express his ideas on the philosophical aspects of the interpretation to his own satisfaction, as

is evidenced by the vicissitudes of the second version of the *Dialectica* paper...."

3.4 *The limits of finitism revisited.* This is an appropriate point to return to the question of Gödel's unsettled views as to the exact upper bound of finitary reasoning that was discussed above in connection with letter 40 and that surfaced several times in the extended correspondence concerning the second version of *1958*. In the opening paragraph of *1972*, finitary mathematics is now defined as "the mathematics of *concrete intuition*," instead of "the mathematics in which evidence rests on what is *intuitive*," as it appeared in *1958*. And the relevant footnote 4 now reads: "Note that an adequate proof-theoretic characterization of concrete intuition, in case this faculty is *idealized* by abstracting from the practical limitation, will include induction procedures which *for us* are *not* concretely intuitive and which could very well yield a proof of the inductive inference for ϵ_0 or larger ordinals. Another possibility of extending the original finitary viewpoint for which the same comment holds consists in considering as finitary any abstract arguments which only reflect... on the content of finitary formalisms constructed before, and iterate this reflection transfinitely, using only ordinals constructed in previous stages of this process. A formalism based on this idea was given by G. Kreisel [in his *1960*]."[bg] Again, since the ordinal limit of that formalism is ϵ_0, the latter part of this quotation shows Gödel would put finitary reasoning well below PA in strength. But the distinction between concrete intuition in a practical sense and that in an idealized sense made in the first part of this footnote would seem to allow concrete intuition to go far beyond PA in strength; only the qualification, that such an idealization is not concrete intuition "*for us*", apparently saves Gödel from a contradiction. In that respect, compare the quote from Gödel's letter 61 above, "that I am now convinced that ϵ_0 is a bound on Hilbert's finitism, not merely in practice [but] in principle" and his concern in letters 68a and 68b with the question as to how close Bernays' proof of the well-foundedness of ϵ_0 comes to finitism. Perhaps Gödel wanted it seen as one of the values of his work in *1958* and *1972* that the step to the notions and principles of the system T would be just what is needed to go beyond finitary reasoning in order to capture arithmetic.

[bg] Of the variant formulation of the ordinal logic proposed by Kreisel in *1965* to characterize finitist proof, Gödel wrote in note f to this footnote of *1972*, p. 274, "Kreisel wants to conclude from [the fact that the limit of his procedure is ϵ_0] that ϵ_0 is the exact limit of idealized concrete intuition. But his arguments would have to be elaborated further to be fully convincing."

3.5 *Side topics.* Several other side matters appeared in this third period of the Bernays–Gödel correspondence, the first two more or less in conjunction with the preceding long story.

3.5.1 *A "written" symposium on the foundations of mathematics.* There is indirect reference in letters 61–63 to a symposium on the foundations of mathematics to be conducted in writing and to which Gödel had been invited to contribute. The intended proceedings were evidently those eventually published in a series of articles for *Dialectica 24* (1970) and *25* (1971). In letter 61 of January 1967, Gödel wrote that he had already clearly expressed his positions on recent developments, adding that "[p]erhaps I could collect those passages and send them to you. For I thought about these formulations very carefully, and thus in a new essay I could not do much more than repeat them." What he ended up proposing for inclusion were "Some remarks on the undecidability results", including "the best formulation" of the theorem on unprovability of consistency (cf. letters 60 and 72). When the proof sheets for the revision of *1958* were prepared, they contained these remarks in a kind of appendix but due to the fate of the revision were never published as such; they eventually appeared in volume II of these *Works* as *1972a*, for which see the introductory notes thereto.[bh] As to the written symposium itself, Bernays provided a full summary of the contributions in his article *1971* for *Dialectica*.

3.5.2 *Gentzen's "original" consistency proof.* The first mention of this in the exchange appears in letter 69 of January 1970 from Bernays, in connection with his new proof for *Hilbert and Bernays 1970* of the wellfoundedness of (the natural ordering for) ϵ_0 discussed above: "The way in which choice sequences are used in [this proof] is similar to that used in Gentzen's *original* consistency proof, which at the time I thought implicitly used the *fan theorem*...That gave rise to opposition to the proof, on the basis of which Gentzen then withdrew it and replaced it with the one by induction up to ϵ_0. Today I no longer think that Gentzen's original proof requires the fan theorem." Bernays referred to his then forthcoming article *1970* for the 1968 conference on intuitionism and proof theory held in Buffalo, in which he gave a simplified presentation of that proof with some historical background. A translation by

[bh]Bernays does not seem to have thought of asking Gödel to detach the remarks (*1972a*) on the undecidability results and send the proofs of them back separately; if he had at the right time, maybe Gödel would have complied, since they had nothing to do with his worries over the notes in *1972*. Another version of the third remark, also dating from 1972, was communicated to Hao Wang and appeared on pp. 325–326 of *Wang 1974*; it is reprinted in Appendix B of volume V of these *Works*.

M. Szabo of the relevant passages in the original proof is to be found as the appendix to *Gentzen 1936* in *Gentzen 1969*.

3.5.3 *Proof theory.* In letter 61 of January 1967, Gödel asked whether Bernays had seen Takeuti's consistency proof for the Π_1^1 comprehension axiom, and what he thought about its significance. (Bernays did not respond with an opinion.) The axiom in question is a scheme in the language of second order arithmetic which asserts the existence of sets defined by Π_1^1 formulas, i.e. is of the form $(\exists X)(n)[n \in X \leftrightarrow (\forall Y)A(n, X, Y)]$ where A is an arithmetical formula which may contain number and set parameters besides n, X, Y. Takeuti proved the consistency of such a system in his paper *1967*, by transfinite induction on an ordering of what he called *ordinal diagrams.* This was the first step toward ordinally informative consistency proofs of impredicative subsystems of analysis; cf. *Takeuti 1987*, section 27 for an exposition of this result and some extensions, as well as the appendices to that work by W. Pohlers and the undersigned for a discussion of subsequent work sparked by it.

3.5.4 *Set theory.* In letter 60 (May 1966), in one of his first apologies for not forwarding the promised corrected translation of *1958*, Gödel said that "two things have quite unexpectedly kept me very much occupied." The first concerned his "best" formulation of the incompleteness theorem mentioned above. The second concerned "a new and *quite astonishingly simple and elegant* proof for the independence of the continuum hypothesis... given by Dana Scott." The proof in question is by the method of Boolean-valued models of set theory as an alternative to Paul Cohen's method of forcing, discovered independently by Dana Scott, Robert Solovay and Petr Vopěnka. Scott's proof was presented in his *1967*; see *Jech 1997* for references and a comprehensive exposition of the method and its applications to independence proofs.

A second matter in set theory was brought up by Gödel in letter 68b, concerning the possible relationship between the proposed "solution" of the continuum problem in *Hilbert 1926* and Gödel's proof of its consistency with set theory via the model in the constructible sets. Hilbert had indicated a general metamathematical lemma directly before Lemma 1 on p. 181 of his paper. Gödel asked Bernays whether Hilbert meant that "[o]ne can consistently assume that every number sequence is recursively definable. ... In that case the structure of the intended proof of the continuum hypothesis would really be *very* similar to my proof. One would then need only to replace Hilbert's transfinite recursive functions by my 'constructible' ones." Then he suggested that, if so, this should be added to his remarks quoted in the introductory

note to the translation of *Hilbert 1926* in *van Heijenoort 1967*, p. 369 (where the connection had been rather minimized). In his response (letter 70), Bernays affirmed that "Hilbert actually did mean by that lemma that one can assume without contradiction that by admitting variables of higher type every number-theoretic function can be recursively defined." It's not clear that this was directly responsive to Gödel's question. In any case, it does not appear that Gödel pursued the matter any further. For a discussion of the tenuous relations between Hilbert's ideas and (a variant of) Gödel's model in the constructible sets see the introductory note to *1939b* and *1940a* in volume III of these *Works*.

A third reference to set theory appeared in letter 74 of December 1970 from Gödel, in which he had written about the delay over returning the second proofs of the *Dialectica* article due to the difficulties with notes (c) and (k). "I will however submit the proofs very soon after the New Year [sic!]. Under no other circumstances do I then want to make further changes, if for no other reason than that I want to devote myself now entirely to a most important set-theoretic question." There is no elaboration of what that is, but presumably the reference is to Gödel's efforts around 1970 to settle the continuum problem by means of new axioms concerning scales of functions, for whose fate see the introductory note to *1970a*, *b* and *c* in volume III of these *Works*. As explained in that note, that effort ran into serious difficulties and Gödel veered from trying to prove that the power of the continuum is $\aleph_2$ to trying to prove that it is $\aleph_1$. But the matter continued to preoccupy Gödel: four years later, in his next to last letter to Bernays (number 83, December 1974), he wrote "I am making progress in my work on the true power of the continuum."

3.5.5 *Three individuals.* In Gödel's last letter to Bernays, number 84 of 12 January 1975, he asked about the logical work of Jorge Emmanuel Ferreira Barbosa, Karl Popper and Eduard Wette. In his direct response (and the last letter of the available correspondence), Bernays' evaluations of the work of which he had some knowledge, namely that of Popper and Wette, are quite balanced in the former case and interestingly frank in the latter. Writing then at the age of 86, three years before his death, Bernays' views are of undiminished clarity.

* * *

In chapter XII of his biography (*1997*) of Gödel, John Dawson paints a depressing picture of the last decade of Gödel's life, much of which was consumed by illness; a few items in this and related correspondence

testify to this.[bi] In May 1968, to excuse the delay at that point in his
work on the revision of the *Dialectica* article, Gödel wrote that "at the
beginning of April I fell ill and I'm still under the care of a physician."
In December of that year he wrote (letter 66) apologizing for taking so
long to respond to a letter of Bernays since "[u]nfortunately my state of
health was rather bad in recent months." In 1970, in his unsent letter
1970c to Alfred Tarski, Gödel wrote, by way of explanation for the
problems that had emerged with his proposed proof in *1970a* that the
power of the continuum is $\aleph_2$, "[u]fortunately my paper, as it stands, is
no good. I wrote it in a hurry after I had been ill, had been sleeping
very poorly and had been taking drugs impairing the mental functions."

[bi]Bernays and Gödel traded information about heart medications in letters 78
and 79 of 1972.

1. Bernays to Gödel

Berlin-Wilmersdorf, 24. XII. 30.

Sehr geehrter Herr Dr. Gödel!

Zunächst möchte ich Ihnen für die freundliche Zusendung Ihrer Abhandlung über die Vollständigkeit des logischen Funktionenkalkuls[a] meinen Dank sagen. Ich habe diese Abhandlung mit grossem Interesse gelesen. Damals hatte ich gerade die Thèse von Herrn Herbrand[b] in der
Lektüre, in welcher auch die Frage der Vollständigkeit des Funktionenkalkuls behandelt wird; und mir schien, dass die dort angestellten recht
mühsamen Betrachtungen durch eine Heranziehung | des Skolemschen
Verfahrens sich vereinfachen lassen müssten. Diese Vereinfachung fand
ich nun von Ihnen mit bestem Erfolge durchgeführt.

Ich habe mich auch nicht daran gestossen, dass Sie an zwei Stellen—
jedenfalls mit vollem Wissen—eine nicht-finite Betrachtung anwenden,
einmal bei der Aufstellung der Alternative "1., 2." (auf S. 355) sowie bei
der Bildung der Folge $S_1, S_2, \ldots$ (auf S. 356). Man kann ja hier die Betrachtung ohne weiteres finit gestalten, wenn man sich mit einer beschei-
deneren Fassung des | Ergebnisses begnügt,—oder aber: man kann dieses
Ergebnis im Sinne der Beweisbarkeit innerhalb der formalisierten Zahlentheorie (der ersten Stufe) deuten.

[a] *1930*.

Gödel's extended difficulties with the *Dialectica* article detailed above may also be credited in part to his poor health from 1968 on.[bj]

Solomon Feferman[bk]

A complete calendar of the correspondence with Bernays appears on pp. 552–555 of this volume.

[bj]Cf., in that respect, p. 219 of the introductory note to *1958* and *1972* in volume II of these *Works*.

[bk]I am indebted to J. W. Dawson, Jr., C. Parsons and W. Sieg for helpful comments on a draft of this note, and I wish to thank S. Buss, E. Engeler, P. Mancosu, E. Menzler-Trott, G. Müller, R. Parikh, P. Pudlák and T. Strahm for providing helpful information.

---

1. Bernays to Gödel

Berlin-Wilmersdorf, 24 December 1930

Dear Dr. Gödel:

First of all, I would like to express my thanks to you for the kind forwarding of your paper on the completeness of the functional calculus of logic.[a] I read that paper with great interest. At the time, I had just been reading Mr. Herbrand's thesis,[b] in which the question of the completeness of the functional calculus is also treated; and it seemed to me that the quite difficult considerations employed there should admit of being simplified through recourse to Skolem's procedure. That simplification I found now to have been carried out by you most successfully.

I also did not object that at two places—in any case with full knowledge—you make use of a non-finitary consideration: once in the establishment of the alternative "1., 2." (on p. 355), and again in the formation of the sequence $S_1, S_2, \ldots$ (on p. 356). Here, of course, without further ado, one can frame the consideration finitarily, if one is satisfied with a more modest version of the result—or, in other words: this result can be interpreted in the sense of provability within formalized number theory (of the first order).

---

[b]*Herbrand 1930.*

Was das Skolemsche Verfahren betrifft, (dessen deduktive Fassung
ich mir auch einmal—nach der Art Ihrer Betrachtung zum Beweise des
Satzes IV—(gelegentlich einer Vorlesung im vorigen Winter) zurechtgelegt habe), so möchte ich beiläufig erwähnen, dass man, unter Vermeidung des Begriffes der "Erfüllbarkeit"[,] das Ergebnis dieses Verfahrens
so formulieren kann:
Jeder Ausdruck (nach Ihrer Bezeichnung) ist "*deduktionsgleich*" mit
einem Ausdruck von der Form

$$(E\mathfrak{x})(\mathfrak{y})\mathfrak{A}(\mathfrak{x};\mathfrak{y}),$$

4 | —wenn "deduktionsgleich" zwei Ausdrücke genannt werden, von denen
jeder aus dem anderen mittels des Funktionenkalkuls ableitbar (beweisbar) ist.
Von Prof. Courant und Prof. Schur hörte ich, dass Sie neuerdings zu
bedeutsamen und überraschenden Ergebnissen im Gebiete der Grund-
lagen-Probleme gelangt sind und dass Sie diese demnächst publizieren
wollen.[c]—Würden Sie die Liebenswürdigkeit haben, mir, wenn es Ihnen
möglich ist, von den Korrekturbogen ein Exemplar zu schicken. Meine
Adresse ist bis zum 4.I. (einschliesslich) "Berlin-Wilmersdorf, Holsteinischestr. 38[I]", von da an wieder "Göttingen, Gaussstr. 1[I]".

Mit bestem Gruss empfiehlt sich Ihnen

Ihr

Paul Bernays

---

[c]Presumably Bernays refers here to *1931*.

2. Bernays to Gödel

Göttingen, 18. I. 31
Gaussstr. 1[I].

Sehr geehrter Herr Dr. Gödel!

Haben Sie vielen Dank für Ihr freundliches Antwortschreiben vom
31.XII. mit dem beigelegten Sonderdruck, ⟨das ich noch in Berlin erhielt,⟩
sowie für die Zusendung des Korrektur-Exemplars von Ihrer neuen Abhandlung.[a]

---

[a]Presumably *1931*. See the previous letter.

As to Skolem's procedure (whose deductive version I too once justified to myself—along the lines of your consideration in the proof of theorem IV—on the occasion of a lecture last winter), I would like to mention in passing that, by avoiding the notion of "satisfiability", the result of that procedure can be formulated as follows:

Every expression (according to your designation) is "*deductively equivalent*" to an expression of the form

$$(E\mathfrak{x})(\mathfrak{y})\mathfrak{A}(\mathfrak{x};\mathfrak{y}),$$

— if two expressions are said to be "deductively equivalent" in case each is provably derivable from the other by means of the functional calculus.

From Professor Courant and Professor Schur I heard that you have recently succeeded in obtaining significant and surprising results in the domain of problems in laying foundations, and that you intend to publish them shortly.[c] Would you be so kind as to send me, if you possibly can, a copy of the proof sheets? Until January 4 (inclusive), my address is "Berlin-Wilmersdorf, Holsteinischestr. 38[I]", from then on once again "Göttingen, Gaussstr. 1[I]".

Bidding you goodbye with best regards,

yours truly

Paul Bernays

---

2. Bernays to Gödel

Göttingen, 18 January 1931
Gaussstr. 1[I].

Dear Dr. Gödel,

Thank you very much for your friendly reply of 31 December with the enclosed offprint, which I received while still in Berlin, as well as for sending the copies of the galleys of your new paper.[a]

Wenn ich Ihnen noch nicht früher geschrieben habe, so geschah es, weil
ich erst den von Ihnen mir in Aussicht gestellten Empfang der Korrekturbogen abwarten wollte, ehe ich mich zu der in Ihrem Briefe vorkommenden grundsätzlichen Frage äusserte.

Nun habe ich in diesen Tagen Ihre Abhandlung (die mir am 14.⟨I.⟩
2 von Ihnen zuging) | mir ⟨gründlich⟩ zu Gemüte geführt.

Diese Lektüre war für mich sehr interessant und sehr lehrreich. Das ist
wirklich ein erheblicher Schritt vorwärts in der Erforschung der Grundla-
genprobleme, den Sie da getan haben.

Ihre Ergebnisse haben für mich noch eine über ihre allgemeine Bedeutsamkeit hinausgehende besondere Aktualität ⟨da⟩durch, dass sie ein Licht
werfen auf eine kürzlich von Hilbert vorgenomme[ne] Erweiterung des
üblichen Rahmens der Zahlentheorie.

Unter der "Zahlentheorie" will ich (so wie Sie) das System verstehen,
welches durch die Peanoschen Axiome, das Schema der Rekursion und
3 die logischen Regeln | der ersten Stufe (einschliesslich der Axiome der
Gleichheit) abgegrenzt ist. (Zu bemerken ist hier freilich, dass zufolge des
Rekursionsschemas das Axiom

$$x' = y' \supset x = y \quad \text{(nach Ihrer Schreibweise)}$$

entbehrlich ist, und
$$\sim (x' = 0)$$
durch
$$\sim (0' = 0)$$
ersetzt werden kann.)

In diesem Formalismus der Zahlentheorie kann auch noch das ι-Symbol,
d. h. die Formalisierung des Begriffes "dasjenige welches" aufgenommen
werden. Es lässt sich zeigen—allerdings nicht so mühelos, wie man es
zunächst denken sollte—, dass aus dem Beweise einer Formel, die das ι-
4 Symbol nicht enthält, dieses Symbol ganz eliminiert werden kann. | Doch
dieses nur nebenbei.

Die Hilbertsche Erweiterung besteht nun in folgender Regel: Wenn

$$\mathfrak{A}(x_1, x_2, \ldots, x_n)$$

eine (nach Ihrer Bezeichnung) *rekursive* Formel ist, von der sich finit
zeigen lässt, dass sie für beliebig gegebene Zahlwerte $x_1 = z_1$, $x_2 = z_2 \ldots x_n = z_n$ eine numerische Identität ergibt, so darf die Formel

$$(x_1)(x_2)\ldots(x_n)\,\mathfrak{A}(x_1 \ldots x_n)$$

als Ausgangsformel (d. h. als Axiom) benutzt werden.

That I haven't yet written to you earlier happened because I first wanted to await receipt of the proof-sheets you promised me before I spoke to the foundational questions occurring in your letters.

In these [last few] days I have now thoroughly digested your paper (which reached me from you on 14 January).

For me this was very interesting and very instructive reading. What you have done is really an important step forward in the investigation of the foundational problems.

Your results have moreover a special topical interest for me that goes beyond their general significance, in that they cast light on an extension of the usual framework for number theory recently undertaken by Hilbert.

By "number theory" (like you) I wish to understand the system delimited by the Peano axioms, the schema of recursion and the logical rules of the first order (including the axioms of equality). (Here, though, it is to be noted that, in consequence of the recursion schema, the axiom

$$x' = y' \supset x = y \quad \text{(in your notation)}$$

is dispensable, and

$$\sim (x' = 0)$$

can be replaced by

$$\sim (0' = 0).)$$

In this formalism for number theory the ι-symbol, i.e., the formalization of the notion "the one which", can also be assumed as well. It may be shown—not so effortlessly, I admit, as one should at first think—that this symbol can be completely eliminated from the proof of a formula that does not [itself] contain the ι-symbol. But this, after all, is only incidental.

Hilbert's extension now consists in the following rule: If

$$\mathfrak{A}(x_1, x_2, \ldots, x_n)$$

is a *recursive* formula (according to your designation) which may be shown, finitarily, to yield a numerical identity for arbitrarily given numerical values $x_1 = z_1$, $x_2 = z_2$, ..., $x_n = z_n$, the formula

$$(x_1)(x_2)\ldots(x_n)\, \mathfrak{A}(x_1 \ldots x_n)$$

may be used as an initial formula (i.e., as an axiom).

Hilbert zeigt nun durch eine einfache Überlegung, dass jede Formel

$$(x_1)(x_2)\ldots(x_n)\,\mathfrak{A}(x_1\ldots x_n),$$

bei welcher

$$\mathfrak{A}(x_1\ldots x_n)$$

5   eine rekursive Formel ist,~~φa,~~ und | welche ⟨(durch finite Überlegung)⟩ als *widerspruchsfrei* mit dem gewöhnlichen System der Zahlentheorie— nennen wir es kurz 3—erwiesen ist, in dem durch die neue Regel erwei- terten System—es heisse 3*—*beweisbar* ist.

Für 3* gilt daher, dass jede als widerspruchsfrei erweisbare ⟨(d. h. finit erweisbare)⟩ Formel

$$(x_1)\ldots(x_n)\,\mathfrak{A}(x_1\ldots x_n),$$

worin $\mathfrak{A}(x_1,\ldots,x_n)$ kein All- oder Seinszeichen enthält, auch beweisbar ist.

Die Widerspruchsfreiheit der neuen Regel folgt aus der Methode des Ackermannschen (oder auch des v. Neumannschen) Nachweises für die Widerspruchsfreiheit von 3. ⟨(Hierauf machte zuerst Herr *R. Schmidt* aufmerksam.)⟩—Hier lassen sich nun ohne weiteres Ihre Überlegungen anknüpfen: Sei

6 | $$Q(x,y) \equiv \overline{xB_3Sb(y^{19}_{Z(y)})}$$

und sei p die Nummer der Formel $(x)Q(x,y)$ (wir brauchen hier ja nicht den Übergang zu dem Relationszeichen q, weil $Q(x,y)$ selbst zu 3 ge- hört); dann zeigen Sie, dass die Formel

$$R(x) \equiv Q(x,p)$$

welche ja rekursiv ist,⟨—(aufgrund der Widerspruchsfreiheit von 3)—⟩die Eigenschaft hat, dass für jeden gegebenen Zahlwert n die Formel

$$R(n)$$

eine numerische Identität, also gewiss in 3 beweisbar ist, während

$$(x)R(x)$$

nicht in 3 beweisbar ist.

Nehmen wir nun die Tatsache der Widerspruchsfreiheit von 3* hinzu,
7   so | folgt—da ja $(x)R(x)$ in 3* beweisbar ist—, dass die Formel

$$(Ex)\overline{R(x)}$$

nicht in 3*, also auch nicht in 3 beweisbar ist.

Hilbert now shows, by a simple consideration, that every formula

$$(x_1)(x_2) \ldots (x_n) \, \mathfrak{A}(x_1 \ldots x_n),$$

in which

$$\mathfrak{A}(x_1 \ldots x_n)$$

is a recursive formula which (through finitary consideration) is shown to be *consistent* with the usual system of number theory—we name it 3 for short—is *provable* in the system—call it 3*—extended by the new rule.

For 3* it therefore holds that every (finitarily) demonstrably consistent formula

$$(x_1) \ldots (x_n) \, \mathfrak{A}(x_1 \ldots x_n),$$

in which $\mathfrak{A}(x_1, \ldots, x_n)$ contains no universal or existential quantifier, is also provable.

The consistency of the new rule follows from the method of Ackermann's (or also of von Neumann's) demonstration of the consistency of 3. (To which Herr *R. Schmidt* first drew attention.)—Now here, without further ado, your considerations may be added on: Let

$$Q(x,y) \equiv \overline{xB_3 Sb(y^{19}_{Z(y)})}$$

and let p be the number of the formula $(x)Q(x,y)$ (here of course we don't need the passage to the relation symbol q, because $Q(x,y)$ itself belongs to 3); you then show that the formula

$$R(x) \equiv Q(x,p),$$

—which is of course recursive (on the basis of the consistency of 3)—has the property that for every given numerical value n the formula

$$R(n)$$

is a numerical identity, hence certainly provable in 3, whereas

$$(x)R(x)$$

is not provable in 3.

If we now adjoin the fact of the consistency of 3*, it follows — since $(x)R(x)$ is of course provable in 3*— that the formula

$$(Ex)\overline{R(x)}$$

is not provable in 3*, hence also not in 3.

Es ergibt sich somit, dass das System 3 deduktiv *unabgeschlossen* ist. Zugleich damit folgt, dass das System 3* mehr beweisbare Formeln liefert als 3, d. h. es gibt elementare Sätze (betreffend die Allgemeingültigkeit rekursiver Zahlbeziehungen), die sich im finiten Sinne begründen lassen, aber nicht formal in 3 beweisbar sind.

8   Was nun die allgemeineren formalen Systeme betrifft, auf welche sich Ihre Betrachtung erstreckt, nämlich alle die, welche Ihren Voraussetzungen "1.,2." (S. 18) | genügen, so ⟨er⟩hält man jedenfalls—wie man sich auch hinsichtlich der Frage der Formalisierbarkeit von beliebigen finiten Überlegungen durch das System P entscheiden mag—aus Ihrem *Satz VI* folgende Konsequenz: Wenn ein formales System S, welches Ihren Voraussetzungen "1.,2." genügt, auf finitem Wege als widerspruchsfrei erwiesen werden kann, so gibt es jedenfalls einen Satz, betreffend die Allgemeingültigkeit einer rekursiven Zahl-Beziehung, der finit beweisbar, aber nicht formal in S beweisbar ist; es können also dann auch gewiss nicht alle finiten Überlegungen in S formalisiert werden.

9   Nimmt man also, wie v. Neumann ⟨es tut⟩, als | sicher an, dass jedwede finite Überlegung sich im Rahmen des Systems P formalisieren lässt—ich halte das, ebenso wie Sie, keineswegs für ausgemacht—, so kommt man zu der Folgerung, dass ein finiter Nachweis der Widerspruchsfreiheit von P unmöglich ist. (Diese Folgerung ergibt sich also bereits aufgrund Ihres Satzes VI.)

Aber auch ohne die Hinzunahme der v. Neumannschen Vermutung ergibt sich für jedes formale System, das Ihrer Voraussetzung "1." genügt, die Alternative: entweder ist das System in seiner Beweiskraft unzulänglich, oder seine Widerspruchsfreiheit lässt sich nicht finit erweisen.

Man wird so dazu hingedrängt, die Voraussetzung "1." fallen zu lassen. Das System 3* bildet ein Beispiel eines Systems, welches | nicht der Voraussetzung "1." genügt, das aber doch auf finitem Wege als widerspruchsfrei erwiesen werden kann.

10   Die Erweiterung des Systems 3 zum System 3* zeigt sich auch dadurch als erfolgreich, dass—(wie ich mir überlegt habe)—jede Formel des Funktionenkalkuls, von der sich (metamathematisch finit) nachweisen lässt, dass sie im Funktionenkalkul nicht ableitbar ist, innerhalb 3* *formal widerlegbar* ist.

Ein gewisser Schönheitsfehler haftet dem System 3* insofern an, als die Erweiterungs-Regel und das Induktions-Axiom als zwei Axiome von ganz verschiedener Form, aber annähernd der gleichen Intention entsprechend, nebeneinander treten. Man kann aber diese beiden Axiome

11   zu einem schärferen Axiom vereinigen, | welches sich als Regel etwa so aussprechen lässt: Ist

$$\mathfrak{A}(x_1, \ldots, x_n)$$

eine ⟨(*nicht notwendig rekursive*)⟩ Formel, in welcher als freie Individuen-

It therefore emerges that the system 3 is *not* deductively *closed*. At the same time, along with that, it follows that the system 3* provides more provable formulas than does 3, i.e., there are elementary propositions (concerning the validity of recursive number-relations) that may be justified in the finitary sense but that are not formally provable in 3.

Now, as concerns the more general formal systems to which your study is extended, namely all those that satisfy your assumptions "1.,2." (p. 18), one obtains in any case from your *Theorem VI*—however one may decide the issue of the formalizability of arbitrary finitary considerations within the system *P*—the following consequence: If a formal system *S* that satisfies your assumptions "1., 2." can be shown to be consistent by finitary means, there is in any case a proposition, concerning the validity of a recursive number-relation, that is finitarily provable, but not formally provable in *S*; so then, also, certainly not all finitary considerations can be formalized in *S*.

Thus if, as von Neumann does, one takes it as certain that any and every finitary consideration may be formalized within the framework of the system *P*—like you, I regard that in no way as settled—one comes to the conclusion that a finitary demonstration of the consistency of *P* is impossible. (This consequence thus already emerges on the basis of your Theorem VI.)

But even without the addition of von Neumann's conjecture, for every formal system that satisfies your assumption "1.", there results the alternative: Either the system is inadequate in its proof strength, or its consistency cannot be shown finitarily.

One is thus impelled to drop the assumption "1.". The system 3* forms an example of a system that does not satisfy the assumption "1.", but that nevertheless can be shown to be consistent by finitary means. The extension of the system 3 to the system 3* also appears to be successful, in that—(as I have shown)—every formula of the functional calculus which can be shown (by metamathematical finitary means) not to be derivable in the functional calculus is *formally refutable* within 3*.

A certain aesthetic defect inheres in the system 3*, insofar as the extension rule and the induction axiom stand next to one another as two axioms of quite different form, but corresponding to roughly the same intention. One can, however, unite these two axioms into one sharper axiom, which can perhaps be enunciated as the rule: If

$$\mathfrak{A}(x_1, \ldots, x_n)$$

is a (*not necessarily recursive*) formula in which only $x_1, \ldots, x_n$ occur

Variablen nur $x_1, \ldots, x_n$ auftreten und welche bei der Einsetzung von
irgend welchen Zahl-Werten anstelle von $x_1, \ldots, x_n$ in eine solche Formel
übergeht, die aus den ⟨formalen⟩ Axiomen und den bereits abgeleiteten
Formeln durch die logischen Regeln ableitbar ist, so darf die Formel

$$(x_1) \ldots (x_n) \, \mathfrak{A}(x_1, \ldots, x_n)$$

zum Bereich der abgeleiteten Formeln hinzugenommen werden.

Für diese Regel "R" habe ich mir (wenn auch nur überschlagsweise)
den Nachweis der Widerspruchsfreiheit überlegt, der sich wiederum aus
demjenigen für das System 3 ergibt.

12   Mit Hilfe der Regel R kann man das | Induktionsaxiom ableiten, und
ferner ergibt sich aus R als spezielle Anwendung die vorherige Erweite-
rungsregel (für das System 3^*).

Ob das durch die Hinzunahme der Regel R und Weglassung des In-
duktions-Axioms aus 3 entstehende System 3^{**} die Eigenschaft hat:

$$(n)[Sb(a_{Z(n)}^v) \in \text{Flg}(3^{**})] \rightarrow [(v \text{ Gen } a) \in \text{Flg}(3^{**})],$$

welche man, in Analogie zu Ihrer "ω-Widerspruchsfreiheit", als "ω-Abge-
schlossenheit" bezeichnen kann, bleibt noch zu untersuchen.—Von beson-
derem Interesse war mir in Ihrer Arbeit auch Ihre Betrachtung zum Satz
VII. Diese zeigt, soviel ich sehe, noch mehr, als Sie explizite formulieren.

13   Nämlich Ihr Satz VII | besagt ja, dass jede rekursive Relation inbezug
auf das System P äquivalent ist mit einer aus Addition, Multiplikation
und den logischen Verknüpfungen aufgebauten Relation. Aus Ihrem Beweise des Satzes VII ergibt sich aber darüber hinaus, dass zur deduktiven
Entwicklung der Zahlentheorie (entsprechend dem System 3) die Rekursionen bis auf diejenigen für die Addition und Multiplikation *entbehrlich*
sind. Man braucht ja, um dieses an Hand Ihres Verfahrens zu zeigen, nur
die Formel

$$(En, d)\{S([n]_{d+1}, x_2, \ldots, x_n) \,\&\, (k)\,[k < x_1 \rightarrow$$
$$T([n]_{1+d\cdot(k+2)}, k, [n]_{1+d\cdot(k+1)}, x_2, \ldots, x_n)]\}$$

so abzuleiten, dass man keine anderen Rekursionen als die für die Addi-
14   tion und Multiplikation benutzt. Dieses gelingt, indem man | die Anwen-
dung der Funktion $l!$ ersetzt durch den Beweis des Satzes

$$(k)(El)(n) \,\{n < k \rightarrow (n < l \,\&\, l \equiv 0 \,(\text{mod } n))\}.$$

Dieses Ergebnis—(ich nehme an, dass ich mich nicht versehen habe)—
ist deshalb besonders auffallend, weil bei der Beschränkung auf die Addition (sogar mit Hinzufügung der Subtraktion und der negativen Zahlen) ein vollkommen metamathematisch beherrschbares System entsteht,

as free variables and which, through the substitution of any numerical values whatever in place of $x_1, \dots, x_n$, is transformed into a formula such as is derivable from the formal axioms and the formulas already derived, the formula

$$(x_1) \dots (x_n) \, \mathfrak{A}(x_1, \dots, x_n)$$

may be adjoined to the domain of the derived formulas.

I have (even if only in rough outline) thought about the demonstration of the consistency of this rule "R", which again results from that for the system 3.

With the help of the rule R one can derive the induction axiom, and furthermore, there results from R, as a particular application, the previous extension rule (for the system 3*).

Whether the system 3** arising from 3 by the adjunction of the rule R and the omission of the induction axiom has the property:

$$(n)[Sb(a^v_{Z(n)}) \in \mathrm{Flg}(3^{**})] \to [(v \; \mathrm{Gen} \; a) \in \mathrm{Flg}(3^{**})],$$

which, by analogy with your "ω-consistency", can be designated as "ω-closure", still remains to be investigated.—Also of particular interest to me in your work was your discussion on *Theorem VII*. That shows, as far as I see, still more than you explicitly formulate. Namely, your Theorem VII of course asserts that every recursive relation is equivalent with respect to the system P, to one built up from addition, multiplication, and the logical connectives. But above and beyond that, from your proof of Theorem VII it follows that in the deductive development of number theory (corresponding to the system 3) the recursions except for those for addition and multiplication are *dispensable*. In light of your procedure, in order to show this one in fact need only derive the formula

$$(En, d)\{S([n]_{d+1}, x_2, \dots, x_n) \; \& \; (k)\,[k < x_1 \to$$
$$T([n]_{1+d\cdot(k+2)}, k, [n]_{1+d\cdot(k+1)}, x_2, \dots, x_n)]\}$$

in such a way that no other recursions are used than those for addition and multiplication. This is achieved by replacing the use of the function $l!$ by the proof of the proposition

$$(k)(El)(n) \, \{n < k \to (n < l \; \& \; l \equiv 0 \, (\mathrm{mod} \; n))\}.$$

This result—(I assume I have not erred)—is therefore especially striking, because with the restriction to addition (even with adjunction of subtraction and the negative numbers) there arises a completely meta-mathematically tractable system, whose completeness (i.e., deductive

von welchem *M. Presburger* die Vollständigkeit (d. h. ⟨die deduktive⟩
Abgeschlossenheit und auch die *Entscheidbarkeit* einer jeden zum System
gehörigen Frage) nachgewiesen hat.[b]—

Auf die Resultate Ihrer weitergehenden Untersuchungen über die Beweisbarkeit ⟨bezw. Widerlegbarkeit⟩ der vorher unentscheidbaren Sätze
durch | Hinzunahme höherer Variablen-Typen bin ich besonders deshalb
sehr gespannt, weil hier ein Paradoxon vorzuliegen scheint: Einerseits
sollte man denken, dass, solange nur abzählbar viele Typen hinzugenommen werden, Ihr Satz VI in Kraft bleibt, dass also dann immer noch unentscheidbare Sätze vorhanden sind. Andrerseits kann ⟨aber⟩ doch nicht
die ganze zweite Zahlenklasse verbraucht werden, da ja die rekursiven
Relationen abzählbar sind.—

Zum Thema der Axiomatisierung der Mengenlehre möchte ich schliesslich noch erwähnen, dass ich mir von der v. Neumannschen Mengenlehre
eine modifizierte Fassung zurechtgelegt habe, welche vor allem eine engere Beziehung zu den üblichen logischen Prozessen der Mengenlehre-
⟨bildung⟩ herstellt, | ferner verschiedene unnötige Abweichungen von dem
Zermeloschen System beseitigt und die Formulierung der Axiome leichter
fasslich macht. (Ich habe darüber im letzten Sommer in der Göttinger
mathem. Gesellschaft vorgetragen.[c])—

Nun bin ich etwas ausführlich geworden. Ich denke aber, Sie werden
das meinem Interesse für den Gegenstand zugute halten.

Mit bestem Gruss empfiehlt sich

Ihnen Ihr   Paul Bernays

---

[b]*Presburger 1930.*
[c]The reports from the Göttingen Mathematical Society list a lecture by Bernays
on "Zur Axiomatik der Mengenlehre" on 24 June 1930. When writing the letter he

3. Gödel to Bernays

Wien VIII, Josefstädterstr. 43, 2./IV 1931.

Sehr geehrter Herr Professor!

Meinen allerbesten Dank für Ihr ausführliches Schreiben vom 18./I.
Wenn ich dasselbe so spät beantworte, so geschieht dies deswegen, weil

closure, and also the *decidability* of any question belonging to the system) was established by *M. Presburger.*[b]

I am therefore especially agog at the results of your further-going investigations about the provability, [[or]] respectively, refutability, of the previously undecidable propositions by means of the adjunction of higher variable-types, for here a paradox seems to present itself: On the one hand, one should think that as long as only denumerably many types are adjoined, your Theorem VI remains in force, and thus that undecidable propositions do still occur. But on the other hand, the entire second number class certainly can't be used up, since, after all, the recursive relations are denumerable.

In conclusion, on the topic of the axiomatization of set theory, I would still like to mention that I have laid out a modified version of von Neumann's set theory which, first of all, establishes a closer relation to the ordinary logical processes of set formation, and furthermore eliminates various unnecessary deviations from Zermelo's system and makes the formulation of the axioms more easily understandable. (I lectured on it last summer before the Göttingen Mathematical Society.)[c]

I've now been somewhat prolix. But I think you will attribute this to my interest in the subject.

Bidding you goodbye with best regards,

Yours truly,   Paul Bernays

---

was obviously concerned with a new lecture on "Neueres and Neuestes aus der Grundlagentheorie" delivered on 10 February 1931. The source is *Jahresbericht der deutschen Mathematiker-Vereinigung 41*, 2. Abt., 14. We are grateful to Volker Peckhaus for bringing this and other items to our attention.

3. Gödel to Bernays

Vienna VIII, Josefstädterstrasse 43, 2 April 1931

Dear Professor,

My most sincere thanks for your detailed letter of 18 January. If I am answering it so late, that is because I first wanted to reflect precisely on

ich mir zunächst dessen Inhalt (besonders was die Systeme 3*, 3** betrifft) genau überlegen wollte und andrerseits gerade in den letzten Wochen ziemlich stark beschäftigt war.—Die Resultate meiner Überlegungen möchte ich Ihnen nun (soweit das in einem Briefe möglich ist) mitteilen.

Zunächst kann man zeigen, daß auch die Systeme 3*, 3** nicht deduktiv abgeschlossen sind, unter einer (sehr plausiblen) Voraussetzung, die ich zunächst formulieren will. Eine Formel der Gestalt $(P)A(x_1 \ldots x_n)$ (wobei $A(x_1 \ldots x_n)$ eine rekursive Relation zw. nat. Zahlen ist u. durch das Präfix (P) sämtliche Variable $x_1 \ldots x_n$ gebunden sind) will ich "finit bewiesen" nennen, wenn man für jede Variable x_j, die in (P) durch E-Zeichen gebunden ist, eine finit konstruierte zahlentheor. Funktion f_j

2   angegeben hat [wobei f_j als Argumente die sämtlichen | durch Allzeichen gebundenen Variablen hat, welche in (P) *vor* x_j stehen] und wenn man die Formel, welche durch Ersetzen der x_j durch f_j entsteht finit bewiesen hat. Das formale System, welches aus 3 durch Adjunktion der Variablen für Klassen natürlicher Zahlen ⟨u. der entsprechenden Axiome⟩ entsteht, will ich S nennen. Man kann nun zeigen, daß es in 3* und 3** sicher dann unentscheidbare Sätze gibt, wenn für keine finit beweisbare Formel $(P)A(x_1 \ldots x_n)$ ihre Negation $(\overline{P})A(x_1 \ldots x_n)$ in S formal beweis-

f   bar ist, d. h. man kann eine ⟨Satz⟩Formel aus 3 angeben, deren Entscheidung in 3* (bzw. 3**) zur Folge hätte, daß man für eine gewisse (angebbare) Formel einerseits einen finiten Beweis und andrerseits eine formale Widerlegung in S angeben könnte. Der Beweis ist etwa der folgende: Man kann in 3 ein (nicht rekursives) Klassenzeichen $A(x)$ angeben, so daß $A(x)$ in der gewöhnlichen (naiven) Weise interpretiert besagt: x ist ein *rekursives* ~~Klassen~~⟨*Relations*⟩*zeichen*, das bei *Einsetzung* beliebiger *Zahlzeichen* an Stelle der *freien Variablen* in eine in 3 *beweisbare Formel*

3   übergeht. [Die unter⟨strichenen[a]⟩ | metamath. Termini haben die Bedeutung der kursivgedruckten in meiner Arbeit[b] vgl. S.7]. Ferner kann man ein ~~e~~ (ebenfalls nicht rekursives) Klassenzeichen $B(x)$ ⟨aus 3⟩ angeben, welches besagt: x ist aus den durch die *Formelklasse* $\hat{x}A(x)$ erweiterten Axiomen von 3 (nach den Schlußregeln von 3) *formal beweisbar*. Nun kann man sich leicht folgendes klar machen: "Eine *Formel* v aus 3 ist in 3* *beweisbar*" bedeutet genau dasselbe wie: "$B(v)$" ist finit beweisbar". Nun bilde ich die Formel $\overline{B}[x; x]^1$ und das ihr zugeordnete *Klassenzeichen* p (d. h. die Nummer von $\overline{B}[x; x]$) und ferner die *Satzformel* $[p; p]$. Dann kann man im System S die beiden folgenden Formeln beweisen: $\overline{B}[p; p]$ und $\overline{B}(\mathrm{Neg}[p; p])$. Die Methode, mit deren Hilfe ich dies zeige, ist

---

^{1}Statt $Sb\left(x^{17}_{Z(y)}\right)$ schreibe ich kurz: $[x; y]$.

---

[a]Underlining is here rendered in italics.

its content (especially concerning the systems 3^*, 3^{**}); and on the other hand, just in the last few weeks I've been rather busy.—Now I would like to tell you the results of my reflections (as far as that is possible in a letter).

To begin with, one can show that the systems 3^*, 3^{**} are also not deductively closed, under a (very plausible) assumption, which I want to formulate first. A formula of the form $(P)A(x_1 \ldots x_n)$ (in which $A(x_1 \ldots x_n)$ is a recursive relation among natural numbers and all the variables $x_1 \ldots x_n$ are bound by the prefix (P)) I will call "finitarily proven" if for each variable x_j bound in (P) by existential quantifiers, a finitarily constructed number-theoretic function f_j has been specified [where f_j has as arguments all the variables bound by universal quantifiers that stand *before* x_j in (P)], and if the formula that results by replacing the x_j by f_j has been finitarily proven. The formal system that results from 3 by adjoining the variables for classes of natural numbers and the corresponding axioms I will call S. One can show now that there certainly are undecidable propositions in 3^* and 3^{**}, in case for no finitarily provable formula $(P)A(x_1 \ldots x_n)$ one can formally prove in S its negation $(\overline{P})A(x_1 \ldots x_n)$, i.e. one can specify a formula of 3 whose decision [proof or refutation] in 3^* (or, respectively, 3^{**}) would imply that for a certain (specifiable) formula one could give, on the one hand, a finitary proof, and, on the other, a formal refutation, in S. The proof is roughly the following: In 3 one can specify a (non-recursive) class sign $A(x)$ such that $A(x)$, interpreted in the usual (naive) way, says: x is a *recursive class sign* that is transformed into a *provable formula* in 3 by insertion of any *number signs* whatever in place of the *free variables*. [The underlined[a] metamathematical terms have the meaning of the cursively printed [symbols] in my paper;[b] see p. 7]. Moreover one can specify a (likewise non-recursive) class sign $B(x)$ of 3 which says: x is *formally provable* (according to the inference rules of 3) from the axioms of 3 extended by the *class of formulas* $\hat{x}A(x)$. Now one can easily make clear to oneself: "a *formula* v of 3 is *provable* in 3^*" means exactly the same as: "$B(v)$ is finitarily provable". Now I construct the formula $\overline{B}[x;x]^1$ and the *class sign* p assigned to it (i.e., the number of $\overline{B}[x;x]$) and furthermore the *formula* $[p;p]$. Then one can prove in the system S the following two formulas: $\overline{B}[p;p]$ and $\overline{B}(\text{Neg}[p;p])$. The method with whose help I show this is the same one that also leads to the decision of

---

[1]Instead of $Sb\left(x_{Z(y)}^{17}\right)$ I write for short: $[x;y]$.

---

[b]See *1931*.

dieselbe, welche auch zur Entscheidung der vorher unentscheidbaren
Sätze mittels höherer Variablentypen führt. Sie besteht in folgendem:
4   Man kann in den Symbolen von S | eine Klasse natürlicher Zahlen de-
finieren, welche (grob gesprochen) aus allen denjenigen Zahlen besteht,
die richtigen Satzformeln aus 3 zugeordnet sind. (Analog für irgend zwei
⟨formale⟩ Systeme, deren zweites in der Typenbildung über das erste hin-
ausgeht). Präzise ausgedrückt heißt das in unserem Fall folgendes: Man
kann ein Klassenzeichen $W(x)$ [~~mit der Bedeutung "x ist ein wahrer Satz
aus 3"~~] aus S angeben, das folgenden Bedingungen genügt:
    1. Ist V eine beliebige Satzformel aus 3 und v ihre Nummer (d. h. die
entspr. *Satzformel*), dann ist in S beweisbar:

$$W(v) \sim V \quad (1)$$

    2. In S ist beweisbar:

$$(x)\{B(x) \to W(x)\} \quad (2)$$

    3. Aus 1. folgt insbes., daß für jede *Satzformel* v aus 3 in S beweisbar
ist:

$$W[\mathrm{Neg}(v)] \sim \overline{W(v)} \quad (3)$$

Auf Grund von (1) ist nun in S beweisbar:

$$W[p;p] \sim \overline{B}[p;p] \quad (4)$$

5 | denn p ist die Nummer von $\overline{B}[x;x]$ folglich $[p;p]$ die von $\overline{B}[p;p]$.
Ferner ist auf Grund von (2) in S beweisbar:

$$B[p;p] \to W[p;p] \quad (5)$$

Schließlich auf Grund von (2) und (3):

$$B(\mathrm{Neg}[p;p]) \to \overline{W}[p;p] \quad (6)$$

Aus (4) (5) (6) ergibt sich nun sofort (durch indirekten Schluß):

$$\overline{B}[p;p] \quad (7) \qquad \text{sowie} \qquad \overline{B}(\mathrm{Neg}[p;p]) \quad (8),$$

welche also in S beweisbar sind. Wäre nun entweder $[p;p]$ oder $\mathrm{Neg}[p;p]$
in 3* *beweisbar*, dann wäre $B[p;p]$ bzw. $B(\mathrm{Neg}[p;p])$ finit beweisbar,
was wegen (7), (8) u. der gemachten Voraussetzung unmöglich ist. Bei
3** verläuft der Beweis ganz analog. Ich halte es übrigens auch für sehr
unwahrscheinlich daß die Systeme 3*, 3** ω-abgeschlossen sind. Unter
gewissen weitergehenden (auch noch ziemlich plausiblen) Voraussetzun-
6   gen | kann man nämlich zeigen, daß sie es nicht sind. Bemerken möchte
ich noch, daß man, statt unter S das System zu verstehen, welches aus

the previously undecidable propositions by means of higher variable types. It consists in the following: In the symbols of S one can define a class of natural numbers which (roughly speaking) consists of all those numbers that are assigned to correct formulas of 3 (analogously for any two formal systems the second of which surpasses the first in the formation of types). Expressed precisely, that means in our case the following: one can specify a class sign $W(x)$ of S that satisfies the following conditions:

1. If V is any formula of 3 and v its number (i.e., the corresponding *formula*), then

$$W(v) \sim V \quad (1)$$

is provable in S.
   2.

$$(x)\{B(x) \to W(x)\} \quad (2)$$

is provable in S.

3. From 1. it follows in particular that for every *formula* v of 3,

$$W[\text{Neg}(v)] \sim \overline{W(v)} \quad (3)$$

is provable in S.
On the basis of (1),

$$W[p; p] \sim \overline{B}[p; p] \quad (4)$$

is now provable in S, for p is the number of $\overline{B}[x; x]$, and consequently $[p; p]$ is that of $\overline{B}[p; p]$.

Furthermore, on the basis of (2),

$$B[p; p] \to W[p; p] \quad (5)$$

is provable in S.

Finally, on the basis of (2) and (3):

$$B(\text{Neg}[p; p]) \to \overline{W}[p; p] \quad (6)$$

From (4), (5) and (6) there now immediately follows (by indirect inference):

$$\overline{B}[p; p] \quad (7) \qquad \text{as well as} \qquad \overline{B}(\text{Neg}[p; p]) \quad (8),$$

which are thus provable in S. Now, if either $[p; p]$ or $\text{Neg}[p; p]$ were *provable* in 3*, then $B[p; p]$, or, respectively, $B(\text{Neg}[p; p])$, would be finitarily provable, which is impossible on account of (7), (8) and the assumption that was made. The proof for 3** proceeds quite analogously. Moreover, I take it to be very improbable that the systems 3*, 3** are ω-closed. Namely, under certain further (but still plausible) assumptions one can show that they are not. In addition I would like to note that, instead of taking S to be the system which arises from 3 by adjunction of class

3 durch Adjunktion der Klassenvariablen entsteht, man darunter auch
ein System verstehen kann, welches durch Hinzunahme gewisser höherer
Rekursionsschemata (~~Rekur n~~simultane Rek. nach mehreren Variablen)
⟨aus 3⟩ hervorgeht, ohne daß an den oben aufgestellten Behauptungen
sich das geringste ändert. Das Prinzip, nach welchem die Klasse $W(x)$
definiert wird, ist nämlich rekursiv; ich definiere zunächst was W für die
allereinfachsten Sätze (Zahlgleichungen etc.) bedeutet u. gehe dann zu
den komplizierteren Sätzen etwa nach folgendem Schema über:

$$W[\text{Neg}(a)] =_{\text{Df}} \overline{W}(a)$$
$$W(a \text{ Dis } b) =_{\text{Df}} W(a) \vee W(b)$$
$$W(v \text{ Gen } a) =_{\text{Df}} (x)W[Sb(a^v_{Z(x)})]$$

(wobei in der letzten Formel a ein *Klassenzeichen* mit der *freien Variablen v* ist)[.] Natürlich ist das Verfahren damit nur ganz vage skizziert.
7 Die Idee, | den Begriff "wahrer Satz" auf diesem Wege zu definieren, hat
(wie ich einem Gespräch entnehme) gleichzeitig und unabhängig von mir
Herr A. Tarski entwickelt (allerdings zu anderen Zwecken).

Übrigens glaube ich, daß man sich (auch abgesehen von ihrer deduktiven Unabgeschlossenheit) bei den Systemen 3*, 3** als einer befriedigenden Begründung der Zahlentheorie nicht beruhigen kann u. zw. vor
allem deswegen, weil in ihnen der sehr komplizierte und problematische
Begriff "finiter Beweis" ohne nähere mathem. Präzisierung vorausgesetzt
wird (bei Angabe der Axiomenregel).

Was die Entscheidbarkeit der unentscheidbaren Sätze in höheren Systemen betrifft, so ergibt sich diese sofort aus den Eigenschaften des Begriffs $W(x)$ [der sich immer im höheren System in Bezug auf das engere definieren läßt]. Denn aus (2), (3) Seite 4. dieses Briefes folgt, daß
die Widerspruchsfreiheit des engeren Systems (in diesem Fall 3*) im
weiteren (d. h. 3**) formal beweisbar ist. Die Paradoxie, die Sie vermuten, klärt sich folgendermaßen auf: Da die Anzahl der aus 3 durch
8 Typen~~erhöhung~~⟨erweiterung⟩ hervorgehenden Systeme $\aleph_1$ ist, können |
nicht alle diese Systeme rekursiv definierbare *Axiomenklassen* haben.
Von einer gewissen Stelle an werden daher die nach meiner Methode konstruierten unentscheidbaren Sätze (in die ja die Axiomenklasse wesentlich
eingeht) nicht mehr die Form $(x)F(x)$ (mit rekursivem F) haben und
man hat auch keinen Anhaltspunkt dafür, daß schließlich alle "rekursiven" Probleme (der Form $(x)F(x)$) lösbar werden, denn es könnte ja
das Überspringen auf unentscheidbare Sätze anderer Art eintreten, bevor
sie alle ausgeschöpft sind.

Bez. Satz VII hatte ich auch schon eine ähnliche Überlegung angestellt, wie Sie sie vorschlagen, aber in meiner Arbeit nicht publiziert, weil
sie ja für das dort verfolgte Ziel irrelevant ist.

variables, one can also take it to be a system which arises from 3 by addition of certain higher recursion schemas (simultaneous recursion on several variables), without changing in the least the assertions made above. Namely, the principle according to which the class $W(x)$ is defined is recursive; I first define what W means for the simplest propositions of all (numerical equations, etc.) and then go on to the more complicated propositions, say according to the following schema:

$$W[\mathrm{Neg}(a)] =_{\mathrm{Df}} \overline{W}(a)$$
$$W(a \text{ Dis } b) =_{\mathrm{Df}} W(a) \vee W(b)$$
$$W(v \text{ Gen } a) =_{\mathrm{Df}} (x)W[Sb(a^v_{Z(x)})]$$

(where, in the last formula, a is a *class sign* with the *free variable v*). Of course, the procedure is only vaguely sketched by that. Simultaneously and independently of me (as I gathered from a conversation), Mr. Tarski developed the idea of defining the concept "true proposition" in this way (for other purposes, to be sure).

By the way, I don't think that one can rest content with the systems 3^*, 3^{**} as a satisfactory foundation of number theory (even apart from their lack of deductive closure), and indeed, above all because in them the very complicated and problematical concept "finitary proof" is assumed (in the statement of the rule for axioms) without having been made mathematically precise.

As concerns the decidability of the undecidable propositions in higher systems, that results immediately from the properties of the concept $W(x)$ [which may always be defined in the higher system with respect to the narrower one]. For from (2) and (3) on page 4 of this letter it follows that the consistency of the narrower system (in this case 3^*) is formally provable in the more extensive one (i.e. 3^{**}). The paradox that you conjecture is resolved as follows: since the number of systems arising from 3 by type extension is $\aleph_1$, not all these systems can have recursively definable *classes of axioms*. From a certain point on, therefore, the undecidable propositions constructed by my method (in which the class of axioms enters in an essential way) no longer have the form $(x)F(x)$ (with recursive F), and one also has no indication that ultimately all "recursive" problems (of the form $(x)F(x)$) will be solvable, for it could be that the leap to undecidable propositions of another kind occurs before they are all exhausted.

With respect to Theorem VII, I had also already employed a consideration similar to that you propose, but did not publish it in my paper because it is irrelevant for the goal pursued there.

Ihre Verbesserung am Axiomensystem der Mengenlehre würde mich
sehr interessieren und ich wäre Ihnen für Zusendung eines Manuskripts
Ihres Vortrags (falls ein solches vorhanden ist) sehr dankbar.

Anbei übersende ich Ihnen zwei Separata meiner Abhandlung und bin
mit den besten Grüssen

Ihr ergebener   Kurt Gödel

P.S. Aus den Eigenschaften von $W(x)$ ergibt sich u. a., daß man in
Satz VI meiner Arbeit die Vorauss. der ω-Widerspruchsfreiheit ersetzen
kann durch Widerspruchsfreiheit des nächsten Typs.

4. Bernays to Gödel

Göttingen, 20. IV. 31
Gaußstr. 1$^\text{I}$.

Sehr geehrter Herr Gödel!

Haben Sie schönen Dank für die Zusendung der beiden Exemplare
Ihrer Abhandlung (für Herrn Prof. Hilbert und für mich), sowie auch
für Ihren eingehenden Antwortbrief!

Ihre Ergebnisse, deren Nachweis Sie mir skizziert haben, sind ja
wiederum sehr ⟨überraschend und⟩ bedeutsam und zeigen in noch ver-
stärktem Masse die grosse Tragweite Ihrer Methoden.

Über einen Umstand würde ich gern von Ihnen noch Aufklärung erhalten. Nämlich ich erkenne nicht, warum man zur Formalisierung des
Begriffes "richtige Formel aus 3", durch das Klassenzeichen $W(x)$, nötig
hat, über das System 3 hinauszugehen. Man braucht doch $W(x)$ nur für
Satzformeln in der Normalform ("normale Satzformeln") als "gültig"
2   zu definieren. Dann aber | scheint folgende Rekursion innerhalb von 3
ausführbar zu sein:

$W_0(x)$ soll besagen (im Sinne der gewöhnlichen Deutung): "x ist eine
Satzformel, welche keine Variable enthält und welche bei der Auswertung
aller vorkommenden rekursiven Funktionen—die auftretenden Negationen
und Disjunktionen sollen zuvor ins Arithmetische übersetzt sein—in eine
identische Gleichung übergeht".

Your improvement of the axiom system of set theory would interest me very much, and I would be very grateful to you for sending me a manuscript of your talk (in case such a thing is available).

I am sending you herewith two offprints of my paper and am, with best wishes,

Yours faithfully,   Kurt Gödel

P.S. From the properties of $W(x)$ it follows, among other things, that in Theorem VI one can replace ω-consistency by consistency of the next type.

4. Bernays to Gödel

Göttingen, 20 April 1931
Gaussstr. 1$^\mathrm{I}$.

Dear Mr. Gödel,

Many thanks for sending the two copies of your paper (for Professor Hilbert and for me), and also for your detailed letter of reply!

Your results, whose proof you sketched to me, are once again very surprising and significant and show in an even stronger way the great reach of your methods.

On one point I would like to receive further clarification from you. Namely, I don't see why, in the formalization of the notion "true formula of 3" by means of the class sign $W(x)$, one must necessarily go beyond the system 3. Indeed, $W(x)$ is needed only to define "valid" for *formulas in normal form* ("normal formulas"). But then the following recursion seems to be realizable within 3:

$W_0(x)$ is supposed to say (in the sense of the usual interpretation): "x is a formula which contains no variable and which becomes an identity after the evaluation of all the recursive functions that occur ⟦in it⟧ (the negations and disjunctions that appear are first supposed to be translated into arithmetic operations)".

Hierzu tritt die Formel

$W_{n+1}(x)$

$\sim (Ek)(Ey)\{kFry \,\&\, \overline{k\ \text{Geb}\ y} \,\&\, [[x = k\ \text{Gen}\ y \,\&\, (u)W_n(Sb(y^k_{Z(u)}))]$

$\lor [x = \text{Neg}(k\ \text{Gen}\ (\text{Neg}\ y)) \,\&\, (Eu)W_n(Sb(y^k_{Z(u)}))]]\},$

welche W_{n+1} durch W_n ausdrückt.

Und nun kann

$$W(x) \sim (En)(n \leqq l(x) \,\&\, W_n(x))$$

definiert werden.

Andrerseits aber erscheint mir, dass | diese Überlegung einen Fehler
enthalten muss. Denn wenn $W(x)$ in 3 definiert ist, so kann man die
*"Beweisbarkeit von x in dem durch die Formelklasse $\hat{z}W(z)$ erweiterten
System 3"* ebenfalls innerhalb von 3 durch eine Klassenzeichen $B^*(x)$
darstellen.

Ferner ist dann für jede normale Satzformel $\mathfrak{N}$ aus 3, mit der Nummer
n, die Formel

(1) $\qquad\qquad\qquad W(n) \sim \mathfrak{N}$

sowie auch

(2) $\qquad\qquad\qquad B^*(n) \to W(n)$

formal (wenn nicht in 3, so doch jedenfalls in S) beweisbar.

Dazu tritt nun noch, aufgrund der Definition von $B^*(x)$:

$$W(n) \to B^*(n),$$

sodass wir im ganzen

$$\mathfrak{N} \sim B^*(n)$$

erhalten.

Wenden wir aber diese Äquivalenz | auf eine solche normale Satzformel
$\mathfrak{N}$ an, die wir durch Umformung der Formel $\overline{B^*}[p;p]$, mit der Nummer
$[p;p]$, erhalten, so gelangen wir, aufgrund von

$$\mathfrak{N} \sim \overline{B^*}[p;p]$$
$$B^*(n) \sim B^*[p;p]$$

zu dem Widerspruch

$$B^*[p;p] \sim \overline{B^*}[p;p].—$$

Here the formula

$$W_{n+1}(x)$$

$$\sim (Ek)(Ey)\{kFry \ \& \ \overline{k \ \text{Geb} \ y} \ \& \ [[x = k \ \text{Gen} \ y \ \& \ (u)W_n(Sb(y^k_{Z(u)}))]$$

$$\lor [x = \text{Neg}(k \ \text{Gen} \ (\text{Neg} \ y)) \ \& \ (Eu)W_n(Sb(y^k_{Z(u)}))]] \},$$

which expresses W_{n+1} in terms of W_n , comes into play.

And now

$$W(x) \sim (En)(n \leq l(x) \ \& \ W_n(x))$$

can be defined.

But on the other hand, it appears to me that this consideration must contain a mistake. For if $W(x)$ is defined in 3, the *"provability of x in the system 3 extended by the formula class $\hat{z}W(z)$"* can likewise be represented within 3 by a class sign $B^*(x)$.

Furthermore, for each normal formula $\mathfrak{N}$ from 3, with the number n, the formula

(1) $$W(n) \sim \mathfrak{N}$$

as well as, also,

(2) $$B^*(n) \to W(n),$$

is then formally provable (if not in 3, then in any case certainly in S).

Now in addition, on the basis of the definition of $B^*(x)$, it emerges that

$$W(n) \to B^*(n),$$

so that, overall, we obtain

$$\mathfrak{N} \sim B^*(n)$$

But if we apply this equivalence to such a normal formula $\mathfrak{N}$, which we obtain through transformation of the formula $\overline{B^*}[p;p]$ with the number $[p;p]$, we achieve, on the basis of

$$\mathfrak{N} \sim \overline{B^*}[p;p]$$

[and]

$$B^*(n) \sim B^*[p;p]$$

the contradiction

$$B^*[p;p] \sim \overline{B^*}[p;p].$$

Sie sind gewiss in der Lage, diese Sachlage aufzuklären. Die hier für
mich bestehende Unklarheit steht vermutlich auch im Zusammenhang
mit derjenigen, die ich inbetreff des *Ackermann*schen Beweises für die
Widerspruchsfreiheit der Zahlentheorie (System 3) einstweilen nicht beheben kann.

Diesen Beweis,—über den Hilbert in seinem Hamburger Vortrag über
"die Grundlagen der Mathematik" (Abh. aus d. Hamburger Math. Se-   D
minar 1928 Band VI Heft 1/2),[a] mit dem von mir hinzugefügten Zusatz
5   referiert hat— | habe ich mir wiederholt überlegt und für richtig befun-
den. Aufgrund Ihrer Ergebnisse muss man doch nun folgern, dass dieser
Beweis nicht innerhalb des Systems 3 formalisiert werden kann; ja dieses
muss sogar auch dann gelten, wenn man das als widerspruchsfrei zu er-
weisende System dadurch einschränkt, dass man von den rekursiven Definitionen nur die für die Addition und Multiplikation beibehält. Andrerseits sehe ich nicht, durch welche Stelle des Ackermannschen Beweises die
Formalisierung innerhalb von 3 unmöglich werden soll, besonders wenn
man die genannte Beschränkung ⟨des Problems vornimmt⟩.

—Beiläufig möchte ich hier folgendes erwähnen: Herr Herbrand, mit
dem ich in Berlin zu sprechen Gelegenheit hatte, war so freundlich, mir
einen Durchschlag von dem Brief zu senden, den er unlängst an Sie gerichtet hat. Was die Stelle dieses Briefes betrifft, wo Herr Herbrand über
6   den Beweis der Widerspruchsfreiheit des Systems | I + 2 + 3'' spricht,[b]
so hat er mich insofern missverstanden, als ich diesen Beweis nicht mir,
sondern Herrn Ackermann zugeschrieben habe. Es handelt sich hier um
den eben erwähnten Beweis.—

Von meinem Vortrag über die Axiome der Mengenlehre habe ich nur
ein stenographisches Konzept (Gabelsberger). Aber ich will Ihnen gern
nächstens die Axiome aufschreiben. Heute möchte ich nur die Absendung
dieser Zeilen nicht verzögern.

Nochmals vielen Dank und beste Grüsse von Ihrem

Paul Bernays

P.S. Herr Prof. Heinrich Scholz in Münster (i. Westfalen), [Adr.: Philosophisches Seminar der Univ., Abteilung B], der sich seit mehreren Jahren intensiv mit mathem. Logik befasst und die Entwicklung der Grundlagenforschung eifrig verfolgt, möchte gern Sonderdrucke von Ihren Abhandlungen (insbes. den von der neuen Abhandlung) erhalten.

---

[a] *Hilbert 1928.*

You are certainly in the position to clear up this state of affairs. What remains unclear to me here is presumably connected with the unclarity I am unable to resolve concerning *Ackermann*'s proof for the consistency of number theory (the system 3).

That proof—to which Hilbert referred in his Hamburg lecture on "The foundations of mathematics" (Abh. aus d. Hamburger Math. Seminar 1928 Band VI Heft 1/2),[a] with the addendum appended by me — I have repeatedly considered and viewed as correct. On the basis of your results one must now conclude, however, that that proof cannot be formalized within the system 3; this must in fact also hold true even if the system to be proved consistent is restricted so that, of the recursive definitions, only those for addition and multiplication are retained. On the other hand, I don't see at which place in Ackermann's proof the formalization within 3 should become impossible, especially if one presupposes the aforementioned restriction of the problem.

In passing I would here like to mention the following: Mr. Herbrand, with whom I had the opportunity to speak in Berlin, was so kind as to send me a carbon copy of the letter that he addressed to you not long ago. Concerning the place in that letter where Mr. Herbrand speaks about the proof of the consistency of the system $I+2+3''$,[b] he has misunderstood me, insofar as I did not ascribe the proof to me, but rather to Mr. Ackermann. It is the proof just mentioned that is in question here.

I have only a stenographic draft (Gabelsberger) of my talk on the axioms of set theory. But I will be happy to write out the axioms for you next time. Only today I would like not to delay sending off these lines.

Once again, many thanks and best regards from yours truly,

Paul Bernays

P.S. Professor Heinrich Scholz in Münster (Westphalia) [Address: Philosophisches Seminar der Univ., Abteilung B], who has been intensively involved with mathematical logic for several years and avidly follows the development of foundational research, would like very much to receive offprints of your papers (especially one of the new paper).

---

[b]See the correspondence with Jacques Herbrand, these *Works*, vol. V.

5. Bernays to Gödel

Göttingen, 3. V. 31.

Gaussstr. 1^{I}

Sehr geehrter Herr Gödel!

Endlich komme ich dazu, Ihnen den versprochenen Nachtrag zu meinem neulich an Sie gerichteten Brief zu bringen.

Zuvor möchte ich noch sagen, dass die Frage, die ich Ihnen neulich vorlegte, sich mir inzwischen aufgeklärt hat. Die aufgestellte Rekursion für $W_n(x)$ enthält ja in dem Ausdruck für $W_{n+1}(x)$ die Bestandteile $(m)W_n(Sb(y^k_{Z(m)}))$ und $(Em)W_n(Sb(y^k_{Z(m)}))$,—(Nota bene: ich weiss nicht mehr genau, ob ich die Benennungen ganz ebenso gewählt habe, aber das macht Ihnen jedenfalls nichts aus)—, und so ist es erklärlich, dass sie nicht im System 3 formalisierbar ist.—Auch betreffs des Ackermannschen Beweises für die Widerspruchsfreiheit der Zahlentheorie glaube ich jetzt ins Klare zu kommen.

2   Es scheint mir die Aufklärung des | Sachverhaltes darin zu bestehen, dass Rekursionen vom Typus:

$$\varphi(a,0) = \alpha(a)$$
$$\varphi(0,n) = \beta(n)$$
$$\varphi(a+1, n+1) = \varphi(a, \gamma(n, \varphi(a+1,n))),$$

(worin α, β, γ bereits definierte elementare Funktionen sein sollen), im allgemeinen *nicht innerhalb* des Systems 3 *formalisierbar* sind.—

Um nun zu den Axiomen der Mengenlehre zu kommen, so schicke ich voraus, dass ich mich hier auf den *inhaltlichen* Standpunkt stellen will, d. h. ich will die logische Symbolik nur *zur Mitteilung* anwenden und auch nur insoweit, als es mir zur Erleichterung der Übersicht geraten erscheint.

Eine *vollständige* Formalisierung, und zwar im Rahmen der *ersten Stufe*, ist ohne Schwierigkeit ausführbar.—

Das *System der Dinge* enthält zwei Gattungen: "~~Elemente~~⟨Mengen⟩" und "~~Mengen~~⟨Klassen⟩". Ich schreibe kleine Variable a, b, ... für ~~Elemente~~ ⟨Mengen⟩, grosse Variable A, B, M,... für ~~Mengen~~⟨Klassen⟩.

3 | Als Grundbeziehungen haben wir:

      "$a \, \epsilon \, b$":   "a ist in b enthalten",

               "b enthält a"

               "a ist *Element von b*"

      "$a \, \eta \, M$":   "a gehört zu M"

               "a ist *Element von M*".

5. Bernays to Gödel

Göttingen, 3 May 1931
Gaussstr. 1[I]

Dear Mr. Gödel,

At last I come to the matter of sending you the promised postscript to the letter I recently addressed to you.

First, I would like to add that the [answer to the] question I recently posed to you has become clear to me in the meantime. The recursion set up for $W_n(x)$ of course contains, in the expression for $W_{n+1}(x)$, the constituents $(m)W_n(Sb(y^k_{Z(m)}))$ and $(Em)W_n(Sb(y^k_{Z(m)}))$,—(N.B.: I no longer know exactly whether I chose the notation just so, but in any case that doesn't matter to you)—and that explains why it is not formalizable in the system 3.—Also, I think I have now reached a clear understanding with regard to Ackermann's proof of the consistency of number theory.

It seems to me that the explanation of the matter lies in the fact that recursions of the type:

$$\varphi(a, 0) = \alpha(a)$$
$$\varphi(0, n) = \beta(n)$$
$$\varphi(a + 1, n + 1) = \varphi(a, \gamma(n, \varphi(a + 1, n))),$$

(in which α, β, γ are supposed to be elementary functions already defined) are in general *not formalizable* within the system 3.

Turning now to the axioms of set theory, I mention at the outset that I will here adopt the *contentual* standpoint, i.e., I will employ the logical symbolism only *as a means of communication* and also only insofar as it appears advantageous to me for facilitating the overview.

A *complete* formalization, and in fact in a *first-order* framework, can be carried out without difficulty.

The *system of things* contains two kinds of entities: "sets" and "classes". I write small variables a, b, ... for sets, large variables A, B, M, ... for classes. As primitive relations we have:

"$a \, \epsilon \, b$":   "a is contained in b",
          "b contains a"
          "a is *an element of* b"
"$a \, \eta \, M$":   "a belongs to M"
          "a is *an element of* M".

Die Beziehung $a \,\epsilon\, b$ findet zwischen Mengen, die Beziehung $a\,\eta\,M$ zwischen einer Menge und einer Klasse statt.

Definition:

$$a\,\sigma\,M \sim_{\mathrm{Df}} (x)(x\,\epsilon\,a \rightleftarrows x\,\eta\,M)$$

"$a\,\sigma\,M$" wird gelesen "die Menge a ist der Klasse M zugeordnet" oder "M ist durch a repräsentiert."— Die *Identität* will ich hier nicht als eigentliche Beziehung auffassen, sondern als etwas schon mit dem System der Dinge Gegebenes (entsprechend wie in der Geometrie). Zur Formalisierung soll, wie üblich, das Symbol = verwendet werden.

(Diese Auffassung ist nicht wesentlich. Man könnte, nach Fraenkels Methode, die | Identität von Mengen durch die ϵ-Beziehung definieren, und entsprechend die Identität von Klassen durch die η-Beziehung. Auch brauchte man von Identität bei Klassen überhaupt nicht zu reden.)—

Zur Formalisierung der Identität dienen die Formeln:

$$a\llbracket ? \rrbracket a$$

$$a = b \to (a\,\epsilon\,c \rightleftarrows b\,\epsilon\,c)$$
$$a = b \to (c\,\epsilon\,a \rightleftarrows c\,\epsilon\,b)$$
$$a = b \to (a\,\eta\,C \rightleftarrows b\,\eta\,C)$$
$$A = B \to (c\,\eta\,A \rightleftarrows c\,\eta\,B)$$

(Die weiteren Formeln für Symmetrie usw. brauchen wir nicht.)

Nun folgen die *Axiome*.

I. Axiome der Bestimmtheit:

1) $(x)(x\,\epsilon\,a \rightleftarrows x\,\epsilon\,b) \to a = b$

2) $(x)(x\,\eta\,A \rightleftarrows x\,\eta\,B) \to A = B$.

(Das Axiom I2) ist, wie schon aus dem zuvor Bemerkten hervorgeht, *nicht wesentlich*, sondern nur der einfacheren Auffassung halber eingeführt.)

II. Elementare Mengen-Axiome.

1) Es gibt eine Menge, die kein Element enthält.

2) Ist b eine nicht in a enthaltene Menge, | so gibt es eine Menge c, welche aus a durch *Hinzunahme von b* (*als Element*) entsteht.

Folgerungen u. Bezeichnungen.

Aus I1), II1) folgt, dass es eine eindeutig bestimmte "Nullmenge" gibt, sie werde mit O bezeichnet.

Aus I1), II2) folgt weiter: Zu jeder Menge a gibt es eine eindeutig bestimmte Menge "(a)", welche a als einziges Element hat. Und zu zwei verschiedenen Mengen a, b gibt es eine eindeutig bestimmte Menge "(a, b)", welche a und b, und auch nur diese Mengen zu Elementen hat.

Es seien jetzt a, b irgend welche Mengen (es braucht nicht $a \neq b$ zu sein). Jedenfalls ist $O \neq (b)$ und $(a) \neq (O, (b))$.

The relation $a \, \epsilon \, b$ obtains between sets, the relation $a \, \eta \, M$ between a set and a class.

Definition:

$$a \, \sigma \, M \sim_{\text{Df}} (x)(x \, \epsilon \, a \rightleftarrows x \, \eta \, M)$$

"$a \, \sigma \, M$" is read "the set a is assigned to the class M" or "M is represented by a".

Here I do not want to view *identity* as a proper relation, but rather as something already given with the system of things (as it is done in geometry). As usual, the symbol $=$ is supposed to be used in the formalization.

(This conception is not essential. One could define identity of sets in terms of the ϵ-relation, according to Fraenkel's method, and correspondingly define identity of classes in terms of the η-relation. One also need not speak of identity of classes at all.)

The [following] formulas serve for the formalization of identity:

$$a = b \rightarrow (a \, \epsilon \, c \rightleftarrows b \, \epsilon \, c)$$
$$a = b \rightarrow (c \, \epsilon \, a \rightleftarrows c \, \epsilon \, b)$$
$$a = b \rightarrow (a \, \eta \, C \rightleftarrows b \, \eta \, C)$$
$$A = B \rightarrow (c \, \eta \, A \rightleftarrows c \, \eta \, B)$$

(We don't need the further formulas for symmetry, etc.) Now follow the *axioms*.

I. Axioms of definiteness [extensionality]:

    1) $(x)(x \, \epsilon \, a \rightleftarrows x \, \epsilon \, b) \rightarrow a = b$

    2) $(x)(x \, \eta \, A \rightleftarrows x \, \eta \, B) \rightarrow A = B$.

(Axiom *I2*) is *not essential*, as follows already from what was remarked at the outset, but is only introduced for the sake of a simpler conception.)

II. Elementary set-axioms.

    1) There is a set that contains no element.

    2) If b is a set not contained in a, then there is a set c that is obtained from a by *adjunction of b* (*as an element*).

Consequences and notation

From I1) and II1) it follows that there is a uniquely determined "null set"; it will be denoted by O. From I1) and II2) it follows further: For each set a there is a uniquely determined set "(a)" that has a as its only element. And for two different sets a, b there is a uniquely determined set "(a, b)" that has a and b, and only those sets, as elements.

Now let a and b be any sets whatever (it need not be that $a \neq b$). In any case, $O \neq (b)$ and $(a) \neq (O, (b))$.

Wir definieren als das "*geordnete Paar a, b*" die Menge

$$((a), (O, (b)))$$

und schreiben dafür kurz "$\langle a, b \rangle$". Die Eigenschaft einer Menge, ein geordnetes Paar zu sein, lässt sich mit Hilfe der ϵ-Beziehung (und der Identität) ausdrücken. | Für jedes Paar $\langle a, b \rangle$ ist eindeutig a als "erstes Element" und b als "zweites Element" bestimmt. D. h. wenn c ein geordnetes Paar ist, so gibt es nur eine ~~Ding~~⟨Menge⟩ a und nur eine Menge b, sodass

$$c = \langle a, b \rangle.$$

Aus I1) und II2) folgt zu a und b gibt es stets eine eindeutig bestimmte Menge c, sodass

$$(x)(x \,\epsilon\, c \rightleftarrows x \,\epsilon\, a \lor x = b).$$

Diese Menge werde mit "$a + (b)$" bezeichnet. Wenn b Element von a ist, so ist $a + (b) = a$. Aus den Axiomen II folgt bereits, dass es unbegrenzt viele Mengen gibt. Nämlich

$$O, (O), (O) + ((O)), (O) + ((O)) + ((O) + ((O))), \dots$$

sind lauter verschiedene Mengen.

III. Elementare Axiome für Klassen.

A) 1) Jede Menge repräsentiert eine Klasse, d. h. zu jeder Menge a gibt es eine Klasse, deren Elemente die Elemente von a sind.

2) Zu jeder Klasse $\langle A \rangle$ gibt es eine "*Komplementärklasse*", deren Elemente diejenigen Mengen sind, die nicht zu A gehören.

3) Zu zwei Klassen A, B gibt es einen "*Durchschnitt*", d. h. eine Klasse deren Elemente die gemeinsamen Elemente von A und B sind.

|Folgerungen: Es gibt eine (eindeutig bestimmte) "*Nullklasse*", zu der keine Menge gehört. Es gibt eine "*Allklasse*"[,] zu der jede Menge gehört.

Zu zwei Klassen A, B gibt es eine "*Vereinigungsklasse*", welche diejenigen Mengen enthält, die zu mindestens einer der Klassen A, B gehören. (Bemerkung: Anstelle des Axioms IIIA1) könnte man auch folgendes Axiom nehmen: "Zu jeder Menge a gibt es eine Klasse, die a als einziges Element hat". Aufgrund der nachfolgenden Axiome leistet dieses Axiom dasselbe wie IIIA1).)

B) Vorbemerkung: Die jetzt folgenden Axiome betreffen die Existenz von *Klassen von Paaren*, wobei jedesmal die Eindeutigkeit aus I2) folgt. Der Einfachheit halber soll die Formulierung so gewählt werden, dass die Behauptung der Eindeutigkeit gleich in den Wortlaut des Axioms eingeschlossen ist.

1) Es existiert die ~~Menge~~⟨Klasse⟩ aller Paare $\langle a, a \rangle$

2) Es existiert die Klasse aller der Paare $\langle a, b \rangle$, für welche $a \,\epsilon\, b$ gilt.

We define as the "*ordered pair a, b*" the set

$$((a), (O, (b)))$$

and, for short, we write "$\langle a, b \rangle$" for it. The property of a set's being an ordered pair may be expressed with the help of the ϵ-relation (and of identity). For every pair $\langle a, b \rangle$ a unique a is determined as "first element" and [a unique] b as "second element". I.e., if c is an ordered pair, there is only one set a and only one set b such that

$$c = \langle a, b \rangle.$$

From I1) and II2) it follows that for a and b there is always a uniquely determined set c such that

$$(x)(x \epsilon c \rightleftarrows x \epsilon a \lor x = b).$$

This set will be denoted by "$a + (b)$". If b is an element of a, then $a + (b) = a$. From the axioms II it already follows that there are unboundedly many sets. Namely,

$$O, (O), (O) + ((O)), (O) + ((O)) + ((O) + ((O))), \ldots.$$

are all different sets.

III. Elementary axioms for classes.

A) 1) Every set represents a class, i.e., for every set a there is a class whose elements are the elements of a.

2) For every class A there is a "*complementary class*", whose elements are those sets that do not belong to A.

3) For two classes A, B there is an "*intersection*", i.e. a class whose elements are the common elements of A and B.

Consequences: There is a (uniquely determined) "null class", to which no set belongs. There is a "universal class" to which every set belongs.

For two classes A, B there is a "union class" which contains those sets that belong to at least one of the classes A, B.

(Remark: Instead of the axioms IIIA1) one could also take the following axiom: "For every set a there is a class that has a as its only element". On the basis of the succeeding axioms this axiom accomplishes the same as IIIA1).)

B) Prefatory remark: The axioms that now follow concern the existence of *classes of pairs*, for which the uniqueness follows in every case from I2). For the sake of simplicity the formalization should be so chosen that the assertion of uniqueness is contained directly within the symbolic expression of the axiom.

1) There exists the class of all pairs $\langle a, a \rangle$.

2) There exists the class of all the pairs $\langle a, b \rangle$ for which $a \epsilon b$ holds.

8 | 3) Zu jeder Klasse A existiert die Klasse der Paare $\langle a, b \rangle$, für welche
$a \, \eta \, A$ gilt.

4) Zu jeder Klasse $\langle C \rangle$ von Paaren existiert der "*Vorbereich*", d. h. die
Klasse derjenigen Mengen a, für welche gilt:

$$(Ey)(\langle a, y \rangle \, \eta \, C)$$

5) Zu jeder Klasse C von Paaren existiert die "*konverse Klasse*", deren
Elemente aus den Elementen von C durch Vertauschung des ersten mit
dem zweiten Paar-Element hervorgehen.

6) Zu jeder Klasse, deren Elemente die Form $\langle \langle a, b \rangle, c \rangle$ haben, existiert
die Klasse der entsprechenden Elemente $\langle a, \langle b, c \rangle \rangle$.—

Mit Hilfe dieser Axiome kann man zeigen, dass jede "definite Aussage"
(im Sinne von Zermelo) $\langle$über eine Menge $a \rangle$ sich auf die Form $a \, \eta \, C$
bringen lässt, wo C eine gewisse durch die Aussage bestimmte Klasse
9   ist. | So kann man z. B. beweisen: Zu jeder Menge m gibt es die "Ver-
einigungsklasse", d. h. die Klasse derjenigen Mengen, welche in den Elementen von m enthälten sind. Nämlich sei M die (nach IIIA1) vorhan-
dene) durch m repräsentierte Klasse. A sei die *konverse* Klasse zu der
Klasse aller der Paare $\langle a, b \rangle$, für die $a \, \eta \, M$ gilt. B sei der Durchschnitt
von A mit der Klasse der Paare $\langle a, b \rangle$, für die $a \, \epsilon \, b$ gilt. Dann besteht B
aus denjenigen Paaren $\langle c, d \rangle$, für welche

$$c \, \epsilon \, d \; \& \; d \, \eta \, M$$

bezw., was damit gleichbedeutend ist:

$$c \, \epsilon \, d \; \& \; d \, \epsilon \, m.$$

Der *Vorbereich* von B ist also die Vereinigungsklasse von m.

In ähnlicher Weise zeigt man, dass es zu jeder Menge m eine *Klasse*
gibt, deren Elemente die *Teilmengen von m* sind, (gemäss der üblichen
Definition der "Teilmenge").

Definitionen. Der Begriff der *Teilklasse* wird wie üblich definiert: "A ist
Teilklasse von B" besagt

$$(x)(x \, \eta \, A \to x \, \eta \, B)$$

10 |Eine Klasse C von Paaren heisst eine "*Funktion*", wenn

$$(x)(y)(z)(\langle x, y \rangle \, \eta \, C \; \& \; \langle x, z \rangle \, \eta \, C \to y = z).$$

Eine Klasse C von Paaren heisst eine "*umkehrbar eindeutige Abbildung*",
wenn C sowie auch die konverse Klasse zu C eine Funktion ist.

3) For every class A there exists the class of pairs $\langle a, b \rangle$ for which $a \eta A$ holds.

4) For every class C of pairs there exists the "*domain*", i.e., the class of those sets a for which

$$(Ey)(\langle a, y \rangle \, \eta \, C)$$

holds.

5) For every class C of pairs there exists the "*converse class*", whose elements are obtained from the elements of C by switching the first element of each pair with the second.

6) For every class whose elements have the form $\langle \langle a, b \rangle, c \rangle$ there exists the class of the corresponding elements $\langle a, \langle b, c \rangle \rangle$.—

With the help of these axioms one can show that every "definite proposition" (in Zermelo's sense) over a set a may be brought into the form $a \, \eta \, C$, where C is a certain class determined by the proposition. So *e.g.* one can prove: For every set m there is the "union class", i.e., the class of those sets which are contained in the elements of m. Namely, let M be the class (occurring according to IIIA1) represented by m. Let A be the *converse* class to the class of all the pairs $\langle a, b \rangle$ for which $a \, \eta \, M$ holds. Let B be the intersection of A with the class of pairs $\langle a, b \rangle$ for which $a \, \epsilon \, b$ holds. Then B consists of those pairs $\langle c, d \rangle$ for which

$$c \, \epsilon \, d \; \& \; d \, \eta \, M$$

respectively, which is equivalent to:

$$c \, \epsilon \, d \; \& \; d \, \epsilon \, m.$$

The *domain* of B is thus the union class of m.

One shows in a similar way that for every set m there is a *class* whose elements are the *subsets of m* (in accordance with the usual definition of "subset").

Definitions. The concept of *subclass* is defined as usual: "A is a subclass of B" means

$$(x)(x \, \eta \, A \to x \, \eta \, B)$$

A class C of pairs is called a "*function*" if

$$(x)(y)(z)(\langle x, y \rangle \, \eta \, C \; \& \; \langle x, z \rangle \, \eta \, C \to y = z).$$

A class C of pairs is called a "*one-to-one mapping*" if C as well as the class converse to C is a function.

Zwei Klassen A, B heissen *"gleichzahlig"*, wenn es eine umkehrbar eindeutig Abbildung gibt, für welche A der Vorbereich und B der "Nach-
bereich", d. h. der Vorbereich der konversen Menge[a] ist.

Mit diesen Begriffen können wir nun die drei Hauptaxiome der *allgemeinen Mengenlehre* formulieren:

IV 1) Axiom der *Aussonderung*: Ist A Teilklasse von B und ist B durch
eine *Menge* repräsentiert, so ist auch A durch eine Menge repräsentiert.

2) Axiom der *Ersetzung*: Ist A mit B gleichzahlig und ist B durch
eine Menge repräsentiert, so auch A.

11 | 3) *Auswahlaxiom*: Zu jeder Klasse C von Paaren gibt es eine *Funktion*,
deren Vorbereich mit dem von C übereinstimmt.

Zur Vereinfachung der Theorie fügen wir noch folgendes zuerst von v.
Neumann aufgestellte Axiom hinzu:

4) In jeder nicht leeren Klasse $\langle K \rangle$ (d. h. jeder Klasse ausser der Nullklasse) gibt es mindestens ein Element, von dem kein Element zu K gehört.—

Mit anderen Worten: Wenn eine von der Nullklasse verschiedene Klasse K mit jedem Element $a \neq O$, welches sie enthält zugleich auch ein
Element von a enthält, so enthält sie die *Nullmenge* als Element.

Durch dieses Axiom werden die zirkelhaften Mengen ausgeschlossen.
Es folgt daraus insbesondere:

$$\overline{(Ex)}\ (x \,\epsilon\, x)$$
$$\overline{(Ex)}(Ey)(x \,\epsilon\, y \ \&\ y \,\epsilon\, x).—$$

Mit Hilfe der bisherigen Axiome kann bereits die allgemeine Mengenlehre
12 entwickelt | werden, d. h.: die Theorie der Gleichzahligkeit (z. B. Bern-
steinscher Satz), ferner die Theorie der Wohlordnung u. der Ordnungszahlen. Man beweist insbes., dass jede *Menge* umkehrbar eindeutig auf
eine Ordnungszahl abbildbar und somit auch wohlordenbar ist. Ferner
beweist man im Bereich der Ordnungszahlen die Gültigkeit der *transfiniten Induktion* und die *Zulässigkeit der rekursiven Definition*.

Der Begriff der "Ordnungszahl" wird am einfachsten mittels des Begriffs der "Vollzähligkeit" definiert: Eine Menge heisst *"vollzählig"*, wenn
sie zugleich mit jedem ihrer Elemente a auch die Elemente von a enthält.
Und eine Menge ist eine "Ordnungszahl", wenn sie selbst, sowie jedes
ihrer Elemente vollzählig ist. (Es existiert die *Klasse* aller Ordnungszahlen.)

Der Begriff der *endlichen* Ordnungszahl kann auf verschiedene Arten
eingeführt werden. Man gelangt damit zur Aufstellung der *Zahlentheorie*.—

---

[a]It would appear that "Klasse" is intended.

Two classes A, B are said to be *"equinumerous"* if there is a one-to-one mapping for which A is the domain and B is the "range", i.e., the domain of the converse set.[a]

With these concepts one can now formulate the three principal axioms of *general set theory*:

IV 1) Axiom of *separation*: If A is a subclass of B and B is represented by a *set*, A is also represented by a set.

2) Axiom of *replacement*: If A is equinumerous with B and B is represented by a set, A is also.

3) *Axiom of choice*: For every class C of pairs there is a *function* whose domain coincides with that of C.

In order to simplify the theory we adjoin in addition the following axiom, first stated by *von Neumann*:

4) In every non-empty class K (i.e., every class except the null class) there is at least one element, no element of which belongs to K.

In other words: If a class K is distinct from the null class and, for every element $a \neq O$ that it contains, contains an element of a as well, then it contains the *null set* as an element.

By means of this axiom the circular sets are excluded. In particular, it follows therefrom:

$$\overline{(Ex)}\ (x \in x)$$
$$\overline{(Ex)}(Ey)(x \in y\ \&\ y \in x).$$

With the help of the aforementioned axioms general set theory can already be developed, i.e.: the theory of equinumerosity (e.g. Bernstein's Theorem), and furthermore the theory of well-ordering and of ordinal numbers. In particular, one proves that every *set* can be mapped one-to-one onto an ordinal number and so is thereby also well-orderable. In the domain of the ordinal numbers one further proves the validity of *transfinite induction* and the *admissibility of definition by recursion*.

The concept of "ordinal number" is defined most easily by means of the concept of "transitivity": A set is said to be *"transitive"* if for each of its elements a it contains at the same time the elements of a as well. And a set is an "ordinal number" if it itself, as well as each of its elements, is transitive. (The *class* of all ordinal numbers exists.)

The concept of *finite* ordinal number can be introduced in various ways. One thereby succeeds in setting up *number theory*.

Nun kommen noch die höheren Mengen-Axiome:

13 |V. 1) Unendlichkeits-Axiom (im Sinne von Zermelo): Es gibt eine *Men-
ge*, die O enthält, und die zugleich mit a auch $a + (a)$ enthält.

Dieses Axiom gestattet die Entwicklung der *Analysis* und der Theorie der *Zahlen der zweiten Zahlenklasse*. (Es liefert das "System S" nach
Ihrer Bezeichnung.)

Dagegen kann hiermit noch nicht die Wohlordenbarkeit des Kontinuums bewiesen werden. Nämlich das Kontinuum tritt hier nur als *Klasse*,
nicht als Menge auf.

2) Für jede Menge m ist die Klasse ihrer Teilmengen durch eine Menge
repräsentiert.

Mit diesem Axiom hat man das System der "Principia Mathematica".
Aber man kann noch nicht die Existenz einer Menge von der Mächtigkeit
$\aleph_\omega$ beweisen.

3) Für jede Menge m ist die Vereinigungs-Klasse durch eine Menge
repräsentiert.

14 —Ich glaube, dass man sich bei dieser | Zusammenstellung die Axiome
verhältnismässig leicht merken kann.[1]—

Für heute möchte ich hiermit schliessen, zumal da meine Ausführungen schon recht lang geworden sind.

Mit den besten Grüssen empfiehlt sich Ihnen Ihr

Paul Bernays

P.S. Aus dem Gesamt-System I–V kann man nach v. Neumann beweisen, dass jede *nicht* durch eine Menge repräsentierte Klasse gleichzahlig ist mit der Allklasse und daher auch mit der Klasse aller Ordnungszahlen, sodass in diesem Sinne auch *jede Klasse wohlgeordnet* werden kann.

[1]Ich gedenke diese Sache bald einmal zu publizieren.

6. Gödel to Bernays

Bremen 19./VI. 1939.

Lieber Herr Bernays!

Ich habe im Wintersemester 1938 in Princeton eine Vorlesung über die
Widerspruchsfreiheit des Kontinuumsatzes gehalten u. dabei Ihr Axiomensystem der Mengenlehre [wie Sie es mir vor einigen Jahren brieflich

Still to come now are the higher set axioms:

V. 1) Axiom of infinity (in Zermelo's sense): There is a *set* that contains *O* and at the same time as *a* also contains $a + (a)$.

This axiom allows the development of *analysis* and of the theory of the *ordinals of the second number class.* (It yields the "system *S*" according to your notation.)

On the the other hand, the well-orderability of the continuum still cannot be proven hereby. Namely, the continuum here appears only as a class, not as a set.

2) For every set *m* the class of its subsets is represented by a set.

With this axiom one has the system of "Principia Mathematica". But one still cannot prove the existence of a set of power $\aleph_\omega$.

3) For every set *m* the union class is represented by a set.

I think that with this grouping the axioms can be kept in mind relatively easily.[1]

Here I would like to stop for today, especially since my remarks have become quite lengthy already.

I commend myself to you with best wishes.

Paul Bernays

P.S. From the entire system I–V one can prove, in the manner of von Neumann, that every class that is *not* represented by a set is equinumerous with the universal class, and therefore also as the class of all ordinal numbers, so that in this sense also *every class* can be *well-ordered*.

---

[1] I intend to publish this matter sometime soon.

6. Gödel to Bernays

Bremen, 19 June 1939

Dear Mr. Bernays:

In the winter semester of 1938, in Princeton, I gave a lecture ⟦course⟧ on the consistency of the continuum hypothesis in which I took your axiom system for set theory (as you imparted it to me a few years ago

mitteilten] zu Grunde gelegt. Meine Vorlesung soll jetzt mimeographiert
werden u. ich habe leider übersehen, dass sich das System, welches Sie
im Journ. symb. Logik[a] publizierten, in einem Axiom[1] von dem mir mitgeteilten unterscheidet. Hoffentlich haben Sie gegen eine Publikation in
der früheren Form (selbstverständlich mit Erwähnung Ihrer Autorschaft)
nichts einzuwenden. Ferner wollte ich Sie bezüglich des Fundierungsaxioms ~~in der Form~~ ⟨(das ich auch verwende)⟩:

$$\sim (\exists x)\{x \neq 0 \,.\, (y)[y \,\epsilon\, x \supset (\exists z)(z \,\epsilon\, x \,.\, y)]\}$$

[oder was auf dasselbe hinausläuft: $(x)[x \neq 0 \supset (\exists y)(y \,.\, x = 0)]$] fragen,
ob diese Form der Idee nach auf v. Neumann zurückgeht u. von Ihnen
bloss für das Axiomensystem mit der ϵ-Relation als Grundbegriff umfor-
muliert wurde.

2 | Ich wäre Ihnen für eine umgehende Antwort ⟨nach Wien⟩ ausseror-
dentlich dankbar, weil ich die Vervielfältigung meiner Vorlesungen nicht
weiter verzögern möchte.

Kommen Sie nicht bald wieder einmal nach Princeton?

Mit besten Grüssen

Ihr   Kurt Godel

Wien XIX. Himmelstrasse 43.

[1] Sie benutzten statt der Existenz der Klasse aller $\{x\}$ ein zweites Inversionsaxiom
für 3-stellige Relationen.

---

[a] *Bernays 1937.*

7. Bernays to Gödel

Zürich 7, 21. VI. 39.
Eidmattstr. 55.

Lieber Herr Gödel!

Auf Ihren freundlichen Brief, den ich heute empfing—es war mir sehr
erfreulich, einmal wieder direkt von Ihnen zu hören—möchte ich Ihnen
sogleich antworten.

Was das Axiom von der Existenz der Klasse aller Mengen (c) betrifft,
so hatte ich früher an dessen Stelle das Axiom von der Existenz einer

by letter) as the basis. My lecture⟦s⟧ are now supposed to be mimeo-graphed, and I unfortunately overlooked that the system you published in the *Journal of Symbolic Logic*ᵃ differs from the one imparted to me in one axiom.[1] I hope you have no objection to a publication of the earlier form (with mention of your authorship, of course). Furthermore, with regard to the axiom of foundation (that I also use):

$$\sim (\exists x)\{x \neq 0 \,.\, (y)[y \,\epsilon\, x \supset (\exists z)(z \,\epsilon\, x \,.\, y)]\}$$

[or what amounts to the same thing: $(x)[x \neq 0 \supset (\exists y)(y \,.\, x = 0)]$] I wanted to ask you whether this form of the idea goes back to von Neu-mann and was recast by you merely for the axiom system with the ϵ-relation ⟦taken⟧ as a primitive notion.

I would be especially grateful to you for a reply by return mail to Vienna, because I would like not to delay further the duplication of my lectures.

Aren't you coming to Princeton again soon?

With best regards,

Yours truly,

Kurt Gödel
Vienna XIX, Himmelstrasse 43

[1] Instead of the existence of the class of all $\{x\}$ you used a second axiom of inversion for 3-place relations.

---

7. Bernays to Gödel

Zurich 7, 21 June 1939
Eidmattstr. 55

Dear Mr. Gödel:

I would like to reply at once to your friendly letter, which I received today—it was very pleasing to me to hear from you directly once again.

As to the axiom of the existence of the class of all sets (c), I formerly had in place of it the axiom of the existence of a class of all pairs $\langle c, c \rangle$.

Klasse aller Paare $\langle c, c \rangle$. Dieses ist sogar in gewisser Hinsicht natürlicher (wie ich es in der Anm. 11 auf S. 70 der Arbeit im Journal vermerkte).[1] Ein zweites Inversionsaxiom für 3-stellige Relationen ausser demjenigen, dass zu jeder Klasse von Paaren der Form $\langle a, \langle b, c \rangle \rangle$ die Klasse der entsprechenden Paare $\langle \langle a, b \rangle, c \rangle$ existiert, ist nicht nötig. Ich hatte anstelle dieses Axioms anfangs zwei andere Axiome, von denen ich aber sehr bald $\langle$(in einer Vorlesung 1929/30)$\rangle$ feststellte, dass sie auf dieses Axiom zurückführbar sind.

Falls Sie in Ihrer Vorlesungs-Ausarbeitung die Fassung des Axiomensystems mit zwei Inversionsaxiomen benutzen, so können Sie ja vielleicht in einer Bemerkung auf die Ersetzung durch das eine Axiom verweisen.

Den Gedanken des Fundierungsaxioms (wie Sie es bezeichnen) habe ich ganz und gar von v. Neumann übernommen. Übrigens dürfte die

2   von Ihnen erwähnte | Fassung, die wohl meiner damaligen brieflichen Formulierung entspricht, dem v. Neumannschen Axiom nicht gleichwertig sein, und es scheint mir, dass sie für die Haupt-Anwendung, die v. Neumann von dem Axiom macht, nicht ausreicht. Ich habe deshalb seit längerem die stärkere, dem v. Neumannschen Axiom $\langle$enger$\rangle$ angepasste Formulierung gewählt: "In jeder nicht leeren *Klasse A* gibt es ein Element, das zu *A* elementenfremd ist."—

Nun möchte ich auch Sie etwas fragen. Im vorigen August hörte ich von Herrn Mostowski, der sich auf eine Mitteilung v. Neumanns berief, Sie hätten etwas über die Unabhängigkeit des Kontinuumssatzes bewiesen. In Ihrer neuen Publikation in den Proceedings[a] finde ich aber keinen derartigen Satz. Ebenso enthielt auch die Mitteilung, die Sie im Winter an Herrn Gonseth sandten, keine dahingehende Behauptung. Nun würde ich gern von Ihnen erfahren, ob jene Mitteilung auf einem Missverständnis beruhte.

Jedenfalls sind Ihre neuen Ergebnisse sehr imponierend und $\langle$inhaltlich$\rangle$ befriedigend! Sobald ich mich wieder in die axiom. Mengenlehre begebe, (was ich bald tun will) werde ich mir Ihre Proceedings-Arbeit genauer vornehmen.—

Leider scheint fürs nächste keine Aussicht für mich zu bestehen, wieder nach Princeton eingeladen zu werden.—

Mit dem Wunsche, dass es Ihnen gut gehe, und mit freundlichem Gruss Ihr

Paul Bernays

---

[1] Mir ist es daher jedenfalls ganz ebenso recht, wenn Sie dieses Axiom anstatt jenes anderen nehmen.

---

[a] *1939a.*

In a certain respect the latter is even more natural (as I noted in remark 11 on page 70 of the work in the journal).[1] A second axiom of inversion for 3-place relations, in addition to the one that to every class of pairs of the form $\langle a, \langle b, c \rangle \rangle$ the class of the corresponding pairs $\langle \langle a, b \rangle, c \rangle$ exists, is not necessary. In place of that axiom I originally had two other axioms, but very soon (in lectures [of] 1929/30) I established that they are reducible to that axiom.

In case, in your lecture notes, you use the version of the axiom system with two axioms of inversion, you can perhaps refer in a remark to the[ir] replacement by the one axiom.

The conception of the axiom of foundation (as you designate it) I took over wholly from von Neumann. However, the version mentioned by you, which probably corresponds to my formulation in the letter of that time, might not be equivalent to von Neumann's axiom, and it seems to me that it does not suffice for the principal use that von Neumann makes of the axiom. I have therefore long since chosen the stronger formulation that conforms more closely to von Neumann's axiom: "In every nonempty *class* A there is an element that is disjoint from A."

Now I would also like to ask you something. Last August I heard from Mr. Mostowski, who referred to a communication of von Neumann, that you had proved something about the independence of the continuum hypothesis. But in your new publication in the Proceedings[a] I find no theorem of that sort. Likewise, the communication that you sent to Mr. Gonseth in the winter also contains no assertion along those lines. Now I would like to hear from you whether that report was based on a misunderstanding.

In any case your new results are very impressive and contentually satisfying! As soon as I take up axiomatic set theory again (which I want to do soon), I will look more closely at your Proceedings article.

Unfortunately, for the time being there appears to be no prospect of my being invited to Princeton again.

In the hope that things are going well with you, and with friendly greeting[s],

Yours truly,

Paul Bernays

---

[1] Therefore in any case it is quite all right with me if you take this axiom instead of that other one.

P.S. Für den Fall, dass Sie meine Arbeit über d. Mengenlehre nicht zur Hand haben, sende ich Ihnen ⟨noch⟩ ein Exemplar davon.

8. Gödel to Bernays

Wien 20./VII. 1939.

Lieber Herr Bernays!

Vor allem recht herzlichen Dank für die rasche Beantwortung meines Briefes u. die Zusendung des Separatums.—Ich verwendete in meiner Vorlesung zwei Inversionsaxiome für 3-stellige Relationen, aber dafür weder das Axiom von der Existenz der Klasse der (c) noch der $\langle c, c \rangle$. Ich war der Meinung, dass dies mit der mir seinerzeit von Ihnen brieflich mitgeteilten Formulierung übereinstimmt, was sich aber jetzt an der Hand Ihres Briefes von 1932, den ich in Amerika ⟨nicht⟩ mit hatte, als Irrtum   1 herausgestellt hat.

Das Fundierungsaxiom für Mengen scheint mir doch mit dem für Klassen äquivalent zu sein[1], denn es folgt aus dem Fund.-ax. für Mengen, dass jede Menge x in der transfiniten Hierarchie der Typen vorkommt
2 [einen Typus hat]. Um dies zu zeigen, | bilde man die "vollzählige Hülle" $\bar{x}$ [d. h. $\bar{x} = x + \gamma(x) + \gamma\gamma(x) + \cdots .$] u. die Menge m der "typenlosen" Elemente von $\bar{x}$. Wäre $m \neq 0$, so gäbe es ein $y \,\epsilon\, m$, $y \cdot m = 0$; d. h. alle Elemente von y hätten einen Typus, y selbst aber nicht, was unmöglich ist.[2]

Bezüglich der Unabhängigkeit der Kontinuumhypothese habe ich wohl einige Ansätze, bin aber zu einem endgültigen Ergebnis noch nicht gelangt. Die Mitteilung von Mostowski kann nur auf einem Missverständnis beruhen. Von Mostowski hörte ich übrigens, dass er Ihnen meine Definition der "konstruierbaren" Mengen [in einer von meiner Proceedings-Arbeit etwas abweichenden, aber äquivalenten Formulierung] erzählte u. dass Sie behauptet hätten, die Existenz von $\aleph_1$ konstruierbaren Mengen
3 natürlicher Zahlen beweisen zu | können. Dieses Resultat würde mich ausserordentlich überraschen, da ich vermute, dass die Annahme der Existenz von nur abzählbar vielen konstr. Mengen nat. Zahlen widerspruchsfrei ist. Handelt es sich vielleicht auch bei dieser Sache um ein Missverständnis?

Ich hoffe[,] dass es Ihnen recht gut geht u. bin mit den besten Grüssen

Ihr   Kurt Gödel

---

[1]allerdings nur auf Grund des Unendlichkeits- u. Ersetzungsaxioms.

[2]Ich habe übrigens in meiner Vorlesung das Fund.-axiom für Klassen angenommen.

P.S. In case you don't have my work on set theory at hand, I am sending you another copy of it.

8. Gödel to Bernays

Vienna 20 July 1939

Dear Mr. Bernays:

First of all, most hearty thanks for the prompt answering of my letter and for the forwarding of the offprint.—In my lecture I used two axioms of inversion for 3-place relations, but neither the axiom of the existence of the class of [[all]] (c) nor of [[all]] $\langle c,c \rangle$. I was of the opinion that that agree[[d]] with the formulation you communicated to me by letter at the time, which, however, in light of your letter of 1932 (which I did not have with me in America) has now turned out to be mistaken.

The axiom of foundation for sets still seems to me to be equivalent to that for classes,[1] for it follows from the axiom of foundation for sets that every set x occurs in the transfinite hierarchy of types (has a type). In order to show that, one forms the "complete hull" $\bar{x}$ [i.e. $\bar{x} = x + \gamma(x) + \gamma\gamma(x) + \cdots$.] and the set m of the "typeless" elements of $\bar{x}$. Were $m \neq 0$, there would be a $y \in m$, $y \cdot m = 0$: i.e., all elements of y would have a type, but not y itself, which is impossible.[2]

With regard to the independence of the continuum hypothesis I have indeed [[made]] a few starts, but I've not yet succeeded in obtaining a final result. Mostowski's report can only be based on a misunderstanding. By the way, I heard from Mostowski that he told you my definition of the "constructible" sets (in a formulation somewhat different from that of my Proceedings article, but equivalent to it) and that you had claimed to be able to prove the existence of $\aleph_1$ constructible sets of natural numbers. I would find that result exceedingly surprising, since I conjecture that the assumption of the existence of only countably many constructible sets of natural numbers is consistent. Is this matter perhaps also a case of a misunderstanding?

I hope that things are going quite well for you. With best regards,

Yours truly,

Kurt Gödel

---

[1] To be sure, only on the basis of the axioms of infinity and replacement.

[2] Anyway, in my lectures I assumed the axiom of foundation for classes.

9. Bernays to Gödel

Zürich, den 28. Sept. 39.
Eidmatt Str. 55.

Lieber Herr Gödel!

Obwohl ich ungewiss bin, ob Sie sich zur Zeit in Wien befinden, will ich doch versuchsweise diese Zeilen an Ihre Wiener Adresse richten.

Ueber den Empfang Ihres freundlichen Briefes vom 20. 7. habe ich mich sehr gefreut. Auch danke ich Ihnen noch vielmals für die Zusendung Ihrer neuen Publikation.

Ihre Bemerkung über die Ersetzbarkeit des Fundierungsaxioms für Klassen durch dasjenige für Mengen (bei Benutzung der übrigen mengentheoretischen Axiome) hat mich sehr interessiert. Im Anschluss daran fiel mir ein, dass mit Hilfe des Auswahlaxioms, des Unendlichkeitsaxioms und des Ersetzungsaxioms das Fundierungsaxiom für Klassen auch folgendermassen auf dasjenige für Mengen zurückgeführt werden kann: Ist A eine Klasse, die mit jedem ihrer Elemente ein Element gemeinsam hat, so lässt sich rekursiv mit Anwendung des Auswahlaxioms eine Folge $m_1, m_2, \ldots$ von Elementen von A definieren, worin

$$m_{n+1} \, \epsilon \, m_n \quad (\text{für } n = 1, 2, \ldots).$$

Mittels des Unendlichkeits- und des Ersetzungsaxioms ergibt sich die Existenz einer Menge s, welche die Elemente m_n ($n = 1, 2, \ldots$) und nur diese enthält; und diese Menge s hat dann mit jedem ihrer Elemente mindestens ein Element gemeinsam.—

Mit Ihrer Abhandlung "Consistency-Proof for the generalized Continuum-Hypothesis" habe ich mich des Näheren beschäftigt. Die Vereinigung der engeren ("verzweigten") Stufentheorie mit der Ordinaltheorie der allgemeinen Mengenlehre, die Sie ja bereits in Ihren | Wiener Vorlesungen über Mengenlehre einführten (von denen mir seinerzeit Herr Mostowski erzählte) und jetzt in modifizierter Weise ausgesch⟨t⟩altet haben, ist jedenfalls ein sehr durchschlagender und leistungsfähiger Ansatz. Die Durchfürung ist ja freilich nichts weniger als leicht, und man empfindet die ansehnliche mathematische Kraftleistung, die hinter dem dargebotenen Gedankengang der⟨Ihrer⟩ Beweisführung steht.

Dieser Gedankengang hat ja eine gewisse Analogie mit demjenigen des Hilbertschen Ansatzes zum Beweis des Kontinuumssatzes: dem Formalismus der Variablen-Gattungen (Variablen-Typen) entspricht die zugrunde gelegte Ordinaltheorie, dem A Lemma I der Satz A und dem Lemma II Ihr Theorem II ⟨2⟩. Dass es sachgemäss sei, die Aufgabe des Nachweises der Widerspruchsfreiheit des Kontinuumssatzes anhand der axiomatischen Mengenlehre zu präzisieren und in Angriff zu nehmen,

9. Bernays to Gödel

Zurich, 28 September 1939
Eidmatt Strasse 55

Dear Mr. Gödel:

   Although I'm uncertain whether you are presently in Vienna, I will nevertheless try to direct these lines to your Viennese address.

   I was very pleased to receive your friendly letter of 20 July. Also, many thanks once again for the sending of your new publication.

   Your remark about the replaceability of the axiom of foundation for classes by that for sets (with the use of the other set-theoretic axioms) interested me very much. In connection with that it occurred to me that with the help of the axiom of choice, the axiom of infinity and the axiom of replacement, the axiom of foundation for classes can also be reduced to that for sets as follows: If A is a class that has an element in common with each of its elements, then by application of the axiom of choice a sequence $m_1, m_2, \ldots$ of elements of A may be defined recursively, in which

$$m_{n+1} \, \epsilon \, m_n \qquad (\text{for } n = 1, 2, \ldots).$$

By means of the axiom of infinity and the axiom of replacement there emerges the existence of a set s which contains the elements m_n ($n = 1, 2, \ldots$) and only those; and this set s then has at least one element in common with each of its elements.

   I have been concerned in greater detail with your article "Consistency-Proof for the generalized Continuum-Hypothesis". The fusion of the restricted ("ramified") theory of levels with the theory of ordinals of general set theory, which you already introduced in your Vienna lectures on set theory (of which Mr. Mostowski told me at the time) and have now worked out in a modified manner, is in any case a very penetrating and efficient approach. Indeed, the execution is nothing less than easy, and one perceives the considerable mathematical tour de force that lies behind the train of thought displayed in your proof.

   That train of thought does have a certain analogy with that of Hilbert's approach to the proof of the continuum hypothesis: the underlying theory of ordinals corresponds to the species of variables (variable types), proposition A to lemma I, and your theorem 2 to lemma II. For a long time I was of the opinion that axiomatic set theory would be suitable for making precise the problem of demonstrating the consistency of

war seit langem meine Meinung, doch habe ich keinen deutlichen Weg zur Durchführung dieses Gedankens gesehen. Das Theorem ~~II~~ ist insbesondere ~~im Hinblick darauf~~⟨darum⟩ sehr erfreulich, weil es den berechtigten Teil von dem Inhalt des Reduzierbarkeitsaxioms aufdeckt.

Eine genauere Darstellung Ihres Beweisganges werden Sie doch noch veröffentlichen. Im Hinblick darauf habe ich mir von den Teilen der Ueberlegung, die Sie bloss programmatisch angedeutet haben, das meiste nur überschlagsweise plausibel gemacht. Es sind aber doch zwei grundsätzliche Punkte, über die ich Sie gern befragen möchte.

1. Ich sehe nicht, wie im Rahmen der axiomatischen Mengenlehre die rekursive Definition der M_α ohne Benutzung des Ersetzungsaxioms möglich ist.

2. Bei der Folgerung des Theorems 10 aus Th. 8 und 9 habe ich in Bezug auf den Satz C eine gewisse Schwierigkeit: Wenn ich recht | verstehe, so folgt aus Th. 8 zunächst nur, dass, falls für das zugrunde gelegte System der Mengenlehre, mittels dessen die M_α definiert werden, Satz A gilt, dann für die Modelle M_{ω_ω} und M_Ω die Sätze R und C gelten. Aus Theorem 9 ergibt sich dann weiter, dass wenn man, ~~ausgeht~~⟨hend⟩ von irgendeiner axiomatisch angesetzten Mengenlehre, die M_α definiert, dann M_{ω_ω} und M_Ω solche (nennen wir sie "ausgezeichnete") Modelle bilden, für die A gilt. Um die beiden Ergebnisse zu kombinieren, muss man doch nun die M_α mittels eines ausgezeichneten Modells definieren. Bei dieser Aenderung des zugrunde gelegten Systems der Mengenlehre wird freilich der Begriff "konstruierbar" gemäss dem Theorem 9 nicht verändert. Aber es könnte sich, wie mir scheint, der Begriff der Gleichmächtigkeit verändern, da in dem ursprünglichen System eine Gleichmächtigkeit auf einer nicht-konstruierbaren Abbildung beruhen kann. Hiernach käme die Möglichkeit in Betracht, dass die Rolle einer Anfangszahl ω_α bei der Aenderung des zugrunde gelegten Systems von einer kleineren Ordnungszahl ω_α^* übernommen wird und dass ω_α, $\omega_{\alpha+1}$ nun nicht mehr als aufeinanderfolgende Anfangszahlen zu gelten haben. Es brauchte also dann der Satz C nicht für M_{ω_ω} und M_Ω zuzutreffen, wohl aber würde er für $M_{\omega_\omega^*}$ und M_{Ω^*} zutreffen.

Vielleicht sind Sie in der Lage, den genannten Fall als nicht in Betracht kommend zu erweisen. Doch Sie selbst sprechen ja in Ihrem Brief die Vermutung aus, dass die Menge aller konstruierbaren Teilmengen von ω ohne Widerspruch als abzählbar angenommen werden kann. Das würde aber doch in sich schliessen, dass die Anfangszahl ω_1 in einem nur aus konstruierbaren Mengen bestehenden Modell der Mengenlehre durch eine solche Ordnungszahl repräsentiert wird, die vom Standpunkt eines umfassenderen Systems der Mengenlehre als Zahl der zweiten Zahlenklasse zu gelten hat.

| Freilich folgt aus Ihrem Th. 10, dass die Existenz einer Abzählung aller konstruierbaren Zahlenmengen nicht im Rahmen der axiomatischen

the continuum hypothesis and for tackling it; yet I saw no clear way of carrying out that idea. For that reason in particular, theorem 2 is very gratifying, because it reveals the part of the content of the axiom of reducibility that is justified.

No doubt you are still going to publish a more precise presentation of how your proof goes. In view of that, I have convinced myself only in a rough way of the plausibility of most of the parts of the consideration that you indicated merely in outline. Still, there are two basic points I would like to ask you about.

1. I don't see how the recursive definition of the M_α is possible within the framework of axiomatic set theory without the use of the axiom of replacement.

2. In the deduction of theorem 10 from theorems 8 and 9, I have a certain difficulty with respect to proposition C: If I understand correctly, from theorem 8 it follows directly only that if proposition A holds for the underlying system of set theory by means of which the M_α are defined, then the propositions R and C hold for the models M_{ω_ω} and M_Ω. From theorem 9 it then follows further that if, proceeding from any axiomatically given set theory, one defines the M_α, then M_{ω_ω} and M_Ω form such models (let us call them "distinguished") for which A holds. In order to combine the two results, one must, however, now define the M_α by means of a "distinguished" model. To be sure, according to theorem 9 the concept "constructible" is not changed by this change of the underlying system of set theory. But it seems to me that the concept of cardinal equivalence could be altered, since in the original system a cardinal equivalence can depend on a non-constructible mapping. From that would arise the possibility that in the change of the underlying system the role of an initial ordinal ω_α is taken over by a smaller ordinal number ω_α^*, and that ω_α and $\omega_{\alpha+1}$ now no longer have to be successive initial ordinals. So then proposition C would not have to be true for M_{ω_ω} and M_Ω, but would indeed be true for $M_{\omega_{\omega^*}}$ and M_{Ω^*}.

Perhaps you are in the position to show that the case mentioned does not arise. Yet in your letter you yourself express the conjecture that the set of all constructible subsets of ω can be assumed to be countable without contradiction. But that would imply that, in a model of set theory consisting only of constructible sets, the initial ordinal ω_1 is represented by an ordinal which, from the standpoint of a more comprehensive system of set theory, has to be a number of the second number class.

To be sure, it follows from your theorem 10 that the existence of a countable enumeration of all constructible sets of integers cannot be

Mengenlehre, sofern diese widerspruchsfrei ist, erwiesen werden kann,
da sich ja andernfalls mittels des Diagonalverfahrens die Existenz einer
nicht-konstruierbaren Zahlenmenge und damit ein Widerspruch gegen
den Satz A ergäbe. Demnach würde im Falle des Zutreffens Ihrer Annahme die Aussage der Abzählbarkeit der Menge aller konstruierbaren
Zahlenmengen einen in der axiomatischen Mengenlehre unentscheidbaren
Satz bilden.

Sie erwähnten, dass Ihnen Herr Mostowski über eine Behauptung von
mir betreffend die Ueberabzählbarkeit der Menge der konstruierbaren
Mengen von natürlichen Zahlen berichtet habe. Welches des Genaueren
die Aeusserung war, die ich Herrn Mostowski gegenüber zu dieser Frage
machte, habe ich nicht mehr in Erinnerung. Jedenfalls steht das, was
ich in dieser Richtung behaupten kann, nicht im Widerspruch zu Ihrer
geäusserten Vermutung.

Immerhin besteht ja von dem "halb-intuitionistischen" Standpunkt,
d. h. dem Standpunkt der Beschränkung des tertium non datur auf die
Gattung der natürlichen Zahlen (und solche Gattungen, die sich auf die
der natürlichen Zahlen zurückführen lassen), die Ueberabzählbarkeit der
Menge aller Mengen natürlicher Zahlen—wobei es sich ja nur um konstruierbare Mengen handelt—insofern in einem nicht trivialen Sinne, als
bei einem jeden Formalismus der halb-intuitionistischen Analysis, sofern
dieser die Anwendung Ihrer Methode der Nummernzuordnung gestattet, diejenigen nicht mehr in dem Formalismus darstellbaren Zahlenmengen, die sich mit Hilfe der semantischen Begriffe (wie "Wahrheit" einer
satzdarstellenden Formel, "Wert" eines zahlbestimmenden Ausdrucks)
definieren lassen, doch noch konstruierbare Mengen sind, sodass jeweils
die Metamathematik eines solchen Formalismus weitere konstruierbare
5   Zahlenmengen liefert. | Dieser Sachverhalt ist analog demjenigen wie
er beim Begriff der berechenbaren Funktion vorliegt: dass eine effek-
tive metamathematische Abzählung von Definitionsgleichungs-Systemen
berechenbarer Funktionen stets neue berechenbare Funktionen liefert.
(Sie haben ja seinerzeit bei der Kleeneschen Annalen-Abhandlung[a] über
die allgemein rekursiven Funktionen gerade auf diese Konsequenz hinge-
wiesen.)—

Ich wollte Ihnen noch mitteilen, dass kürzlich Herr Gentzen sich überlegt hat, dass die Methode der Interpretation der klassischen Aussagen-
und Prädikatenlogik im Rahmen der intuitionistischen sich über den Bereich des zahlentheoretischen Formalismus hinaus auch unschwer auf die
einfache Stufentheorie ausdehnen lässt. Natürlich bedeutet dieses nicht
eine intuitionistische Interpretation der einfachen Stufentheorie; denn es

---

[a] *Kleene 1936.*

proved within the framework of axiomatic set theory, so long as that [theory] is consistent, since otherwise the existence of a non-constructible set of integers, and thereby a contradiction to proposition A, would arise by means of the diagonal procedure. Accordingly, in case your assumption turned out to be true, the statement of the countability of the set of all constructible sets of integers would constitute a proposition undecidable in axiomatic set theory.

You mentioned that Mr. Mostowski reported to you about an assertion of mine concerning the uncountability of the set of constructible sets of natural numbers. What, more precisely, the remark was that I made to Mr. Mostowski *vis-à-vis* that question I no longer recall. In any case, what I can assert in that direction does not stand in contradiction to your expressed conjecture. All the same, from the "semi-intuitionistic" standpoint, i.e., the standpoint of the restriction of *tertium non datur* to the type of the natural numbers (and such types as may be reduced to that of the natural numbers), the uncountability of the set of all sets of natural numbers—where it is a question only of constructible sets— persists in a non-trivial sense, insofar as in each formalism of semi-intuitionistic analysis, so long as it permits the application of your method of assigning numerals, those sets of numbers that are no longer representable in the formalism—which may be defined with the help of semantic concepts (like the "truth" of a formula representing a proposition, [or] the "value" of an expression denoting a number)—are nevertheless still constructible sets, so that the metamathematics of such a formalism always furnishes further constructible sets of [natural] numbers. This state of affairs is analogous to what happens with the concept of computable function: that an effective metamathematical enumeration of systems of defining equations for computable functions always furnishes new computable functions. (At the time of Kleene's Annalen article[a] about the general recursive functions you pointed to just this consequence.)

In addition I wanted to let you know that Mr. Gentzen showed that the method of interpreting classical propositional and predicate logic within the framework of intuitionistic [propositional and predicate logic] may also be extended without difficulty beyond the domain of the number-theoretic formalism to the simple theory of types. Of course, this does not constitute an intuitionistic interpretation of simple type theory;

werden ja⟨dabei⟩ die vom Intuitionismus nicht zugelassenen Arten der imprädikativen Definitionen und Existenzbeweise nicht eliminiert.—

Hoffentlich haben Sie trotz der gegenwärtigen Kriegslage die Möglichkeit, Ihre Arbeiten weiterzuführen.

Mit den besten Wünschen für Ihr Ergehen und freundlichen Grüssen.

Ihr   Paul Bernays

10. Bernays to Gödel

Zürich 7, 12. XII. 39
Eidmattstr. 55

Lieber Herr Gödel!

Würden Sie so nett sein, mir Bescheid zu geben, ob Sie den Brief, den ich Ihnen Ende September sandte, bekommen haben. Bei den heutigen erschwerten Bedingungen ist ja die Möglichkeit, dass er verloren gegangen ist, in Betracht zu ziehen. Ich schrieb Ihnen zunächst, um Ihnen noch für Ihren freundlichen Brief u. für die Separata zu danken, dann erwiderte ich etwas auf Ihre mich sehr interessierende Bemerkung über das Fundierungsaxiom und vor allem hatte ich verschiedene Fragen betreffend Ihren Beweis der Widerspruchsfreiheit d. Auswahlaxioms u. der verall-

2   gem. Kontinuumshypoth., | mit dem ich mich des näheren beschäftigt hatte u. über den ich übrigens auch aufgefordert worden bin eine Review zu schreiben.

Falls d. Brief nicht eingetroffen ist, so könnte ich Ihnen noch einen Durchschlag schicken. Mit den besten Grüssen   Ihr

Paul Bernays

11. Gödel to Bernays

Wien, 29./XII. 1939.

Lieber Herr Bernays!

Bitte entschuldigen Sie, dass ich erst heute dazukomme, Ihren freundlichen Brief vom 28. Sept. zu beantworten. Ich hatte in letzter Zeit eine Menge wegen meiner Ausreise nach U.S.A. zu tun u. hoffe jetzt, dass ich

for the kinds of impredicative definitions and existence proofs not permitted by intuitionism are not eliminated thereby.

Hopefully, despite the present war situation you have the opportunity to carry on your works.

With best wishes for your welfare, and with friendly greetings,

Yours truly,

Paul Bernays

10. Bernays to Gödel

Zurich 7, 12 December 1939
Eidmattstr. 55

Dear Mr. Gödel:

Would you be so kind as to let me know whether you received the letter that I sent you at the end of September? In today's troubled conditions the possibility that it was lost has to be considered. I wrote you first of all in order to thank you once more for your friendly letter and for the offprints, then I replied somewhat to your remark about the axiom of foundation, which was very interesting to me, and above all I had various questions concerning your proof of the consistency of the axiom of choice and of the generalized continuum hypothesis, with which I had busied myself in detail and about which, moreover, I had also been asked to write a review.

In case the letter did not arrive, I could still send you a carbon copy.

With best regards,

Yours truly,

Paul Bernays

11. Gödel to Bernays

Vienna, 29 December 1939

Dear Mr. Bernays:

Please excuse that only today do I get around to replying to your friendly letter of 28 September. Recently I had a lot of things to do on account of my departure to the U.S.A. and I now hope that I will still be

doch noch im Jänner über Russland u. Japan werde fahren können. Ihre
Karte vom 12./XII. hat mich auch erst verspätet am 22./XII. erreicht.

   Bezüglich Ihrer Fragen kann ich Sie zunächst auf meine Princetoner
Vorlesungen vom Herbst 1938 verweisen, welche jetzt bald vervielfältigt
herauskommen sollen.—Wie das Ersetzungsaxiom bei der Konstruktion
des Modells M_{ω_ω} zu vermeiden ist, habe ich ganz kurz in Fussnote 12
der Proceedingsarbeit[a] angedeutet. In dem modifizierten Modell sind
2 die "Mengen" die sämtlichen Ordinalzahlen $< \omega_\omega$ und | die "ϵ-Relation"
ist eine gewisse durch transfinite Rekursion definierte Relation R zwischen diesen Ordinalzahlen, wobei die Definition des R so konstruiert
ist, dass R mit der auf die Menge M_{ω_ω} beschränkten (gewöhnlichen) ϵ-
Relation isomorph ist.—Ihre zweite Frage (bez. der Ableitung von Th.
10 aus Th.8 u. 9) erledigt sich dadurch, dass die Implikationen $A \to R$
und $A \to C$ formal aus den Axiomen der Mengenlehre beweisbar sind u.
daher R und C in *jedem* Modell gelten, in dem A gilt, daher auch in den
Modellen M_{ω_ω} u. M_Ω. Beim Übergang zu den Modellen M_{ω_ω} (bzw. M_Ω)
kann selbstverständlich eine Änderung des Gleichmächtigkeitsbegriffs
eintreten; das würde aber nur bedeuten, dass bereits gewisse echte An-
fangsabschnitte von M_{ω_ω} (bzw. M_Ω) als Modelle für Mengenlehre +
Kontinuum-Hyp. verwendbar sind, nicht aber dass M_{ω_ω} (M_Ω) keine sol-
3 chen Modelle sind. Die Existenz gewisser | sogar abzählbarer Anfangs-
abschnitte von M_{ω_ω} (M_Ω), die bereits Modelle für Mengenlehre + Kont.
Hyp. sind, folgt ja übrigens auch schon nach dem Skolemschen Verfahren
zur Konstruktion abzählbarer Modelle.

   Was die intuitionistische Interpretation der klassischen Analysis betrifft, so ist mir bekannt, dass diese mittels gewisser Begriffe von ungefähr derselben Art von "Konstruktivität" wie die der Begriffe des Heyting-
schen Aussagenkalküls in trivialer Weise möglich ist. Doch scheint mir
das von keinem besonderen Interesse zu sein, weil eben die Konstruktivität der verwendeten Begriffe sehr problematisch ist.

   Ich habe erst jetzt bemerkt, dass im Beweis von Th. 2 meiner Proceedingsarbeit der Buchstabe α bald als Konstante bald als Variable verwendet wird, u. sende Ihnen daher noch ein Separatum, in dem dieser Irrtum
korrigiert ist.

   Mit besten Grüssen

                    Ihr   Kurt Gödel

4 | P.S. Meine Adresse hat sich geändert u. ist jetzt: Wien I, Hegelgasse 5.

---

[a] *1939a.*

able to go in January via Russia and Japan. Also, your card of 12 December reached me only belatedly, on 22 December.

With regard to your questions I can first of all refer you to my Princeton lectures from autumn 1938, which should now soon come out in mimeographed form.

In footnote 12 of the Proceedings article[a] I indicated quite briefly how the axiom of replacement is to be avoided in the construction of the model M_{ω_ω}. In the modified model the "sets" are all the ordinal numbers $< \omega_\omega$ and the "ϵ-relation" is a certain relation R between those ordinal numbers that is defined by transfinite recursion, where the definition of R is so constructed that R is isomorphic to the (usual) ϵ-relation restricted to the set M_{ω_ω}.

Your second question (regarding the derivation of theorem 10 from theorems 8 and 9) is disposed of by the fact that the implications $A \to R$ und $A \to C$ are formally provable from the axioms of set theory and therefore R and C hold in *every* model in which A does, hence also in the models M_{ω_ω} and M_Ω . In the passage to the model M_{ω_ω} or, respectively, M_Ω, a change in the concept of cardinal equivalence can of course happen; but that would only mean that certain proper initial segments of M_{ω_ω} (or, respectively, M_Ω) are suitable as models for set theory + the continuum hypothesis, not that M_{ω_ω} (M_Ω) are not such models. The existence of certain initial segments of M_{ω_ω} (M_Ω), even countable ones, that are already models for set theory + the continuum hypothesis also follows after all by Skolem's procedure for the construction of countable models.

As concerns the intuitionistic interpretation of classical analysis, it is well known to me that that is possible in a trivial way by means of concepts of approximately the same sort of "constructivity" as that of Heyting's propositional calculus. Yet that seems to me to be of no particular interest, because exactly the constructivity of the concepts used [[in this interpretation]] is very problematical.

I have just now noticed that in the proof of theorem 2 of my Proceedings article the letter α is used both as a constant [[and]] as a variable, and I am therefore sending you another offprint in which that error is corrected.

With best regards,

Yours truly,

Kurt Gödel

P.S. My address has changed; it is now: Wien I, Hegelgasse 5.

12. Gödel to Bernays

Princeton, 16./I. 1942

Lieber Herr Bernays!

Ich habe mit grossem Interesse Ihren Artikel in den Entretiens de
Zurich vom Jahre 1938[a] gelesen; bloss das was Sie auf p. 152 Zeile 8–11
sagen ist mir unverständlich. Würde das nicht auf ein Aufgeben des formalistischen Standpunktes hinauslaufen? Ich würde mich sehr freuen,
wieder einmal von Ihnen zu hören. Die letzte Nachricht die ich von Ihnen
habe ist Ihr Brief vom Jänner 1940,[b] über den mir mein Bruder berich-
tete. Wie geht es Ihnen u. was gibt es wissenschaftlich bei Ihnen Neues?
Wir haben hier ausser einer allgemeinen Preissteigerung vom Krieg noch
fast nichts zu spüren bekommen. Ich habe im vergangenen Jahre meine
Arbeit über das Kontinuumproblem durch Anwendung eines sehr allgemeinen Begriffs von rekursiven Ordinalzahlfunktionen in eine dem
Hilbertschen Ansatz näherstehende Form gebracht. Leider kann man
~~darüber~~ ohne logistische Symbole darüber nichts Näheres schreiben.

Mit besten Grüssen

Ihr   Kurt Gödel

---

[a] *Bernays 1941a.*

13. Bernays to Gödel

Zürich 7, 7. Sept. 42.
Eidmattstrasse 55.

Lieber Herr Gödel!

Es war mir sehr erfreulich, einmal wieder von Ihnen direkt zu hören.
Herr Prof. Church, den ich bat, Ihnen vorerst meinen schönen Dank für
Ihren Brief vom 16.I. auszurichten, war ja so freundlich, meine Bestellung zu übermitteln. Wenn ich nicht selbst gleich Ihren Brief (den ich
im März empfing) erwidert habe, so geschah es, weil ich mit der Antwort
ein paar Bemerkungen zu Ihrer Abhandlung über die Widerspruchsfreiheit des Auswahlaxioms und der Kontinuums-Hypothese zu verbinden
gedachte und hierfür gern erst noch einmal einige Abschnitte Ihrer Abhandlung durchgehen wollte,—wozu ich erst jetzt in den Sommerferien
gekommen bin.
    Was ich hier bemerken will, wird inhaltlich für Sie kaum Neues enthal-
ten; ich möchte nur zum Ausdruck bringen, dass sich mir diese Gesichtspunkte im Zusammenhang mit Ihren Ergebnissen aufgedrängt haben. Es

12. Gödel to Bernays

Princeton, 16 January 1942

Dear Mr. Bernays:

I read your article in the Entretiens de Zurich from the year 1938[a] with great interest; only what you say on p. 152, lines 8–11 is not comprehensible to me. Wouldn't that be tantamount to giving up the formalist standpoint? I would be very pleased to hear from you once again. The last news that I have from you is your letter of January 1940,[b] about which my brother reported to me. How are you, and what's new with you scientifically? Here, apart from a general price inflation we have still gotten to feel almost nothing from the war. In the past year, through application of a very general concept of recursive functions of ordinal numbers, I have brought my work on the continuum problem into a form that is closer to Hilbert's approach. Unfortunately, without logical symbols none of the details of it can be written down.

With best regards,

Yours truly,

Kurt Gödel

---

[b]This letter has not been found.

13. Bernays to Gödel

Zurich 7, 7 September 1942
Eidmattstrasse 55

Dear Mr. Gödel:

I was very pleased to hear directly from you once again. Professor Church, whom I asked to express to you, for the time being, my thanks for your letter of 16 January, was kind enough to convey my message. That I myself did not at once reply to your letter (which I received in March) happened because I intended to combine with the answer a few remarks on your work about the consistency of the axiom of choice and of the generalized continuum hypothesis, and for that I really wanted first to go over a few sections of your work once more—which I had a chance to [do] only now in the summer vacation.

What I want to note here will, as to its content, hardly contain anything new for you; I would only like to state that this point of view pressed itself upon me in connection with your results. It is a question,

handelt sich in erster Linie um die Möglichkeit, das Axiom C3, betref-
fend die Potenzmenge, aus der gesamten Betrachtung auszuschalten.
Tatsächlich kommt, soviel ich sehe, dieses Axiom für die Aufstellung des
Modells Δ und für die Beweisführung Ihres Kapitels VII nur insofern zur
Anwendung, als es zum Beweis von $\mathfrak{M}(x \times y)$ (Satz 5.18) benutzt wird.
Dieser kann ja aber auch ohne C3, durch Verwendung von C2 und C4
geführt werden. (Ich denke an die Beweisart, wie ich sie in meiner Ar-
beit über die Mengenlehre, ~~Teil II~~ Teil II, S. 4–5[a] angewendet habe.)—
Allerdings ist in dem hier betrachteten Deduktionszusammenhang der
Satz $\mathfrak{M}(\omega)$ (8.51) nicht enthalten, bei dessen Beweis Sie das Axiom C3
zwecks der Vermeidung des Auswahlaxioms heranziehen. Aber dieser Anlass zur Verwendung des Axioms tritt ja nur auf, weil Sie das Unendlich-
2 keitsaxiom in der v. *Neumann*schen Fassung | zugrunde legen, während
Zermelos Unendlichkeitsaxiom oder auch das an die Dedekindsche Un-
endlichkeitsdefinition knüpfende (Axiom VI in meiner Aufstellung) den
Beweis von $\mathfrak{M}(\omega)$ ohne C3 und ohne das Auswahlaxiom gestattet.

Was ferner die verallgemeinerte Kontinuumshypothese betrifft, so
kann diese ja bei Abwesenheit des Axioms C3 so formuliert werden: "Für
jede unendliche Kardinalzahl κ ist $\mathfrak{P}(\kappa)$ gleichmächtig der Klasse derjenigen Ordnungszahlen ν, für welche $\bar{\bar{\nu}} = \kappa$." Ihre Beweisführung aus
Kapitel VIII liefert ohne Weiteres auch den Beweis dieser Behauptung
für das Modell Δ (d. h. auf Grund von $V_l = L_l$), ohne Benutzung von
C3 ~~nämlich zufolge~~⟨in Verfolgung⟩ der Alternative, dass entweder κ die
grösste Kardinalzahl ist oder dass es zu κ eine nächstgrössere Kardinalzahl gibt.[1]

Die Elimination des Axioms C3 erscheint mir insbesondere darum von
Interesse, weil sie die Anwendung Ihrer Ergebnisse auf manche nicht-
trivialen Ausgangsmodelle der Mengenlehre ermöglicht, die bei Anwendung des Axioms C3 von vornherein ausgeschlossen ~~wären~~⟨werden⟩. So
haben Sie doch selbst, wie mir Herr Mostowski einmal erwähnte, in Ihren
Vorlesungen von 1937 einen Beweis der Widerspruchsfreiheit der Kontinuumshypothese im Rahmen der Analysis geführt. Ich kenne diesen Beweis zwar nicht; aber ich nehme an, dass er sich von Ihrer jetzigen Beweisführung aus gewinnen lässt, indem man Ihre Methode für ein auf die
Zwecke der Analysis eingeschränktes Axiomensystem der Mengenlehre
zur Anwendung bringt. (Dabei gestaltet sich ja der Beweis des Kontinu-
3 ums-Satzes für das Modell Δ wesentlich einfacher als Ihr Beweis in |
Kapitel VIII, weil das Modell Δ so eingerichtet werden kann, dass darin

---

[1]Zugleich ergibt sich dabei, dass, wenn es eine Kardinalzahl $> \kappa$ gibt, dann $\mathfrak{M}(\mathfrak{P}(\kappa))$
gilt. Und dieser Satz tritt gewissermassen dann an die Stelle des Axioms C3.

---

[a]*Bernays 1941*.

in the first place, of the possibility of eliminating the axiom C3, concerning the power set, from the whole consideration. Actually, so far as I see, in setting up the model Δ and in carrying out the proof of your chapter VII this axiom comes into play only insofar as it is used in the proof of $\mathfrak{M}(x \times y)$ (theorem 5.18). But that can also be proved without C3, by making use of C2 and C4. (I'm thinking of the sort of proof that I applied in my paper on set theory,[a] part II, pp. 4–5.)—To be sure, the theorem [about] $\mathfrak{M}(\omega)$ (8.51) is not included in the deductive context considered here; in its proof, in order to avoid the axiom of choice, you have recourse to axiom C3. But this reason for making use of the axiom only comes about because you take as a basis von Neumann's version of the axiom of infinity, whereas Zermelo's axiom of infinity, or even that to which Dedekind's definition of infinity refers (axiom VI in my setup) permits the proof of $\mathfrak{M}(\omega)$ without C3 and without the axiom of choice.

Furthermore, as concerns the generalized continuum hypothesis, in the absence of the axiom C3 it can be formulated in this way: "For every infinite cardinal number κ, $\mathfrak{P}(\kappa)$ has the same power as the class of those ordinal numbers ν for which $\bar{\bar{\nu}} = \kappa$." Your line of proof from chapter VII directly furnishes the proof of this assertion for the model Δ (i.e., on the basis of $V_l = L_l$), without use of C3, when pursuing the alternative that either κ is the largest cardinal number or else there is a cardinal number next larger than κ.[1]

The elimination of the axiom C3 seems to me to be of particular interest, because it makes possible the application of your results to many non-trivial ground models of set theory that are excluded at the outset by the application of the axiom C3. You yourself, as Mr. Mostowski once told me, in your 1937 lectures gave a proof of the consistency of the continuum hypothesis in the framework of analysis. I am not, I admit, familiar with that proof; but I assume that it may be derived from your present line of proof, in which your method is brought to bear on a set-theoretical axiom system that is restricted to the goals of analysis. (Indeed, the proof of the continuum hypothesis for the model Δ turns out to be much easier than your proof in chapter VIII, because the model Δ can

---

[1]From that it follows at once that if there is a cardinal number $> \kappa$, then $\mathfrak{M}(\mathfrak{P}(\kappa))$ holds. And that proposition to some extent takes the place of the axiom C3.

ω die grösste Kardinalzahl ist, sodass von der obigen Alternative nur der erste Fall in Betracht kommt.)—

Man könnte daran denken, entsprechend wie C3 auch das Unendlichkeitsaxiom auszuschalten; aber das würde $\langle$sich$\rangle$ nicht lohnen, weil man ja bei Auslassung dieses Axioms ein viel simpleres Modell als Ihr Modell Δ zur Verfügung hat.

Inbezug auf das Axiom D schliesst sich hier noch folgende Bemerkung an: Der Beweis v. Neumanns dafür, dass das Axiomensystem mit dem Axiom D widerspruchsfrei ist, sofern es ohne dieses widerspruchsvoll$\langle$frei$\rangle$ ist, beruht ja wesentlich darauf, dass das Axiom der Potenzmenge unter den Axiomen auftritt. Ihre Methode der Herstellung des Modells Δ er-ermöglicht aber den Nachweis, dass im Falle der Widerspruchsfreiheit der Axiome A, B, C1*, 2, 4,—(C1* bedeute eine der beiden oben genannten Formen des Unendlichkeitsaxioms)—auch bei Hinzufügung von D die Widerspruchsfreiheit erhalten bleibt. Dieses ist einfach eine Folge davon, dass für Ihre Konstruktion von Δ in Kapitel V die Definition der Ordinalzahl durch eine solche ersetzt werden kann, bei welcher die Begründung der Ordinalzahl-Theorie, anstatt mit Hilfe von Axiom D, mit Hilfe des Aussonderungs-Satzes 5.12 erfolgen kann. (Dabei ergibt sich die Gültigkeit von D_l ja daraus, dass unter den Elementen einer konstruktiblen Klasse $\overline{A}$ solche Elemente $\overline{x}$ sind, für welche $Od^{\prime}\overline{x}$ die kleinste in $Od^{\prime\prime}\overline{A}$ vorkommende Ordinalzahl ist; und die Gültigkeit des Unendlichkeitsaxioms für das Modell Δ folgt aus dem Umstand, dass $\mathfrak{F}^{\prime}\omega$ die Nullmenge und mit a auch stets $\{a\}$ als Element enthält.)

Ob sich auf Grund des modificierten Ordinalzahl-Begriffes auch der Satz $On_l = On$ beweisen lässt, ist mir zweifelhaft; aber für den betrachteten Nachweis der relativen Widerspruchsfreiheit von D ist das ja nicht nötig.

4   (Durch Verbindung der Bemerkung über D mit der vorherigen, | betreffend die Eliminierbarkeit des Axioms C3, ergibt sich noch, dass, falls die Axiome A, D, C1*, 2, 4, widerspruchsfrei sind, auch bei Z $\langle$Hinz$\rangle$unahme von D, E und $V = L$ die Widerspruchsfreiheit erhalten bleibt.)—

Im vorigen Herbst habe ich auf einem Kongress in Basel über Ihre Ergebnisse inbetreff des Auswahlaxioms und der verallgemeinerten Kontinuums-Hypothese vorgetragen.—Sie schrieben mir, dass Sie neuerdings Ihre Untersuchung über das Kontinuumsproblem in eine neue, dem Hilbertschen Ansatz nähere Gestalt gebracht haben. Es würde mir natürlich sehr erwünscht sein, diese neue Form Ihrer Behandlung des Gegenstandes näher kennen zu lernen; insbesondere denke ich mir, dass diese Untersuchung auch Direktiven liefern wird für die weitere Ausgestaltung des konstruktiven Standpunktes über den finiten Standpunkt hinaus.

be set up in such a way that ω is the largest cardinal in it, so that only the first case of the alternative above comes into consideration.)

One could also think of eliminating, similarly to C3, the axiom of infinity; but that would not be worthwhile, because by leaving out that axiom one has a much simpler model than your model Δ.

With regard to the axiom D, the following remark is yet to be added: von Neumann's proof that the axiom system with the axiom D is consistent as long as it is consistent without it, rests essentially on the fact that the power set axiom appears among the axioms. But your method of constructing the model Δ makes possible the proof that in the case of the consistency of the axioms A, B, C1 , 2, and 4 (C1 denotes one of the two stated forms of the axiom of infinity), consistency is preserved when adjoining D. This is simply a consequence of the fact that for your construction of Δ in chapter V, the definition of ordinal number can be replaced by one which allows the grounding of the theory of ordinal numbers with the help of the separation theorem 5.12 instead of axiom D. (Thereby, the validity of D_l results from the fact that among the elements of a constructible class $\overline{A}$ are such elements $\overline{x}$ for which $Od'\overline{x}$ is the least ordinal number occurring in $Od''\overline{A}$; and the validity of the axiom of infinity for the model Δ follows from the circumstance that $\mathfrak{F}'\omega$ contains the null set and, together with a, also always contains $\{a\}$ as an element.)

Whether the theorem $On_l = On$ can also be proved on the basis of the modified notion of ordinal number is doubtful to me; but that is not necessary for the demonstration of the relative consistency of D that is being considered.

(By combining the remark about D with the preceding one concerning the eliminability of the axiom C3, it further results that, if the axioms A, D, C1*, 2 and 4 are consistent, consistency is also preserved when adjoining D, E and $V = L$.)

Last fall I lectured at a congress in Basel on your results concerning the axiom of choice and the generalized continuum hypothesis. You wrote me that you recently cast your investigation of the continuum problem in a new form closer to Hilbert's approach. Of course, I would very much like to become better acquainted with this new form of your treatment of the subject; in particular, I think that that investigation will also provide directions for the further working out of the constructive standpoint, beyond the finitary standpoint.

Ihre Verwunderung über den Passus aus meinem Exposé in den Zürcher "Entretien⟨s⟩" von 1938[b] ist sehr begreiflich. Die Stelle ist in der Ausarbeitung wirklich zu kurz geraten. Was ich hier sagen wollte, ist, dass es nicht als angemessen erscheint, in einem absoluten Sinne einen methodischen Standpunkt als den schlechthin evidenten und die davon abweichenden Standpunkte als bedenklich bzw. als nur technisch gerechtfertigt hinzustellen; diese Art der Gegenüberstellung hat man auch gar nicht nötig, um den methodischen Unterschieden gerecht zu werden, sofern man sich entschliesst, verschiedene Schichten und Arten der Evidenz zu unterscheiden. Es kommt hier noch der Gesichtspunkt hinzu, dass die Sicherheit eines Gedankensystems (einer ~~Dialektique~~ "dialectique" im Sinne von *Gonseth*) nicht von vornherein gegeben ist, sondern im Gebrauch, an Hand einer Art "geistiger Erfahrung" erworben wird. (Für die Einführung des Begriffs der geistigen Erfahrung habe ich mich in der Diskussion zu dem Vortrag von *Fréchet* ausgesprochen; vgl. "Entretien⟨s⟩", S. 78.) Eine solche erworbene Sicherheit und Evidenz ist aus⟨c⟩h der Analysis zuzuerkennen. Das hindert aber nicht, dass man den Methoden der Analysis eine Betrachtungsweise von mehr elementarer | Evidenz und spezifischer arithmetischem Charakter gegenüberstellt. Die Aufgabe, die innere Einstimmigkeit der Analysis von einem solchen Standpunkt elementarer Evidenz an Hand der Formalisierung der Schlussweisen der Analysis als eine syntaktische Notwendigkeit zu erweisen, erhält damit ihre methodische Bedeutsamkeit.

Diese Auffassungen entsprechen allerdings nicht einem strikt "formalistischen" Standpunkt; aber einen solchen habe ich niemals eingenommen, insbesondere habe ich mich in meinem (Sommer 1930 geschriebenen) Aufsatz "Die Philosophie der Mathematik und die Hilbertsche Beweistheorie"[c] deutlich davon distanziert, und noch mehr dann in dem (Ihnen wohl bekannten) Vortrag "Sur le platonisme dans les mathéma-   é
tiques".[d]

Hier an den Hochschulen nimmt das akademische Leben weiter seinen Gang, wenn auch etliche Studenten und manche Assistenten und Dozenten zeitweise im militärischen Dienst beansprucht sind. Ich selbst las in den letzten Semestern über verschiedene geläufige Gebiete der Mathematik, wobei ich mich den Vorschlägen der Fakultät anpasste.

In der allgemeinen Lebenshaltung sind hier die B⟦e⟧schränkungen durch die Kriegsumstände einstweilen noch recht glimpflich; man macht sich aber Sorgen für den Winter wegen der Knappheit des Heizmaterials.

---

[b] *Bernays 1941a.*
[c] *Bernays 1930.*

Your astonishment about the passage from my exposé in the Zurich "Entretiens" of 1938[b] is very understandable. The passage turned out to be too short in its elaboration. What I intended to say here is that it doesn't seem appropriate, in an absolute sense, to posit one methodological standpoint as plainly evident and the standpoints differing from it as dubious or, respectively, as only technically justified. That sort of opposition is also not at all necessary in order to do justice to the differences, so long as one resolves to distinguish different layers and kinds of evidence. Here the point of view is to be added that the certainty of a conceptual system (a "dialectic" in the sense of Gonseth) is not given beforehand, but is acquired through use, in the light of a kind of "intellectual experience". (I spoke in favor of the introduction of the notion of intellectual experience in the discussion of Fréchet's lecture; cf. "Entretiens", page 78.) Such an acquired certainty and evidence is also to be accorded to analysis. But that doesn't prevent one from contrasting the methods of analysis with an approach based on more elementary evidence and of a more specifically arithmetical character. The task of establishing the internal coherence of analysis, as a syntactic necessity, from such a standpoint of more elementary evidence by means of the formalization of the inference methods of analysis thereby retains its methodological significance.

To be sure, these views do not correspond to a strict "formalist" standpoint. But I've never taken such a standpoint; in particular, in my essay "Die Philosophie der Mathematik und die Hilbertsche Beweistheorie"[c] (written in the summer of 1930) I clearly distanced myself from it, and even more so in the lecture "Sur le platonisme dans les mathématiques"[d] (with which you may be acquainted).

At the *Hochschulen* here academic life continues its course, even though some students and many assistants and docents are from time to time called to military service. I myself lectured last semester on various common parts of mathematics, conforming to the proposals of the faculty.

In [regard to] the general mode of living, the restrictions here due to the war are for the time being still quite mild; but people are worried about the winter because of the scarcity of fuel.

---

[d] *Bernays 1935.*

Hoffentlich haben Sie sich in Princeton jetzt gut eingelebt und ist das
eheliche Leben und die damit verbundene Häuslichkeit Ihnen für Ihr
gesundheitliches und gemütsmässiges Befinden, und dadurch auch für
Ihre wissenschaftliche Arbeit recht zuträglich.

Möchten Sie Ihrer Ge[m]ahlin unbekannterweise eine Empfehlung von
mir ausrichten.

> Seien Sie selbst freundlich gegrüsst

> > von Ihrem   Paul Bernays

14. Gödel to Bernays[a]

Lieber Herr Bern.

Vielen Dank für Ihren Brief vom 7. Sept. Ihre interessante Bemerkung über das Pot. mengenaxiom regt die Frage an, wie weit man den in
meinen Vorlesungen gegebenen Beweis wörtlich auf die schwächeren Systeme (insbesondere das System der *Principia Math.*) übertragen kann.
Ich hatte ursprünglich die Absicht, meine Arbeit auf einer derartigen all-
gemeineren Basis aufzubauen, habe diese aber dann wieder aufgegeben.
Möglicherweise stellt sich die Verallgemeinerung nachträglich als ganz
leicht heraus. Wegen des Ersetzungsaxioms müsste man sich auf den Fall
beschränken, in dem ~~in dem System den hochsten "Typ" gibt die *Pot*~~
~~Operation~~ Ordinalzahl des Systems [in dem [die] Sinne der Höhe der Iteration der Pot.mengenbildung] eine isolierte Zahl ist. Ein anderer Weg
wäre, die Widerspruchsfreiheit derartiger Systeme als mit der Widerspruchsfreiheit der Existenz von Mengen gewißer höherer Mächtigkeiten
als äquivalent nachzuweisen. Es sollte irgend jemand einmal [diese?][b]
Fragen durchdenken.

Die Gleichung $On_l = On$ läßt sich ⟨*bei entsprechender Definition der
Ordinalzahl,*⟩ so scheint mir, auch ohne Ersetzungsaxiom leicht beweisen—
aber ~~es est fraglich ich glaube~~ [zumindest das Schwächere bekommt es, in
der Ordnen geben kann] diese Fragen scheinen mir weniger interessant,
da das Fundierungs Ax. doch in allen vernünftigen Systemen erfüllt

---

[a]Undated, presumably unsent reply to Bernays' letter of 7 September 1942, transcribed from a shorthand draft.

Hopefully you are now well settled in Princeton and married life and the domesticity associated with it is quite salubrious for your physical and emotional health, and thereby also for your scientific work.

Would you please convey my respects to your wife, even though I am not personally acquainted with her.

Friendly greetings to you yourself.

Yours truly,

Paul Bernays

14. Gödel to Bernays[a]

Dear Mr. Bernays:

Many thanks for your letter of 7 September. Your interesting remark about the power set axiom raises the question how far the proof given in my lectures can be carried over verbatim to the weaker systems (in particular, the system of Principia Mathematica). I originally had the intention of building up my work on such a more general basis, but then gave it up again. It is possible that the generalization will subsequently turn out to be quite easy. Because of the axiom of replacement, one would have to restrict to the case in which the ordinal of the system [in the sense of the height of the iteration of the power set formation] is an isolated ordinal. Another way would be to show the consistency of systems of that kind to be equivalent to the consistency of the existence of sets of certain higher cardinalities. Someone should really look into these[b] questions.

It seems to me that the equation $On_l = On$ with the corresponding definition of ordinal number may also easily be proved without the axiom of replacement (one gets at least the weaker [result], in which one can give [unintelligible word])—but these questions seem to me less interesting, since the axiom of foundation ought surely to be satisfied in all

---

[b] The shorthand symbols for "zu" and "diese" are rather similar. The symbol used here more closely resembles that for "zu", but "diese" better fits the context.

2 sein sollte. ~~Ich wurde Ihnen gern~~ φ_1 Über meinen Begriffen der rekur-
siven Ordinalzahlfunktionen | werde ich Ihnen vielleicht nächstens näher
schreiben können.

Ich habe schliesslich noch eine Bitte an Sie—könnten Sie vielleicht
meiner Mutter in [Brünn Č. S. Mähren Pelicog 8ª] schreiben, daß Sie
Nachrichten von mir haben und es mir und meiner Frau gut geht? Ich
habe schon versucht, durch das Rotkreuz ihr Nachricht zukommen zu lassen, weiß aber nicht, ob diese angekommen ist.

In der Hoffnung, daß es Ihnen auch weiterhin gut geht und Sie die
Möglichkeit zur wissenschaftlichen Arbeit haben, verbleibe ich mit herzlichen Grüssen

Unbekannterweise auch Grüsse von meiner Frau

15. Bernays to Gödel

Zürich 2, 28. XII. 1956
Bodmerstr. 11.

Lieber Herr Gödel!

Zum neuen Jahr sende ich Ihnen meine besten Wünsche! Es war sehr
schön, dass ich Sie im Frühjahr bei meinen Besuchen in Princeton mehrmals sehen und mich ausführlicher mit Ihnen unterhalten konnte.

Haben Sie Ihre damaligen Überlegungen zur Interpretation der Arithmetik mit Hilfe von Funktionalen hernach zur Darstellung gebracht?—
2 | Ich schulde Ihnen noch die versprochene Zusendung der Briefe, in de-
nen von dem Lebensende Gerhard Gentzens berichtet wurde. Es ist ein
Brief ⟨an mich⟩ von Dr. habil. Fritz Kraus, (bei dem die Mitteilung erst
auf der *zweiten* Seite beginnt) und ein andrer von Dr. F. Krammer an
M P Dr. H̷. F̷inl (in einer Abschrift, die Herr Dr. Kraus damals beigelegt
hatte), ausserdem noch ein Lebenslauf von Gentzen, den seine Mutter
verfasst hatte. Ich schicke Ihnen diese Schriftstücke mit gewöhnlicher
Post.ª—
3 Möchten Sie Ihrer Frau ein[[e]] Empfehlung | von mir ausrichten.

Mit herzlichen Grüssen

Ihr   Paul Bernays

ªThe cover letter accompanying them is letter 16 herein.

reasonable systems. I will perhaps be able to write you in more detail next time about my notions of the recursive functions of ordinal numbers.

I will finally ask you for a favor —could you perhaps write to my mother (in Brno, Moravia, Pellicogasse 8ᵃ [and tell her] that you have news from me and that things are going well for me and my wife? I have already tried to have a report sent to her by the Red Cross, but I don't know whether it was received.

In the hope that things are also going well with you and that you have the opportunity for scientific work, I remain, with kindest regards

[unsigned]

Though she is not personally acquainted with you, my wife also sends her regards.

15. Bernays to Gödel

Zurich 2, 28 December 1956
Bodmerstr. 11

Dear Mr. Gödel,

I send you my best wishes for the New Year! It was very nice that during my visit in Princeton last year I was able to see you several times and converse with you at greater length.

Did you subsequently present your considerations on the interpretation of arithmetic by means of functionals?

I still owe you the letters I promised to send, which reported on the end of Gerhard Gentzen's life. There is a letter to me from Dr. habil. Fritz Kraus (in which the information only begins on the *second* page) and another from Dr. F. Krammer to Dr. M. Pinl (in a copy that Herr Dr. Kraus had enclosed at that time), as well as a biographical sketch of Gentzen that his mother had composed. I am sending you these documents by regular mail.ᵃ

Please convey my regards to your wife.

With kindest regards,

Yours truly,

Paul Bernays

16. Bernays to Gödel

31. XII. 1956

Lieber Herr Gödel!

Beiliegend befinden sich die Ihnen versprochenen Schriftstücke über
Gerhard Gentzen.

Mit dem Wunsche, dass Sie das Jahr 1957 gut begonnen haben, und
herzlichem Gruss

Ihr Paul Bernays

17. Gödel to Bernays

Princeton, 6./II. 1957.

Lieber Herr Bernays:

Ich danke Ihnen herzlich für die Neujahrswünsche u. die Zusendung
der Dokumente zu Gentzen's Lebenslauf. Die letzteren schicke ich gleichzeitig mit eingeschriebener Post an Sie zurück. Der Inhalt ist ja leider
sehr deprimierend. Der Nachlaß Gentzens muss sich also in den Händen
der tschechischen Behörden befinden, falls er nicht in dem damaligen
Chaos abhanden gekommen ist. Man sollte glauben, dass jetzt, wo doch
die kommunistischen Regime etwas liberaler geworden sind, vielleicht
doch eine Auskunft aus Prag zu erhalten wäre.

Auch ich habe mich sehr gefreut, dass ich im vergangenen Frühjahr
wieder einmal Gelegenheit hatte, ⟨mit Ihnen⟩ über Grundlagenfragen
mit Ihnen zu sprechen, u. bedaure nur, dass nicht öfter die Möglichkeit
dazu besteht. Meine Interpretation der intuitionistischen Logik mit Hilfe
von Funktionalen habe ich bisher nicht publiziert. Ich habe sie übrigens
im Jahre 1941 in einer Vorlesung am Institut ausführlich dargestellt,
aber es existiert davon keine Nachschrift. Es würde mich interessieren,
ob Sie es für möglich halten, durch Erweiterung diese Interpretation
für Funktionstypen genügend hoher (aber durch Zulassung von Funktionalen, die durch Induktion nach genügend grossen (aber noch konstruktiven) Ordinalzahlen definiert sind, zu einer Interpretation der Analysis zu machen (wobei natürlich die Begriffe der klassischen Logik in der
in Erg. Math. Koll. Heft 4, p. 34[a] angegebenen Weise zu interpretieren
sind).

2

---

[a] *1933e.*

16. Bernays to Gödel

31 December 1956

Dear Mr. Gödel,

Enclosed are the documents about Gerhard Gentzen that I promised you.

With the wish that you have begun the year 1957 well, and with cordial greetings,

Yours truly,

Paul Bernays

17. Gödel to Bernays

Princeton, 6 February 1957

Dear Mr. Bernays,

I thank you cordially for the New Years wishes and for sending the documents concerning Gentzen's life. I am returning the latter to you simultaneously by registered mail. Unfortunately, the content is very depressing. Gentzen's *Nachlaß* must thus be in the hands of the Czech authorities, if it was not lost then in the chaos. One should think that now, when the Communist regimes have become somewhat more liberal, information from Prague could perhaps be obtained.

I was also very pleased that last spring I once again had the opportunity to speak with you about foundational questions, and only regret that this possibility does not exist more often. Up to now I have not published my interpretation of intuitionistic logic with the help of functionals. I did present it in detail in a lecture at the Institute in 1941, but no transcript of it exists. I would be interested [to know] whether you think it possible to make this interpretation into an interpretation of analysis by admitting functionals that are defined by induction over sufficiently large (but still constructive) ordinal numbers (in which, of course, the concepts of classical logic are to be interpreted in the way stated in Erg. Math. Koll. vol. 4, p. 34[a]).

Es fällt mir übrigens ein, dass ich vom ersten Teil Ihrer kürzlich in Dialectica erschienen Arbeit über Kant[b] kein Separatum besitze. Ich wäre Ihnen sehr dankbar, wenn Sie mir eines zuschicken würden.

Mit herzlichen Grüssen u. besten Wünschen

Ihr   Kurt Gödel

P.S. Meine Frau lässt ebenfalls bestens grüssen.

---

[b] *Bernays 1955.*

18. Gödel to Bernays

Princeton, Nov. 20, 1957.

Lieber Herr Bernays:

Ich habe vor kurzem erfahren, dass Sie an dem Symposium ueber die axiomatische Methode in Berkeley teilnehmen, und nachher noch einige Zeit in den Vereinigten Staaten bleiben werden. Ich wuerde mich ausserordentlich freuen, wenn Sie bei dieser Gelegenheit einige Wochen am Institute for Advanced Study verbringen moechten und wir unsere Diskussionen vom letzten Fruehjahr fortsetzen koennten. Zweifellos wuerde es auch fuer andere Mitglieder des Institutes von Interesse sein, mit Ihnen ueber Grundlagenfragen zu sprechen. Ich denke an einen Aufenthalt von etwa drei Wochen, wofuer Ihnen das Institut eine Remuneration von $1000 zur Verfuegung stellen koennte. Falls Sie einen oder mehrere Vortraege halten wollten, wuerde gewiss auch dafuer Interesse vorhanden sein. Jedoch bleibt das ganz Ihnen ueberlassen und soll durchaus nicht der Zweck Ihres Besuches sein.

Ich hoffe, es wird Ihnen moeglich sein, diese Einladung anzunehmen, und freue mich darauf Sie wiederzusehen.

Mit herzlichen Gruessen

Ihr   Kurt Gödel

P.S. Ich moechte Sie noch fragen, ob Sie in letzter Zeit etwas von, oder ueber Schuette gehoert haben. Ist er noch immer Dozent in Marburg? Ich finde seine Arbeit ueber die Widerspruchsfreiheit der unendlichen

By the way, it occurs to me that I do not own an offprint of the first part of your paper on Kant that appeared recently in *Dialectica*.[b] I would be very grateful if you would send me one.

With kindest regards and best wishes,

Yours truly,

Kurt Gödel

P.S. My wife likewise sends best regards.

---

18. Gödel to Bernays

Princeton, November 20, 1957

Dear Mr. Bernays,

I recently learned that you are taking part in the symposium in Berkeley on the axiomatic method and that you will remain in the United States for some time afterwards. I would be exceedingly glad if on this occasion you could spend a few weeks at the Institute for Advanced Study and we could continue our discussion from last spring. Undoubtedly it would also be of interest to other members of the Institute to speak with you about foundational questions. I'm thinking of a stay of about three weeks, for which the Institute could offer you a remuneration of $1000. In case you wanted to give one or more lectures, there would certainly also be interest in that. However, that is entirely up to you and should not at all be the purpose of your visit.

I hope it will be possible for you to accept this invitation and look forward to seeing you again.

With kindest regards,

Yours truly,

Kurt Gödel

P.S. I would like to ask you whether you have heard anything recently from or about Schütte. Is he still a docent in Marburg? I find his paper

Induction[a] ganz ausgezeichnet und moechte eine Einladung ans Institut fuer ihn vorschlagen. Es ist allerdings fraglich ob das schon für 1958/59 möglich sein wird.

---

[a] *Schütte 1951.*

19. Bernays to Gödel

Zürich 2, den 4. XII. 1957
Bodmerstrasse 11

Lieber Herr Gödel!

Über Ihren Brief und Ihre so freundliche Einladung habe ich mich sehr gefreut. Die Möglichkeit zu einem Aufenthalt am Institute und insbesondere zu eingehenden Aussprachen mit Ihnen würde mir in hohem Masse erwünscht sein, nur ist die Voraussetzung, dass ich nach dem Symposium in Berkeley noch einige Zeit in den U.S.A. zu bleiben gedenke, nicht erfüllt. Ich habe die Einladung zu dem Symposium im Hinblick darauf angenommen, dass dadurch meine Wintervorlesung hier nicht wesentlich unterbrochen wird, und habe auch keinen Urlaub dafür beantragen wollen, nachdem ich schon vorher um einen Urlaub für das kommende Frühjahr nachgesucht hatte. Auch zwei andere freundliche Einladungen nach Notre Dame und nach Lincoln musste ich deshalb leider ablehnen. Übrigens würde der Ausfall meiner Vorlesungen den Januar über bereits den Abbruch der Vorlesung überhaupt bedeuten.—Es sind jetzt nur noch zwei und ein halbes Semester, in denen ich hier am Poly noch lese. Vielleicht bestünde für die anschliessende Zeit (nach dem März 1959) eine Möglichkeit, mir einen Aufenthalt am Institute zu arrangieren, was ich dann sehr begrüssen würde.

Was Ihre Frage betreffs Herrn Schütte angeht, so finde auch ich, dass er in Anbetracht seiner vielen bedeutsamen Publikationen schon längst eine Professur verdiente. Zur Zeit ist er aber noch Dozent an der Universität Marburg. Er arbeitet übrigens an einem Buch für Springer, welches die neuen Fortschritte der Beweistheorie darlegen soll.[a]

Es tut mir sehr leid, dass ich Ihnen nicht zustimmend antworten kann. Gern aber will ich in Aussicht nehmen, dass | ich auf der Rückreise von

2

---

[a] *Schütte 1960.*

on the consistency of infinite induction[a] first-rate and would like to propose an invitation to the Institute for him. It is however doubtful whether that will already be possible for 1958/59.

---

19. Bernays to Gödel

Zurich 2, 4 December 1957
Bodmerstrasse 11

Dear Mr. Gödel,

I am very pleased by your letter and your friendly invitation. The possibility of a stay at the Institute and especially of detailed conversations with you would be desirable to me to the greatest degree; only the presumption that I am thinking of remaining in the U.S. for some time after the symposium in Berkeley is not fulfilled. I accepted the invitation to the symposium mindful that my winter course here will not be appreciably interrupted, and I didn't want to request a leave for it after I had applied for a leave for the coming spring. Therefore I also had to turn down two other friendly invitations, to Notre Dame and to Lincoln [Nebraska?]. After all, cancelling my lectures for all of January would mean ending the course altogether. There are now only two and a half semesters left in which I am still lecturing here at the Poly[technic Institute]. Perhaps a possibility might exist to arrange a stay for me at the Institute for the ensuing time (after March 1959); I would welcome that then very much.

As to your question regarding Mr. Schütte, I too feel that in view of his many significant publications he should have received a professorship a long time ago. At present, however, he is still a docent at Marburg. He is working, by the way, on a book for Springer, which is supposed to present the new advances in proof theory.[a]

I'm very sorry that I can't answer affirmatively. But, if it suits you, I will be pleased to consider stopping in Princeton on my way back from

Berkeley am 6. und 7.I, wenn es Ihnen passt, in Princeton Aufenthalt mache.[1]

Mit noch vielem Dank für Ihr so nobles Anerbieten und mit herzlichen Grüssen

Ihr   Paul Bernays

[1]Wenn Sie so gut sein wollen, mir hierüber nach Berkeley Bescheid zu geben, so ist meine dortige Adresse "Hotel Carlton, 2338 Telegraph Avenue".

20. Bernays to Gödel

Zürich 2, 5. März 1958
Bodmerstr. 11

Lieber Herr Gödel!

Sie werden das versprochene Fries-Separatum,[a] das ich Ihnen mit noch ein paar anderen Abhandlungen zuschickte, bekommen haben.

Mein Aufenthalt in Princeton steht mir noch in sehr schöner Erinnerung, und ich danke Ihnen noch vielmals für die so freundliche Aufnahme, die Sie mir boten!

An unser Gespräch anknüpfend wollte ich Sie noch fragen, wer der Autor ist, von dem Sie mir ein Ergebnis zur Frage des Kontinuumproblems erzählten. Soviel ich mich entsinne, nannten Sie Herrn Hajnal und sagten, dass sich die Arbeit in den Berichten des Cornell-Kongresses 1957[b] befinde.

Jedoch ich habe sie unter diesen nicht vorgefunden, und so werde ich mich sicherlich geirrt haben.

Würden Sie so gut sein, meiner mangelnden Erinnerung in dieser Sache abzuhelfen.

Mit herzlichen Grüssen

Ihr Paul Bernays

---

[a]*Bernays 1953.*

Berkeley on January 6 and 7.[1]

With many thanks once again for your generous offer, and with kindest regards,

Yours truly,

Paul Bernays

[1]Please be so kind as to let me know about this; my address there is "Hotel Carlton, 2338 Telegraph Avenue".

20. Bernays to Gödel

Zurich 2, 5 March 1958
Bodmerstrasse 11

Dear Mr. Gödel,

You will have received the promised Fries offprint,[a] which I sent you along with a few other papers.

My stay in Princeton is still a very beautiful memory for me, and I thank you again very much for the friendly reception you offered me!

With reference to our conversation, I still wanted to ask you who the author is [[who]], you told me, [[had]] a result on the continuum question. So far as I recall, you named Mr. Hajnal and said that the work was to be found in the reports of the 1957 Cornell Congress.[b]

However, I have not found it in those [[reports]], and so I was surely mistaken.

Would you be so good as to assist my failing memory in this matter?

With kindest regards,

Yours truly,

Paul Bernays

---

[b]*Hajnal 1956.*

21. Gödel to Bernays

Princeton, 14./III. 1958.

Lieber Herr Bernays!

Besten Dank für die Übersendung der Separata, sowie Ihren Brief vom
5./III. Ich freue mich sehr, dass Sie mit Ihrem Aufenthalt in Princeton
zufrieden waren, u. hoffe, dass wir Sie im Herbst 1959 einmal für längere
Zeit bei uns haben werden.

Ich habe mit grossem Interesse Ihre Arbeit über Fries gelesen. Man
bekommt durch die Zitate u. beigefügten Erklärungen ein sehr deutliches
Bild von den Friesschen Ideen. Die modernen Auffassungen, die Sie dann
auseinandersetzen, scheinen mir allerdings mit dem Kern der Friesschen
Lehre nicht verträglich zu sein, da ja nach Fries die *unbewusste* Erkenntnis die unvergleichlich vollkommenere ist. Auch Ihr Vortrag über Syntax u. Philosophie der Wissenschaften[a] hat mich sehr interessiert. Ich
stimme in den meisten Punkten mit Ihnen überein. Was die Existenz der
Begriffe betrifft, so glaube ich, dass man wohl in einem gewissen Sinn |
sagen kann, die Raum-Zeit-Dinge hätten ein anderes Sein. Aber das sind
doch bloss zwei Spezies eines allgemeinen Genus des "Seins", welches von
beiden ausgesagt werden kann.

Die Arbeit von Hajnal über das Kontinuum Problem ist in Zs. f. math.
Log. u. Grundl. d. Math. vol. 2 (1956) p. 131[b] erschienen. Daran knüpft
an eine Arbeit von A. Levy in der Pariser Comptes Rendus, 4. Nov. 1957.[c]
Ich habe bemerkt, dass Sie in Ihrem neuen Buch über Mengenlehre[d]
einen gewissen Tsen[[g]] Ting-Ho[e] zitieren. Ist diese Arbeit interessant?

Für Schütte konnte ich für 1958/59 leider nichts arrangieren.

Mit herzlichen Grüssen

Ihr Kurt Gödel

P.S. Ich habe zu meinem Bedauern bemerkt, dass ich Sie irrtümlich um
ein Separatum der Arbeit über Fries ersucht habe, da ich ein solches
schon besass. Ich retourniere es zusammen mit den Resumés der Sym-
posionvorträge, die Sie mir geliehen haben. Besten Dank.

---

[a] *Bernays 1957.*
[b] *Hajnal 1956.*
[c] *Lévy 1957.*

21. Gödel to Bernays

Princeton, 14 March 1958

Dear Mr. Bernays,

Thanks very much for sending along the offprints, as well as for your letter of 5 March. I'm very pleased that you were satisfied with your stay in Princeton, and I hope we will have you with us for a longer time in the fall of 1959.

I read your work on Fries with great interest. Through the quotations and appended explanations one gets a very clear picture of the Friesian ideas. However, the modern conceptions that you then present seem to me to be incompatible with the core of the Friesian theory, since according to Fries the *unconscious* comprehension is the incomparably more complete one. Also, your lecture on syntax and philosophy of science[a] interested me very much. I agree with you on most points. As to the existence of concepts, I think one can say in a certain sense that the spatiotemporal entities have another being. But they are nevertheless only two species of a general genus of "being", which can be predicated of both.

The work of Hajnal on the continuum problem appeared in Zeitschrift für mathematische Logik und Grundlagen der Mathematik, vol. 2 (1956), p.131.[b] Connected with it is a paper by A. Levy in the Paris Comptes Rendus of 4 November 1957.[c] I noticed that in your new book on set theory[d] you cite a certain Tsen⟦g⟧ Ting-Ho.[e] Is that paper interesting?

Unfortunately I wasn't able to arrange anything for Schütte for 1958/59.

With kindest regards,

Yours truly,

Kurt Gödel

P.S. I noticed to my regret that I asked you mistakenly for an offprint of the work on Fries, since I already own one. I am returning it together with the resumés of the symposium lectures that you lent to me. Thanks very much.

---

[d] *Bernays and Fraenkel 1958.*
[e] *Tseng Ting-Ho 1938.*

22. Bernays to Gödel

Zürich 2, den 2. Juni 1958
Bodmerstrasse 11

Lieber Herr Gödel!

Haben Sie vielen Dank für Ihren freundlichen Brief vom 14. III und
für die Rücksendung der Exposés sowie des Fries-Separatums.

Bezüglich der Fries'schen Abhandlung haben Sie Recht, dass bei den
von mir vorgeschlagenen modifizierenden Auffassungen sehr Wesentliches
von der Fries'schen Erkenntnistheorie verloren geht. Es wäre die Frage,
wie man die Fries'sche Lehre mit stärkerer Bewahrung ihrer methodischen Gedanken so ausgestalten könnte, dass eine befriedigende Erkenntnistheorie der heutigen theoretischen Physik resultiert,—die wohl kaum
nach dem Schema des unterordnenden Schlusses (unter vorgängige feste
Prinzipien) verstanden werden kann.

Ich danke Ihnen auch noch für die Angabe der Arbeit von Hajnal,
sowie derjenigen von A. Levy. Von Herrn Levy bekam ich inzwischen
seine Dissertation zugesandt, die Sie wohl auch von ihm werden empfangen haben. Die hierin in der Einleitung formulierten Ergebnisse sind ja
sehr interessant.

Das Zitat einer Abhandlung von Tsen⟦g⟧ Ting Ho, das Sie in der Mono-
graphie über die Mengenlehre fanden, stammt nicht von mir. Die Bibliographie enthält zugleich die von Herrn Fraenkel angeführte Literatur.
Ich kenne jene Abhandlung nicht, aber vielleicht sollte ich sie mir einmal
vornehmen.

Mit herzlichen Grüssen

Ihr Paul Bernays

23. Gödel to Bernays

Princeton, 30. Sept. 1958

Lieber Herr Bernays:

Besten Dank für Ihren Brief vom 2. Juni. Von den in der Einleitung
der Dissertation Levy's angekündigten Resultaten scheint mir das Interessanteste das über das Ackermannsche System der Mengenlehre zu sein.
Das sieht ja sehr überraschend aus. Haben Sie es kontrolliert?

22. Bernays to Gödel

Zurich 2, 2 June 1958
Bodmerstrasse 11

Dear Mr. Gödel,

Many thanks for your friendly letter of 14 March and for sending back the resumés as well as the Fries offprint.

With respect to the Friesian essay you are right that in the modified conception proposed by me something very essential of the Friesian epistemology is lost. The question would be, how could one work out the Friesian theory with stricter adherence to its methodical ideas so that a satisfying epistemology of contemporary physics results—which probably can hardly be understood according to the schema of the subordinate inference (among previously established principles).

I also thank you once more for the information on Hajnal's paper, as well as that of A. Levy. In the meantime I received from Mr. Levy his dissertation, which you probably will also have received from him. The results formulated in its introduction are certainly very interesting.

The reference to a paper by Tsen⟦g⟧ Ting Ho that you found in the monograph on set theory did not stem from me. The bibliography contains also the literature cited by Mr. Fraenkel. I don't know that paper, but perhaps I should take a look at it sometime.

With kindest regards,

Yours truly,

Paul Bernays

23. Gödel to Bernays

Princeton, 30 September 1958

Dear Mr. Bernays,

Thanks very much for your letter of 2 June. Of the results announced in the introduction to Levy's dissertation, the most interesting seems to me to be that on Ackermann's system of set theory. That really looks very surprising. Have you verified it?

Kreisel erzählte mir, dass Sie in Ihren Vorträgen in England den kombinatorischen Mengenbegriff näher besprochen haben. Ich habe sehr bedauert, dass darüber nichts im Druck erscheinen wird. Begriffliche Untersuchungen dieser Art sind ja heute äusserst selten. A. Wittenberg scheint einer der wenigen zu sein, die über solche Fragen nachdenken. Ich hatte leider noch keine Gelegenheit, mir sein kürzlich erschienenes Buch[a] näher anzusehen, werde das aber bald tun. Die Argumente gegen den Platonismus, die er in den Dialectica gab, scheinen mir zwar durchaus nicht überzeugend, aber seine Einstellung den Problemen gegenüber gefällt mir sehr gut.

Haben Sie Kreisel's Vortrag über die Formalisierung der finiten Mathematik gesehen?[b]

Ich möchte jetzt auf meinen Vorschlag zurückkommen, dass Sie einmal dem Institute for Advanced Study einen längeren Besuch abstatten. Sie schrieben mir seinerzeit, dass Sie bis Ende März 1959 in Zürich beschäftigt sind. Ich möchte Sie also, zunächst inoffiziell, fragen, ob es Ihnen recht wäre, das Wintersemester 1959 in Princeton zu verbringen. Die

2 übliche Remuneration würde in Ihrem Fall $3500 sein. Es dürfte | kaum Schwierigkeiten machen, eine solche Einladung für Sie zu arrangieren, und ich würde mich ausserordentlich freuen, wenn Sie annehmen könnten.   S
Ich hoffe, für 1959/60 auch ein Institutsstipendium für Schütte erwirken zu können.

Mit herzlichen Grüssen

Ihr Kurt Gödel

---

[a] *Wittenberg 1957.*

24. Bernays to Gödel

Zürich 2, 12. X. 1958
Bodmerstr. 11

Lieber Herr Gödel!

Vielen Dank für Ihren Brief, den ich in diesen Tagen empfing. Im Besonderen danke ich Ihnen herzlich, dass Sie Ihre Einladung an mich zu einem längeren Aufenthalt in Princeton erneuern. Ich würde sehr gern den Winter 1959/60 in Princeton verbringen und freue mich in dem Gedanken, dass ich, wenn es sich macht, mit Ihnen ausgiebig werde über

Kreisel told me that in your lectures in England you discussed the combinatorial concept of set in detail. I very much regret that nothing about that has appeared in print. Conceptual investigations of that sort are extremely rare today. A. Wittenberg seems to be one of the few who ponders such questions. Unfortunately I still had no opportunity to look more closely at his book that recently appeared,[a] but I will do so soon. The arguments against platonism he gave in Dialectica seem to me not at all convincing, but I like his attitude toward the problems very much.

Have you seen Kreisel's lecture on the formalization of finitary mathematics?[b]

Now I would like to come back to my proposal that you pay a longer visit to the Institute for Advanced Study one of these days. At that time you wrote me that you are busy in Zurich through the end of March 1959. So I would like to ask, at first unofficially, whether you would like to spend the winter semester of 1959 in Princeton. The standard remuneration would in your case be $3500. There will be hardly any difficulty in arranging such an invitation for you, and I would be especially pleased if you could accept it. I hope also to be able to secure an Institute scholarship for Schütte for 1959/60.

With kindest regards,

Yours truly,

Kurt Gödel

---

[b] *Kreisel 1960.*

24. Bernays to Gödel

Zurich 2, 12 October 1958
Bodmerstrasse 11

Dear Mr. Gödel,

Many thanks for your letter, which I received in these [last few] days. In particular, I thank you heartily for renewing your invitation to me for a longer stay in Princeton. I would like very much to spend the winter of 1959/60 in Princeton and am pleased by the thought that, if it comes

grundlagentheoretische und philosophische Fragen ~~sprieh~~sprechen kön-nen[1]

Wenn, wie Sie planen, auch Herr Schütte am Princetoner Institute sein wird, so werde ich das natürlich auch sehr begrüssen.—

Dass Sie Herrn Wittenbergs Buch ein lebhafteres Interesse entgegen bringen, freute mich zu hören. In meiner Korrespondenz mit ihm dis-putiere ich noch mit ihm über verschiedenes Grundsätzliche aus dem Buch.

Herrn Kreisel's Vortrag, den Sie erwähnen (über die Formalisierung der finiten Mathematik), habe ich bisher noch nicht gesehen. Als er das letzte Mal in Zürich war, sprach er mir unter anderm davon, dass er den Begriff des Finiten genauer zu bestimmen suche, als es bisher geschah; | der Vortrag steht gewiss wohl mit dieser Problemstellung in Zusammen-hang.

Was die Dissertation von Azriel Lévy betrifft, so werde ich durch Ihre Frage, ob ich das auf Ackermann's System der Mengenlehre bezügliche Ergebnis kontrolliert habe, darauf aufmerksam, dass es ja für eine solche Kontrolle eventuell genügt, das bei A. Lévy erwähnte Abstract von Vaught[a] nachzusehen und dann den in der Einleitung der Dissertation dargelegten Gedankengang zu verfolgen. Ich will mich bald einmal daran machen.

Zurzeit bin ich mit Wittgenstein's "Remarks on the Foundations of Mathematics"[b] (für den Zweck einer Besprechung)[c] beschäftigt. In die-sem Buch ist auch verschiedentlich von Ihrem Unableitbarkeits-Theorem die Rede. Freilich sieht Wittgenstein dabei seine Aufgabe nicht darin, über Ihren Beweis zu sprechen, sondern, wie er selbst erklärt, "an ihm vorbei zu reden".—Das Buch ist im Ganzen nicht uninteressant, aber in vieler Hinsicht unbefriedigend.

Mit herzlichen Grüssen

Ihr Paul Bernays

---

[1] Wie ist jetzt am Institute die Semester-Abgrenzung?

---

[a] *Vaught 1956.*
[b] *Wittgenstein 1956.*

about, I will be able to speak with you a great deal about foundational and philosophical questions.[1]

If, as you plan, Mr. Schütte will also be at the Princeton Institute, I will of course also welcome that very much.

It pleased me to hear that you have such a lively interest in Mr. Wittenberg's book. In my correspondence with him I'm still discussing with him various fundamental [points] from the book.

Mr. Kreisel's lecture you mentioned (about the formalization of finitary mathematics) I have not yet seen. The last time he was in Zurich he told me, among other things, that he seeks to fix the concept of what is finitary more precisely than has been done hitherto; the lecture is undoubtedly connected with posing this problem.

As concerns the dissertation of Azriel Lévy, your question whether I've verified the result for Ackermann's system of set theory makes me aware that for such a verification it suffices perhaps to examine the abstract of Vaught[a] mentioned by A. Lévy and then to follow the reasoning presented in the introduction of the dissertation. I want to do that sometime soon.

At the moment I'm busy with Wittgenstein's "Remarks on the Foundations of Mathematics"[b] (for the purpose of a review).[c] Your underivability theorem is mentioned several times. To be sure, in that regard Wittgenstein does not view it as his task to speak about your proof, but rather, as he himself declares, "to talk past it". The book is not entirely uninteresting, but is unsatisfying in many respects.

With kindest regards,

Yours truly,

Paul Bernays

---

[1] [Written vertically in the left margin of page 1:] What are the dates for the semesters at the Institute now?

---

[c] *Bernays 1959.*

25. Gödel to Bernays

Princeton, 30./X. 1958.

Lieber Herr Bernays!

Besten Dank für Ihren Brief vom 12./X. Ich freue mich, Ihnen nach Rücksprache mit meinen Kollegen mitteilen zu können, dass Ihrem Besuch in Princeton im Wintersemester 1959 nichts im Wege steht. Falls Sie einverstanden sind, können wir Ihnen die offizielle Einladung binnen kurzem zusenden. Falls Ihnen irgend ein anderes Arrangement besser konvenieren sollte, so bitte ich Sie, mir das mitzuteilen, damit ich nach Möglichkeit Ihren Wünschen entsprechen kann. Die Semesterabgrenzung ist im Jahre 1959/60 wie folgt: Wintersemester: 28. Sept.–18. Dez., Sommersemester: 11 Jan–8 Apr.

Auch bezüglich einer Einladung Herrn Schütte's für das Jahr 1959/60 haben sich keinerlei Schwie|rigkeiten ergeben. Ich habe Herrn Schütte gestern davon Mitteilung gemacht. Ich hoffe, dass es Ihnen sowie auch Herrn Schütte möglich sein wird, diese Einladungen anzunehmen. Ich bin sicher, dass der Kontakt mit Ihnen für die andern hier anwesenden Logiker interessant u. fruchtbar sein würde.

Was das Wittgensteinsche Buch über die Grundlagen der Mathematik betrifft, so habe auch ich einiges daraus gelesen. Es schien mir damals, dass es hauptsächlich dadurch Nutzen stiften kann, dass es die Falschheit der in ihm[1] aufgestellten Behauptungen zeigt. Zumindest galt das von den Stellen, die ich gelesen habe.

Ich würde mich sehr freuen, einmal über derartige Fragen ausführlich mit Ihnen sprechen zu können.

Mit herzlichen Grüssen

Ihr Kurt Gödel

---

[1] u. im Tractatus[a] (das Buch selbst enthält ja sehr wenige Behauptungen)

---

[a] *Wittgenstein 1922.*

25. Gödel to Bernays

Princeton, 30 October 1958

Dear Mr. Bernays,

Thanks very much for your letter of 12 October. After consultation with my colleagues, I am pleased to be able to report to you that nothing stands in the way of your visit to Princeton in the winter semester of 1959. If you are agreed, we can send you the official invitation within a short time. In case some other arrangement should be more convenient for you, I ask you to let me know so that I can comply as far as possible with your wishes. The semester dates for the year 1959/60 are as follows: winter semester: 28 Sept.–18 Dec., summer semester: 11 Jan.–8 Apr.

No sort of difficulties arose either regarding an invitation to Mr. Schütte for the year 1959/60. I informed Mr. Schütte about it yesterday. I hope it will be possible for you as well as for Mr. Schütte to accept this invitation. I am sure that the contact with you would be interesting and fruitful for the other logicians who are here.

As to Wittgenstein's book on the foundations of mathematics, I also read parts of it. It seemed to me at the time that the benefit created by it may be mainly that it shows the falsity of the assertions set forth in it.[1] At least that was true of the passages that I read.

I would be very pleased to be able to speak with you in detail about such questions.

With kindest regards,

Yours truly,

Kurt Gödel

---

[1]and in the *Tractatus*.[a] (The book itself really contains very few assertions.)

26. Bernays to Gödel

Zürich 2, 24. November 1958.

Lieber Herr Gödel!

Es ist mir sehr leid, dass meine Antwort an Sie sich so lange verzögert hat. Wollen Sie dieses mit Nachsicht entschuldigen!

Ich möchte zuerst Ihnen und Ihrer Frau von Herzen für Ihren telegraphischen Glückwunsch[a] danken, der mir eine Überraschung bereitete und mich sehr erfreute.

Ferner danke ich Ihnen vielmals für Ihre Beteiligung an der Festschrift.[b] Sie haben zu dieser ein wertvolles Schmuckstück beigetragen. Auch werden es, unabhängig von dem Anlass, sehr viele ⟨Mathematiker und Logiker⟩ begrüssen, dass Sie die Idee und die Methode der Durchführung Ihres neuen Widerspruchsfreiheitsbeweises, der ja eine neue Forschungslinie der Beweistheorie eröffnet, zur Darstellung gebracht haben.

Nunmehr komme ich zum Thema Ihres Briefes, für den ich Ihnen auch herzlich danke! Sie stellen mir bezüglich Ihrer Einladung so freundlich anheim, Ihnen etwaige Wünsche hinsichtlich des zeitlichen Arrangements mitzuteilen. So möchte ich hiervon Gebrauch machen und Sie fragen, ob es Ihnen gleich gut passen und sich einrichten lassen würde, dass mein Besuch offiziell für des Sommersemester 1960 (d. h. 11. Jan.–8. April)

2   angesetzt würde; ich könnte dann de facto schon im früheren | Dezember, oder auch schon Ende November kommen.

Ich weiss wohl, dass in Princeton die Zeit des "Wintersemesters" mehr sommerlich ist als die des "Sommersemesters" und dass Sie mir somit das Schönere anbieten. Aber im Hinblick auf meine hiesige Einteilung würde ich gern meine Abreise nach den U.S.A. etwas später (d. h. nicht vor November) legen.

Seien Sie sehr herzlich gegrüsst

von Ihrem

Paul Bernays

---

[a]The occasion for the congratulatory telegram (no longer extant) was probably Bernays' 70th birthday (17 October 1958).

26. Bernays to Gödel

Zurich 2, 24 November 1958

Dear Mr. Gödel,

I am very sorry that my reply to you has been so long delayed. Would you please excuse this with leniency!

I would like first of all to thank you and your wife with all my heart for your telegraphed good wishes,[a] which surprised and pleased me very much.

Furthermore, many thanks to you for your participation in the *Festschrift*.[b] You contributed a valuable gem to that. Also, independently from the occasion, very many mathematicians and logicians will welcome that you presented the idea of your new consistency proof and the method of carrying it out, which opens a new direction of research in proof theory.

I come now to the main issue of your letter, for which I also heartily thank you! With regard to your invitation, you are kind enough to leave it to me to tell you of possible wishes concerning the temporal arrangement. So I would like to make use of this [opportunity] and ask you whether it would suit you equally well and be possible to arrange that my visit be officially scheduled for the summer semester of 1960 (i.e., 11 Jan.–8 April); I could then in fact come already in early December or even towards the end of November.

I am well aware that in Princeton the time of the "winter semester" is more summery than that of the "summer semester" and that you are therefore offering me the more beautiful [alternative]. But in view of my local arrangement[s] I would like to depart for the U.S. somewhat later (i.e., not before November).

The very kindest regards

from your

Paul Bernays

---

[b] *Gödel 1958.*

27. Gödel to Bernays

Princeton, 7./I. 1959

Lieber Herr Bernays!

Besten Dank für Ihren Brief vom 24. XI. sowie auch die freundlichen Neujahrswünsche, die ich auf das herzlichste erwidere. Ich freue mich sehr, dass Sie Gefallen an meinem Beitrag zur Festschrift fanden. Die Hauptfrage wäre ja nun, ob man ~~vielleicht~~ durch die am Schlusse meines Aufsatzes erwähnten Erweiterungen des Systems T vielleicht die Widerspruchsfreiheit der Analysis beweisen kann.

Was Ihren Besuch in Princeton betrifft, so kann Ihren Wünschen ohne weiteres entsprochen werden, auch in der Weise, dass die offizielle Einladung auf $1\frac{1}{2}$ Semester ausgedehnt wird. Ein Brief des Direktors in diesem Sinn dürfte Ihnen ja bereits zugegangen sein. Ich freue mich sehr, dass Sie längere Zeit in Princeton bleiben wollen.

Mit herzlichen Grüssen

Ihr Kurt Gödel

2 | P.S. Besten Dank im voraus für die freundliche Zusendung Ihres neuen Buches.[a] Was das Buch Herrn Wittenbergs[b] betrifft, so scheint es mir den philosophischen Grundlagenproblemen gegenüber im allgemeinen recht pessimistisch eingestellt zu sein. Aber ich habe die Behandlung der einzelnen Fragen noch nicht gelesen.

---

[a] *Bernays and Fraenkel 1958.*

28. Bernays to Gödel

Zürich 2, den 22. I. 1959
Bodmerstrasse 11

Lieber Herr Gödel!

Vielen Dank für Ihren Brief vom 7.I. und besonders auch dafür, dass Sie mir nun die Einladung nach Princeton für anderthalb Semester erwirkt haben. Ich erhielt inzwischen das offizielle Schreiben des Direktors und habe ihm meinen Dank und meine Zusage ausgesprochen.

27. Gödel to Bernays

Dear Mr. Bernays,

Thank you very much for your letter of 24 November as well as for the friendly New Years wishes, which I reciprocate most heartily. I'm very glad that you enjoyed my contribution to the *Festschrift*. The main question now would be whether one can prove the consistency of analysis by means of the extensions of the system T mentioned at the end of my article.

As to your visit to Princeton, your wishes can also be accommodated without further ado by expanding the official invitation to $1\frac{1}{2}$ semesters. A letter from the director to that effect may already have reached you. I am very pleased that you want to stay in Princeton for a longer time.

With kindest regards,

Yours truly,

Kurt Gödel

P.S. Thanks very much in advance for kindly sending me your new book.[a] As to Mr. Wittenberg's book,[b] it seems to me in general to take quite a pessimistic view of the philosophical problems of the foundations. But I've still not read the treatment of the individual questions.

---

[b] *Wittenberg 1957.*

28. Bernays to Gödel

Dear Mr. Gödel,

Many thanks for your letter of 7 January and also especially for your having now procured the invitation to Princeton for one and a half semesters. In the meantime I received the official word from the director and have expressed to him my thanks and my acceptance.

Mit Ihrer Methode des Widerspruchsfreiheitsbeweises mittels der Funktionale wollen wir uns demnächst in unserem Logikseminar beschäftigen.   ä
Freilich zu Versuchen einer Ausdehnung der Methode auf die Analysis werden wir schwerlich vordringen.

Mit herzlichen Grüssen

Ihr Paul Bernays

29. Gödel to Bernays

Princeton, 8./X. 1959.
129 Linden Lane

Lieber Herr Bernays!

Wie ich höre, beabsichtigen Sie Anfang November hier einzutreffen, u. ich freue mich sehr darauf, Sie bald hier begrüssen zu können. Ich danke Ihnen auch bestens für die freundliche Zusendung eines Exemplars Ihres neuen Buches über die Mengenlehre.[a] Ich glaube, es ist bei der heutigen Einstellung vieler Mathematiker sehr wünschenswert, die *infinitistischen* Methoden zu pflegen. Es wurde mir mitgeteilt, dass Sie auch auf dem diesbezüglichen Kongress in Warschau Interessantes gesagt haben. Das Buch von Wittenberg[b] habe ich jetzt einigermassen gelesen. Leider finde ich darin fast gar keine sachliche Stellungnahme zum Inhalt der Mathe-
2   matik, was man doch | auf Grund seiner Arbeit in den Dialectica[c] erwartet hätte. Ich freue mich sehr darauf, bald mit Ihnen über Grundlagenfragen sprechen zu können.

Mit herzlichen Grüssen u. besten Wünschen für eine angenehme Überfahrt,

Ihr Kurt Gödel

---

[a] *Bernays and Fraenkel 1958.*
[b] *Wittenberg 1957.*

In our logic seminar we will soon be concerned with your method of proving consistency by means of functionals. However, we shall hardly get to the point of attempting to extend the method to analysis.

With kindest regards,

Yours truly,

Paul Bernays

29. Gödel to Bernays

Princeton, 8 October 1959
129 Linden Lane

Dear Mr. Bernays,

As I heard it, you intend to arrive here at the beginning of November, and I am very pleased to be able to welcome you here soon. Thanks very much also for kindly sending me a copy of your new book on set theory.[a] I think it is very desirable, given the contemporary attitude of many mathematicians, to cultivate the *infinitary* methods. It was reported to me that you also said interesting things at the Congress in Warsaw that was devoted to that theme. I have now read the book of Wittenberg[b] to some extent. Unfortunately, I find in it almost no objective position as to the content of mathematics, which one would certainly have expected on the basis of his paper in Dialectica.[c] I am very pleased to be able to speak with you soon about foundational questions.

With kind regards and best wishes for a pleasant crossing,

Yours truly,

Kurt Gödel

---

[c] *Wittenberg 1953.*

30. Bernays to Gödel

Zürich 2, 9. X. 1959
Bodmerstr. 11

Lieber Herr Gödel!

Es ist an der Zeit, dass Ich Ihnen über meine Reisedisposition für den November schreibe.

Das Visum für die U.S.A. habe ich im September erhalten; und vor kurzem habe ich mir einen Platz für den Flug am 9. November abends mit der Swissair reservieren lassen. Das Flugzeug soll am 10. Nov. morgens um 8^{30} in New York eintreffen.

Ich hoffe, dass Ihnen die Zeitwahl so recht ist; andernfalls könnte ich die Bestellung noch ändern lassen.

Im Anfang September habe ich an dem Symposium in Warschau teilgenommen, zu dem Sie ja auch eine Einladung erhielten. Es war eine sehr anregende Veranstaltung. Übrigens verspürte man gegenüber der Tagung vor zwei Jahren in Amsterdam, trotz der Gegensätzlichkeit der angesetzten Themata (damals: "constructivity in mathematics", dieses Mal: "infinitistic methods"), keinen stärkeren methodischen Abstand.

Herr Schütte ist ja wohl bereits einige Zeit in Princeton, und Sie haben gewiss mit ihm schon mancherlei überlegt.

Im Ausblick auf das bald kommende Zusammensein mit Ihnen grüsst Sie sehr herzlich

Ihr Paul Bernays

⟦Written vertically in left margin:⟧ Möchten Sie bitte Ihrer Frau eine Empfehlung von mir ausrichten.

31. Bernays to Gödel

Zürich 2, 11. Mai 1960

Lieber Herr Gödel!

Jetzt sind schon über zwei Wochen verstrichen, seit ich von Princeton abreiste. Es war besonders freundlich von Ihnen, dass Sie mir zum Abschluss meines Princetoner Aufenthaltes das Zusammensein mit Ihnen und die angenehme Fahrt nach Princeton-Junction boten.—In Syracuse (wo ich von Herrn Baum eingeladen war) hatte ich es sehr hübsch; und die beiden Vorträge dort und in Rochester fanden, wie mir schien, guten

30. Bernays to Gödel

Zurich 2, 9 October 1959
Bodmerstrasse 11

Dear Mr. Gödel,

The time has come to write you about my travel plans for November.
I received the visa for the U.S. in September, and a short time ago I
reserved a place for myself on the Swissair flight the evening of 9 November. The plane should arrive in New York around 8:30 on the morning of
the 10th.

I hope the choice of time is all right with you; otherwise I could still
change the reservation.

At the beginning of September I participated in the symposium in
Warsaw to which you had also been invited. It was a very stimulating
event. By the way, one hardly noticed a methodological difference when
comparing it with the meeting two years ago in Amsterdam, despite the
contrast between the official themes (then: "constructivity in mathematics"; this time: "infinitistic methods").

Mr. Schütte has probably been in Princeton for some time already,
and you have surely reflected upon all sorts of things with him.

Looking ahead to our getting together soon, [I send my] kindest regards.

Yours truly,

Paul Bernays

[Written vertically in the left margin:] Please convey my respects to
your wife.

31. Bernays to Gödel

Zurich 2, 11 May 1960

Dear Mr. Gödel,

More than two weeks have now already passed since I left Princeton.
It was especially kind of you that, at the end of my stay in Princeton,
you offered me the [chance of] being together with you and the pleas-
ant ride to Princeton Junction.—In Syracuse (where I was invited by
Mr. Baum) I had a very nice time, and the two lectures there and in
Rochester were, it seemed to me, well received. (In the one I spoke about

Anklang. (In dem einen sprach ich über Direktiven zur Axiomatisierung
der Mengenlehre, in dem anderen über die zweite Zahlenklasse und insbes. über die Veblen–Schütte'sche Methode zur konstruktiven Gewinnung eines beträchtlichen Anfangsstückes.)

Am 29. IV. traf ich hier, wie beabsichtigt, in Zürich ein.—Natürlich
stehe ich sehr unter dem Eindruck des Mannigfachen und Reichhaltigen,
das ich in der schönen Zeit am Institute erlebte, und der vielen Anregungen, die ich empfing.

In Anknüpfung an unsere Unterhaltungen über Fries habe ich mir hier
die "Metaphysik" von Fries[a] wieder angesehen (von der Sie bemerkten,
dass sie in der Princetoner Bibliothek vorhanden ist). Diese gibt einen
recht guten Überblick über seine Philosophie. (Insbesondere findet sich
da auch ein Abschnitt "Metaphysik der inneren Natur oder metaphysische Lehre von unserer wissenschaftlichen | Erkenntnis des Geistes".)

Von dreien meiner älteren Arbeiten, die ich Ihnen erwähnte, habe
ich Ihnen Sonderdrucke geschickt (die Sie als Ihre Exemplare behalten
möchten). In der Arbeit "Über den transzendentalen Idealismus"[b] ist
die Darstellung der Antinomienlehre noch unter der Voraussetzung der
Verbindlichkeit der Euklidischen Geometrie für die Physik gegeben. In
einem anderen Punkte erfuhr seinerzeit die Abhandlung eine (freilich nur
mündlich mir gegenüber geäusserte) Kritik von Seiten des ⟨Experimental-⟩Psychologen G. E. Müller, der den Passus in der Anmerkung auf S.
22 beanstandete, dass "durch keine einzige der uns aus der Wahrnehmung bekannten Qualitäten ein dreidimensionales Raumgebiet erfüllt
werden kann". Die Behauptung, dass ein dreidimensionales Raumgebiet
nicht durch eine Farbe ausgefüllt sein könne, erklärte er als "Schreibtisch-
Philosophie".—

Ein ⟨ebenfalls älterer⟩ Aufsatz von mir "Über Nelsons Stellungnahme
in der Philosophie der Mathematik"[c] kam mir bei einer Durchsicht von
Sonderdrucken in die Hände; ich hätte Ihnen diesen mitgeschickt, doch
habe ich nur noch ein sehr zerfledertes Exemplar. Übrigens steht er in
der Zeitschrift "Die Naturwissenschaften" (16. Jahrgang, Heft 9, 1928),
und diese ist ja, wie ich glaube, in der Instituts-Bibliothek vorhanden.
Unter anderm wird hier Nelsons Einwand gegen die Auffassung erwähnt,
dass unsere Anschauung nicht völlig genau sei und dass die genauen geometrischen Gesetze aus dem anschaulich Gegebenen durch einen Idealisierungsprozess gewonnen werden. Nelson macht geltend, dass ohne die
reine Anschauung für die Idealisierung gar keine Norm gegeben sei. Auf

2

[a] *Fries 1924.*
[b] *Bernays 1913.*

directions for the axiomatization of set theory, in the other about the second number class and in particular about the Veblen–Schütte method for constructively building up a considerable initial segment.)

On 29 April I arrived here in Zurich, as planned.—Of course, the impression of the manifold and rich things that I experienced during the beautiful time at the Institute, and of the many stimuli that I received, is still fresh in my mind.

In connection with our conversation about Fries, I looked here again myself at the "Metaphysik" of Fries[a] (which, you noted, is available in the Princeton library). That gives quite a good overview of his philosophy. (In particular, also to be found there is a chapter "Metaphysics of the inner nature, or, metaphysical theory of our scientific knowledge of the mind".)

I have sent you offprints of three of my older papers that I mentioned to you (which you may wish to keep as your own copies). In the paper "On transcendental idealism"[b] the presentation of the theory of antinomies is still given under the assumption that we are obliged to use Euclidean geometry for physics. In another point the essay was criticized (just in an oral remark to me) by the experimental psychologist G. E. Müller, who objected to the passage in the remark on p. 22 that "a three-dimensional region of space cannot be filled by any single quality known to us from perception". He declared that the assertion that a three-dimensional region of space could not be filled up by a color was "arm-chair philosophy".

I came across another older essay of mine, "On Nelson's position in the philosophy of mathematics"[c] while looking through offprints; I would have sent that to you, but I have only a very tattered copy. Anyway, it's in the journal *Die Naturwissenschaften* (16th year, volume 9, 1928), and that, I believe, is available in the Institute library. Among other things mentioned there is Nelson's objection to the view that our intuitive perception is not completely precise and that the precise geometrical laws are obtained from what is intuitively given by means of a process of idealization. Nelson maintains that without pure intuition no norm at all

---

[c] *Bernays 1928.*

3　dieses Argument könnte | man aber wohl entgegnen, dass die Gesichtspunkte der begrifflichen Einfachheit (Durchgängigkeit, Permanenz) hinlängliche Direktiven für die Idealisierung bieten. Das Buch von Herrn Schütte[d] ist ja jetzt kürzlich erschienen. Haben Sie sich es schon näher angesehen? Gewiss hat Herr Schütte Ihnen bereits das Hauptsächliche der Anlage und der Ergebnisse mündlich erzählt.

Zu der Arbeit für die Neu-Ausgabe des Grundlagenbuchs[e] bin ich vorerst noch nicht gekommen, da erst einiges Zurückgebliebene noch nachzuholen ist; doch denke ich, dass dieses provisorische Stadium bald abgeschlossen ist ⟨sein wird⟩.

In dankbaren Gefühlen grüsst Sie sehr herzlich

Ihr Paul Bernays

---

[d] *Schütte 1960.*

32. Bernays to Gödel

Zürich 2, 20. XII. 1960
Bodmerstr. 11

Lieber Herr Gödel!

Zu den Weihnachtstagen und zum neuen Jahr sende ich Ihnen, und auch Ihrer Frau, meine herzlichen Wünsche!

Hoffentlich konnten Sie mit Ihrem gesundheitlichen Ergehen in der letzten Zeit zufrieden sein.

Von Herrn Wette hörte ich im Herbst, dass er anschliessend an den Stanforder Kongress Gelegenheit hatte, mit Ihnen über seine Ansätze und Überlegungen zum Kontinuumproblem zu sprechen,—welche ja, wenn sie gelingen sollten, auch eine Art von konstruktivem Wider-

2　spruchsfreiheitsbeweis für die Mengenlehre in sich schliessen würden. | Welches war denn Ihr Eindruck hiervon?—

Meine Überlegungen, die ich in Princeton anknüpfend an das Axiomen-Schema von Azriel Lévy anstellte, habe ich hier fortgesetzt und eine

would be given for the idealization. But to that argument one might object that the point of view of conceptual simplicity (universality, permanence) affords adequate directives for the idealization.

The book by Mr. Schütte[d] has now recently appeared. Have you already taken a closer look at it? Surely Mr. Schütte has already told you orally the principal features of its structure and results.

I have not yet begun work on the new edition of the book on foundations,[e] as I had to take care of some older matters first; nevertheless, I think that this provisional stage will soon be completed.

In gratitude, with very kindest regards to you,

Yours truly,

Paul Bernays

---

[e] *Hilbert and Bernays 1968* and *1970*.

32. Bernays to Gödel

Zurich 2, 20 December 1960
Bodmerstrasse 11

Dear Mr. Gödel,

I send you and your wife my cordial wishes for the Christmas holidays and the New Year.

I hope your state of health has been satisfactory of late.

In the autumn I heard from Mr. Wette that following the Stanford Congress he had the opportunity to speak with you about his approaches to and deliberations on the continuum problem—which, if they should succeed, would also entail a sort of constructive consistency proof for set theory. What was your impression of it?

I've continued here the work I started in Princeton in connection with the axiom schema of Azriel Lévy, and I've submitted a paper on the

Arbeit darüber[a] für den Fraenkel-Festband eingesandt.

Für den kommenden spring term habe ich eine Einladung nach Philadelphia angenommen. So hoffe ich, dass ich in 1961 Gelegenheit haben werde, Sie wiederzusehen.

Mit herzlichen Grüssen

Ihr Paul Bernays

---

[a] *Bernays 1961.*

33. Gödel to Bernays

Princeton, 21./XII. 1960.

Lieber Herr Bernays!

Ich muss mich vor allem entschuldigen, dass ich mich noch immer nicht für die 3 mir freundlicherweise zugesandten Separata bedankt habe, was ich hiemit auf das herzlichste tue. Ihr Inhalt hat mich sehr interessiert u. ich hatte mir schon lange vorgenommen, ausführlich sachlich darauf einzugehen, bin aber dann immer wieder daran gehindert worden, nicht zuletzt durch meinen Gesundheitszustand, der das ganze vergangene Jahr leider sehr schlecht war. Ihre Behandlung der Antinomien des Kontinuums u. die Zusammenstellung mit den Zenonischen fand ich sehr schön u. berechtigt. Nur *hier* scheinen mir Antinomien in einem einigermassen strikten Sinn vorzuliegen. Ein Ding, von dem nichts übrig bleibt,
2   wenn man alle Zusammensetzung | weglässt (die Punkte des Anschauungsraumes sind ja nachweislich nur Grenzen) scheint tatsächlich eine Unmöglichkeit zu sein. Die anderen Antinomien sollte man heutzutage durch die mengentheoretische der Totalität aller Dinge u. die Plancksche der Willensfreiheit ersetzen. Bez. der 4-ten müsste man einen modernen Theologen fragen. Sehr originell u. interessant fand ich auch Ihre Behandlung des Moralgesetzes. Aber diese gehört wohl nur Ihnen u.
3   nicht | der Friesschen Schule an. Ich finde übrigens in den Philosophiegeschichten Beneke immer als einen Fortsetzer von Fries hingestellt was mir *gänzlich* unberechtigt scheint. Eher könnte man das vielleicht von Volkelt sagen. Ist die Philosophiegeschichte von Nelson wirklich erschienen? Da ich vermute, dass Sie keine Duplikate der 3 Separata haben, werde ich sie demnächst retournieren.

subject to the Fraenkel *Festband.*[a]

I've accepted an invitation to Philadelphia for the coming spring term. So I hope that I shall have the opportunity to see you again in 1961.

With best regards,

Yours truly,

Paul Bernays

---

33. Gödel to Bernays

Princeton, 21 December 1960

Dear Mr. Bernays,

First of all I must beg your pardon that I have still not thanked you for the three offprints you kindly sent me, which I hereby do most cordially. Their content interested me very much, and I had intended for a long time to explore the matter thoroughly, but then was prevented from it again and again, last but not least by my state of health, which unfortunately was very bad for the whole past year. I found your treatment of the antinomies of the continuum side by side with those of Zeno to be very beautiful and justified. Only *here*, it seems to me, do antinomies arise in a relatively strict sense. A thing of which nothing remains when one omits all compositions (the points of perceptual space are demonstrably only limits) actually seems to be an impossibility. The other antinomies should be replaced nowadays by the set-theoretic one of the totality of all things and the Planckian one of free will. With respect to the fourth one would have to ask a modern theologian. I also found your treatment of the moral law very original and interesting. But that probably belongs only to you and not to the Friesian school. By the way, in the histories of philosophy I always find Beneke made out to be a disciple of Fries, which seems to me *entirely* unjustified. Rather, one could perhaps say that of Volkelt. Has Nelson's history of philosophy actually appeared? Since I suspect that you don't have duplicates of the three offprints, I will return them soon.

Ich hoffe, dass es Ihnen in jeder Hinsicht gut geht, u. wünsche Ihnen ein recht glückliches Neues Jahr.

Mit der herzlichsten Grüssen

Ihr Kurt Gödel

Beste Grüsse auch von meiner Frau

34. Gödel to Bernays

Princeton, 16./III. 1961

Lieber Herr Bernays!

Besten Dank für Ihre Weihnachtsgrüsse. Ich nehme an, dass Sie jetzt schon längere Zeit in Philadelphia sind, u. würde mich natürlich ausserordentlich freuen, Sie einmal hier in Princeton zu sehen. Ich würde auch sehr gerne über den Inhalt der mir freundlicherweise überlassenen Separata mit Ihnen sprechen. Haben Sie Lust für 1–2 Wochen herzukommen? Das Institut könnte Ihnen die Kosten vergüten. Ich nehme an, $300.– würden für etwa 10 Tage hinreichen.

Bitte lassen Sie mich also wissen, ob u. wann Ihnen ein solcher Besuch passen würde. Ich hoffe, dass Sie sich wohlbefinden u. dass Sie auch mit Ihrem diesjährigen Besuch in U.S.A. zufrieden sind, u. würde mich sehr freuen, Sie bald wiederzusehen.

Mit herzlichen Grüssen

Ihr Kurt Gödel

2 | P.S. Zu meinem grossen Bedauern hörte ich, dass Julius Kraft gestorben ist. Wie steht es jetzt mit der Herausgabe von Nelson's Geschichte der Philosophie?[a]

---

[a] *Nelson 1962.*

I hope that things are going well for you in every respect, and I wish you a very happy New Year.

    With kindest regards,

               Yours truly,

               Kurt Gödel

Best wishes also from my wife.

34. Gödel to Bernays

                           Princeton, 16 March 1961

Dear Mr. Bernays,

    Thank you very much for your Christmas greetings. I assume you've now already been in Philadelphia for some time, and I would be especially pleased to see you here in Princeton. I would also like very much to speak with you about the content of the offprints [[you]] kindly forwarded to me. Do you have the inclination to come here for 1–2 weeks? The Institute could reimburse you. I assume $300 would suffice for about 10 days.

    Please let me know whether and when such a visit would suit you. I hope that you are in good health and are also satisfied with your visit this year in the U.S., and I would be very pleased to see you again soon.

    With best regards,

               Yours truly,

               Kurt Gödel

    P.S. To my great sorrow, I heard that Julius Kraft died. How do things stand now with the editing of Nelson's history of philosophy?[a]

35. Bernays to Gödel

Philadelphia 4, Pa., 23. März 1961.
220 S. De Kalb Str.

Lieber Herr Gödel!

Gerade war ich im Begriff, Ihnen zu schreiben, um Sie zu fragen, ob ich in der kommenden Woche—hier sind die Osterferien vom 27. III. (bezw. 25. III.) bis zum 3. IV.—Gelegenheit hätte, Sie in Princeton zu sprechen.

Nun empfange ich Ihre freundliche Aufforderung, die mich sehr erfreut. (Ich erhielt Ihren Brief wohl dadurch mit etwas Verzögerung, dass Sie begreiflicherweise mich beim Mathematics Department vermuteten, während ich beim Philosophy Department bin.)

Haben Sie vielen Dank für Ihr splendides Anerbieten! Gewiss habe ich Lust, einige Tage wieder in Princeton zu verbringen, und die Osterferien geben ja eine Möglichkeit dafür. Nur weiss ich noch nicht, ob ich nicht vielleicht einen durchreisenden Besuch bekomme.

Wollen wir vorerst so vereinbaren, dass ich am Dienstag, 28. III. nach Princeton komme und wir das Weitere dann mündlich verabreden. Ich kann mit dem Zug um 9^{59} in Princeton sein und, wenn es Ihnen recht ist, um 11 in Ihrem Instituts-Zimmer | vorsprechen.

Ihren freundlichen Neujahrsbrief empfing ich, nachdem ich Ihnen kurz zuvor meinen Neujahrsgruss gesandt hatte. Darf ich Ihnen nachträglich noch vielmals dafür danken. Über die Themata, die Sie darin berührten, werde ich mich gern mit Ihnen unterhalten.

Von dem leider so frühen Hinscheiden des Herrn Julius Kraft erfuhr ich auch. (Ich hatte ihn noch im Oktober 1960 in München gesehen.) Die Herausgabe der Nelson'schen Vorlesungen über die Geschichte der Philosophie hatte er, wie ich glaube, so weit gebracht, dass sie nicht gefährdet ist.

Im Januar empfing ich von einem Nelson-Schüler L. H. Grunebaum (in Scarsdale, N.Y.) einen Brief, in dem er mir mitteilte, dass er vor einigen Jahren eine Leonard Nelson Foundation gestiftet hat. Aus dieser sind einige Publikationen finanziert worden sowie Vorlesungen und Seminare, die Prof. Stephan Körner im vorigen Jahr in Yale und an anderen Orten hielt. (Mich forderte er auch zu Vorträgen auf; ich habe ihm erst kürzlich geantwortet.)

Nun auf Wiedersehen am Dienstag. (Es wird wohl am besten sein, wenn ich gleich nach der Ankunft bei Ihnen anrufe.)

Mit herzlichen Grüssen Ihr

Paul Bernays

35. Bernays to Gödel

Philadelphia 4, Pa., 23 March 1961
220 S. De Kalb Street

Dear Mr. Gödel,

I was about to write to you, in order to ask you whether I would have the opportunity to speak with you in Princeton during the coming week— the Easter vacation here is from 27 March (or, respectively, 25 March) until 3 April.

Now I receive your cordial invitation, which pleases me very much. (I probably received your letter with some delay because, as might be expected, you presumed I was in the mathematics department, whereas I'm in the philosophy department.)

Many thanks for your splendid offer! Certainly I am inclined to spend a few days in Princeton again, and the Easter vacation provides a possibility for it. But I still don't know whether I might not perhaps have a visit from someone passing through.

For the time being, let's agree that I'll come to Princeton on Tuesday, 28 March, and we'll then make further arrangements orally. I can be in Princeton on the 9^{59} train and, if it is convenient for you, call around 11 at your office at the Institute.

I received your cordial New Year's letter after I had, shortly before, sent my New Year's greeting to you. May I, belatedly, still thank you very much for it. I will gladly talk with you about the topics that you mentioned in it.

I also learned of the too early death of Mr. Julius Kraft. (I had just seen him in Munich in October 1960.) I think he had gotten so far with the editing of Nelson's lectures on the history of philosophy that [the project] is not at risk.

In January I received a letter from a student of Nelson, L. H. Grunebaum (in Scarsdale, N.Y.), in which he informed me that some years ago he had founded a Leonard Nelson Foundation. From that [source] a few publications have been financed, as well as lectures and seminars that Professor Stephan Körner delivered in the past year at Yale and at other places. (He invited me to [give] lectures also; I just recently replied to him.)

Goodbye now until Tuesday. (It will probably be best if I call you upon arrival.)

With best regards,

Yours truly,
Paul Bernays

[Written vertically in left margin:] Möchten Sie bitte Ihrer Frau eine Empfehlung von mir ausrichten.

36. Bernays to Gödel

Philadelphia 4, Pa. 20. April, 1961

Lieber Herr Gödel!

Sie sollten eigentlich schon ⟨viel⟩ früher einen Brief von mir bekommen, doch es liegt immer so vielerlei vor. Ich gedachte auch schon, Ihnen vorzuschlagen, dass ich an diesem Samstag nach Princeton komme. Jedoch ich habe meine Vorbereitung für einen Vortrag in Yale, den ich für den 24. April zugesagt habe, noch nicht unter Dach.

Nun möchte ich Ihnen zuerst noch einmal herzlich danken, dass Sie mir den Aufenthalt um Ostern in Princeton so angenehm und schön gestalteten. Vielen Dank ferner für die noble Zusendung, die Sie mir durch das Institut zukommen liessen!

Von den Manuskripten, die Sie mir mitgaben, habe ich mir verschiedenes angesehen, und insbesondere mit der Abhandlung von Takeuti[a] mich etwas näher abgegeben. Darf ich Ihnen angeben, was ich daraus entnommen habe. Takeuti betrachtet Mengen-Theorien bestehend aus Formeln gebildet mit den Mitteln des Prädikaten-Kalküls und dem Grundprädikat $\in$ ⟨ϵ⟩, wobei eine vollständige und widerspruchsfreie Einteilung in gültige und ungültige gegeben ist (eine Formel ist dann und nur dann gültig, wenn Ihre Negation ungültig ist); ferner: Jede beweisbare Formel der Zermelo–Fraenkel'schen Mengenlehre ist gültig. Eine solche Mengenlehre wird "positiv definit" genannt, wenn erstens die Bedingung der Konstruktivität jeder Menge erfüllt ist, das heisst, wenn, nach Ihrer Bezeichnung, die Gleichung $V = L$ gültig ist (die Gleichheit wird hier mittels ϵ definiert) und wenn ferner für die Mengenlehre ein "reguläres Modell" existiert[,] das heisst ein solches, bei dem keine im Sinne der ϵ-Beziehung absteigende Folge existiert. (Diese Bedingung ist nicht selbstverständlich, obwohl ja aus der Gleichung $V = L$ das Fundierungsaxiom beweisbar ist. Nämlich der Begriff der Folge ist hier nicht derjenige innerhalb der betreffenden Theorie, sondern derjenige der Modell-Theorie, in der die gesamte Überlegung angestellt wird.)

---

[a]See letter 44, 30 July 1962, third paragraph.

[Written vertically in the left margin:] Would you please extend my compliments to your wife.

36. Bernays to Gödel

Philadelphia 4, Pa. 20 April, 1961

Dear Mr. Gödel,

You really should already have received a letter from me much earlier, but ever so many things come up. I already thought too of proposing to you that I come to Princeton this Saturday. However, I still haven't completed the preparation for a lecture at Yale, which I promised for 24 April.

I would like now, first of all, to thank you cordially yet again for organizing the Easter visit in Princeton so agreeably and excellently. Moreover, many thanks for the generous remittance you had sent to me by the Institute!

I've looked at several of the manuscripts that you sent along with me, and in particular I've concerned myself somewhat more closely with the paper by Takeuti.[a] Let me tell you what I got out of it. Takeuti considers set theories consisting of formulas formed by means of the predicate calculus and the primitive predicate ϵ, in which a complete and consistent division into valid and invalid [formulas] is given (a formula is valid if and only if its negation is invalid). Moreover: Every provable formula of Zermelo–Fraenkel set theory is valid. Such a set theory is called "positive definite" if, first of all, the condition of the constructivity of each set is satisfied—that is if, according to your denotation, the equation $V = L$ is valid (equality is defined here in terms of ϵ—and if, further, a "regular model" exists for the set theory, that is, one in which no descending sequence in the sense of the ϵ-relation exists. (This condition is not self-evident, even though the axiom of foundation is provable from the equation $V = L$, because the concept of sequence here is not that within the theory under consideration, but rather that of the model theory in which the whole deliberation is made.)

Takeuti zeigt nun zunächst folgendes: (ich formuliere es der Einfachheit wegen ein wenig anders, als es in der Abhandlung steht, indem ich
in der Theorie auch Kennzeichnungen (ι Terme) zulasse. Dabei bleibt ja
die Vollständigkeit erhalten, sofern man für jedes Prädikat $A(c)$, welches
nicht die Existenz und Eindeutigkeitbedingungen erfüllt, die Gleichheit
von $\iota_x A(x)$ mit der Nullmenge ⟨in der Theorie⟩ festsetzt.) Für eine positiv definite Theorie bilden die ι Terme ein reguläres Modell ⟨M⟩, wenn
als Elementbeziehung $a\,\epsilon^*\,b$ genommen wird: "Die Formel $a\,\epsilon\,b$ ist gültig".

Dieses Modell M ist, wie sich leicht ergibt, ebenfalls regulär. Hieraus
folgt insbesondere, dass für jede Ordnungszahl a aus M die Ordnung der
Elemente von a durch die ϵ^*-Beziehung eine Wohlordnung auch im Sinne
der Modell-Theorie ist. Der Typus dieser | Wohlordnung wird mit "$\hat{a}$"
bezeichnet. Es gilt dabei: Wenn a und b Ordnungszahlen aus M sind, so
ist $\hat{a} = \hat{b}$ dann und nur dann, wenn $a = b$; $\hat{a}$ ist kleiner als $\hat{b}$ dann und nur
dann, wenn $a\,\epsilon^*\,b$; zu jeder Ordnungszahl β in der Modell-Theorie, welche
kleiner als eine Ordnungszahl $\hat{a}$ ist, gibt es in M eine Ordnungszahl b, so
dass $\beta = \hat{b}$.

Die Mengen des Modells M sind (als ι-Terme) in der Modell-Theorie
abzählbar, und somit bestimmen die den Ordnungszahlen a in M entsprechenden Ordnungszahlen $\hat{a}$ der Modell-Theorie einen Wohlordnungs-
Typus der zweiten Zahlenklasse, welchen Takeuti als Typus der betrachteten Mengen-Theorie bezeichnet. Bedeutet F sowohl für M wie auch innerhalb der Modell-Theorie die von Ihnen definierte Funktion, welche die
Ordnungszahlen auf die sämtlichen konstruktiblen Mengen abbildet, so
beweist Takeuti—(das habe ich freilich nicht mehr genau nachgeprüft)—
dass für irgend zwei Ordnungszahlen a, b, aus M, $F'a\,\epsilon^*\,F'b$ dann und
nur dann besteht, wenn $F'\hat{a}$ Element von $F'\hat{b}$ ist. Hiernach ist das Modell M durch den Typus der Mengentheorie, zu der es gehört, eindeutig
bestimmt. Nun wird als "Cantor'sches Absolut" von Takeuti eine solche
positiv definite Mengentheorie erklärt, welche sich nicht in eine andere
solche Mengentheorie "einbetten" lässt, und diese Bedingung wird auf die
einfachere zurückgeführt, dass der Typus der Mengentheorie ein maxima-
ler ist, so dass das Cantor'sche Absolut charakterisiert ist als eine solche
positiv definite Mengentheorie, zu der es keine von höherem Typus gibt.

Diese Charakterisierung müsste sich—so folgert Takeuti—(weil der
Typus eine Zahl der zweiten Zahlenklasse ist)—in dem Cantor'schen Absolut ausdrücken lassen, und das ergäbe einen Widerspruch gegen den allgemeinen Satz über Nicht-Definierbarkeit aus "Undecidable Theories".[b]
(Es fehlt da in dem Manuskript eine Nummernangabe; ich nehme an, es
handelt sich um Theorem 1 aus Teil II, Seite 46.)

---

[b] *Tarski, Mostowski and Robinson 1953.*

Takeuti now first shows the following: (for simplicity, I formulate it a little differently than it is in the paper, in that I also permit designators (ι-terms) in the theory. Completeness is thereby preserved, as long as one stipulates the equality of $\iota_x A(x)$ with the null set in the theory for every predicate $A(c)$ which doesn't satisfy the existence and uniqueness conditions.) For a positive definite theory the ι-terms form a regular model M if "The formula $a \, \epsilon \, b$ is valid" is taken as the element relation $a \, \epsilon^* \, b$.

As is easily seen, this model M is likewise regular. Hence it follows in particular that for every ordinal number a from M the ordering of the elements of a by the ϵ^*-relation is also a well-ordering in the sense of the model theory. The type of this well-ordering is denoted by "$\widehat{a}$" . Thereby it holds: If a and b are ordinal numbers from M, then $\widehat{a} = \widehat{b}$ if and only if $a = b$; $\widehat{a}$ is less than $\widehat{b}$ if and only if $a \, \epsilon^* \, b$; [and] to each ordinal number β in the model theory that is less than an ordinal number $\widehat{a}$, there is an ordinal number b in M such that $\beta = \widehat{b}$.

The sets of the model M are (as ι-terms) countable in the model theory, and thus the ordinal numbers $\widehat{a}$ of the model theory that correspond to the ordinal numbers a in M determine a well-ordering type of the second number class, which Takeuti designates as the type of the set theory under consideration. If, in M as well as in the model theory, F denotes the function defined by you which maps the ordinal numbers onto the totality of constructible sets, then Takeuti proves—(I admit I haven't verified that in more detail)—that for any two ordinal numbers a and b from M, $F'a \, \epsilon^* \, F'b$ holds if and only if $F'\widehat{a}$ is an element of $F'\widehat{b}$. Accordingly, the model M is uniquely determined by the type of the set theory to which it belongs. Now "Cantor's absolute" is declared by Takeuti to be such a positive definite set theory as may not be embedded in another such set theory. And this condition is reduced to the simpler one that the type of the set theory be maximal, so that Cantor's absolute is characterized as such a positive definite set theory for which there is none of higher type.

This characterization itself—so Takeuti concludes—would have to be expressible in Cantor's absolute (because the type is a number of the second number class), and that would yield a contradiction to the general theorem on undefinability from [the monograph] "Undecidable Theories".[b] (A numerical designation is lacking in the manuscript; I assume it is Theorem 1 from part II, page 46, that is in question here.)

Dieses letzte Stück der Takeuti'schen Beweisführung ist mir gar nicht ersichtlich. Abgesehen aber davon erscheint es doch als keineswegs sachgemäss, das Cantor'sche Absolut mit einer in standard logic formalisierten Mengenlehre zu identifizieren, welche von einer umfassenderen Modelltheorie aus betrachtet wird.

⟦The typescript ends here. The remainder of the letter is handwritten in pen.⟧

5. Mai 1961.

3   Der Brief hat mehrmalige Unterbrechungen erfahren. Ich hatte | inzwischen ⟨—wie anfangs schon erwähnt—⟩einmal beabsichtigt, nach Princeton zu kommen; doch die letzten Wochen waren sehr ausgefüllt. Am vorletzten Wochenende⟨—oder eigentl. Wochen-Anfang—⟩ trug ich in Yale vor, und am letzten hatte ich auswärtigen Besuch von Freunden.

An dem Symposium in New York habe ich wenigstens partiell teilgenommen. Dort traf ich (unter ⟨den⟩ vielen Bekannten) auch Herrn Mostowski, der mir ein Exemplar der gerade im Erscheinen begriffenen Berichte über das Warschauer Symposium von 1959[c] gab. Darin findet sich unter anderm ein ausführlicher Artikel von Esenin Volpin "Le programme ultra-intuitionniste des fondements des mathématiques",[d] welcher von der Arbeit, die er Ihnen seinerzeit zusandte, wesentlich verschieden ist. Sie haben vielleicht inzwischen ein Exemplar des Berichtes bekommen.

Ich bin Ihnen noch die nähere Angabe von dem Haupt-Ergebnis von Surányi schuldig. Er beweist, dass das allgemeine Entscheidungsproblem des Prädikatenkalkuls der ersten Stufe (Frage der Erfüllbarkeit) sich zurückführen lässt auf dasjenige für Formeln von der Gestalt

$$(x)(y)(z)M_1(x,y,z) \ \& \ (x)(y)(Ez)M_2(x,y,z),$$

mit quantorenfreien M_1, M_2 und mit höchstens einer zweistelligen, aber beliebig vielen einstelligen Prädikaten-Variablen. (Sein Buch heisst: "Reduktionstheorie des Entscheidungsproblems", Verlag d. ungarischen Akademie der Wissensch., Budapest 1959.[e])—

4   Der Herr W. W. Tait ist ja offenbar ein höchst begabter Mann. | Sie gaben mir von ihm zur Ansicht: *zwei* Abhandlungen über die first order number theory, eine Abh. über die Analysis (p⟦p⟧. 1–15; das ist wohl nicht das Ganze?), und eine Abh. "Nested recursion".[f]

[c] *Polish Academy of Sciences 1961.*
[d] *Esenin-Vol'pin 1961.*

This last bit of Takeuti's proof demonstration is not at all clear to me. But aside from that, it seems in no way appropriate that Cantor's absolute be identified with a set theory formalized in standard logic, which is considered from a more comprehensive model theory.

⟦The typescript ends here. The remainder of the letter is handwritten in pen.⟧

5 May 1961

The letter was interrupted several times. In the meantime I had intended—as already mentioned at the beginning—to come to Princeton; but the last few weeks were very busy. The weekend before last—or really the beginning of the week—I lectured at Yale, and last weekend I had out-of-town friends visiting.

I participated at least partially in the symposium in New York. There (among the many acquaintances) I also met Mr. Mostowski, who gave me a copy of the report on the 1959 Warsaw Symposium,[c] just now being published. In it is to be found, among other things, a detailed article by Esenin-Vol'pin, "Le programme ultra-intuitionniste des fondements des mathématiques",[d] which is essentially different from the paper that he sent you at the time. Perhaps you've received a copy of the report in the meantime.

I still owe you the more precise statement of Surányi's main result. He proves that the general decision problem for the predicate calculus of first order (the question of satisfiability) may be reduced to that of formulas of the form

$$(x)(y)(z)M_1(x, y, z) \ \& \ (x)(y)(Ez)M_2(x, y, z),$$

with quantifier-free M_1 and M_2 and with at most one two-place but arbitarily many one-place predicate variables. (His book is entitled: "Reduktionstheorie des Entscheidungsproblems", Hungarian Academy of Sciences Press, Budapest 1959.[e])

Mr. W. W. Tait is obviously a most highly talented man. You gave me ⟦the following⟧ of his to look over: *two* papers on first-order number theory, one paper on analysis (p⟦p⟧. 1–15; is that perhaps not the whole thing?), and one paper "Nested recursion".[f]

---

[e] *Surányi 1959.*

[f] *Tait 1961.*

Ich hoffe im Laufe dieses Monats nochmals nach Princeton zu kommen und will Ihnen dann auch diese Manuskripte wiedergeben. Das Konzept von Herrn Takeuti und seine Begleitzeilen schicke ich Ihnen mit diesem Brief.—

Über die Frage des Verhältnisses von Strafe und Freiheit, auf die wir zu sprechen kamen—(wir mussten die Unterhaltung unterbrechen)—, habe ich noch nachgedacht.

Im Ausblick darauf, Sie in ein paar Wochen wiederzusehen, grüsst Sie sehr herzlich

Ihr Paul Bernays

Möchten Sie bitte Ihrer Frau einen schönen Gruss von mir bestellen.

P.S. Die Besprechung des Buches von Herrn Wittenberg erschien in dem *Züricher* Blatt "Die Tat".

37. Gödel to Bernays

Princeton, 11./V. 1961.

Lieber Herr Bernays!

Vor allem meinen besten Dank für den genauen Bericht über Takeuti u. Suranyi. Ich freue mich sehr, dass ich voraussichtlich bald das Vergnügen haben werde, Sie hier wiederzusehen. Bitte schreiben Sie mir den genauen Tag Ihrer Ankunft, damit ich ein Zimmer im Institut reservieren lassen kann. Tait schrieb an Spector über die Fortsetzung seiner Arbeit über die Analysis,[1] dass diese Dinge noch zu sehr im Flusse seien[,] um Ihnen eine Kopie zu schicken. Das bezog sich übrigens auch auf seine andern unpublizierten Arbeiten, von denen er nur an Kreisel, Spector u. mich Kopien schickte.[a]

Ich habe mir kürzlich die Einleitung in die Kritik der Urt. Kr.[b] angesehen u. war sehr erstaunt, zu finden, dass die Urteilskraft ⟨bei Kant⟩ eigentlich das fundamentale Erkenntnisvermögen ist. Das beantwortet

---

[1]ich weiss übrigens nicht[,] ob davon überhaupt schon ein Manuskript existiert

---

[a]See the introductory note, section 2.3.

I hope in the course of this month to come to Princeton once more, and I will then also return these manuscripts to you. I am sending you Mr. Takeuti's draft and his accompanying commentary with this letter.

I've thought more about the question of the relation of punishment and freedom that we came to speak about (we had to interrupt the conversation).

In the expectation of seeing you again in a few weeks, I send you my very kindest regards.

Yours truly,

Paul Bernays

Please give your wife a nice greeting from me.

P.S. The review of the book by Mr. Wittenberg appeared in the *Zurich* newspaper "Die Tat".

37. Gödel to Bernays

Princeton, 11 May 1961

Dear Mr. Bernays,

Above all, my most sincere thanks for the precise report on Takeuti and Surányi. I'm very pleased that I will probably soon have the pleasure of seeing you here again. Please write me the exact day of your arrival, so that I can have a room reserved at the Institute. Tait wrote to Spector about the continuation of his work on analysis[1] that these things are still too much in flux to send you a copy. That applies also, by the way, to his other unpublished works, of which he sent copies only to Kreisel, Spector and me.[a]

I recently looked at the introduction to the *Critique of Judgment*[b] and was very astonished to find that judgment is for Kant really the fundamental faculty of knowledge. That also answers my question about the

---

[1] I don't know, by the way, whether a manuscript of it exists at all.

[b] *Kant 1790.*

auch meine Frage über die Kantsche Theorie der Begriffsbildung. Be-
2 friedigend | scheint mir die Einteilung des höheren Erkenntnisvermögens
in Verstand, Vernunft u. Urteilskraft allerdings nicht zu sein. Überhaupt
finde ich, dass die Kantsche Systematik an vielen Inconsequenzen leidet.
Warum gibt es nur 3 statt 4 Vernunftideen? Die vierte sollte offenbar die
"materia prima" sein. Andrerseits sollte es nur 3 Paralogismen geben (da
die Unterteilungen bei Kant doch 3gliedrig sind). Auch sollte man erwarten das⟦s⟧ es kosmologische, theologische etc. "Paralogismen" gäbe.
Ferner sollte es eine praktische Seite des Verstandes u. der Sinnlichkeit u.
eine ästhetische der Sinnlichkeit geben.

Ich freue mich darauf Sie bald wiederzusehen u. verbleibe mit herzlichen Grüssen

Ihr Kurt Gödel

P.S. Beste Grüsse auch von meiner Frau.

Dass Kreisel die zweite Hälfte des Juni in Princeton verbringen wird,
dürften Sie ja wissen.

38. Bernays to Gödel

Philadelphia 4, Pa., 19. V. 1961.

Lieber Herr Gödel!

Haben Sie vielen Dank für Ihren freundlichen Brief!
Ich hätte Ihnen schneller geantwortet; nur hatte ich für meine Einteilung einen Bescheid abzuwarten, den ich erst heute empfing.
Die eigentlichen Vorlesungen schliessen hier mit dieser Woche; es kom-
2 men dann | noch die examinations.
Anschliessend an einen Wochenend-Aufenthalt in New York gedenke
ich im Laufe des *Montag 22.* nach Princeton zu kommen und bis Donners-
tag mittag zu bleiben.

Im Ausblick auf das Zusammensein mit Ihnen grüsst Sie sehr herzlich

Ihr Paul Bernays

Freundliche Grüsse auch an Ihre Gemahlin.

Kantian theory of concept formation. To be sure, the division of the higher faculty of knowledge into understanding, reason, and judgment doesn't seem to me to be satisfying. In general I find that the Kantian systematization suffers from many inconsistencies. Why are there only 3 instead of 4 ideas of reason? The fourth should obviously be the "materia prima". On the other hand there should be only 3 paralogisms (since the subdivisions in Kant are always tripartite). One should also expect that there would be cosmological, theological, etc. "paralogisms". Moreover there should be a practical side of the understanding and of sensibility and an aesthetic one of sensibility.

I'm happy to be seeing you again soon and remain, with kindest regards,

Yours truly,

Kurt Gödel

P.S. Kindest regards also from my wife.

You may be aware that Kreisel will be spending the second half of June in Princeton.

38. Bernays to Gödel

Philadelphia 4, Pa., 19 May 1961

Dear Mr. Gödel,

Many thanks for your friendly letter!

I would have replied more quickly, but I had to await a decision on my schedule, which I only received today.

The actual lectures conclude here this week; then there are still the examinations.

Following a weekend stay in New York I'm thinking of coming to Princeton in the course of *Monday the 22nd* and staying until Thursday noon.

In the expectation of being together with you, very kind regards,

Yours truly,

Paul Bernays

Cordial regards also to your wife.

39. Bernays to Gödel

Zürich 2, den 10. Juli 1961.
Bodmerstr. 11

Lieber Herr Gödel!

Es war meine Absicht gewesen, Ihnen noch von Philadelphia aus zu schreiben und Ihnen für die schönen Tage zu danken, die ich durch Sie in Princeton hatte. Es waren aber die letzten Wochen meines Aufenthaltes in den U.S.A. sehr gedrängt ausgefüllt. Es kam dann noch hinzu, dass ich Ihnen gern zugleich über einen sachlichen Punkt schreiben wollte, über den ich in Anknüpfung an unsere Unterhaltung etwas nachgedacht hatte, und wofür ich das Nötige nicht zur Hand hatte.

Wir sprachen ja unter anderem von der Fries'schen Auffassung von der Strafe und so wie ich Ihnen diese formuliert hatte, schien es Ihnen, dass es sich dabei nur um das Erforderniss eines sozusagen zivilrechtlichen Ausgleiches handle. Tatsächlich aber unterscheidet Fries von diesem Erforderniss ausdrücklich noch ein weitergehendes. Um den Text bei Fries genauer nachzusehen, habe ich mir hier in der Zentralbibliothek seine "Philosophische Rechtslehre und Kritik aller positiven Gesetzgebung"[a] ausgeliehen. Ich schicke Ihnen beiliegend eine Abschrift einiger Stellen aus dem Buch (in der ursprünglichen Orthographie). Diejenigen aus dem ersten Abschnitt betreffen die allgemeine Ansicht vom Ethischen und Rechtlichen, die aus dem dritten Abschnitt die speziellen Fragen des Strafgesetzes. Gegenüber diesen Ausführungen drängt sich zunächst der Einwand auf, dass die Betrachtung zu einseitig auf Schädigungen an Eigentum abgestellt sei; aber vermutlich hat diese Spezialisierung hier nur den Sinn einer schematisierenden Vereinfachung.[1]

Von Herrn Kreisel haben Sie gewiss gehört, dass ich ihn in Paris am Flughafen sprach. Sie hatten gewiss mit ihm während seines kürzlichen Princetoner Aufenthaltes angeregte Unterhaltungen.

2 | Nun möchte ich Ihnen nochmals für den zweimalig mir gespendeten Aufenthalt in Princeton von Herzen danken, der mir so anregende Unterhaltungen mit Ihnen ermöglichte.

Seien Sie, und auch Ihre Gemahlin, herzlich gegrüsst

von Ihrem Paul Bernays

---

[1] [Written vertically in the left margin of page 1:] Ein gewichtigerer Einwand richtet sich aber wohl gegen die ~~etwas~~ Drakonische Art der Argumentation.

---

[a] *Fries 1803* (Erster Theil, 1. Abschnitt and 3. Abschnitt), pp. 31, 33–35, 58–60, 61–62; reprinted in *Fries 1968-*, vol. 9, pp. 51, 53–55, 78–80, 81–82.

39. Bernays to Gödel

Zurich 2, 10 July 1961
Bodmerstrasse 11

Dear Mr. Gödel,

It had been my intention to write you again from Philadelphia and to thank you for the lovely days I had in Princeton through you. But the last week of my stay in the U.S. was very crowded. In addition to that, I wanted at the same time to write you about a substantive point on which I had reflected somewhat in connection with our conversation, and for which I didn't have the necessary [items] at hand.

We spoke among other things of the Friesian conception of punishment, and as I had formulated it to you it seemed to you that it was only a question of the requirement of a so-to-speak "civil legal reconciliation". Actually, however, Fries expressly distinguishes from that requirement another farther-reaching one. In order to look into Fries' text in greater detail, I checked out from the central library here his *Philosophische Rechtslehre und Kritik aller positiven Gesetzgebung.*[a] I enclose a copy of a few passages from the book (in the original orthography). Those from the first chapter concern the general view of ethics and legal matters, those from the third chapter the particular questions of criminal law. Confronted with this exposition the objection that comes to mind first of all is that the consideration is too narrowly focused on damage to personal property; but presumably this specialization has here only the sense of a schematizing simplification.[1]

You have surely heard from Mr. Kreisel that I spoke to him at the airport in Paris. I'm sure you had stimulating conversations with him during his recent stay in Princeton.

Now I would like to thank you yet again from the heart for the stay in Princeton [you] twice bestowed upon me, which made possible such stimulating conversations with you.

Cordial greetings to you, and also to your wife from

Yours truly,

Paul Bernays

---

[1][Written vertically in the left margin of page 1:] A weightier objection, however, might be leveled against the Draconian sort of argumentation.

40. Gödel to Bernays

Princeton, 11. VIII. 1961.

Lieber Herr Bernays!

Ich danke Ihnen herzlich für Ihren Brief vom 10./VII. sowie den beigelegten Auszug aus ~~den~~ Friesschen. Ich bin mit vielem darin gar nicht einverstanden. Insbesondere scheint es mir *gänzlich* falsch, dass auf Grund des Naturrechts die kleinste Beleidigung das Rechtsverhältnis völlig auflöst u. den Beleidiger rechtlos macht. (Übrigens würde doch ⟨daraus⟩, da nach Fries dann die Gewalt an Stelle des Rechtsverhältnisses tritt, sich die noch absurdere Folgerung ergeben, dass gegen eine weitere Schadenszufügung seitens des Beleidigers ⟨moralisch⟩ nichts mehr einzuwenden ist.) Ich verstehe nicht, mit welcher Begründung man so etwas behaupten kann. Die Behauptung über das richtige Strafausmass scheint mir vernünftig zu sein, aber die Begründung sehr schlecht. Da würde mir noch die folgende mehr zusagen: Die Strafe ist auch eine Schadenersatz-
2   leistung, nämlich an die organisierte Gesellschaft für die Störung | der Ordnung, in der sie (die org. Gesellschaft) besteht. Da der Zweck dieser Ordnung in dem Schutz der einzelnen vor Schaden besteht, so ist offenbar der dem einzelnen verursachte Schaden das richtige Mass für die Grösse der Störung.—Ich weiss nicht, ob sie schon gehört haben, dass Spector am 29./VII. plötzlich gestorben ist; wie man mir sagt, an Leukämie, obzwar er anscheinend bis kurz vor seinem Tod keine grösseren Beschwerden hatte. Es ist wirklich erstaunlich, wie viele hochbegabte Logiker jung gestorben sind (Nicod, Herbrand, Gentzen, Spector, Ramsey etc.)

Tait hat die Widerspruchsfreiheit der Analysis bisher scheinbar nur mit nicht konstruktiven Mitteln bewiesen. Mit Kreisel hatte ich interessante Diskussionen. Er scheint jetzt wirklich in mathematisch befriedigender Weise gezeigt/ zu haben, dass die erste ϵ-Zahl die genaue Grenze des Finitismus ist. Ich finde dieses Resultat sehr schön, wenn es auch
3   vielleicht eines phänomenologischen Unterbaus bedürfen | wird, um voll zu befriedigen.—Es freut mich, dass Sie mit Ihrem Besuch in Princeton zufrieden waren. Noch mehr würde es mich freuen, wenn wir Sie hier für das II. Semester des Jahres 1961/62 (d. h. Mitte Jänner bis Mitte April 1962) begrüssen könnten. Ich glaube, dass das II. Semester dem I. gegenüber vorzuziehen wäre, da Tait wahrscheinlich dann schon präzisere Resultate über die Widerspruchsfreiheit der Analysis haben wird, die Sie vermutlich interessieren werden. Tait ~~wird~~⟨dürfte⟩ übrigens auch 1962/63 noch in Princeton sein.

Jetzt habe ich noch eine Bitte bez. zweier Literaturangaben:

1. Wo hat Ackermann bewiesen (vielleicht in einer Rezension?), dass im System der typenfreien Logik gewisse Aussagefunktionen in nicht trivialer Weise (d. h. nicht von der Form $P \cdot \varphi(x)$ wo P keinen Sinn hat, oder

40. Gödel to Bernays

Princeton, 11 August 1961

Dear Mr. Bernays,

I thank you cordially for your letter of 10 July as well as for the enclosed extract from Fries. I am not at all in agreement with much therein. In particular, it seems to me *entirely* false that on the basis of natural law the smallest offense totally dissolves the legal relationship and leaves the offender without rights. (Incidentally, from that follows—since, according to Fries, power then replaces the legal relationship—the still more absurd consequence that morally nothing more can be objected to a further infliction of damage by the offender.) I don't understand on what grounds one can assert such a thing. The assertion about the just measure of punishment seems to me to be reasonable, but the grounding [for it] very bad. Here the following would be more agreeable to me: punishment is also a rendering of compensation, namely to the organized society for the disruption of the order in which it (the organized society) consists. Since the purpose of this order consists in the protection of the individual from damages, the damages caused to the individual are obviously the proper measure for the magnitude of the disruption.

I don't know whether you've already heard that Spector suddenly died on 29 July; of leukemia, so I was told, although he apparently had no more serious complaints until shortly before his death. It is really astonishing how many highly gifted logicians died young (Nicod, Herbrand, Gentzen, Spector, Ramsey etc.).

So far Tait has apparently proved the consistency of analysis only with non-constructive means. I had interesting discussions with Kreisel. He now really seems to have shown in a mathematically satisfying way that the first ϵ-number is the precise limit of what is finitary. I find this result very beautiful, even if it will perhaps require a phenomenological substructure in order to be completely satisfying.

I'm pleased that you were satisfied with your visit in Princeton. It would please me still more if we could welcome you here for the second semester of the year 1961/62 (i.e., mid-January until mid-April 1962). I believe that the second semester would be preferable to the first, since by then Tait will probably have more precise results on the consistency of analysis, which will presumably be of interest to you. Tait might moreover still be in Princeton [during] 1962/63.

Now I still have a favor to ask you concerning two references:

1. Where did Ackermann prove (perhaps in a review?) that in the system of type-free logic certain propositional functions [of a] non-trivial [sort] (i.e., [ones] not of the form $P \cdot \varphi(x)$, where P has no meaning, or

ähnliches) für *keinen* Wert des Arguments einen Sinn haben. Das könnte
man als eine Widerlegung | des in meiner Arbeit im Schilpp-Band über
Russell[a] (auf p. 150 oben) beschriebenen Lösungsversuchs der Antinomien
ansehen. Allerdings würde das nur eine geringfügige Modifikation dieses Lösungsversuchs nötig machen, falls *1.* nur *wenige* gänzlich sinnlose
Funktionen existieren *2.* eine Aussagefunktion, die für mindestens 1 Argument einen Sinn hat, fast überall sinnvoll ist. Natürlich sind diese
beiden Fragen sehr unpräzise, aber es könnte *2.* in einem präzisen Sinn
widerlegbar sein.

2. Könnten Sie mir mitteilen, wo Ihre Arbeit über die Unendlichkeitsaxiome[b] erscheinen wird u. wo die von Lévy[c] erschienen sind? Sonst hat
ja wohl niemand darüber publiziert (abgesehen von Tarski's, Keisler's, u.
Hanf's Arbeiten über den Zusammenhang mit der kleinsten "messbaren"
Kardinalzahl).[d]

Ich erinnere mich übrigens, dass Sie in Ihrer Rezension meiner Arbeit
über die Russellsche Logik[e] | auch Einwände gegen den Inhalt der Fuss-
note 47 (auf p. 151) hatten. Was ich meinte, war, dass mathematische
Sätze vielleicht als Identitäten von Bedeutungen (Wahrheitswerten, Klassen, Eigenschaften etc.) *auf Grund des Sinns*, nicht aber als Sinnidenti-
täten interpretierbar sind. Nur das letztere (was ⟨z. B.⟩ aus ihrer Ableit-
barkeit aus Definitionen im Sinn von Abkürzungen folgen würde) hätte
ihre "Inhaltslosigkeit" zur Folge.

Ich hoffe, dass es Ihnen in jeder Hinsicht gut geht, u. verbleibe

mit herzlichen Grüssen

Ihr Kurt Gödel

P.S. Die Geschichte der Philosophie von Nelson ist wohl noch immer
nicht erschienen?

| P.S. Die Stelle in der Bibel, an der expressis verbis gesagt wird, dass
Gott auch das Übel erschaffen hat, ist: Isaias 45,7. Aus dem Zusammenhang folgt, dass das in einem sehr allgemeinen Sinn gemeint ist.

P.S. Sicherlich muss man in dem oben erwähnten Lösungsversuch der
Antinomien den Ausdruck "except for certain singular points" in einem

---

[a] *1944.*
[b] *Bernays 1961.*
[c] *Lévy 1960.*

similar ones) have meaning for *no* value of the argument? That could
be regarded as a refutation of the method for the resolution of the antinomies described in my paper in the Schilpp volume on Russell[a] (top of
p. 150). To be sure, that would necessitate only a minor modification
of that resolution method if *1.* only *a few* entirely meaningless functions
exist [and] *2.* a propositional function that has a meaning for at least
one argument is meaningful almost everywhere. Both these questions are
of course very imprecise, but the second could be refutable in a precise
sense.

2. Could you inform me where your paper on the axiom of infinity[b]
will appear and where those of Lévy[c] appeared? Probably no one else
has published about it (except for Tarski's, Keisler's, and Hanf's works
on the connection with the least "measurable" cardinal number).[d]

I recall, by the way, that in your review of my paper on Russell's
logic[e] you also objected to the content of footnote 47 (on p. 151). What
I meant was that mathematical propositions are perhaps interpretable
as identities of signification (truth values, classes, properties, etc.) *on
the basis of the sense*, but not as sense-identities. Only the latter (which
would follow e.g. from their derivability from definitions in the sense of
abbreviations) would have their "contentual emptiness" as a consequence.

I hope that things are going well for you in every respect, and remain

With kindest regards,

Yours truly,

Kurt Gödel

P.S. Nelson's history of philosophy has probably still not appeared?

P.S. The place in the Bible at which it is said *expressis verbis* that
God also created evil is Isaiah 45:7. From the context it follows that that
is meant in a very general sense.

P.S. Certainly in the attempt to solve the antinomies mentioned above
the expression "except for certain singular points" must be understood in

---

[d] *Erdös and Tarski 1961, Hanf 1964* and *Keisler and Tarski 1964.*
[e] *Bernays 1946.*

ziemlich liberalen Sinn verstehen, da ja, z. B., für den Begriff der Ordinalzahl alle aus diesem Begriffe durch Anwendung von Rechenoperationen definierten grösseren Ordinalzahlen singuläre Punkte sein müssen.

41. Bernays to Gödel

Zürich 2, den 12. Oktober 1961.
Bodmerstr. 11

Lieber Herr Gödel,

Es tut mit sehr leid, dass meine Antwort auf Ihren freundlichen Brief vom August sich so verspätet hat. Ich hatte die Zeit nach unserer Rückkunft von unserem Ferienaufenthalt in Mürren sehr ausgefüllt, hauptsächlich dadurch, dass ich die beabsichtigten Ergänzungen für die neunte Auflage von Hilbert's "Grundlagen der Geometrie"[a] fertigzustellen hatte. (Ich hatte die Einsendung des Manuskripts aufgrund eines bereits längeren Aufschubes für Ende September zugesagt.)

Um an den Beginn Ihres Briefes anzuknüpfen, worin Sie sich über die Fries'sche Ansicht von der Strafe äusserten, so bin ich natürlich ebensowenig wie Sie mit dieser Auffassung einverstanden (und habe das ja auch ⟨in dem Brief⟩ durch meine seitliche Randbemerkung kundgegeben). Inzwischen fand ich übrigens beim Nachsehen in Nelson's Rechtslehre ("System der philosophischen Rechtslehre und Politik", Leipzig 1924, Verlag Der neue Geist),[b] dass Nelson die Fries'sche Argumentation im Wesentlichen übernimmt. Ich schicke Ihnen wiederum beiliegend eine Abschrift der betreffenden Stelle. Ihr Argument gegen die Behauptung, dass jede Uebertretung ⟨des Gesetzes⟩ bereits das Rechtsverhältnis völlig aufhebt, finde ich sehr schlagend. Die Motivierung der These ist bei Nelson ja etwas eingehender als bei Fries, aber überzeugend ist sie doch nicht. Zumal, wenn zuletzt Nelson sich auf "das Gebot der vergeltenden Gerechtigkeit" beruft, so scheint gerade die Formulierung und Begründung dieses Gebotes dafür zu sprechen, dass die Bereitschaft, eine über die Vergeltung einer begangenen Gesetzesübertretung hinausgehende Bestrafung zu erleiden, gerechtermassen nicht zugemutet werden kann.

---

[a] *Hilbert 1962.*

a rather liberal sense, since, e.g., for the concept of ordinal number all greater ordinal numbers defined from this concept by application of computational operations must be singular points.

41. Bernays to Gödel

Zurich 2, 12 October 1961
Bodmerstrasse 11

Dear Mr. Gödel,

I'm very sorry that my reply to your friendly letter of August has been delayed so much. The time after our return from our vacation in Mürren I had filled up, chiefly because I had to finish the intended supplements to the ninth edition of Hilbert's *Grundlagen der Geometrie*.[a] (As a result of an already longer delay, I had promised to submit the manuscript at the end of September.)

To start at the beginning of your letter, where you expressed your opinion of the Friesian view of punishment, I am of course as little in agreement with that conception as you (and declared that also in the letter through my marginal comment). In the meantime, by the way, I found by looking into Nelson's theory of law (*System der philosophischen Rechtslehre und Politik*, Verlag Der neue Geist, Leipzig, 1924)[b] that Nelson adopts the Friesian argumentation in its essentials. I am enclosing again for you a copy of the relevant passages. Your argument against the assertion that every violation of the law already totally abolishes the legal relationship I find to be very compelling. The motivation for the thesis is given with somewhat more detail in Nelson than in Fries, but it is still not convincing. Especially when, in the end, Nelson refers to "the precept of compensatory justice", the very formulation and grounding of that precept seems to argue that the willingness to suffer a punishment that goes beyond the compensation for violating the law cannot justly be demanded.

---

[b] *Nelson 1924, Rechtslehre*, 2. Abteilung, 3. Kapitel; reprinted in *Nelson 1970–74*, vol. 6, pp. 104–106.

2 | Die Nachricht von dem plötzlichen Tode Clifford Spectors empfing ich
zugleich von Ihnen und von Herrn Kreisel. Dieser Verlust ist wirklich
sehr zu beklagen. Dass ich im Frühjahr noch mit Herrn Spector näher
bekannt wurde, schätze ich nun um so mehr.—
    Was Ihre Erkundigungen zur Literatur betrifft so sind die Abhandlun-
gen von Azriel Lévy[,] die er im Anschluss an seine auf Hebräisch abge-
fasste Dissertation (mit einem englischen Summary) publizierte[,] die
beiden folgenden:
    1. "Axiom schemata of strong infinity in axiomatic set theory", Pacific
       Journal of Mathematics, Vol. 1, 10, No. 1, 1960, pp. 223–238.[c]
    2. "On Ackermann's set theory", The Journal of symbolic logic, Vol.
       24, No. 2, ⟨1959⟩, pp. 154–166.[d]
Seitdem ist, wie ich glaube, eine neuere Abhandlung von ihm erschienen.
(Von einer etwas älteren Arbeit von ihm "Indépendence conditionelle de
$V = L$ et d'axiomes qui se rattachent au système de M. Gödel", Comptes
rendus Paris, vol. 245 (1957), pp. 1582–83[e] finden Sie eine Besprechung
durch Shepherdson[f] im Journal of symb. logic, Vol. 24, No. 3, p. 225.)
Meine Ueberlegungen über die Verstärkung des ursprünglichen Schemas
von Azriel Lévy,[g] wovon ich Ihnen im Frühjahr erzählte, sollen in dem
Festband für Fraenkel[h] erscheinen.
    Ein Beweis von Ackermann dafür, dass in der typenfreien Logik Aus-
sagefunktionen existieren, die (in nicht-trivialer Weise) für keinen Wert
des Argumentes sinnvol sind, ist mir nicht gegenwärtig. Ich habe auf Ihre
diesbezügliche Frage hin allerlei bei Ackermann nachgesehen, aber keine
solche Stelle gefunden.
    Was die Methode des Lösung des Russell'schen Paradoxons durch Zu-
lassung singulärer Stellen von Aussagefunktionen anbelangt, so hat man
ja dabei das folgende Dilemma: Entweder der Sachverhalt der Singu-
larität eines Argumentwertes für ein gewisses Prädikat wird nur metamathematisch konstatiert; dann ist das typenfreie System gewissermassen
nicht die ganze Logik. Oder man führt eine Formalisierung des Begriffes
"sinnlos" in das System ein, wie es Behmann in verschiedenen Abhandlungen vorschlägt[1]; dann besteht wiederum die Gefahr eines formalen
Widerspruches⟨, was eine "Korrektur der logischen Formeln" nötig macht⟩.

---

[1][Handwritten in the left margin of page 2:] Die hauptsächliche hierauf bezügliche
Abhandlung von Behmann ist "Der Prädikatenkalkül mit limitierten Variablen ...",
Journal of symb. logic, vol. 24, No. 2, 1959, pp. 112–140.[i]

---

[c] *Lévy 1960.*

[d] *Lévy 1959.*

[e] *Lévy 1957.*

I received the report of the sudden death of Clifford Spector simultaneously from you and from Mr. Kreisel. This loss is really very much to be lamented. That I became more closely acquainted with Mr. Spector in the spring I now value all the more.

As concerns your inquiries about the literature, the papers of Azriel Lévy that he published in connection with his dissertation (written in Hebrew, with an English summary) are the following two:

1. "Axiom schemata of strong infinity in axiomatic set theory", *Pacific Journal of Mathematics*, Vol. 10, No. 1, 1960, pp. 223–238.[c]

2. "On Ackermann's set theory", *The Journal of Symbolic Logic*, Vol. 24, No. 2, 1959, pp. 154–166.[d]

Since then, I believe, a newer paper by him has appeared. (You may find a review by Shepherdson of a somewhat earlier paper of his, "Indépendence conditionelle de $V = L$ et d'axiomes qui se rattachent au système de M. Gödel", *Comptes rendus Paris*, vol. 245 (1957), pp. 1582–83,[e] in the *Journal of Symbolic Logic*, Vol. 24, No. 3, p. 225.[f]) My deliberations about the strengthening of the original schema of Azriel Lévy,[g] which I mentioned to you in the spring, are supposed to appear in the *Festband* for Fraenkel.[h]

I don't recall a proof by Ackermann that propositional functions exist in type-free logic which (in a non-trivial way) are meaningful for no value of the argument. With regard to your question about that, I checked many papers by Ackermann but found no such passage.

As far as the solution of Russell's Paradox by the admission of singular points of propositional functions is concerned, one is faced with the following dilemma: Either the fact of the singularity of an argument value for a certain predicate is only metamathematically established, in which case the type-free system is, as it were, not the whole of logic. Or else one introduces into the system a formalization of the concept "meaningless", as Behmann proposes in various papers,[1] in which case the danger of a formal contradiction persists, which necessitates a "revision of the logical formulas".

---

[1] [Handwritten in the left margin of page 2:] The principal paper of Behmann relevant to this is "Der Prädikatenkalkül mit limitierten Variablen ...", *Journal of symbolic logic*, vol. 24, No. 2, 1959, pp. 112–140.[i]

[f] *Shepherdson 1959*.

[g] *Bernays 1961*.

[h] *Bar-Hillel et alii 1961*.

[i] *Behmann 1959*.

3 | Sie kamen noch auf Ihre Fussnote 47 in Ihrer Abhandlung über Russell's mathematischeɤ Logik zu sprechen, sowie auf das, was ich in meiner Review der Abhandlung dazu bemerkte. Sie weisen darauf hin, dass die Alternative für die Interpretation, wie ich sie in dem Passus der Review hinstelle, für die Formulierung dessen, worauf es Ihnen ankam, einer feineren Unterteilung bedarf. Die Gegenüberstellung, auf die es Ihnen ankommt, ist ⟨ja⟩, wenn ich recht verstehe, diejenige von vérités de fait und vérités de raison, wobei natürlich die raison "auf Grund des Sinnes" besteht. Was mich bei Ihrer Formulierung in der Anmerkung ⟨seinerzeit⟩ störte, war wohl dies⟨es⟩ dass Sie die view about analyticity in Zusammenhang brachten mit der Möglichkeit, die mathematischen Sätze auf die Form $a = a$ zu bringen. Sie wollten aber wohl ~~bloss~~ damit ⟨bloss⟩ sagen, dass die Möglichkeit einer solchen Reduktion besteht, ~~und~~ ohne dass damit die mathematischen Sätze zu analytischen Identitäten in einem *engeren* Sinne des Wortes werden. Den Philosophen gegenüber ist es ~~aber~~ vielleicht ganz gut hervorzuheben, dass die Möglichkeit jener Reduktion einen rein technischen Charakter hat und uns eigentlich keinerlei erkenntnistheoretischen Aufschluss gibt, (~~z. B.~~⟨Als Beispiel diene etwa⟩ die Transformation der Behauptung, dass es unendlich viele Primzahlen gibt, in diejenige, dass die Klasse der natürlichen Zahlen identisch ist mit der Klasse derjenigen natürlichen Zahlen, zu denen es eine grössere Primzahl gibt.) ⟨Die Überführung eines Satzes in die Form $a = a$ ist ja im allgemeinen nicht eigentlich eine Reduktion, sondern eine zusätzliche Konstruktion.—⟩

Kürzlich fiel mir ein, dass ich Ihnen bei einer unserer Unterhaltungen von dem Buch von Andreas Speiser[j] gesprochen hatte, in welchem er die Hegel'sche Philosophie zu einer neuartigen Darstellung bringt. Ich habe Ihnen ein Exemplar dieses Buches zugehen lassen. Ich brauche Ihnen gewiss nicht zu sagen, dass ich mich nicht in Versuchung fühle, zu dieser Art des Philosophierens bekehrt zu werden. Aber viele Bemerkungen in dem Buch sind interessant.

Für Ihre freundliche Aufforderung zum kommenden spring term bin ich Ihnen von Herzen dankbar! Zurzeit jedoch bin ich mit so vielen Dingen im Rückstande; und ich muss erst einmal mit meinen Arbeiten etwas mehr im Reinen sein, damit ich mit gutem Gewissen und geeigneter Empfänglichkeit ⟨wieder⟩ eine Reise nach Princeton unternehmen kann.

---

[j] *Speiser 1952.*

You also mentioned your footnote 47 of your essay on Russell's mathematical logic and, in addition, what I had remarked in my review of the essay. You point out that the alternative for the interpretation, as I present it in the passage of the review, requires a finer subdivision for the formulation of what was of concern to you. The contrast that is of concern to you is, after all, if I understand correctly, between vérités de fait and vérités de raison, where, of course, the raison subsists "on the basis of sense". What disturbed me at the time about your formulation in the note was probably this, that you connected the view about analyticity with the possibility of bringing mathematical statements into the form $a = a$. But in doing so, you probably only intended to say that the possibility of such a reduction exists, not that the mathematical statements thereby become analytic identities in a *narrower* sense of the word. With regard to the philosophers, it is perhaps important to emphasize that the possibility of that reduction has a purely technical character and gives us no real epistemological information. (The transformation of the assertion that there are infinitely many prime numbers into the assertion that the class of natural numbers is identical with the class of those natural numbers for which there exists a greater prime number may perhaps serve as an example.) The conversion of a statement into the form $a = a$ is of course in general not really a reduction, but rather an additional contruction.

It occurred to me recently that in one of our conversations I had spoken to you of the book by A[n]dreas Speiser[j] in which he offers a novel account of Hegel's philosophy. I have sent you a copy of this book. I surely needn't tell you that I do not feel tempted to become converted to this sort of philosophizing. But many remarks in the book are interesting.

I'm heartily grateful to you for your cordial request for the coming spring term! Just now, however, I am in arrears with so many things; and I must first get my papers somewhat more in order before I can undertake a trip to Princeton again with a clear conscience and proper receptivity.

Mit vielen herzlichen Grüssen und der Bitte um eine Empfehlung an Ihre Gemahlin

Ihr Paul Bernays

⟦Written vertically in the margin:⟧ Schönen Dank auch noch für Ihre Angabe der Bibelstelle!

42. Gödel to Bernays

Princeton, 15./XII. 1961

Lieber Herr Bernays,

Meinen herzlichsten Dank für Ihren Brief vom Okt. u. den Auszug aus Nelson, sowie vor allem für das interessante Buch von Speiser. Über das letztere u. andere philos. Fragen will ich Ihnen demnächst ausführlicher schreiben. Ich habe von Ihnen noch den Band der Abh. der Friesschen Schule, den Sie mir bei Ihrem letzten Aufenthalt freundlicherweise über-
2   lassen haben, u. will ihn bald | retournieren. Ich habe einiges daraus gelesen u. wieder die grosse Klarheit der Nelsonschen Darstellungsweise bewundert. Was seine Kantkritik betrifft, so scheint es mir allerdings, dass er Kant allzu primitive Fehler in die Schuhe schiebt. Es tut mir sehr leid, dass Sie dieses Jahr nicht nach Princeton kommen können. Hof-fentlich wird es später einmal möglich sein.

Ich hoffe, es geht Ihnen gut, u. wünsche Ihnen das Beste für Weih-nachten u. das kommende Jahr. Mit herzlichen Grüssen

Ihr Kurt Gödel

Beste Wünsche u. herzliche Grüsse auch von meiner Frau.

With many kind regards and the request that you give my compliments to your wife,

Yours truly, Paul Bernays

[Handwritten in the left margin of page 3:] Thank you very much, too, for your citing of the place in the Bible.

42. Gödel to Bernays

Princeton, 15 December 1961

Dear Mr. Bernays,

My most cordial thanks for your letter of October and the extract from Nelson, as well as, above all, for the interesting book by Speiser. I want to write you about the latter and other philosophical questions in more detail in the near future. I still have your volume of Abhandlungen der Fries'schen Schule that you kindly lent me during your last stay [here], and I want to return it soon. I have read a few [passages] from it and have again admired the great clarity of Nelson's method of presentation. As concerns his critique of Kant, it seems to me, though, that he charges Kant with all too primitive mistakes. I'm very sorry that you can't come to Princeton this year. Hopefully it will be possible once again later.

I hope things are going well for you, and I wish you the best for Christmas and the coming year.

With kind regards,

Yours truly, Kurt Gödel

Best wishes and kind regards also from my wife.

43. Bernays to Gödel

Zürich 2, 31. XII. 1961
Bodmerstr. 11

Lieber Herr Gödel

Haben Sie vielen Dank für Ihre guten Wünsche und für Ihre freundlichen Zeilen!

Den von Ihnen angekündigten Ausführungen über das Speisersche Buch und über noch anderes Philosophische sehe ich mit viel Neugier entgegen.—

Gegenwärtig bin ich mit einer für die "Synthese" versprochenen Arbeit[a] (über Carnap's Philosophie) beschäftigt, die schon längst fertig sein sollte.—

2   Der Fraenkel-Festband ist | anscheinend noch nicht herausgekommen.—

Dass Herr Schütte in diesem Winter und im kommenden Sommer hier Herrn Specker vertritt, wissen Sie wohl.—

Nun wünsche ich Ihnen für das kommende Jahr von Herzen ein gutes Ergehen, möglichst frei von Behinderung im Gesundheitlichen. Möchte es auch für Ihre im Gange befindlichen Arbeiten eine fruchtbare Zeit werden!

Mit herzlichen Grüssen

Ihr Paul Bernays

Die ⟨freundlichen⟩ Grüsse und die Neujahrswünsche Ihrer Frau erwidere ich herzlich!

---

[a] *Bernays 1961a.*

44. Gödel to Bernays

Princeton, 30./VII. 1962.

Lieber Herr Bernays!

Es tut mir leid, dass ich bisher nicht dazu kam, meine Absicht, Ihnen ausführlich zu schreiben, auszuführen. Im letzten Sommersemester[1]

[1] ich meine das im Jan. beginnende

43. Bernays to Gödel

Zurich 2, 31 December 1961
Bodmerstrasse 11

Dear Mr. Gödel,

Many thanks for your good wishes and for your friendly letter!

I look forward with much curiosity to the observations you announced on Speiser's book and on other philosophical matters.

I am presently busy with a paper (on Carnap's philosophy) promised for *Synthese*,[a] which should have been ready a long time ago.

The Fraenkel *Festband* has apparently still not been published.

You probably know that Mr. Schütte is filling in for Mr. Specker here this winter and during the coming summer.

Now, from the heart, I wish that during the coming year things will go well for you, as free as possible from health problems. May it also be a fruitful time for your investigations that are currently underway!

With kind regards,

Yours truly,

Paul Bernays

I heartily return your wife's cordial greeting and New Year's wishes!

---

44. Gödel to Bernays

Princeton, 30 July 1962

Dear Mr. Bernays,

I'm sorry that up to now I did not get around to carrying out my intention of writing you in detail. Last summer semester[1] there were a few

---

[1] I mean the one beginning in January.

waren einige unerwartete u. zeitraubende Angelegenheiten in der Fakultät zu erledigen. Dadurch bin ich mit allem etwas zurückgeblieben.—
Ich danke Ihnen herzlich für den l. Brief, den Sie mir um die Jahreswende schickten.—Was das Buch von Speiser betrifft, so finde ich die
Einleitung sehr schön, besonders auch deswegen, weil ein präziser Formalismus angegeben wird⟨.⟩ ~~was~~ Aber die Durchführung im einzelnen, soweit ich sie gelesen habe, finde ich durchaus nicht klarer als Hegel selbst.
Was Speiser auf p. 20 unten u. p. 21 sagt, ist mir unverständlich. Die
angedeutete Aufklärung des Unterschiedes zwischen Arithmetik u. Ge-
2   ometrie, selbst | wenn sie durchgeführt wäre (u. *wo wurde sie* durchgeführt?)
wäre doch kein Erfolg in der Math., oder eine Überschreitung unseres
Wissens in der Math.—Ich sende gleichzeitig mit diesem Brief den mir
geliehenen Band der Abh. d. Friesschen Schule mit bestem Dank zurück.
Da die Beschäftigung mit den nicht-ethischen phil. Problemen, die gegenwärtig
mein Hauptinteresse sind, meine Zeit vollauf in Anspruch nimmt, so konnte ich leider nur einen kleinen Bruchteil des Buches lesen. Es scheint mir
unberechtigt, dass Nelson auf p 540 die Kantsche Interpretation des Sollens gänzlich ablehnt. Kann jemand dessen Wollen in keinem Sinn mit
dem Sollen übereinstimmt, das letztere überhaupt erleben oder verstehen? Die gänzliche Irreduzibilität des Sollens (p. 539) scheint mir höchst
unbefriedigend u. unphilosophisch zu sein, ebenso die Beschränkung auf
3   die psychologische | Deduktion, während doch gerade die transcenden-
talen u. metaphys. Dedukti⟨onen⟩ das phil. Interessante sind. Sehr gespannt bin ich auf Nelsons Geschichte der Phil. ⟨seit Kant⟩,[a] die ja jetzt
sehr bald erscheinen soll.

Die Takeutische Arbeit über das Absolute[b] (J. Math. Soc. Jap. 13,
197),[2] über die Sie mir einmal schrieben,[c] habe ich mir jetzt etwas näher
angesehen. Ich stimme mit Ihnen überein, dass sein Resultat für das Absolute nichts bedeutet. Aber ich finde doch manche seiner Def., Lemmata
u. Fragestellungen nicht uninteressant. Mit gewissen Modifikationen
würden sie ⟨sogar⟩ etwas für das Absolute bedeuten. Haben Sie übrigens
seine Arbeit über Unendlichkeitsax. im J. Math. Soc. Jap. 13, p 220[e]
gesehen? Sein "G-inference" (p. 223) ist ja auch ein Relativierungsax.,
das aber die nötige Stärke nicht durch gebundene Klassenvar., sondern
durch einen neuen Grundbegriff G der axiomat. Mengenlehre bekommen

---

[2]ergänzt in: Proc. Jap. Ac. 37, 437[d]

---

[a] *Nelson 1962.*
[b] *Takeuti 1961.*
[c] See letter 36, 20 April 1961.

unexpected and time-consuming affairs to deal with among the faculty. As a result of that I fell somewhat behind with everything.

Thank you very much for your last letter, which you sent me at the beginning of the year.—As to the book by Speiser, I find the introduction very beautiful, particularly because a precise formalism is given. But the detailed execution, as far as I've read it, I find not a bit clearer than Hegel himself. What Speiser says at the bottom of p. 20 and on p. 21 is incomprehensible to me. The indicated clarification of the distinction between arithmetic and geometry, even if it could be carried out (and *where was it carried out?*), would still not be a success within mathematics, nor something going beyond our knowledge within mathematics.—I am sending back to you simultaneously with this letter the volume of papers of the Friesian school you lent me; thanks very much. Since my work on non-ethical philosophical problems, which are at present my principal interest, takes up my time completely, I was unfortunately able to read only a small fraction of the book. It seems unjustified to me that Nelson on p. 540 completely rejects the Kantian interpretation of the *Sollen*. Can anyone whose *Wollen* agrees in no sense with the *Sollen* really experience or understand the latter? The total irreducibility of the *Sollen* (p. 539) seems to me to be most unsatisfying and unphilosophical, as does the restriction to psychological deduction, whereas it is precisely the transcendental and metaphysical deductions that are philosophically interesting. I am very eager for Nelson's history of philosophy since Kant,[a] which should surely appear now very soon.

Takeuti's paper on the absolute[b] (J. Math. Soc. Jap. 13, 197),[2] about which you once wrote,[c] I have now looked at somewhat more closely. I agree with you that his result is insignificant for the absolute. But I still find many of his definitions, lemmas and problems not without interest. With certain modifications they would even be significant for the absolute. By the way, have you seen his paper on the axiom of infinity in the J. Math. Soc. Jap. 13, p. 220?[e] His "G-inference" (p. 223) is of course also an axiom of relativization, which, however, is supposed to obtain the necessary strength not by means of bound class variables, but by means of a new primitive concept G of axiomatic set theory. You have probably

[2]amplified in: Proc. Jap. Ac. 37, 437.[d]

---

[d] *Takeuti 1961a.*

[e] *Takeuti 1961b.*

4   soll. | Mit Kondô's Arbeiten über "namable sets"[f] haben Sie sich ja wohl
nie abgegeben? Takeuti scheint übrigens das Hauptgewicht auf sein Ax.
5 (p 223 oben der eben zitierten Arbeit) zu legen, dessen Bedeutung ich
mir noch nicht klar gemacht habe.

Was die Volpinsche Idee betrifft, so möchte ich gerne *irgendwelche*,
auch nur halbwegs plausible Axiome über den Begriff der "erreichbaren
Zahl" sehen, aus denen die Widerspruchsfreiheit, wenigstens der Analysis,
folgt. Ist Ihnen irgend etwas derartiges bekannt? Es wäre ja auch wirklich überraschend, wenn man die Math. (einschliesslich der Zahlentheorie) aus der Einsicht begründen könnte, dass der Begriff der natürlichen
Zahl unsinnig ist. Wie steht es mit der Neuauflage der "Grundlagen der
Mathematik"?[g] Falls Ihre Arbeit über Carnap schon erschienen ist, wäre
ich Ihnen für einen Sonderdruck dankbar.

Mit den besten Grüssen

Ihr Kurt Gödel

---

[f] *Kondô 1958.*

45. Bernays to Gödel

Zürich 2, den 12. Oktober 1962.
Bodmerstr. 11

Lieber Herr Gödel,

Ueber Ihren freundlichen Brief den Sie mir im Sommer sandten, (ich
empfing ihn während eines Ferienaufenthaltes) habe ich mich sehr gefreut. Gern hätte ich Ihnen bald darauf geantwortet; doch bin ich durch
verschiedene Obliegenheiten längere Zeit nicht dazu gekommen, mir die
verschiedenen philosophischen und mathematischen Dinge, über die Sie
schrieben, ~~mir~~ vorzunehmen.

Nun danke ich Ihnen zunächst noch vielmals für die Rücksendung des
Heftes von den Abhandlungen der Fries'schen Schule.—Sie äusserten
sich unbefriedigt davon, dass Nelson das Sollen als etwas absolut Letz-
tes nimmt. Diese Haltung Nelson's steht im Zusammenhang mit seinem
Begriff der sittlichen *Erkenntnis*, die er ganz in Entsprechung zur sonsti-

never spent time with Kondô's papers on "namable sets"?[f] Incidentally, Takeuti seems to put the principal emphasis on his Axiom 5 (top of p. 223 of the paper just cited), whose significance I've still not made clear to myself.

As to Vol'pin's idea, I would very much like to see *some*, even just halfway plausible axioms about the concept of "accessible number" which imply the consistency at least of analysis. Are you aware of any such thing? It would also be really surprising if one could base mathematics (including number theory) on the insight that the concept of natural number is nonsensical. How do things stand with the new edition of "Grundlagen der Mathematik"?[g] In case your paper on Carnap has already appeared, I would be grateful to you for an offprint.

With best regards,

Yours truly,

Kurt Gödel

[g] *Hilbert and Bernays 1968* and *1970*.

45. Bernays to Gödel

Zurich 2, 12 October 1962
Bodmerstrasse 11

Dear Mr. Gödel,

I very much enjoyed your friendly letter you sent me in the summer. (I received it during my vacation.) I would have liked to answer it soon thereafter, but due to various obligations I didn't get around for a longer time to taking up the various philosophical and mathematical matters you wrote about.

Now, first of all, many thanks for sending back the volume of papers of the Friesian school. You expressed dissatisfaction that Nelson takes the *Sollen* as something absolutely basic. This stance is connected with his concept of moral *knowledge* [Erkenntnis], which he tends to regard as

gen Erkenntnis aufzufassen tendiert. Die "Frage nach dem Grunde der
Verbindlichkeit des Sittengesetzes" lehnt er in demselben Sinne ab wie
jene erkenntnistheoretischen Fragen, gegen die er sich in seinem Buche
"Ueber das sogenannte Erkenntnisproblem"[a] wandte. Was mir an einem
absoluten Sollen als besonders unbefriedigend erscheint, ist der Umstand,
dass doch das Sollen nur in Verbindung mit der Zeitform unseres Erlebens seine echte Bedeutung hat. Wenn wir von einer vergangenen Handlung sagen, sie hätte nicht geschehen sollen, so ist damit doch die Mo-
dalität schon wesentlich geändert; das eigentliche Sollen ist dann gegen-
standslos geworden; es bleibt nur noch die Wertung.

Sie sprechen von Nelson's Beschränkung auf die psychologische Deduk-
tion; ich bin dabei aber nicht sicher, ob Sie sich auch die "Deduktion des
Inhaltes des Sittengesetzes" (Nr. 85, S. 658–663)[b] angesehen haben, bei
der es sich, nach der vorangegangenen Aufweisung eines rein vernünftigen
Ursprunges der ethischen Urteile und Antriebe, darum handelt, aus den
aufgewiesenen Charakteristiken der sittlichen Erkenntnis den Inhalt des
Sittengesetzes zu determinieren. Diese Ueberlegung, die Nelson in seinem
Buche "Kritik der praktischen Vernunft"[c] ausführlicher dargestellt hat,
scheint mir jedenfalls des philosophischen Interesses | sehr würdig zu sein.
(Von dem philosophischen Publikum wurde sie anscheinend überhaupt
nicht zur Kenntnis genommen.)

Nelson's Vorlesungen über "Fortschritte und Rückschritte der Philosophie"[d] sind ja inzwischen erschienen, und Sie haben gewiss auch schon
ein Exemplar davon erhalten.

Im Juli fand in Göttingen im Kreise von Anhängern Nelson's eine
Gedenkfeier, anlässlich des Datums seines 80. Geburtstages statt, bei der
ich auch anwesend war.

Was das Buch von Speiser anbelangt, so finde auch ich, dass die Einleitung ("Das Präludium") viel Anregendes enthält. Dass aber hier
gleichwohl die Gedankengänge stellenweise sehr unbefriedigend sind, haben Sie ja auch an dem Beispiel des Passus auf S. 20–21 vermerkt. Auf
die hierbei angetönte Problematik von Subjekt und Objekt kommt Speiser ja verschiedentlich zu sprechen. Charakteristisch ist z. B. die Stelle
(S. 30–31), wo er zuerst Euler's Bemerkung zitiert, dass die Seele "von
dem wirklichen Dasein der Objekte, die sie ausser sich denkt," überzeugt
ist, und dann im Anschluss an diese "Kritik des einseitigen Idealismus"
gleich anschliessend erklärt "Fichte hat immer darauf das Hauptgewicht

---

[a] *Nelson 1908.*

[b] *Nelson 1917,* Dritter Teil, 4. Abschnitt, 3. Kapitel–6. Kapitel; reprinted in *Nelson
1970–74,* vol. 4, p. 500ff.

in complete analogy to other knowledge. He rejects the "question as to the basis of the obligatory nature of the moral law" in the same sense as those epistemological questions against which he turned in his book *Über das sogenannte Erkenntnisproblem*.[a] What seems to me especially unsatisfying about an absolute *Sollen* is the circumstance that, after all, the *Sollen* only has its proper significance in conjunction with the temporal force of our experience. If we say of a past action that it should not have happened, the modality is thereby already essentially changed; the proper *Sollen* has then become pointless; there remains only the value judgment.

You speak of Nelson's restriction to the psychological deduction; in that regard, however, I'm unsure whether you have also looked at the "Deduktion des Inhalt des Sittengesetzes" (No. 85, pp. 658–663)[b] in which, after the foregoing exhibition of a purely rational origin of ethical judgments and drives, it is a question of determining the content of the moral law from the exhibited characteristics of moral knowledge. This consideration, which Nelson presented in detail in his book *Kritik der praktischen Vernunft*,[c] seems to me in any case to be very worthy of philosophical interest. (It has apparently not become at all well known to the philosophical public.)

Nelson's lectures on "Fortschritte und Rückschritte der Philosophie"[d] have indeed appeared in the meantime, and you surely have also already received a copy of it.

In July a celebration was held in Göttingen among the circle of Nelson's followers, on the occasion of the date of his 80th birthday, at which I was also present.

As far as the book by Speiser is concerned, I also find that the introduction ("Das Präludium") contains much that is stimulating. That, however, the trains of thought here are nonetheless in places very unsatisfying you too have noted in the example of the passage on pp. 20–21. Speiser turns in various places to the discussion of the problematic nature of subject and object suggested here. Characteristic [of those discussions], e.g., is the place (pp. 30–31) where he first cites Euler's remark that the soul is convinced "of the real existence of the objects that it conceives outside of itself", and then, in connection with this critique of the one-sided idealism, declares immediately afterward, "Fichte

---

[c] See *Nelson 1917*.

[d] *Nelson 1962*.

gelegt, dass man ein unzertrennliches Subjekt-Objekt als das Oberste annehmen muss". Diese Anlehnung an eine mystische Haltung, wonach die
eigentliche Erkenntnis in einem Eins-sein mit dem Gegenstande bestehen
soll, erscheint mir ~~als~~ für die Philosophie ⟨als⟩ sehr unzuträglich. Es liegt
doch diesem anscheinend so erhabenen Aspekt eine Art von Physikalis-
mus zugrunde, für welchen eine Erkenntnis von etwas dem Erkenntnis-
prozess nicht Immanentem als eine actio in distans auszuschliessen ist.
Durch eine solche Auffassung wird die spezifische Bedeutungsfunktion
des Denkens (die "reference") verkannt, ebenso wie das Erfordernis einer
Distanz, wie sie ja für das Erkennen analog wie für die visuelle Wahrnehmung besteht. (Dass wir im Denken mit unserem Denken eins sind, besagt ja keineswegs, dass wir dabei das real Wesensmässige des Denkvor-
ganges erkennen.)—

Nun zu den mathematischen Dingen. An der Abhandlung Takeuti's
über "Cantor's Absolutes"[e] finde ich besonders die aus der Proposition
1 (S. 199) sich ergebende Bemerkung interessant, dass für eine "definite Mengenlehre" T die ι-Terme mit der angepassten Elementbeziehung
3   ein | Modell bilden. Diese Feststellung könnte eigentlich in der Abhand-
lung als Korrolar der Proposition 1 (Anwendung auf den Fall $n = 0$)—
natürlich in der Bezeichnungsweise von Takeuti—vermerkt sein; das wäre
gewiss für den Leser eine Erleichterung.

Uebrigens ist ja diese Bemerkung nicht auf den Fall einer Mengenlehre
beschränkt, sondern gilt doch für jede im Rahmen der standard logic formalisierte Theorie, welche die beiden Bedingungen der Definitheit (S. 198
Mitte) erfüllt. Von dieser Erwägung aus kommt man ziemlich direkt zu
der Methode des Henkin'schen Vollständigkeitsbeweises, indem man darauf ausgeht, die Erfüllung jener beiden Bedingungen zu erzwingen, was
ja für die Bedingung der Vollständigkeit durch den Lindenbaum'schen
Prozess gelingt und für die zweite Bedingung durch die Einführung von
Individuensymbolen bewirkt wird, welche die Rolle von ϵ-Termen (im
Sinne des Hilbert'schen ϵ-Schemas) haben.

Wie weit man bei der Anwendung dieser Betrachtungen auf die Mengenlehre von der Kennzeichnung eines "Absoluten" entfernt bleibt, ist
ersichtlich. Das Skolem'sche Paradoxon macht sich ja in vollem Umfange geltend. Dieser Umstand hat wohl auch den Esenin-Vol'pin auf
seinen Gedanken der Verschärfung des konstruktiven Standpunktes gebracht. Wenn es schon möglich ist—so mag die Ueberlegung sein—, die
Uneigentlichkeit in der axiomatischen Behandlung der hohen Unendlichkeiten bei der beweistheoretischen Betrachtung auszuschalten, sollte es
dann nicht auch möglich sein, die zahlentheoretische Unendlichkeit in

---

[e] *Takeuti 1961.*

always put the principal emphasis on the fact that one must assume an inseparable subject-object as supreme." This dependence on a mystical attitude, according to which actual knowledge is supposed to be oneness with the object, seems to me not to be conducive for philosophy at all. Indeed, underlying this apparently so sublime aspect is a sort of physicalism, for which knowledge of something that is not immanent to the cognitive process is to be excluded as an "action at a distance". Through such a conception the specific semantic function of thought (the "reference") is not recognized, just as the requirement of a distance that exists for cognition in analogy to such a requirement for visual perception. (That, in thinking, we are at one with our thinking in no way means that we thereby recognize the real nature of the thinking process.)

Now to mathematical matters. In Takeuti's paper on "Cantor's absolute"[e] I find particularly interesting the observation that emerges from Proposition 1 (p. 199) that for a "definite set theory" T the ι-terms with the appropriate element relation form a model. This realization could actually be noted in the paper as a corollary of Proposition 1 (application to the case $n = 0$)—of course, in Takeuti's notation; that would certainly make things easier for the reader.

By the way, this remark is not restricted to the case of some set theory, but rather holds in fact for every theory formalized within the framework of standard logic which satisfies the two conditions of definiteness (middle of p. 198). Out of this consideration one comes rather directly to the method of Henkin's completeness proof, by aiming to force the satisfaction of those two conditions; this succeeds for the completeness condition through the Lindenbaum procedure, and for the second condition it is effected through the introduction of symbols for individuals, which play the role of ϵ-terms (in the sense of Hilbert's ϵ-schema).

In the application of these considerations to set theory, it is evident how far removed one remains from the characterization of an "absolute". The Skolem paradox asserts itself fully. This circumstance probably also brought Esenin-Vol'pin to his thoughts on sharpening the constructive standpoint. If it is possible—the thinking may have proceeded—to eliminate the impropriety in the axiomatic treatment of high infinities from proof-theoretic considerations, shouldn't it then also be possible to avoid the number-theoretic infinity in proof theory? Here one need not at all

der Beweistheorie zu vermeiden? Dabei braucht man ⟨ja⟩ keineswegs der
Meinung zu sein, dass "der Begriff der natürlichen Zahl unsinnig ist",
ebensowenig wie der Beweistheoretiker den Begriff des Ueberabzählbaren
als unsinnig abzulehnen braucht. Die grundsätzliche Problematik besteht
allerdings darin, dass anscheinend jede deduktive Präzisierung bereits die
zahlentheoretische Unendlichkeit einführt. Die Darstellung von Volpin
zeigt nicht, soviel ich sehe, wie diese Schwierigkeit überwunden werden
kann. Er führt das Postulat *Trad* (S. 205) ein, welches die Existenz einer
"natürlichen Folge" N von der Eigenschaft fordert, dass die Operation 2^n
überall in dieser Folge definiert ist, und diese Voraussetzung wird, wenn
ich recht verstehe, in dem Sinne verwendet, dass, wenn n zu N gehört,
dann auch 2^n zu N gehört. (vgl. S. 206, Zeile 23–24). Die Schwierigkeit,
4 die hierin liegt, hat Volpin natürlich auch bemerkt. | So sagt er ja auf S.
221, dass das ultra-intuitionistische Programm vor allem die Rechtferti-
gung des Postulates Trad erfordere. Mir scheint freilich, dass die gleiche
Schwierigkeit auch schon in der Forderung der Existenz des Nachfolgers
für jedes Element einer "natürlichen Folge" (vgl. S. 203, 3. Absatz) besteht. Es müsste wohl eine Art der Unschärfe, analog der Brouwer'schen
Unschärfe bezüglich des tertium non datur, eingeführt werden. Wie sich
das theoretisch durchführen lassen soll, ist schwer zu sehen, wenngleich
ja faktisch solche Unschärfen in der Abgrenzung konkreter Folgen (z. B.
derjenigen der akustisch wahrnehmbaren Tonstufen) bestehen.

Die Abhandlung über Unendlichkeitsaxiome von Takeuti[f] habe ich
etwas angesehen, aber bis zu einem wirklichen Verstehen bin ich nicht
vorgedrungen. Takeuti könnte es hier sicherlich dem Leser viel leichter
machen; anscheinend bereitet es ihm ein gewisses Vergnügen, seine Leit-
gedanken zu verstecken. Es ist ihm auch wohl nicht so sehr daran gele-
gen, seinen Systemen einen appeal zu verleihen; vor allem kommt es ihm
auf die deduktive Stärke an. Was übrigens die gebundenen Klassenvariablen betrifft, so werden in seinem Klassenbildungs-Schema (Group B′,
S. 222) gebundene Klassenvariablen bei den Klassenbestimmenden Prä-
dikaten $\mathfrak{A}(c)$ doch nicht ausgeschlossen. Er braucht sie allerdings bei
seinen Klassenbildungen fast nirgends, nur einmal bei der Definition von
$\mathfrak{S}(A, \mathfrak{T}, \alpha)$ (auf S. 231).[1]

[1][Handwritten vertically in the left margin of page 4:] Die Abhandlung von
Kondô "Sur les ensembles nommables ..." (Jap. Jour. Math. 28, 1958)[g] hätte ich
für das Journal of Symb. Logic besprechen sollen. Ich setzte mehrmals zu einer
eingehenden Lektüre an, doch kam ich nicht so bald zu einem befriedigenden Überblick,
und die Lektüre ⟨der sehr langen Abh.⟩ wurde immer wieder durch Anderweitiges
unterbrochen. Schliesslich wollte Herr Church nicht länger warten.

---

[f] *Takeuti 1961b.*

have the view that "the concept of natural number is nonsensical", any more than the proof-theorist needs to reject the concept of the uncountable as nonsensical. The foundational problem, however, is this: Every way of making deduction precise apparently introduces already the number-theoretic infinity. Vol'pin's account does not show, so far as I see, how this difficulty can be overcome. He introduces the postulate *Trad* (p. 205), which demands the existence of a "natural sequence" N with the property that the operation 2^n is everywhere defined in this sequence, and this presupposition is used, if I understand correctly, in the sense that if n belongs to N then 2^n also belongs to N (cf. p. 206, lines 23–24). The difficulty that lies therein Vol'pin has of course also noted. Thus on p. 221 he says that the ultra-intuitionistic program requires above all the justification of the postulate *Trad*. It seems to me, though, that the same difficulty exists already in the requirement of the existence of the successor for each element of a "natural sequence" (cf. p. 203, chapter 3). A kind of unsharpness, analogous to the Brouwerian unsharpness with respect to tertium non datur, would probably have to be introduced. It is hard to see how that is to be carried out theoretically, even though in actual fact such unsharpness exists when differentiating concrete sequences (e.g., those of acoustically perceptible degrees).

I have looked somewhat at the paper on axioms of infinity by Takeuti,[f] but I have not advanced to a real understanding. Takeuti could surely make things much easier for the reader here; apparently it gives him a certain pleasure to hide his central ideas. He is also probably not very concerned about giving his systems appeal; for him it is, above all, a question of [their] deductive strength. Incidentally, as concerns the bound class variables, in his class-formation schema (Group B′, p. 222) bound class variables are not excluded in the class-determining predicates $\mathfrak{A}(c)$. He almost never needs them, though, in his formations of classes— only once, in the definition of $\mathfrak{S}(A, \mathfrak{T}, \alpha)$ (on p. 231).[1]

---

[1] [Handwritten vertically in the left margin of page 4:] I was supposed to review Kondô's paper "Sur les ensembles nommables ..." (Jap. Jour. Math. 28, 1958)[g] for the *Journal of Symbolic Logic*. I started out to read it in detail several times, but I didn't arrive very soon at a satisfactory overview, and the reading of the very long paper was interrupted again and again by other things. Finally Mr. Church didn't want to wait any longer.

---

[g] *Kondô 1958.*

Bezüglich des Systems am Ende meiner Arbeit über die Unendlichkeitsschemata (S. 47–48)[h] hat vor einigen Monaten Herr Myhill die
Bemerkung gemacht, dass die Formel $a = b \rightarrow \sigma(a) = \sigma(b)$ ableitbar,
also als Axiom entbehrlich ist. Nun bleiben nur noch die beiden formalen
Axiome $a = b \rightarrow (a \,\epsilon\, c \rightarrow b \,\epsilon\, c)$ (bei Definition der Gleichheit) und

$$a \,\epsilon\, c \rightarrow \sigma(c) \,\epsilon\, c \,\&\, a \,\notin\, \sigma(c)$$

(was ja für die elementaren Teile der Mengenlehre nicht gebraucht wird),
und die beiden Schemata

$$c \,\epsilon\, \{x | A(x)\} \leftrightarrow (Ez)(c \,\epsilon\, z) \,\&\, A(c),$$
$$A \rightarrow (Ey)(\text{Strans}(y) \,\&\, (Ez)(y \,\epsilon\, z) \,\&\, \text{Rel}(y, A)),$$

(mit den Festsetzungen für Rel).

5 | Zur Beschäftigung mit der Neuauflage der "Grundlagen der Mathe-
matik"[i] bin ich immer noch nicht gekommen (ausser dass ich mir eine
Reihe von kleineren Aenderungs- und Einschaltungs-Erfordernissen vermerkt habe). Die Monate waren mit Verschiedenem ausgefüllt. Im Frühjahr war ich mit einem Beitrag für die Popper-Festschrift[j] beschäftigt,
im früheren Sommer mit ein paar Supplementen für die 9. Auflage von
Hilbert's "Grundlagen der Geometrie".[k] Im Anfang Juni nahm ich an
einer Tagung in Clermont-Ferrant teil und im frühen September an einem   d
Kolloquium der Akademie für Philosophie der Wissenschaften in Brüssel.
Ich war auch unter den Referenten für die Habilitationsschrift von Herrn
Dr. Müller.[l] A propos: Seine Probevorlesung (in Heidelberg) steht jetzt
in etwa einem Monat bevor. Vor kurzem hatte ich noch eine aussagenlo-
gische Arbeit von H. Leblanc[m] für das Journal zu besprechen.[n] Nun hoffe
ich, dass ich endlich bald mich den "Grundl. d. Math." werde zuwenden
können. Insbesondere will ich mich dafür auch mit den Arbeiten von
Herrn Tait befassen, die er mir zusandte. Auch die Pariser Vorlesungen
von Herrn Kreisel, deren Ausarbeitung ich von ihm empfing, will ich mir
des Näheren vornehmen.

Von dem Kolloquium in Brüssel wollte ich Ihnen noch sagen: Es wurde allgemein bedauert, dass Sie die Annahme Ihrer Wahl zum Mitglied
der Akademie (die ja schon vor längerer Zeit erfolgte) bisher nicht kundgegeben haben. Prof. Fréchet bat mich speziell, Ihnen zum Ausdruck
zu bringen, dass es lebhaft begrüsst werden würde, wenn Sie die An-

---

[h] *Bernays 1961.*
[i] *Hilbert and Bernays 1968* and *1970.*
[j] *Bernays 1964.*
[k] *Hilbert 1962.*

With respect to the system at the end of my paper on the infinity schemata (pp. 47–48),[h] a few months ago Mr. Myhill made the remark that the formula $a = b \to \sigma(a) = \sigma(b)$ is derivable, and so is dispensable as an axiom. Now only the two formal axioms $a = b \to (a \, \epsilon \, c \to b \, \epsilon \, c)$ (in the definition of equality) and

$$a \, \epsilon \, c \to \sigma(c) \, \epsilon \, c \,\&\, a \notin \sigma(c)$$

(which is not needed for the elementary parts of set theory), and the two schemata

$$c \, \epsilon \, \{x | A(x)\} \leftrightarrow (Ez)(c \, \epsilon \, z) \,\&\, A(c),$$
$$A \to (Ey)(\text{Strans}(y) \,\&\, (Ez)(y \, \epsilon \, z) \,\&\, \text{Rel}(y, A)),$$
(with the specifications of Rel)

remain.

I've still not begun with the work for the new edition of the *Grundlagen der Mathematik*[i] (except that I've noted a series of lesser changes and insertions that are required). The months were filled up with various things. In the spring I was busy with a contribution for the Popper *Festschrift*,[j] in the early summer with a few supplements for the 9th edition of Hilbert's *Grundlagen der Geometrie*.[k] At the beginning of June I took part in a conference at Clermont-Ferrand and in early September in a colloquium of the Akademie für Philosophie der Wissenschaften in Brussels. I was also among the referees for the *Habilitationsschrift* of Dr. Müller.[l] A propos: His trial lecture (in Heidelberg) is now upcoming in about a month. A short while ago I also had to review a paper on propositional logic by H. Leblanc[m] for the Journal.[n] I hope now that I will finally be able to turn soon to the *Grundlagen der Mathematik*. In particular, for that I also want to study the papers of Mr. Tait that he sent me. I also want to read in greater detail Mr. Kreisel's Paris lectures, whose written version I received from him.

Re the colloquium in Brussels I wanted to tell you: it was generally regretted that you have not as yet declared your acceptance of your election as a member of the academy (which took place a long time ago). Professor Fréchet especially asked me to express to you that it would be vigorously welcomed if you wanted to declare your acceptance of your

[l] *Müller 1962.*
[m] *Leblanc 1962.*
[n] *Bernays 1962.*

nahme Ihrer Wahl zum Mitglied erklären wollten. Es befinden sich unter
den Mitgliedern der Akademie etliche Ihrer Kollegen, so Church, Curry,
Kleene, Quine.[2] Die Annahme der Mitgliedschaft würde für Sie keine
Belastungen zur Folge haben. Es würde hierfür auch genügen, wenn Sie
mich bei Gelegenheit davon verständigen, da mir neuerdings das
Präsidium d. Akademie zuerteilt wurde.— Den Sonderdruck von meinem
Carnap-Aufsatz haben Sie gewiss erhalten. Ich schicke Ihnen auch noch
einen solchen von meinem kleinen Beitrag zu dem Stanford-Kongress.[o]

Hoffentlich ist Ihr gesundheitliches Ergehen befriedigend. Mit herzlichen Grüssen, auch an Ihre Gemahlin,

Ihr Paul Bernays

[2][Handwritten vertically in the left margin:] Auch wurde Einstein noch zu seinen
Lebzeiten Mitglied.

[o]*Bernays 1962a.*

46. Bernays to Gödel

Zürich 2, 31. XII. 1962
Bodmerstr. 11

Lieber Herr Gödel!

Ihnen und Ihrer Frau sende ich ⟨zum neuen Jahr⟩ meine besten Wün-
sche!
Hoffentlich konnten Sie im verflossenen Jahr mit Ihrem gesundheit-
lichen Ergehen zufrieden sein.—
Haben Sie sich schon mit den im Spätsommer erschienenen Vorlesun-
gen von Nelson[a] des Näheren befasst?—
Herr Tucker hat wohl auch Ihnen seinen Artikel "Constructivity and
grammar"[b] zugesandt. Im ganzen scheint mir, dass er es sich mit | der
Erledigung der Paradoxien doch zu leicht macht.—

Mit herzlichen Grüssen

Ihr Paul Bernays

[a]*Nelson 1962.*

election as a member. Among the members of the academy are quite a few of your colleagues, such as Church, Curry, Kleene and Quine.[2] The acceptance of membership would entail no burden for you. ⟦The following sentence was added in pen:⟧ For that it would also suffice if you were to inform me at some point, since the presidency of the academy was recently awarded to me.

You have surely received the offprint of my essay on Carnap. I am also going to send you one of my short contribution to the Stanford congress.°

Hopefully your state of health is satisfactory. With best regards, also to your wife,

Yours truly,

Paul Bernays

[2]⟦Handwritten vertically in the left margin:⟧ Einstein became a member still during his lifetime.

---

46. Bernays to Gödel

Zurich 2, 31 December 1962
Bodmerstrasse 11

Dear Mr. Gödel,

I send my best wishes to you and your wife for the New Year!

Hopefully you were able to be satisfied with your state of health during the past year.

Have you already studied in detail the lectures of Nelson[a] that appeared in late summer?

Mr. Tucker probably also sent you his article "Constructivity and grammar".[b] On the whole it seems to me that he dispenses with the paradoxes too easily.

    With best regards,

Yours truly,

Paul Bernays

---

[b] *Tucker 1963.*

47. Gödel to Bernays

Princeton, 9./I. 1963.

Lieber Herr Bernays!

Besten Dank für die übermittelten Neujahrswünsche, die ich auf das herzlichste erwidere. Vielen Dank auch für Ihren freundlichen Brief vom 12./X. u. die beiden Sonderdrucke. In Ihrem Aufsatz "Zur Rolle der Sprache"[a] gehen Sie leider auf die Grundauffassung der Neo-Positivisten über das Wesen der Mathematik (nämlich dass diese gänzlich inhaltslos sei, dass es also keine mathematischen Inhalte gebe) nicht ein. Ich halte diese Auffassung (an der, soviel ich weiss, auch heute noch festgehalten wird) zwar für vollkommen falsch, aber für interessant insofern, als sich gerade bei dem Versuch sie durchzuführen herausstellt, worin die mathematischen Inhalte bestehen. Interessant war für mich, dass Sie auf p[[.]]199 von den "neueren abstrakten Disziplinen der Mathematik" als etwas ausserhalb der Mengenlehre liegendes sprechen. Ich vermute, dass Sie damit auf den Begriff der Kategorie u. ~~seine~~ ⟨die⟩ Anwendbarkeit von Kategorien auf sich selbst anspielen. Es scheint mir aber, dass alles dies in einer Mengenlehre mit endlich iteriertem Klassenbegriff enthal-
2   ten ist, wobei sich die Reflexivität durch eine "typical am|biguity" der Aussagen von selbst ergibt. Ist das nicht auch Ihre Meinung? Ich habe übrigens gehört, dass irgend jemand die Axiome der Mengenlehre mit Hilfe des Begriffs der Kategorie formuliert hat u. dass dies vielleicht sogar publiziert wurde. Wenn Ihnen darüber etwas bekannt ist, wäre ich Ihnen für eine diesbezügliche Mitteilung sehr dankbar.—Die transcendentale Deduktion des Sittengesetzes bei Nelson[b] habe ich mir angesehen. Es kommt mir aber unwahrscheinlich vor, dass man die ganze Moral u. Rechtswissenschaft aus der einzigen Annahme der Invarianz gegen Permutation der moralischen ⋅Subjekte ableiten kann. Die Moral, welche Verbrecher gegeneinander üben, ist doch auch von dieser Art. Jedenfalls ist es interessant, dass Nelson transcendentale Deduktionen auf psychologische aufbaut (so verstehe ich die Sache), während doch bei Kant diese beiden Dinge getrennt sind.—Das Nelsonsche Buch über die Geschichte der Philosophie seit Kant[c] finde ich *sehr* interessant u. anregend, insbesondere durch die ausserordentlich klare Darstellung. Andererseits
3   scheint es mir, dass er der idealistischen Philosophie in seiner Dar|stellung (rein historisch genommen) durchaus nicht gerecht wird. Das Wesentliche bei Hegel, z. B., ist doch der dialektische Prozess, über den

---

[a] *Bernays 1961a.*

47. Gödel to Bernays

Princeton, 9 January 1963

Dear Mr. Bernays,

Thanks very much for the New Year's wishes you conveyed, which I reciprocate most heartily. Many thanks too for your cordial letter of 12 October and the two offprints. In your essay "Zur Rolle der Sprache"[a] you unfortunately do not address the basic conception of the neo-positivists concerning the nature of mathematics (namely that it is entirely without content, that there is consequently no mathematical content). To be sure, I hold that conception (which, so far as I know, is still maintained even today) to be completely false, but interesting insofar as it emerges, when attempting to carry it out, what the mathematical content is. I found it interesting that you speak on p. 199 of the "newer abstract disciplines of mathematics" as something lying outside of set theory. I conjecture that you are thereby alluding to the concept of category and to the self-applicability of categories. But it seems to me that all of this is contained within a set theory with a finitely iterated notion of class, where reflexivity results automatically through a "typical ambiguity" of statements. Isn't that also your opinion? I've heard, by the way, that someone has formulated the axioms of set theory with the aid of the concept of category and that this has perhaps even been published. If you know something about it, I would be very grateful to you for relevant information. I have looked at Nelson's transcendental deduction of the moral law.[b] But it appears improbable to me that all of morality and jurisprudence can be derived from the sole assumption of the invariance under permutation of the moral subjects. The morality that criminals practice toward each other is in fact also of that kind. In any case, it is interesting that (as I understand the matter) Nelson bases transcendental deductions on psychological ones, whereas in Kant those two things are separated.—I find Nelson's book on the history of philosophy since Kant[c] *very* interesting and stimulating, especially by virtue of its extraordinarily clear presentation. On the other hand, it seems to me that in his presentation (taken purely historically) he does not at all do justice to idealistic philosophy. What is essential, e.g., in Hegel is the dialectical

---

[b] *Nelson 1908*, p. 66.
[c] *Nelson 1962*.

nicht viel u. nur im allgemeinen gesprochen wird. Was bei Hegel näher
ausgeführt wird, ist bloss das was ~~dieser~~ ⟨er⟩ mit Aristoteles oder den mittelalterlichen Realisten gemein hat.—

Das Verhältnis Volpins zur klassischen Mathematik scheint mir ein
ganz anderes zu sein als das des Finitismus zu ihr. Denn Volpin nimmt
doch über den Begriff "erreichbar" Axiome an, die nach der klassischen
Mathematik für *jeden* (auch unpräzisen) Begriff falsch sind; es sei denn
dass man den Begriff der Wahrheit aufgibt u. nur von Graden der Wahrheit spricht. Das ~~dürfte~~ scheint mir aber mit der Volpinschen Idee eines
Widerspruchsfreiheitsbeweises der kl. Math inkompatibel zu sein. Ich
bin gegenwärtig leider sehr beschäftigt u. werde Ihnen daher über die
anderen in Ihrem Brief berührten Fragen erst in meinem nächsten Brief
schreiben.

Mit herzlichen Grüssen Ihr

Kurt Gödel

4 | Herzliche Grüsse u. beste Wünsche auch von meiner Frau.

P.S. Mit Rücksicht darauf, dass Volpin für den Widerspruchsfreiheitsbew.[1] seine Axiome über Erreichbarkeit doch nur in Verbindung mit der
Zahlentheorie (oder Kombinatorik) verwenden will, halte ich Folgendes
für äusserst wahrscheinlich:

Wenn man den Sinn seiner Axiome irgendwie (z. B. auch in der von
Ihnen angedeuteten Weise) so abschwächt, dass sie mit der klass. Math.
kompatibel werden, so wird die Existenz (oder Widerspruchsfreiheit mit
der Zahlentheorie) eines solchen Begriffs von Erreichbarkeit in der klass.
Math. beweisbar werden, was einen Widerspruchsfreiheitsbew. für diese
unmöglich macht.

[1] der klass. Math.

48. Bernays to Gödel

Zürich 2, den 23. Februar 1963.
Bodmerstr. 11

Lieber Herr Gödel,

Haben Sie vielen Dank für Ihren freundlichen Brief vom Januar! Wenngleich Sie eine Fortsetzung des Briefes ins Auge fassten, so möchte ich

process, of which not much is said, and ⟦that⟧ only in general terms. What of Hegel is set forth in greater detail is merely what he has in common with Aristotle or the medieval realists.

The relation of Vol'pin to classical mathematics seems to me to be quite different from ⟦the relation⟧ of finitism to it. For Vol'pin assumes axioms about the concept "accessible" that according to classical mathematics are false for *every* (even imprecise) concept, unless one gives up the concept of truth and speaks only of degress of truth. But that seems to me to be incompatible with Vol'pin's idea of a consistency proof for classical mathematics. Unfortunately I am very busy at present and will therefore write you only in my next letter about the other questions mentioned in your letter.

With best regards,

Yours truly,

Kurt Gödel

Cordial regards and best wishes also from my wife.

P.S. With regard to the fact that for the consistency proof[1] Vol'pin nonetheless wants to use his axioms about accessibility only in connection with number theory (or combinatorics), I hold the following to be highly probable:

If one somehow weakens the meaning of his axioms (e.g., also in the way you indicated) so that they become compatible with classical mathematics, then the existence of such a concept of accessibility (or its consistency with number theory) becomes provable in classical mathematics, which makes a consistency proof for it impossible.

[1] of classical mathematics

48. Bernays to Gödel

Zurich 2, 23 February 1963
Bodmerstrasse 11

Dear Mr. Gödel,

Many thanks for your friendly letter of January. Even though you intend to continue the letter, I would still like to reply to you now, since

Ihnen immerhin schon jetzt antworten, da Sie doch auf ziemlich vieles
schon zu sprechen gekommen sind.

Vielleicht beginne ich am besten mit der Nelson'schen Philosophie.
Es erscheint Ihnen als unwahrscheinlich, "dass man die ganze Moral und
Rechtswissenschaft aus der einzigen Annahme der Invarianz gegen Permu-
tationen der moralischen Subjekte ableiten" könne. Das ist aber auch
nach Nelson nicht der Fall. Einerseits werden ja für die Beurteilung einer
Handlung gemäss dem Prinzip der moralischen Wertung, nach Einbezie-
hung der Anliegen ("Interessen") aller von der Handlung betroffenen Per-
sonen, diese Anliegen nicht nach dem Masstab ihrer "zufälligen Stärke"
bemessen, "sondern nach der Stärke[,] die sie bei hinreichender Einsicht in den Wert ihres Gegenstandes hätten". (Ich zitiere hier aus der
Abhandlung "Die kritische Ethik bei Kant, Schiller und Fries".)[a] In der
Beurteilung gemäss dem Moral-Prinzip kommt also implizite eine ausser-
moralische Wertung zur Geltung. Andrerseits spielen bei der Entwicklung der Rechtslehre die "Subsumptionsformeln" eine wesentliche Rolle,
welche allgemeine Bedingungen formulieren, "unter denen vernünftige
Gemeinschaft in der Natur möglich ist". (Sie finden darüber kurz etwas
in den "Fortschritten ..."[b] S. 705–706.)

Uebrigens das Argument, das Sie anführen, dass doch auch die Moral,
welche Verbrecher gegeneinander üben, von solcher Art ist, scheint mir
eher zugunsten des Prinzips zu sprechen, da es doch zeigt, dass selbst
unter Bedingungen, wo keinerlei zusätzliche Stützen des Ethischen durch
den Lebens-Rahmen gegeben sind, dennoch die Art der moralischen Bewertung sich spontan einstellt.

Sie bemerken, dass bei Nelson die Deduktion des Inhalts des Moral-
prinzips auf einer psychologischen Deduktion des Urteils aufgebaut ist,
während das bei Kant nicht der Fall ist. Für Kant fällt eine Deduktion
2 des Ursprungs deshalb weg, weil er schlechtweg behauptet, dass | "das
moralische Gesetz gleichsam als ein Faktum der reinen Vernunft dessen
wir uns a priori bewusst sind, und welches apodiktisch gewiss ist, gegeben"[1] sei. Nach seiner Methodik könnte eine Deduktion des Moral-
prinzips wohl nur in etwas Analogem zu dem in der Kritik der reinen
Vernunft verwendeten Prinzip der Möglichkeit der Erfahrung bestehen;
da aber so etwas hier nicht in Betracht kommt, so hält Kant anscheinend

---

[1] [Handwritten vertically in the left margin of page 2:] Krit. d. prakt. Vernunft,[c]
Elementarlehre, Erstes Buch, Erstes Hauptst. I. Von d. Deduktion ... Hier folgt auch
die andere zitierte Stelle.

---

[a] *Nelson 1914.*
[b] *Nelson 1962.*

you have already broached rather many [topics].

Perhaps it is best that I begin with Nelson's philosophy. It seems improbable to you "that all of morality and jurisprudence can be derived from the sole assumption of the invariance under permutation of the moral subjects". But according to Nelson that is not the case. On the one hand, for the judgment of an act in accordance with the principle of moral valuation, after the inclusion of the concerns ("interests") of all parties affected by the act, those concerns are measured not according to the standard of their "incidental strength", "but rather according to the strength that they would have given sufficient insight into the worth of their objects". (I'm quoting here from the article "Die kritische Ethik bei Kant, Schiller und Fries".[a]) In judging in accordance with the principle of morality an extramoral valuation thus implicitly comes into play. On the other hand, in the development of jurisprudence the "subsumption formulas" play an essential role: they formulate general conditions "under which rational community is possible in nature". (You find something brief about that in the "Fortschritten ...",[b] pp. 705-706.)

Moreover the argument that you put forward that the morality which criminals practice among one another is of such a kind seems to me rather to speak in favor of the principle, since in fact it shows that even under conditions in which no additional support for morality is given by the social setting that kind of moral evaluation arises spontaneously. You note that for Nelson the deduction of the content of the principle of morality is based on a psychological deduction of judgment, whereas that is not the case for Kant. For Kant a deduction of the origin is omitted, because he simply claims that "the moral law is given, as it were, as a fact of pure reason of which we are conscious a priori and which is apodictically certain".[1] According to his methodology a deduction of the principle of morality could at best consist in something analogous to the principle of the possibility of experience employed in the Critique of Pure Reason; since, however, such a thing does not come under consideration here, Kant apparently holds that a proper deduction of the moral law is

---

[1] [Handwritten vertically in the left margin of page 2:] Critique of Practical Reason,[c] Doctrine of elements, first book, first article, I. "Of the deduction ...". The other passages cited also follow here.

---

[c] *Kant 1788*; *Kant 1902–*, vol. V, p. 47.

eine eigentliche Deduktion des moralischen Gesetzes überhaupt nicht für
möglich. So erklärt er auch (in der Kritik der praktischen Vernunft, im
Abschnitt "Von der Deduktion der Grundsätze der reinen praktischen
Vernunft"), es könne "die objektive Realität des moralischen Gesetzes
durch keine Deduktion, durch keine Anstrengung der theoretischen, speku-
lativen oder empirisch unterstützten Vernunft bewiesen" werden.

Bezüglich des Nelson'schen Buches über die Geschichte der Philosophie
seit Kant vermerken Sie, neben den Vorzügen dieses Buches, als einen
Mangel, dass man über den Inhalt der idealistischen Philosophien nur
sehr im Allgemeinen orientiert wird und, dass insbesondere der nähere
Verlauf des dialektischen Prozesses bei Hegel nicht des Genaueren zur
Sprache kommt. Nelson würde dieses nicht bestreiten, es jedoch vom
Standpunkt seiner Aufgabestellung nicht als einen Mangel anerkennen.
Wie er seine Aufgabe auffasst, setzt er in den "Vorbetrachtungen" (S.
27–32) explizite und ausführlich auseinander. Insbesondere sagt er ja da
(S. 29 unten) "Ich werde auf solche Lehren, die sich schon in den Grundlagen als verfehlt erweisen, nicht weiter eingehen, als nötig ist, um uns
von dieser Verfehltheit ihrer Grundlagen zu überzeugen. Es lohnt nicht,
da um Konsequenzen zu rechten, wo die Verfehltheit der Grundlage
schon feststeht; denn wir wissen im Voraus, dass dort nur durch einen
glücklichen Zufall noch etwas Richtiges unterlaufen könnte."

Diese⟨s⟩ Verfahren hat freilich seine Nachteile, da ja bei etlichen Philosophen das offizielle Programm gar nicht das Beste ist und eventuell
die fruchtbare Verwertung ihrer Gedanken ein Abgehen von jenem Programm erfordert. Zu der Hegel'schen Philosophie kann ich freilich auch
im Sinne einer solchen liberaleren Methodik kein Verhältnis gewinnen.
Wohl kann ich den Gedanken des objektiven Geistes, und was sich daran
knüpft, akzeptieren. Man könnte ihn, wie es Herr Lautman tat,[d] für die
Philosophie der Mathematik zu verwerten suchen.—Aber eine generelle
Dynamik der Wechselwirkung zwischen abstrakten Inhalten, in welcher
die Aeusserung mystischen Erlebens, die Begriffswelt der Wissenschaft
und soziologischen | Betrachtung in ein gemeinsames Erkennen zusam-
mengefasst sein sollen, das ist doch etwas recht Horrendes, das einem da
zugemutet wird.

Es scheinen immerhin hinsichtlich der Aeusserung mystischer Ein-
stellungen zwischen etlichen Menschen gewisse Gemeinsamkeiten zu bestehen. So findet sich in dem 1962 posthum herausgegebenen Buche "Erinnerungen, Träume, Gedanken von C. G. Jung"[e] im Anhang der Ab-
druck einer Broschüre, die "Septem Sermones ad Mortuos" (die Jung seinerzeit nur im Privatdruck erscheinen liess). Hier ist Manches, was

---

[d] *Lautman 1938.*

not possible at all. So too he declared (in the Critique of Practical Reason, in the chapter "Of the deduction of the basic principles of pure practical reason") "the objective reality of the moral law cannot be proved by means of deduction, by an effort of the theoretical, speculative or empirically supported reason".

With regard to Nelson's book on the history of philosophy since Kant, in addition to the merits of that book, you note as a defect that [from it] one becomes oriented only in a very general way toward the content of idealistic philosophy and that, in particular, the more detailed course of the dialectical process in Hegel does not come up for more precise discussion. Nelson would not dispute that; however, from the standpoint of what he set out to do it would not be acknowledged as a defect. How he conceives of his task, he discusses explicitly in the "preliminary considerations" (pp. 27–32). In particular, he says there (bottom of p. 29), "I shall not enter further than necessary into theories that have already shown themselves to be misguided in the foundations, i.e., than necessary in order to convince us of this misdirection in their foundations. It is not worthwhile to argue about consequences where the misdirection of the foundations is already established; for we know in advance that only through a lucky accident could something correct still come of it."

This procedure has its shortcomings, I admit, since of course in the case of some philosophers the official program is not at all the best, and perhaps the fruitful utilization of their thoughts requires a departure from that program. To Hegel's philosophy I can, I admit, attain no relation, even in the sense of a more liberal methodology. I can probably accept the ideas of the objective spirit and what is tied in with that. One could, as Mr. Lautman has done,[d] attempt to use it [the objective spirit] for the philosophy of mathematics. But a general dynamics of the interplay between abstract contents, in which the expression of mystical experience, the conceptual world of science and sociological consideration[s] are supposed to be joined into a common act of cognition, that, indeed, is something quite horrendous that is being demanded there.

Nevertheless, with regard to the expression of mystical attitudes there do seem to be certain commonalities among various people. Thus in the appendix in the book *Erinnerungen, Träume, Gedanken von C. G. Jung*,[e] published posthumously in 1962, one finds the reproduction of a brochure, the "Septem Sermones ad Mortuos" (which in his own time Jung allowed to appear only in a private printing). Here there is much

---

[e] *Jung 1962.*

an die Hegel'sche Denk- und Sprechweise erinnert, obwohl es kaum den
Anschein hat, dass Jung von Hegel stärker beeinflusst wurde.

Um auf Nelson zurückzukommen, so meinten Sie, dass er über Hegel
nur solches näher ausführe, was dieser mit Aristoteles oder den mittelalterlichen Realisten gemeinsam hat. Das kann aber doch insofern schwerlich ganz zutreffen, als Nelson (in einem ausführlichen Abschnitt, S. 509–
537) die "Aristotelisch-Kantische und die Neuplatonische Fichte'sche
Logik"[f] einander gegenüberstellt, wobei er natürlich mit der Fichte'schen
auch die Hegel'sche Logik meint.—

Was meinen Carnap-Aufsatz[g] betrifft, so habe ich mich bei diesem be-
müht, möglichst wenig das Gegensätzliche zwischen meinen Ansichten
und denen Carnaps hervorzukehren, insbesondere inbezug auf solche
Punkte, über die ich mich anderwärts schon äusserte. Uebrigens kommt
doch darin das Thema des Inhalts (bezw. der behaupteten Inhaltlosigkeit) der Mathematik wenigstens indirekt, in der Erörterung über
Sinngleichheit (S. 189–192), zur Sprache, worin zur Behebung der
Schwierigkeiten die Verwendung des Begriffes der Vorgängigkeit emp-
fohlen wird. Für die Sinnes-Analyse mathematischer Sätze können natürlich die mathematischen Beziehungen nicht als vorgängig betrachtet
werden.

Sehr interessiert hat mich Ihre Bemerkung, dass sich gerade bei dem
Versuch, die Auffassung von der Inhaltlosigkeit der Mathematik durchzuführen, herausstellt, worin die mathematischen Inhalte bestehen. Freilich obwohl diese Feststellung sehr nach meinem Sinne ist, konnte ich mir
doch nicht ergänzen, wie Sie des Genaueren dieses Durchführen und dieses Sich-Herausstellen meinen.

4 | Dass ich bei der Aufzählung der Gebiete der Mathematik, in denen die
klassischen Methoden zur Verwendung kommen, die "neueren abstrakten Disziplinen" neben Analysis und Mengenlehre genannt habe, war im
Sinne einer Unterscheidung von kategorischer und hypothetischer Mathematik gedacht. Abstrakte axiomatische Topologie und Algebra können
ja in solcher Weise betrieben werden, dass man zwar Begriffe wie die der
natürlichen Zahl und der reellen Zahl sowie mengentheoretische Begriffe
und Sätze benutzt, dagegen eine strikte Einordnung in die Mengenlehre
nicht vornimmt.

Was die mit den "Kategorien" verbundenen Schwierigkeiten angeht,
von denen Herr Mac Lane in seinem Warschauer Vortrag[h] sprach, so
habe ich von den Erfordernissen um die es sich da handelt, nur eine

---

[f] *Nelson 1962*, Zweiter Teil, 4. Kapitel, VII; reprinted in *Nelson 1970–74*, vol. 7, p. 509ff.

[g] *Bernays 1961a*.

that is reminiscent of Hegel's way of thinking and speaking, although it hardly appears that Jung was more strongly influenced by Hegel.

To come back to Nelson, you believe that he treats in detail about Hegel only those things that Hegel has in common with Aristotle or the medieval realists. But that can hardly be entirely true, insofar as Nelson (in a detailed chapter, pp. 509–537) contrasts with each other the "Aristotelian-Kantian and neoplatonist-Fichtean logic",[f] where of course by Fichtean he also means Hegelian logic.

As to my essay on Carnap, I have endeavored to emphasize in it as little as possible the points of disagreement between my views and those of Carnap, especially with regard to such points about which I have already expressed my views elsewhere. Incidentally, the issue of the content (or, respectively, of the claimed lack of content) of mathematics is mentioned there at least indirectly, in the argument about identity of sense[g] (pp. 189–192), where the use of the concept "Vorgängigkeit" [[that which is prior]] is recommended to obviate the difficulties. For the analysis of the sense of mathematical statements the mathematical relations cannot of course be considered as "vorgängig".

I found very interesting your remark that the mathematical content turns out to consist in the very attempt to carry out the conception of the lack of content of mathematics. Although that assessment is very much in conformity with my own sense [[of things]], I admit I still could not piece together how you think of the more precise details of that carrying-out and turning-out.

That, in the enumeration of the domains of mathematics in which the classical methods come to be used, I named the "newer abstract disciplines" in addition to analysis and set theory was conceived in the sense of a distinction between categorical and hypothetical mathematics. Abstract axiomatic topology and algebra can be pursued in such a way that one indeed uses concepts like that of natural number and real number as well as set-theoretical concepts and theorems, without undertaking a strict incorporation into set theory.

As to the difficulties associated with "categories", of which Mr. Mac Lane spoke in his Warsaw address,[h] I have only a rough idea of the requirements in question. All the same, it seems to me that the difficulties rest

[h] *Mac Lane 1961.*

ungefähre Vorstellung. Immerhin scheint mir, dass die Schwierigkeiten
mindestens zu einem Teil darauf beruhen, dass man ausschliesslich die
extensionale Charakterisierung der Kategorien im Auge hat. In analoger
Weise würde z. B., wenn wir die Zahl 5 durch die Klasse der fünfzähligen
Mengen charakterisieren, ein Hinderniss bestehen, eine Klasse mit der
Zahl 5 als Element zu bilden. Die Kategorien sind doch auch effektiv
durch Axiomensysteme gegeben, und die Mannigfaltigkeit der dabei in
Betracht kommenden Axiomensysteme lässt sich doch vermutlich als
eine Menge, gewiss aber als eine Klasse darstellen. Ihr Gedanke, dass
als Rahmen für die Theorie der Kategorien eine Mengenlehre mit endlich
iteriertem Klassenbegriff zulänglich ist, leuchtet mir sehr ein. Doch wenn
man schon die axiomatische Mengenlehre erweitert, so sollten doch auch
die anderweitigen neuen infinitistischen Untersuchungen in dem erweiterten Rahmen Platz finden. Ich bekam gerade kürzlich ein Manuskript-
Exemplar der Dissertation von ⟨William⟩ Hanf[i] zugesandt, worin der Autor in einem Einleitungsabschnitt die Meinung äussert, dass für seine Untersuchungen eine Mengenlehre mit endlich iteriertem Klassenbegriff als
axiomatischer Rahmen nicht ausreiche.

Dass jemand neuerdings die Axiome der Mengenlehre mit Hilfe des
Begriffs der Kategorie formuliert hat, habe ich zum erstenmal aus Ihrem
Briefe erfahren.

Bezüglich der Untersuchungen von Esenin-Vol'pin, sprechen Sie von
den Axiomen, die er über den Begriff "erreichbar" annimmt und die im
5 | Sinne der klassischen Mathematik falsch sind. In der Fassung seiner
Ueberlegungen,[j] die in dem Warschauer Kongressband publiziert ist,
finde ich aber gar nicht solche Axiome, die der klassischen Mathematik
direkt widersprechen, vielmehr nur die Ablehnung mancher geläufigen
Annahmen der klassischen Mathematik. Ferner ist ein Unterschied
gegenüber der sonstigen beweistheoretischen Betrachtung, dass gar *nicht
die Widerspruchsfreiheit schlechthin* bewiesen werden soll, sondern nur
dieses: dass ein Widerspruch sich erst bei einem Beweise von einer gewissen minimalen Länge—die dann nicht mehr als eine konkrete gilt—ein-
stellen könne. Ob ein Widerspruchsfreiheitsbeweis in diesem abgeschwäch-
tem Sinne auch durch Ihr Unableitbarkeitstheorem ausgeschlossen wird,
habe ich mir noch nicht hinlänglich überlegt.—

Im Dezember hat Dr. Gert Müller, der bis 1959 mein Assistent war, in
Heidelberg seine Probevorlesung für die Habilitation gehalten. Er wird
im Sommersemester seine Vorlesungen beginnen. Seine Habilitationsschrift[k] handelte von der "mehrfachen" Rekursion.—Sie haben vermutlich von ihm manche Besprechungen gelesen. Er hat, besonders für das

---

[i] *Hanf 1963.*

at least in part on the fact that it is exclusively the *extensional* characterization of categories that is being considered. In an analogous way, for example, if we characterize the number 5 by means of the class of five-element sets, there is an impediment to forming a class with the number 5 as [an] element. To be sure, categories are also effectively given by means of axiom systems, and the multiplicity of the axiom systems that have to be considered can presumably be represented as a set, and certainly as a class. Your idea that a set theory with a finitely iterated notion of class is suitable as a framework for the theory of categories seems very plausible to me. If axiomatic set theory is being extended in any event, then the other novel infinitistic investigations should also find a place in the extended framework. Just recently I received a manuscript of the dissertation submitted by William Hanf,[i] in which the author expresses the opinion in an introductory chapter that for his investigations a set theory with a finitely iterated notion of class does not suffice as an axiomatic framework.

That someone recently formulated the axioms of set theory with the help of the notion of category I learned for the first time from your letter.

With respect to the investigations of Esenin-Vol'pin, you speak of the axioms that he assumes about the concept "accessible" and that are false in the sense of classical mathematics. But in the version of his deliberations that is published in the Warsaw Congress volume[j] I find no axioms at all that directly contradict classical mathematics; rather, only the rejection of many familiar assumptions of classical mathematics. Furthermore, there is a distinction from the usual proof-theoretic consideration, that it is not *consistency* per se that is to be proved, but only this: that a contradiction can only arise with a proof of a certain minimal length, which then no longer is viewed as a concrete one. Whether a consistency proof in this weakened sense is also excluded by your underivability theorem I have not yet pondered sufficiently.

In December Dr. Gert Müller, who was my assistant up to 1959, delivered his trial lecture for the *Habilitation* in Heidelberg. He will begin his lectures in the summer semester. His *Habilitationsschrift*[k] deals with "multiple" recursion. You have probably read many reviews by him.

---

[j] *Esenin-Vol'pin 1961*.

[k] *Müller 1962*.

Zentralblatt, viele Besprechungen beigetragen, in denen sich seine Gründlichkeit und seine Beschlagenheit zeigt. Er ist sehr versiert in der Grundlagenforschung und auch philosophisch interessiert. In den letzten Jahren sind von ihm die Abhandlungen "Ueber die unendliche Induktion"[1] (in dem Warschauer Kongressband) und "Nicht-Standardmodelle der Zahlentheorie"[m] (Math. Zeitschrift 77) erschienen. Durch verschiedene Umstände hat sich seine Laufbahn verzögert. Es würde sehr wertvoll und erwünscht für ihn sein, wenn er für ein akademisches Jahr eine Einladung an das Princetoner Institute erlangen könnte. Darf ich Sie fragen, ob Sie wohl gewillt und in der Lage wären, ihm eine solche Einladung zu erwirken?—

Hoffentlich geht es Ihnen gesundheitlich befriedigend.—Wenn Sie wieder einmal mehr Zeit haben, so sehe ich der Fortsetzung Ihres Briefes vom 9.1., die Sie mir in Aussicht stellten, mit Vergnügen entgegen.

Mit herzlichen Grüssen an Sie und Ihre Gemahlin

Ihr Paul Bernays

---

[1] *Müller 1961.*

49. Bernays to Gödel

Zürich 2, 14. März 1963.
Bodmerstr. 11

Lieber Herr Gödel!

Sie werden sich ⟨gewiss⟩ wundern, schon wieder einen Brief von mir (nach demjenigen, den ich Ihnen kürzlich sandte) zu bekommen.

Dieses Mal liegt jedoch ein spezieller Anlass vor. Professor Myhill schrieb mir, dass er sich um eine Mitgliedschaft am Institute for Advanced Study für den Sommer oder den fall term dieses Jahres bewerbe und dass er mich als eine der Referenzen angegeben hat.

Es möchte mir scheinen, dass John Myhill hinsichtlich seiner wissenschaftlichen Leistungen kaum einer Empfehlung bei Ihnen bedarf. Sie wissen sicherlich, dass er ein sehr produktiver Forscher ist und dass er,

Especially for *Zentralblatt*, he has contributed many reviews in which his thoroughness and his expertise are displayed. He is very well versed in foundational research and also has philosophical interests. In the last few years he has published "Uber die unendliche Induktion"[1] (in the Warsaw Congress volume) and "Nicht-Standardmodelle der Zahlentheorie"[m] (*Mathematische Zeitschrift 77*. As a result of various circumstances his career has been delayed. It would be very valuable and desirable for him if he could obtain an invitation to the Princeton Institute for one academic year. May I ask you whether you would be willing and in a position to procure him such an invitation?

Hopefully in matters of health things are going satisfactorily for you. When you once again have more time I shall look forward to the continuation of your letter of September 1, which you held out to me in prospect.

With cordial regards to you and your wife,

Yours truly,

Paul Bernays

---

[m] *Müller 1961a.*

49. Bernays to Gödel

Zurich 2, 14 March 1963
Bodmerstrasse 11

Dear Mr. Gödel,

You will certainly be surprised to receive another letter from me already (after the one that I recently sent you).

This time, however, there is a special reason. Professor Myhill wrote me that he has applied for a membership at the Institute for Advanced Study for the summer or fall term of this year and that he named me as one of the references.

It would seem to me that with regard to his scientific accomplishments John Myhill hardly needs to be commended to you. You certainly know that he is a very productive researcher and that, in connection on the

anknüpfend einerseits an die Methoden von Quine, andrerseits an diejenigen von Post und ⟨von⟩ Kleene, mannigfaches sehr Interessante zu den Fragen der Rahmentheorien, zur Behandlung der Metamathematik und zur Untersuchung der rekursiv abzählbaren Mengen beigetragen hat.

Auch über seine gegenwärtigen Forschungsgebiete werden Sie ja durch das Arbeitsprogramm, das Herr Myhill seiner Bewerbung beifügt, orientiert.

Freilich, das offizielle Arbeitsprogramm ist für die Fruchtbarkeit eines Aufenthaltes am Institute for │ Adv. Study gewiss nicht allein massgeblich; diese besteht ja zumeist in der Möglichkeiten der Beeinflussung jenes Programmes durch Gedankenaustausch.

Für derartige fruchtbare Beeinflussung gibt nun die Arbeitsweise von Herrn Myhill besonders gute Vorbedingungen, da sein Forschen im Geiste der Offenheit für verschiedenartige Aspekte erfolgt und da er andrerseits, aufgrund seiner gewandten und gründlichen Beherrschung grundlagentheoretischer Techniken, empfangene Anregungen fruchtbar zu verfolgen versteht.—

Persönlich lernte ich Herrn Myhill—nach einem nur flüchtigen Zusammensein auf dem Amsterdamer Kongress—anlässlich eines Vortrags kennen, den er 1961 im philosophical Departm. der University of Pennsylvania in Philadelphia hielt. Er hat ein sehr freundliches und umgängliches Wesen.

So möchte ich ⟨für⟩ die Bewerbung von Herrn Myhill eine zustimmende Entscheidung mit Wärme befürworten, soweit mir hierüber eine Meinung zusteht.

Mit herzlichen Grüssen

Ihr Paul Bernays

50. Gödel to Bernays

Princeton, 18./XII. 1963

Lieber Herr Bernays!

Ich danke Ihnen bestens für den interessanten Brief, den Sie mir vor längerer Zeit schrieben. Leider war es mir bisher nicht möglich, ihn ausführlich zu beantworten, da ich mit den höchst wichtigen Cohenschen Resultaten, sowie auch mit einer Neuherausgabe einiger meiner Arbeiten u. mit Institutsangelegenheiten so beschäftigt war, dass ich mein Interesse für Philosophie eine Zeit lang zurückstellen musste.

one hand with the methods of Quine and on the other with those of Post and of Kleene, he has contributed a number of very interesting results on the questions of frame theories [Rahmentheorien], on the treatment of metamathematics and on the investigation of the recursively enumerable sets.

Of course, you are also informed about his current areas of research by the program of work that Mr. Myhill enclosed with his application.

To be sure, the official program of work alone certainly does not determine the fruitfulness of a stay at the Institute for Advanced Study; that consists for the most part in the possibilities of that program being influenced by an exchange of ideas.

Mr. Myhill's style of working provides a particularly good basis for that sort of fruitful influence, since his research is pursued in the spirit of openness to various aspects, and since, on the other hand, on the basis of his adroit and thorough mastery of foundational techniques, he knows how to follow up fruitfully on suggestions he receives.

I personally became acquainted with Mr. Myhill—after a fleeting encounter at the Amsterdam congress—on the occasion of a lecture that he gave in 1961 in the philosophy department at the University of Pennsylvania. He has a very friendly and sociable character.

So, insofar as I am entitled to an opinion on the matter, I would like to recommend an affirmative decision on Mr. Myhill's application.

With cordial greetings,

Yours truly,

Paul Bernays

50. Gödel to Bernays

Princeton, 18 December 1963

Dear Mr. Bernays,

I thank you very much for the interesting letter that you wrote me a long time ago. Unfortunately, up to now it was not possible for me to answer it in detail, since I was so busy with the results of Cohen, which are of the highest importance, as well as with a new edition of some of my papers and with Institute affairs, that I had to defer my interest in philosophy for a long time.

Mit herzlichen Grüssen u. besten Wünschen, auch von meiner Frau,

Ihr Kurt Gödel

51. Bernays to Gödel

Zürich 2, 1. I. 1964
Bodmerstr. 11

2 | Lieber Herr Gödel!

Über Ihre freundlichen Neujahrsgrüsse habe ich mich sehr gefreut!
Dass Sie durch die verschiedenen von Ihnen erwähnten Umstände, ins-
besondere durch die Vorgänge in der axiomatischen Mengenlehre, die
sich ja so stark an Ihre Methoden knüpfen, von Ihren philosophischen
Überlegungen abgelenkt wurden, ist mir sehr verständlich.
Hoffentlich können Sie zurzeit mit Ihrem Befinden zufrieden sein.

Mit herzlichen Wünschen für das beginnende Jahr grüsst Sie und
Ihre Frau vielmals

Ihr Paul Bernays

52. Gödel to Bernays

Princeton, 5./II. 1965.

Lieber Herr Bernays!

Besten Dank für Ihren Brief vom 30./I. Ich freue mich sehr, dass es
Ihnen möglich ist, dieses Semester des öfteren nach Princeton zu Besuch
zu kommen. Wenn es Ihnen recht ist, werde ich Ihnen eine offizielle Einladung zugehen lassen, in der Sie zu Besuchen "roughly once a week for
the next two months", mit einer Remuneration von $500 eingeladen wer-
den.
Ich nehme an, es wird Sie interessieren, auch mit andern Logikern hier
zu sprechen. Sie wissen ja wahrscheinlich, dass Hasenjäger dieses Jahr
am Institut arbeitet. Ferner ist W. W. Boone hier, den Sie vielleicht auch
persönlich kennen; ausserdem Solovay und Yates, zwei ausgezeichnete
junge Logiker, von denen der erstere sehr schöne neue Resultate mit der

With cordial greetings and best wishes, also from my wife,

Yours truly,

Kurt Gödel

51. Bernays to Gödel

Zurich 2, 1 January 1964
Bodmerstrasse 11

Dear Mr. Gödel,

I was very pleased by your friendly New Year's greetings!

That you were distracted from your philosophical deliberations by the various circumstances you mentioned, in particular by the events in axiomatic set theory that are so closely tied to your methods, I can well understand.

Hopefully you can be satisfied with your state of health for the time being.

With cordial wishes for the year [now] beginning, I offer many times over greetings to you and your wife.

Yours truly,

Paul Bernays

52. Gödel to Bernays

Princeton, 5 February 1965

Dear Mr. Bernays,

Thanks very much for your letter of 30 January. I'm very pleased that it's possible for you to come to visit Princeton more often this semester. If it is all right with you, I will have an official invitation sent to you, in which you will be invited to visit "roughly once a week for the next two months", with a remuneration of $500 .

I assume it will also be of interest to you to speak with other logicians here. You probably know that Hasenjäger is working at the Institute this year. Moreover W. W. Boone is here, with whom you are also perhaps personally acquainted; in addition [there are] Solovay and Yates, two excellent young logicians, the frst of whom has obtained very beau-

2   Cohenschen | Methode erzielt hat, ~~wh~~ während der letztere eine sehr in-
teressante neue Methode auf dem Gebiet der rekursiven Funktionentheorie entwickelt. Ferner ist ein Schüler Myhill's namens Ellentuck hier, der
über die in den Mostowskischen u. ähnlichen Modellen geltenden Theorien der ⟨im Dedekindschen Sinn⟩ endlichen (d. h. nicht reflexiven) Kardi-
nalzahlen u. damit zusammenhängende Fragen arbeitet, u. der auch sehr
begabt zu sein scheint.

Wenn es Ihnen recht ist, können wir Ihren ersten Besuch auf Freitag
den 12./II ansetzen. Ich würde Sie dann, falls Sie zusagen, um ca. 10^h in
meinem Office erwarten u. wir könnten bis 12^h oder $\frac{1}{2}1^h$ diskutieren. Ich
nehme an, dass Sie um diese Zeit gewöhnlich Ihren Lunch haben.

Indem ich mich darauf freue, Sie bald wiederzusehen, verbleibe ich
mit herzlichen Grüssen

Ihr Kurt Gödel

P.S. Ich bin nicht sicher, ob es mir möglich sein wird Sie *jede* Woche
zu sehen. Aber die genaue Anzahl der Besuche ist ja nicht festgesetzt,
u. ausserdem werden Sie ja wahrscheinlich auch mit anderen Logikern
sprechen wollen.

53. Bernays to Gödel

8002 Zürich, 17. IX. 1965
Bodmerstr. 11.

Lieber Herr Gödel!

Es war mein Vorhaben, Ihnen bald nach meiner Rückkunft von Amerika zu schreiben. Doch meine Korrespondenz hat sich leider sehr verzögert,
vornehmlich dadurch, dass ich im Juli an einem internationalen Kongress
in *London* über Grundlagen- und Methoden-Fragen teilnahm, der von
Herrn *Popper* und seinem Schüler Imre Lakatos veranstaltet wurde und
bei dem ich manche Bekannte aus der mathematischen und philosophischen Grundlagenforschung wiedersah.

Kürzlich habe ich noch ein Symposium der Académie Internat. de
Philosophie des Sciences in Lausanne mitgemacht, das dieses Mal ein
Thema der Kulturphilosophie ("Civilisation technique et humanisme")
behandelte.—

Gern denke ich an unsere Gespräche in den Monaten des spring term
zurück. Ich brauche Ihnen nicht zu sagen, wie erfreulich es für mich war,

tiful new results with Cohen's methods, while the latter has developed a very interesting new method in the domain of recursive function theory. Furthermore, a student of Myhill's named Ellentuck is here, who works on the theories (valid in Mostowski models and similar ones) of cardinal numbers that are finite in the sense of Dedekind (i.e., not reflexive) and on questions connected with them, and who also seems to be very talented.

If it is all right with you, we could schedule your first visit on Friday, the 12th of February. If you agree, I would then expect you in my office around 10 a.m., and we could discuss things until 12 or 12:30 . I assume you usually have your lunch at this time.

Meantime, pleased at ⟦the prospect of⟧ seeing you again soon, I remain, with cordial greetings

Yours truly, Kurt Gödel

P.S. I'm not sure whether it will be possible for me to see you *every* week. But the precise number of visits is of course not specified, and besides, you will probably also want to speak with other logicians.

53. Bernays to Gödel

8002 Zurich, 17 September 1965
Bodmerstrasse 11

Dear Mr. Gödel,

It was my intention to write you soon after my return from America. But unfortunately my correspondence has been very much delayed, principally because in July I took part in an international congress in *London* on foundational and methodological questions, which was organized by Mr. *Popper* and his student Imre Lakatos; and at it I saw again many acquaintances ⟦working in⟧ mathematical and philosophical research on foundations.

Recently I participated in yet another symposium of the Académie Internationale de Philosophie des Sciences in Lausanne, which dealt this time with a theme from the philosophy of culture ("Civilisation technique et humanisme").

I think back with pleasure on our conversations during the months of the spring term. I needn't say to you how enjoyable it was for me that

dass ich in dieser Zeit oftmals mit Ihnen zusammen sein und die philosophischen Fragen der Mathematik besprechen konnte. Als ein kleines
Zeichen meiner dankbaren Erinnerung sende ich Ihnen mit gleicher Post
eine Ausgabe der Schelling'schen Schrift "Über das Wesen der menschlichen Freiheit".[a] Gern hätte ich sie Ihnen in schönerer Ausstattung zu-
2   kommen lassen; doch scheint es, | dass sie zurzeit im Buchhandel nur in
dieser Ausgabe erhältlich ist.—Wir kamen auf diese Schrift einmal zu
sprechen; ich erwähnte Ihnen, dass ich seinerzeit durch einen Artikel
von Robert Corti darauf aufmerksam wurde, dass in der Schrift verschiedene wesentliche Gedanken der Philosophie Schopenhauers bereits
geäussert sind und dass daher ⟨vermutlich⟩ Schopenhauer von den Gedanken Schellings beeinflusst wurde, während andrerseits ja—wie Sie
bemerkten—Schopenhauer immer sehr abschätzig von Schelling gesprochen
hat.

In Anknüpfung an unsere Unterhaltungen bin ich Ihnen noch einige
Angaben schuldig, die ich damals aus dem Stegreif nicht vollständig zu-
sammenbrachte.

Anlässlich der Untersuchung von Herrn *Lawvere* über eine Einordnung
der Mengenlehre in die Theorie der Kategorien fiel mir eine Abhandlung
von *H. Wegel*[b] ein, die ich mir vornahm, Ihnen noch genauer anzugeben.
Diese erschien in den Math. Annalen, Bd. 131 (1956), S. 435–462; der
Titel ist "Axiomatische Mengenlehre ohne Elemente von Mengen". In
dieser Axiomatisierung hat man zwei Sorten von Dingen: Mengen und
Funktionen, und zwei Grundprädikate: $a \subseteq b$ und die Funktionsbeziehung
3   $abcd\varphi e$ ("e ist der Wert der Funktion φ für die Argumente a, b, c, d"); |
als eine Art Ersatz für die Element-Beziehung hat man das *definierbare*
Prädikat "a ist eine Einermenge (unitset), welche Teilmenge von b ist".
Alle Axiome sind *eigentliche* Axiome (nicht Schemata). Der Funktionsbe-
griff wird durch Axiome festgelegt. Metamathematisch tritt dagegen der
Begriff eines "Systems" von Teilmengen einer Menge m auf; ein solches
System wird durch ein Mengen-Prädikat geliefert, das in Anlehnung an
Fraenkel's Verfahren normiert ist. Eine Hauptrolle spielt ein Axiom, das
verlangt, dass zu jedem System von Teilmengen einer Menge m eine "In-
dexmenge" i existiert nebst einer umkehrbar eindeutigen Abbildung der
dem System angehörenden Teilmengen von m auf die Einermengen von i.
(Darüber hinaus wird verlangt, dass die Zuordnung eine gleichmässige ist
inbezug auf m und auf die in dem System auftretenden Parameter, sowie

---

[a] *Schelling 1809*. The edition Gödel received from Bernays was most likely *Schelling
1964*, as that is the version found among his books.

at that time I could often be together with you and discuss the philosophical questions of mathematics. As a small token of my grateful remembrance I am sending you at the same time an edition of Schelling's book *Über das Wesen der menschlichen Freiheit.*[a] I would have liked to send you a more beautiful edition; but it seems that at the moment it is available at the bookdealer only in this edition.—We once had occasion to speak of this book; I told you that I became aware of it at that time through an article by Robert Corti, that in the book various essential ideas of Schopenhauer's philosophy are expressed and that hence, presumably, Schopenhauer was influenced by the ideas of Schelling, whereas on the other hand—as you noted—Schopenhauer always spoke very disparagingly of Schelling.

In connection with our conversations I still owe you a few pieces of information, which at that time, extemporaneously, I couldn't completely put my finger on.

On the occasion of Mr. *Lawvere's* investigation about an incorporation of set theory into the theory of categories, an article by *H. Wegel*[b] came to my mind, which I intended to describe to you still more precisely. It appeared in *Mathematische Annalen*, vol. 131 (1956), pp. 435-462; the title is "Axiomatische Mengenlehre ohne Elemente von Mengen". In this axiomatization one has two sorts of things: sets and functions, and two primitive predicates: $a \subseteq b$ and the functional relation $abcd\varphi e$ ("e is the value of the function φ for the arguments a, b, c, d"); as a sort of substitute for the element-relation one has the *definable* predicate "a is an *Einermenge* (unitset), which is a subset of b". All the axioms are *proper* axioms (not schemas). The concept of function is fixed by means of axioms. In contrast, the concept of a "system" of subsets of a set m occurs metamathematically; such a system is provided by a set-predicate that is normalized on the model of *Fraenkel's* procedure. A principal role is played by an axiom that requires that to every system of subsets of a set m there exists an "index set" i together with a one-to-one mapping of the subsets of m belonging to the system to the unit sets of i. (Beyond that it is required that the assignment be uniform with respect to m and to the parameters occurring in the system, and furthermore that every

[b] *Wegel 1956.*

ferner, dass jede (dem System angehörende) Einermenge von m auf sich
selbst abgebildet wird.)—

Einmal kamen wir auf das Thema der Darstellung einer zweistelligen Funktion durch einstellige Funktionen und die Summe zu sprechen. Dabei ~~fiel mir~~ erwähnte ich eine Untersuchung von *Alexander Ostrowski*, die ich aber nicht genau genug in Erinnerung hatte. Die Abhandlung ist betitelt "Über Dirichletsche Reihen und algebraische Differential-
4 gleichungen";[c] sie steht in der | Math. Zeitschrift, Bd. 8 (1920), S. 241–298. Ostrowski beweist hier unter anderm, dass die Funktion $\zeta(x,s) = \sum_{n=1}^{\infty} \frac{x^n}{n^s}$ als Funktion ihrer beiden Argumente keiner algebraischen partiellen Differentialgleichung genügt,—⟨so⟩wie es Hilbert vermutet hatte. Ferner folgert er aus diesem Ergebnis, dass es unmöglich ist, die Funktion $\zeta(x,s)$ aus beliebigen analytischen Funktionen einer Variablen und beliebigen algebraischen Funktionen mehrerer Variablen durch sukzessive Einschachtelungen zu gewinnen.—

Eine dritte nachzuholende Angabe betrifft eine Schrift von *Albert Lautman*, von der ich Ihnen sagte, dass darin eine Art Hegel'scher Betrachtungsweise auf die Mathematik angewendet werde. Die Schrift[d] erschien 1938 in Paris in der Sammlung "Actualités scientifiques et industrielles", im Verlag Hermann & Cie, mit dem Titel "Essai sur les notions de structure et d'existence en mathématiques". Ich habe sie im Journal of Symbolic Logic, vol 5, Nr 1 (1940), pp. 20–22 besprochen.[e] (Gern hätte ich sie Ihnen auch zugesandt; doch scheint sie schwer erhältlich zu sein.)—

Dass Herr Leo F. Boron in Gemeinschaft mit William Howard Ihre ⟨schöne⟩ Abhandlung in dem Dialectica-Band[f] von 1958 ins Englische übersetzt und diese Übersetzung den Dialectica zur Publikation angeboten hat, ist Ihnen gewiss bekannt. Wenn ich Herrn Boron in seinem
5 Schreiben richtig verstanden habe, so haben | Sie sich ja mit der Veröffentlichung seiner Übersetzung einverstanden erklärt. Auch ist Herr Gonseth sehr dafür, dass die Übersetzung in den Dialectica erscheint.[g] —

Hoffentlich können Sie zurzeit mit Ihrem gesundheitlichen Ergehen zufrieden sein. Mit herzlichen Grüssen und Wünschen für Sie und Ihre Frau, in dankbarem Gefühl

Ihr Paul Bernays

---

[c] *Ostrowski 1920.*
[d] *Lautman 1938.*
[e] *Bernays 1940.*

unit set of m (belonging to the system) is mapped to itself.)

We once came to discuss the topic of the representation of a two-place function by one-place functions and the summation [operation]. On that point I mentioned an investigation by *Alexander Ostrowski*, which, however, I did not recall precisely enough. The paper is entitled "Über Dirichletsche Reihen und algebraische Differentialgleichungen";[c] it is to be found in *Mathematische Zeitschrift*, vol. 8 (1920), pp. 241–298. Here Ostrowski proves, among other things, that as a function of its two arguments the function $\zeta(x, s) = \sum_{n=1}^{\infty} \frac{x^n}{n^s}$ does not satisfy any algebraic partial differential equation—just as Hilbert had conjectured. Moreover, it follows from this result that it is impossible to obtain the function $\zeta(x, s)$ by successive substitutions of arbitary analytic functions of one variable and arbitrary algebraic functions of several variables.

A third item of information to be supplied as promised concerns a paper by *Albert Lautman*, of which I told you that it applies a sort of Hegelian perspective to mathematics. The paper[d] appeared in 1938 in Paris in the collection "Actualités scientifiques et industrielles", published by Hermann & Cie, with the title "Essai sur les notions de structure et d'existence in mathématiques". I reviewed it in the *Journal of Symbolic Logic*, vol. 5, No. 1 (1940), pp. 20–22.[e] (I would like to have sent it to you too; but it seems to be difficult to obtain.)

That Mr. Leo F. Boron, in association with William Howard, is translating your beautiful paper in the *Dialectica* volume of 1958[f] into English, and that this translation was submitted to *Dialectica* for publication is surely known to you. If I correctly understood Mr. Boron in what he wrote, you have agreed to the publication of his translation. Mr. Gonseth is also very much in favor of publishing the translation in *Dialectica*.[g]

Hopefully you can be satisfied for the time being with your state of health.

With cordial greetings and [best] wishes for you and your wife,

With gratitude,

Yours truly,

Paul Bernays

---

[f] *1958.*

[g] For the vicissitudes of this translation that was taken over by Gödel and only published posthumously as *1972* in volume III of these *Works*, see sections 3.1 and 3.3 of the introductory note to this correspondence.

54. Gödel to Bernays

27. Sept. 1965.
145 Linden Lane

Lieber Herr Bernays!

Vor allem vielen herzlichen Dank fuer Ihren freundlichen Brief vom 17. Sept. sowie auch die Abhandlung von Schelling. Die letztere spielt ja zweifellos eine fundamentale Rolle in der Schellingschen Philosophie, da sie doch an der Grenze von deren beiden Hauptphasen steht. Fuer die in Ihrem Brief enthaltenen Zitate danke ich Ihnen noch ganz besonders. Seien Sie versichert, dass unsere Diskussionen im vergangenen Fruehjahr auch fuer mich, nicht nur sehr erfreulich, sondern auch sehr anregend waren.

Leider habe ich augenblicklich nicht die Zeit, Ihren Brief ausfuehrlich zu beantworten, moechte aber, da Sie doch Mitherausgeber der Dialectica sind, etwas ueber die Boron–Howardsche Uebersetzung meiner Arbeit aus Dialectica 12 sagen. Vor allem danke ich Ihnen, dass Sie mich auf deren beabsichtigte Publikation aufmerksam gemacht haben. Ich habe bisher nur eine provisorische Form dieser Uebersetzung gesehen, in der gewisse Verbesserungsvorschlaege von Howard im Text noch nicht beruecksichtigt waren. In dem damaligen Text waren Verbesserungen zweifellos noetig. Nun ist ja Howard ein ausgezeichneter Logiker und es ist durchaus moeglich, dass ich mit der endgueltigen Form der Uebersetzung vollkommen einverstanden waere. Ich moechte Sie aber jedenfalls bitten mir, im Falle die Uebersetzung wirklich im Druck erscheinen soll, eine Kopie des Manuskriptes zukommen zu lassen.

Mit herzlichen Gruessen und nochmaligem besten Dank

Ihr Kurt Gödel

Beste Gruesse auch von meiner Frau.

55. Bernays to Gödel

Lieber Herr Gödel!

8002 Zürich, 8. November 1965.
Bodmerstr. 11.

Vielen Dank für Ihren freundlichen Brief vom 27. September! Ich sende Ihnen zugleich mit diesen Zeilen das Manuskript der englischen

54. Gödel to Bernays

Princeton, 27 September 1965
145 Linden Lane

Dear Mr. Bernays,

First of all, many hearty thanks for your friendly letter of 17 September as well as the treatise by Schelling. The latter undoubtedly plays a fundamental role in Schelling's philosophy, since it stands indeed at the boundary between its two principal phases. I thank you again especially for the citations contained in your letter. Be assured that our discussions during the past spring were for me, too, not only very enjoyable but also very stimulating.

Regrettably I don't have time at the moment to reply to your letter in detail, but, since you are a co-editor of *Dialectica*, I would like to say something about the Boron–Howard translation of my work from vol. 12 of *Dialectica*. Above all, I thank you for having made me aware of its intended publication. Up to now I've seen only a provisional form of this translation, in which certain of Howard's suggestions for improvements in the text had not yet been taken into account. In the text at that time improvements were undoubtedly necessary. Now, Howard is an excellent logician and it is quite possible that I would entirely agree to the final form of the translation. But in any case I would like to ask you, in the event the translation really is supposed to appear in print, to have a copy of the manuscript sent to me.

With cordial greetings and most sincere thanks, once again

Yours truly,

Kurt Gödel

Best wishes also from my wife.

55. Bernays to Gödel

8002 Zurich, 8 November 1965
Bodmerstrasse 11

Dear Mr. Gödel,

Many thanks for your friendly letter of 27 September! I am sending you at the same time as these lines the manuscript of the English

Übersetzung Ihrer Abhandlung in Dialectica 12. Ich habe an dem englischen Text ein paar kleine Korrekturen angebracht; Sie werden vielleicht
noch weitere Änderungen vermerken.—Wenn Sie die Revision | vollzogen
haben, so möchten Sie mir das korrigierte Manuskript wieder zugehen
lassen.

Mit herzlichen Grüssen, auch an Ihre Frau,

Ihr Paul Bernays

56. Gödel to Bernays

Princeton, 2. Dez. 1965.

Lieber Herr Bernays!

Besten Dank fuer die Zusendung der Uebersetzung meiner Dialectica-
Arbeit. Der Text unterscheidet sich fast gar nicht von der ersten (un-
korrigierten) Fassung, ueber die ich Ihnen in meinem letzten Brief schrieb.
Ich glaube daher, dass einige Aenderungen unbedingt noettig und eine
Reihe anderer sehr wuenschenswert waeren. Ferner habe ich bemerkt,
dass man zu den Bezeichnungserklaerungen 1, 2, 3, auf p. 284, 285 unbedingt eine Erlaeuterung hinzufuegen sollte, welche u. a. zu besagen
haette, dass es sich nur darum handelt, sicherzustellen, dass die spaeter
unter 1. bis 6. beschriebenen Formeln sinnvolle Ausdruecke des Systems
T sind. Darf ich fragen, fuer wann die Drucklegung der Uebersetzung in
Aussicht genommen ist und ob auch die andern nicht englischen Artikel
von der Festschrift uebersetzt werden sollen?

Mit herzlichen Gruessen, auch von meiner Frau

Ihr Kurt Gödel

57. Bernays to Gödel

8002 Zürich, 10. Dez. 1965
Bodmerstr. 11.

Lieber Herr Gödel!

Schönen Dank für Ihren Brief vom 2. Dezember. Wenn Sie, wie Sie
vorhaben, an der Uebersetzung Ihrer Abhandlung noch Verbesserungen

translation of your article in vol. 12 of *Dialectica.* I have made a few small corrections to the English text; you will perhaps take note of still other changes.—When you have carried out the revision, would you please have the corrected manuscript sent back to me again.

With cordial greetings, also to your wife,

Yours truly,

Paul Bernays

56. Gödel to Bernays

Princeton, 2 December 1965

Dear Mr. Bernays,

Thanks very much for sending the translation of my *Dialectica* work. The text differs hardly at all from the first (uncorrected) version, about which I wrote you in my last letter. I think, therefore, that a few changes would be absolutely necessary and a series of others very desirable. Furthermore I have noticed that one should undoubtedly add to the notational explanations 1, 2, 3 on pp. 284, 285 an elucidation, which, among other things, would have to say that it is only a question of making sure that the formulas described later under 1. through 6. are meaningful expressions of the system T . May I ask when the printing of the translation is planned and whether the other articles of the *Festschrift* not in English are also supposed to be translated?

With cordial greetings, also from my wife,

Yours truly,

Kurt Gödel

57. Bernays to Gödel

8002 Zurich, 10 December 1965
Bodmerstrasse 11

Dear Mr. Gödel,

Thanks very much for your letter of 2 December. If, as you intend, you still want to make improvements and additions to the translation

und Ergänzungen anbringen ⟨wollen⟩, so sind wir Ihnen gewiss sehr dankbar.

Sie erkundigten sich, wann die Drucklegung in Aussicht genommen ist. Herr Gonseth sagte mir, dass er das Manuskript gegen Ende Januar in Druck geben möchte; bis dahin ist also noch etwas über einen Monat Zeit.

Was Ihre ~~andere~~⟨weitere⟩ Frage betrifft, so ist eine Uebersetzung der anderen auf deutsch verfassten Abhandlungen in dem Dialectica-Band 12 (Heft 3/4), so viel ich weiss, nicht beabsichtigt. Es soll die Uebersetzung Ihrer Abhandlung in den Dialectica mit einer redaktionellen Bemerkung versehen sein, deren Text ich Ihnen beilege. Falls Sie da etwas hinzufügen oder ändern möchten, lassen Sie es mich bitte wissen.

Mit herzlichen Grüssen

Ihr Paul Bernays

Entwurf für den Text der redaktionellen Note:

Diese Uebersetzung der Abhandlung Kurt Gödels "Ueber eine bisher noch nicht benützte Erweiterung des finiten Standpunktes" wurde von ~~W.~~ ⟨Leo⟩ Boron und William Howard im Hinblick auf einen im Kreise der mathematischen Grundlagenforscher rege gewordenen Wunsch angefertigt, es möchte diese Abhandlung, welche einen neuen programmatischen Aspekt für die Beweistheorie eröffnet und auf welche in der neueren grundlagentheoretischen Literatur vielfach Bezug genommen wird, auch in englischer Sprache zugänglich sein. Die deutsche Originalfassung erschien im Bande 12 der Dialectica, und so übernimmt unsere Zeitschrift gern die Publikation der Uebersetzung, in der Meinung, damit auch ihrer Würdigung der Bedeutsamkeit der mathematischen Grundlagenforschung im Problembereich der Philosphie der Wissenschaften Ausdruck zu geben.

58. Gödel to Bernays

Princeton, Jan. 25, 1966.

Lieber Herr Bernays!

Besten Dank fuer Ihre freundlichen Neujahrswuensche, die ich auf das herzlichste erwidere. Ich bedaure, dass ich Ihnen bisher noch nicht die revidierte Uebersetzung meiner Dialecticaarbeit einschicken konnte. Ich

of your paper, we will certainly be very grateful to you.

You inquired when the printing is being planned. Mr. Gonseth told me that he would like to give the manuscript to the printer towards the end of January; until then there is thus still somewhat more than a month's time.

As to your further question, as far as I know a translation of the other papers in vol. 12 (nos. 3/4) of *Dialectica* written in German is not intended. The translation of your paper in *Dialectica* is supposed to be provided with an editorial remark, whose text I enclose for you. In case you would like to add or change something there, please let me know.

With cordial greetings,

Yours truly,

Paul Bernays

Draft for the text of the editorial note:

This translation of Kurt Gödel's paper "Ueber eine bisher nicht benützte Erweiterung des finiten Standpunktes" was prepared by Leo Boron and William Howard in response to a desire that has become keen within the community of researchers in mathematical foundations that this paper, which opened a new programmatic aspect for proof theory and to which many references have been made in the recent foundational literature, also be available in English. The original German version appeared in volume 12 of *Dialectica*, and so our journal is pleased to undertake the publication of the translation, with the intention thereby also of expressing its appreciation of the significance of research in mathematical foundations within the domain of problems of the philosophy of science.

58. Gödel to Bernays

Princeton, 25 January 1966

Dear Mr. Bernays,

Thank you very much for your friendly New Year's wishes, which I cordially reciprocate. I regret that up to now I still was not able to send you the revised translation of my *Dialectica* paper. Namely, in looking

habe naemlich bei genauerem Durchsehen gefunden, dass daran doch wesentlich mehr zu korrigieren ist, als es zuerst den Anschein hatte. Ausserdem halte ich an ein paar Stellen kleine Abaenderungen und Einschaltungen fuer angebracht, um dem Leser das Verstaendnis zu erleichtern. Ich glaube aber, damit jetzt im wesentlichen fertig zu sein, und hoffe, Ihnen das druckfertige Manuskript im̶ spaetestens in einigen Wochen einschicken zu koennen. Ich nehme an, dass es auf das genaue Datum nicht sehr ankommt, da die Dialectica doch alle drei Monate erscheinen[.]

Mit herzlichen Gruessen, auch von meiner Frau,

Ihr Kurt Gödel

59. Bernays to Gödel

8002 Zürich, 24. April 1966.
Bodmerstr. 11

Lieber Herr Gödel!

Recht beschämt fühle ich mich, da ich für die Festschrift zu Ihrem 60. Geburtstag nicht etwas Geeignetes bereit hatte,—dieses um so mehr, als Sie ja seinerzeit bei dem Dialectica-Band zu meinem 70. Geburtstag einen so bedeutsamen Beitrag spendeten.

Nun möchte ich Ihnen jedenfalls meine herzlichen Glückwünsche zu Ihrem vollendeten Lebens-Dezennium aussprechen.

Sie können gewiss im Blick auf die Situation in den Grundlagen-Untersuchungen mit viel Genugtuung feststellen, dass die Entdeckungen und Methoden, die Sie in die Meta-Mathematik brachten, in der heutigen Forschung beherrschend und wegleitend sind. Möge es Ihnen weiterhin auch beschieden sein, den Gang dieser Forschung fruchtbringend zu beeinflussen.—

Doch die Grundlagen der Mathematik sind ja nur eines der Anliegen
2 Ihres Forschens; und ich möchte Ihnen auch wünschen, dass | Ihre philosophischen Überlegungen sich ⟨bald⟩ zu solchen Ergebnissen gestalten mögen, die Ihnen deren Publikation nahelegen.

Last not least, wünsche ich Ihnen noch, dass Ihr allgemeines Ergehen in der nun kommenden Zeit ein möglichst befriedigendes sei. Hof-
s fentlich können Sie Ihren 60. Geburtstag recht schön begehen und ₰ich der gewiss eindrucksvollen Kundgebungen der allgemeinen Würdigung Ihres geistigen Schaffens erfreuen.

Mit sehr herzlichen Grüssen, auch an Ihre Frau,

Ihr Paul Bernays

it over in greater detail I found that substantially more needs to be corrected in it than appeared at first. In addition, to facilitate the reader's understanding I thought it opportune [to make] minor changes and interpolations at a few places. I think, however, I am now essentially finished with it, and I hope to be able to send you the print-ready manuscript in a few weeks at the latest. I assume the exact date doesn't matter very much, since *Dialectica* appears every three months.

With cordial greetings, also from my wife,

Yours truly,

Kurt Gödel

59. Bernays to Gödel

8002 Zurich, 24 April 1966
Bodmerstrasse 11

Dear Mr. Gödel,

I feel quite embarrassed, since I had not prepared anything suitable for the *Festschrift* in honor of your 60th birthday—all the more so, as you made such a significant contribution to the *Dialectica* issues for my 70th birthday.

In any case, I would like now to express my cordial good wishes on your [just] completed decade of life.

In view of the situation in foundational investigations, you can certainly ascertain with much satisfaction that the discoveries and methods that you brought to metamathematics are dominant and leading the way in the research of today. May it be granted to you also in the future to influence the direction of this research in a way that is fruitful.

Yet the foundations of mathematics are of course only one of the concerns of your research; and I would also like to wish that your philosophical reflections may turn into such results that you are induced to publish them.

Last [but] not least, I wish in addition that your general state of health in the coming years be as satisfying as possible. Hopefully you can celebrate your 60th birthday quite beautifully and take pleasure in the certainly impressive statement of the general appreciation of your intellectual work.

With very cordial greetings, also to your wife,

Yours truly,

Paul Bernays

60. Gödel to Bernays

Princeton, 22./V. 1966.

Lieber Herr Bernays!

Ich habe mich sehr über Ihre Glückwünsche zu meinem 60-ten Geburtstag gefreut u. danke Ihnen herzlichst dafür. Ganz besonders danke ich Ihnen für die meine philosophischen Untersuchungen betreffenden Wünsche. Denn diese sind ja seit langer Zeit mein Hauptinteresse. Zu einer Besorgnis ⟨darüber⟩, dass Sie keinen Beitrag zu der beabsichtigten Festschrift geliefert haben, sehe ich nicht den geringsten Anlass.

Auf unsere frühere Korrespondenz zurückkommend, bitte ich Sie, zu entschuldigen, dass ich den in Aussicht gestellten Termin für die Einsendung der korrigierten Übersetzung meiner Arbeit über ⟨eine⟩ Erweiterung des finiten Standpunktes nicht einhalten konnte. Es haben mich in letzter Zeit ganz unerwartet zwei Dinge stark in Anspruch genommen. Erstens hat sich in einer meiner alten Arbeiten (im Menger Kollo-
2 quium Heft 3) ein Versehen in der Formulierung des allgemeinen | Theorems über Unbeweisbarkeit der Widerspruchs[[frei]]heit im selben System herausgestellt (wie ich anlässlich einer Neuherausgabe dieser Note bemerkte).[a] Ich glaube jetzt endlich die beste Formulierung dieses Theorems gefunden zu haben, welche auch einen überraschend einfachen Beweis gestattet. Zweitens wurde ein neuer u. *ganz erstaunlich einfacher u. eleganter* Beweis für die Unabhängigkeit der Kontinuumhypothese von Dana Scott[b] gegeben. Ich möchte aber jetzt die Durchsicht der Übersetzung wieder vornehmen u. hoffe, sie Ihnen dann bald zusenden zu können, da ich im wesentlichen damit schon fertig war u. bloss das Ganze (nach Verlauf einer gewissen Zeitspanne) einer nochmaligen Durchsicht unterziehen wollte.

Mrs Church sagte mir, dass Sie die Absicht hatten[[,]] dieses Frühjahr wieder nach Philadelphia zu kommen, aber, wie ich hörte, haben Sie sich dann leider anders entschlossen.

Ich freue mich immer sehr, von Ihnen zu hören, u. hoffe, dass es Ihnen gut geht.

Mit herzlichen Grüssen, auch von meiner Frau,

Ihr Kurt Gödel

---

[a] *1932b*; see *van Heijenoort 1967*, pp. 616–617, fn. 1 or these *Works*, vol. I, p. 235 and the textual note thereto.

60. Gödel to Bernays

Princeton, 22 May 1966

Dear Mr. Bernays,

I was very pleased by your good wishes on my 60th birthday, and I thank you most cordially for them. Most especially, I thank you for the wishes concerning my philosophical investigations. For they have been my principal interests for a long time. I see not the slightest cause for concern about the fact that you've provided no contribution to the intended *Festschrift*.

Coming back to our earlier correspondence, I apologize that I was unable to adhere to the prospective date for forwarding the corrected translation of my paper on an extension of the finitary standpoint. Of late, two things have quite unexpectedly kept me very much occupied. First, in one of my old papers (in volume 3 of the Menger colloquium [proceedings]) a mistake has turned up in the formulation of the general theorem on the unprovability of consistency within the same system (as I noticed on the occasion of a new edition of that note).[a] I believe now to have found at last the best formulation of this theorem, which also admits a surprisingly simple proof. Second, a new and *quite astonishingly simple and elegant* proof for the independence of the continuum hypothesis was given by Dana Scott.[b] I would like now, however, to take up the examination of the translation again, and I hope then to be able to send it to you soon, since I was already essentially finished with it and merely wanted to reexamine the whole thing once more (after some time had passed).

Mrs. Church told me that you had the intention of coming to Philadelphia again this spring, but, as I heard, you then unfortunately decided not to.

I'm always very pleased to hear from you, and I hope that things are going well for you.

With cordial greetings, also from my wife,

Yours truly,

Kurt Gödel

---

[b]See *Scott 1967*.

61. Gödel to Bernays

Princeton, 24./I. 1967.

Lieber Herr Bernays!

Vor allem besten Dank für die beiden Briefe vom Okt. 1966 u. Jan. 1967, sowie Ihre freundlichen Neujahrswünsche, die ich auf das herzlichste erwidere. Entschuldigen Sie, bitte, dass ich erst heute antworte.— Die Übersetzung meiner Dialectica Arbeit habe ich im Dez. u. Jan. einer nochmaligen Durchsicht unterzogen u. bei dieser Gelegenheit die erkenntnistheoretischen Ausführungen auf den ersten beiden Seiten etwas deutlicher formuliert[1], aber ohne etwas am Inhalt zu ändern. Im übrigen habe ich ausser Verbesserungen der Übersetzung kaum etwas geändert, sondern bloss einige Fussnoten hinzugefügt. Ich bin jetzt mit dem Manuskript ganz zufrieden u. glaube, dass man es so publizieren kann. Ich muss bloss noch die Änderungen u. Ergänzungen in Reinschrift übertragen, was ich bald zu tun hoffe. Denn ich möchte diese Sache endlich zu Ende bringen, u. muss mich entschuldigen, dass ich sie so lange verzögert habe. Meine Ansichten haben sich seit damals kaum geändert, ausser 2   dass | ich jetzt überzeugt bin, dass ϵ_0 eine prinzipielle, nicht bloss praktische, Grenze des Hilbertschen Finitismus ist u. dass man das auch wird überzeugend dartun können. Das schliesst allerdings nicht aus, dass es (im Hilbertschen Sinn) *nicht* finite Beweise geben könnte, die eine ebenso starke Überzeugungskraft haben.—Was das schriftliche Symposium betrifft, so möchte ich sagen, dass ich ja meine Stellungnahme zu den neueren Entwicklungen an mehreren Stellen bereits deutlich ausgesprochen habe. Vielleicht könnte ich diese Stellen sammeln u. Ihnen einschicken. Denn ich habe mir diese Formulierungen sehr genau überlegt u. könnte daher in einer neuen Abhandlung nicht viel mehr tun als sie wiederholen.—Kennen Sie den Takeutischen Widerspruchsfreiheitsbeweis für das Π_1^1-Komprehensionsaxiom[a] u. haben Sie sich schon eine bestimmte Meinung über dessen Bedeutung gebildet? Das Gerücht, dass ich in nächster Zeit nach Europa kommen will, ist *gänzlich* unbegründet.

Mit herzlichen Grüssen, auch von meiner Frau,

Ihr Kurt Gödel

---

[1] [Written across the bottom of both pages:] Es scheint mir auch, dass ich Ihnen eine etwas zu starke Behauptung zugeschrieben habe. Sie haben ja nirgends die *Not-* 2   *wendigkeit* abstrakter Begriffe | behauptet, sondern deren Verwendung bloss als eine *Möglichkeit* der Erweiterung des Finitismus hingestellt.

---

[a] *Takeuti 1967.*

61. Gödel to Bernays

Princeton, 24 January 1967

Dear Mr. Bernays,

First of all, thanks very much for the two letters of October 1966 and January 1967, as well as your friendly New Year's wishes, which I most heartily reciprocate. Please excuse that I am answering only today.

In December and January I reexamined my *Dialectica* paper once again and formulated on this occasion the epistemological considerations of the first two pages somewhat more clearly,[1] but without changing any of the content. Besides [that], apart from improvements of the translation, I have changed hardly anything, but have merely added a few footnotes. I am now quite satisfied with the manuscript and believe it can be published as is. I have only to transcribe the changes and supplementary remarks into fair copy, which I hope to do soon. For I would like at last to bring this matter to an end, and must apologize that I have delayed it for so long. My views have hardly changed since then, except that I am now convinced that ϵ_0 is a bound on Hilbert's finitism, not merely in practice [but] in principle, and that it will also be possible to prove that convincingly. Of course, that does not exclude that there could be (in Hilbert's sense) *non*-finitary proofs that are equally convincing. As to the written symposium [proceedings], I would like to say that I have already clearly expressed my positions on the new developments in several places. Perhaps I could collect those passages and send them to you. For I thought about these formulations very carefully, and thus in a new essay I could not do much more than repeat them.

Do you know Takeuti's consistency proof for the Π_1^1-comprehension axiom,[a] and have you already formed a definite opinion about its significance? The rumor that I will be coming to Europe before long is *entirely* unfounded.

With cordial greetings, also from my wife,

Yours truly,

Kurt Gödel

---

[1][Written across the bottom of both pages:] It seems to me, too, that I ascribed to you somewhat too strong an assertion. You never asserted the *necessity* of abstract concepts, but posited the use of them merely as a *possible way* of extending finitism.

62. Bernays to Gödel

8002 Zürich, 27. X. 1967.
Bodmerstr. 11.

Lieber Herr Gödel!

Seit langem habe ich vor, Sie einmal wieder von mir hören zu lassen, es kam aber immer wieder etwas dazwischen. Ihren freundlichen Brief von Ende Januar habe ich vor mir. Sie schrieben dann, dass Sie die Verbesserungen der Übersetzung Ihrer Dialectica-Abhandlung sowie die Modifikationen und Ergänzungen ausgeführt hätten und nur noch den Text in Reinschrift zu übertragen brauchten, was Sie bald zu tun hofften.

Nun sind Sie aber, wie es scheint, doch hiervon abgehalten worden. Hoffentlich war es nicht durch schlechtes Befinden, vielmehr durch anderweitiges Sachliches, das Sie beschäftigte.[1]

Es wäre schön, wenn Sie es möglich machen könnten, das Manuskript ⟨bald⟩[a] in der von Ihnen verbesserten Fassung—wie Sie es lieber wollen— an mich oder an die Redaktion von Dialectica ⟨d. h. an Herrn Gonseth⟩ (12, chemin du Muveran, 1000 Lausanne, Suisse) gelangen zu lassen. Herr Gonseth würde gern, wenn möglich, die Übersetzung noch in eine Dialectica-Nummer *von 1967* aufnehmen.—

2   Was das schriftliche Symposium angeht, so | begrüsse ich sehr Ihr Anerbieten, dass Sie Ihre verschiedenen Äusserungen, mit denen Sie zu der neueren Entwicklung der mathematischen Grundlagenforschung Stellung genommen haben, sammeln und uns einsenden wollen. Eine solche Zusammenstellung wird gewiss Ihre Auffassung von der Grundlagen-Situation sehr eindrücklich zur Geltung bringen.

Meine rezeptive Verfolgung der Grundlagenforschung war in diesem Jahr ziemlich eingeschränkt. Immerhin hörte ich im Frühjahr in Oberwolfach eine interessante Vortragsreihe von *Dr. Jensen* über die hohen Mächtigkeiten.—Mit der *Neuauflage* der "Grundlagen der Mathematik"[b] ist es jetzt so weit, dass die Fahnen-Korrekturen durchgesehen sind. Ich will aber noch die Wf.-Beweise von Kalmár und Ackermann[c] in einem Supplement zur Darstellung bringen (die ja ganz in den Rahmen des Vorherigen gehören).—

---

[1] ⟦Written vertically in left margin of page 1:⟧ Oder sollte eine Sendung von Ihnen mich nicht erreicht haben?

---

[a] Underlining in this letter may be Gödel's rather than Bernays'. There are also arrows in the margin pointing to each line that has underlining.

[b] *Hilbert and Bernays 1968* and *1970*.

62. Bernays to Gödel

8002 Zurich, 27 October 1967
Bodmerstrasse 11

Dear Mr. Gödel,

For a long time I've been intending to let you hear from me again,
but over and over again something intervened. I have your letter from
the end of January in front of me. You wrote then that you had finished
the improvements to the translation of your *Dialectica* paper, that you
had finished as well the modifications and supplements [to it] and that
you had only to transcribe the text into fair copy, which you hoped to do
soon.

But now, so it seems, you have been held back from that. Hopefully
it was not bad health, but rather other practical matters that kept you
busy.[1]

It would be nice if you could see to it that the manuscript, in the version improved by you, reaches me or the editorial office of *Dialectica* (i.e.,
Mr. Gonseth, 12 chemin du Muveran, 1000 Lausanne, Switzerland)—
whichever you prefer—<u>soon</u>.[a] If possible, Mr. Gonseth would still like to
include the translation in a 1967 issue of *Dialectica*.

As to the written symposium [proceedings], I welcome very much your
offer to collect and send to us your various remarks in which you have
taken a position on the more recent development of research in the foundations of mathematics. Such a compilation will certainly express your
conception of the foundational situation very forcefully.

My reception of foundational research was somewhat hampered this
year. Nevertheless, in the spring I heard in Oberwolfach an interesting
series of lectures by Dr. Jensen on large cardinals. – Things are now so
far along with the new edition of *Grundlagen der Mathematik*[b] that the
galleys have been checked. But in a supplement I still want to present
the consistency proofs of Kalmár and Ackermann[c] (which lie entirely
within the framework of what has gone before).

---

[1] [Written vertically in left margin of page 1:] Or should a mailing of yours not have
reached me?

---

[c] In supplement V to *Hilbert and Bernays 1970*, Bernays presented the consistency
proofs for number theory that had been given by Kalmár (unpublished) and Ackermann
(*1940*).

In Amsterdam habe ich kürzlich viele der Grundlagenforscher wieder-
gesehen.[2] Ich kam freilich nicht zu dem grossen Kongress, sondern nur
zu dem Symposium der Académie internationale ..., das in dem gleichen
Hotel stattfand.

In den Tagen danach nahm ich noch an dem Kolloquium in Tihany
(Ungarn) über rekursive Funktionen teil, das zu Ehren von Rósza Péter
veranstaltet war.—

3   Von dem kürzlich erschienenen Buche von Jean van ⟨Heijenoort⟩ |
"From Frege to Gödel"[d] haben Sie jedenfalls auch ein Exemplar empfangen. Ich finde die an die Abhandlungen angeschlossenen historischen
und sachlichen Ausführungen sehr instruktiv!—

Mit dem Wunsche, dass es Ihnen gut gehe, grüsst Sie und Ihre Frau

sehr herzlich Ihr Paul Bernays

Ihrer Bemerkung, dass ich in dem Vortrag "Sur le Platonisme ..."[e]
nicht die *Notwendigkeit* abstrakter Begriffe behauptet habe, stimme ich
durchaus zu, und so ist es ganz in meinem Sinne, dass Sie den hierauf
bezüglichen Passus Ihrer Dialectica-Abhandlung geändert formulieren.

---

[2] [Written vertically in the left margin of page 2:] Herrn Kreisel habe ich hernach
noch in Zürich gesprochen.

---

[d] *van Heijenoort 1967.*

63. Gödel to Bernays

Princeton, 20./XII. 1967.

Lieber Herr Bernays!

Herzlichen Dank für Ihren frdl. Brief vom 27./X. Die Übersetzung
der Dialectica Arbeit werde ich sofort nach Schluss der Weihnachtsferien
(d. h. anfangs Januar) abtippen lassen. Leider war in den letzten Wochen meine Aufmerksamkeit zu stark durch andere Dinge in Anspruch
genommen. Die Zusammenstellung meiner Äusserungen über die neuere
Entwicklung der math. Grundlagenforschung möchte ich dann auch möglichst bald fertigstellen. Für wann ist denn das Erscheinen des betreffenden Dialectica ~~Nummer~~ Bandes geplant?

In Amsterdam I recently saw many of the researchers in foundations again.[2] I admit that I didn't go to the big congress, but only to the symposium of the international academy ..., which took place in the same hotel.

In the days after the symposium I took part as well in the colloquium in Tihany (Hungary) on recursive functions, which was organized in honor of Rósza Péter.

You too have undoubtedly received a copy of the recently published book *From Frege to Gödel* by Jean van Heijenoort.[d] I find the historical and factual remarks accompanying the papers to be very instructive!

With the wish that things are going well for you, I send greetings to you and your wife.

Very cordially yours,

Paul Bernays

I agree completely with your remark that in the lecture "Sur le Platonism ..."[e] I did not assert the *necessity* of abstract concepts, and it accords perfectly with my view that you reformulate the passage of your *Dialectica* paper relating to it.

[2][Written vertically in the left margin of page 2:] In addition I spoke later on to Mr. Kreisel in Zurich.

---

[e]*Bernays 1935.*

63. Gödel to Bernays

Princeton, 20 December 1967

Dear Mr. Bernays,

Sincere thanks for your friendly letter of 27 October. I will have the translation of the *Dialectica* paper typed up immediately after the end of the Christmas vacation (i.e., at the beginning of January). Unfortunately, in the last few weeks my attention was too much taken up with other things. I would then also like to prepare as soon as possible the compilation of my remarks on the more recent development of research on the foundations of mathematics. When is the issue of *Dialectica* in question planned for publication?

Ich habe mich sehr gefreut, zu hören, dass die Neuauflage des Grundlagenbuches gute Fortschritte macht.

Ich hoffe, dass es Ihnen gesundheitlich u. auch in jeder anderen Hinsicht gut geht u. verbleibe mit besten Wünschen zum Jahreswechsel u. herzlichen Grüssen

Ihr Kurt Gödel

64. Gödel to Bernays

Princeton, 16./V. 1968.

Lieber Professor Bernays!

Herzlichen Dank für Ihren frdl. Brief vom 5./V u. die Ansichtskarte aus Oberwolfach. Es tut mir leid zu hören, dass Ihre Gesundheit nicht ganz zufriedenstellend ist. Meine besten Wünsche für die weitere Entwicklung Ihres Zustandes!

Ich bedaure ausserordentlich, dass ich Ihnen die versprochene Revision der Übersetzung meiner Dialectica Arbeit noch immer nicht einsenden konnte. Als ich das Manuskript vornahm, um es abtippen zu lassen, fand ich in der philosophischen Einleitung (d. h. den ersten $3\frac{1}{2}$ Seiten) in den Dialectica) vieles schon im Urtext so ~~unfe~~ unbefriedigend u. lückenhaft dargestellt,[1] dass ich zahlreiche Ergänzungen u. Änderungen für unbedingt nötig hielt. Schliesslich habe | ich dann diesen Teil gänzlich umgearbeitet u. den Umfang auf mehr als das doppelte ~~erhöh~~ vermehrt. Anfangs April wurde ich dann krank u. bin noch immer in ärztlicher Behandlung. Da aber der philosophische Teil jetzt im wesentlichen fertig ist, besteht immerhin die Möglichkeit, dass ich ihn als meinen Beitrag für das Symposium einschicken könnte. An dem math. Teil will ich nur sehr wenig ändern. Was meine übrigen in den letzten Jahren publizierten Äusserungen über die Grundlagen betrifft, die ja nicht sehr zahlreich sind, so kann ich Ihnen gerne eine Liste einschicken. Es dürfte ja genügen, sie zu zitieren statt neu abzudrucken.

Ich möchte Sie noch fragen, ob Sie in der Neuauflage des Grundlagenbuches die Kreiselsche Ableitung[a] der Induktion nach ϵ_0 (Siehe: Lectures

---

[1] Ich legte ja damals keinen besondern Wert auf das Philosophische, sondern es kam mir hauptsächlich auf das math. Resultat an, während es jetzt umgekehrt ist.

---

[a] See *Kreisel 1965*.

I was very pleased to hear that the new edition of the foundations book is making good progress.

I hope that things are going well for you healthwise and in every other respect, and I remain, with best wishes at the turn of the year and with cordial greetings,

Yours truly,

Kurt Gödel

64. Gödel to Bernays

Princeton, 16 May 1968

Dear Professor Bernays,

Cordial thanks for your friendly letter of 5 May and the picture post-card from Oberwolfach. I am sorry to hear that your health is not quite satisfactory. My best wishes for the further development of your condition!

I'm terribly sorry that I still was not able to send you the promised revision of the translation of my *Dialectica* paper. As I took up the manuscript in order to have it typed, I found in the philosophical introduction (i.e., the first $3\frac{1}{2}$ pages in *Dialectica*) much even in the original text [[that was]] presented in such an unsatisfying and fragmentary way[2] that I considered numerous supplementary remarks and changes as absolutely necessary. In the end, then, I completely rewrote that part and doubled its size. Then at the beginning of April I fell ill, and I'm still under the care of a physician. Since, however, the philosophical part is now essentially finished, the possibility nevertheless exists that I could send it along as my contribution for the symposium. In the mathematical part I want to make only a very few changes. As concerns my other remarks on foundations published in recent years, they are not after all very numerous, so I can readily send you a list [of them]]. It might suffice just to cite them instead of reprinting them.

I would still like to ask you whether in the new edition of the book on foundations you have taken account of Kreisel's derivation of induction up to ϵ_0.[a] (See: *Lectures on modern mathematics*, vol. 3, p. 172; ed. by

---

[2] At that time I placed no particular value on the philosophical matters; rather, I was mainly concerned with the mathematical result, whereas now it is the other way around.

on modern math., vol. 3, p 172; ed. by T. L. Saaty, 1965[b]) berücksichtigt
haben. Sie scheint mir dem Finiten wesentlich näher zu kommen als Ihr
früherer Beweis.

Mit herzlichen Grüssen u. besten Wünschen, auch von meiner Frau

Ihr Kurt Gödel

---

[b] *Saaty 1965.*

65. Bernays to Gödel

8002 Zürich, 20. Juli 1968
Bodmerstrasse 11

Lieber Herr Gödel!

Haben Sie vielen Dank für Ihren Brief vom 16. Mai. Ich bedauerte
sehr zu hören, dass Sie im April erkrankten und auch im Mai noch nicht
wieder wohl waren. Ich möchte hoffen, dass die ärztliche Behandlung erfolgreich war und dass Sie nunmehr von der Krankheit befreit sind.

Ihren Vorschlag, dass Sie die umgearbeitete philosophische Einleitung
Ihrer Dialectica Abhandlung als Beitrag für das Symposium senden wollen, begrüsse ich sehr. Vielleicht macht es Ihnen nicht viel Mühe, wenn
Sie den Beitrag auf deutsch abfassen. (Möglicherweise haben Sie sich
ohnehin die Aenderung der früheren Einleitung zunächst auf deutsch
überlegt.)

Für die von Ihnen angebotene Liste Ihrer publizierten Aeusserungen
über die Grundlagen werden wir auch sehr dankbar sein.

Was die Darstellung der Induktion nach ϵ_0 für die neue Auflage des
Grundlagenbuches betrifft, so habe ich ohnehin vor, in dem Supplement,
welches die Darstellung des Kalmárschen Wf-Beweises und des Ackermannschen (von 1940) bringen soll, einen Beweis jener Induktion zu
geben,[a] der sich enger an die 0-ω-Figuren anschliesst als der Beweis in
§5, der sich ⟨ja⟩ auf die zahlentheoretische Wohlordnung $a \preceq b$ ⟨stützt⟩.
Ich danke Ihnen übrigens ⟨sehr⟩ für den Hinweis auf den Kreiselschen
Artikel in den Lectures of modern mathematics.

---

[a] See *Hilbert and Bernays 1970*, pp. 533–535.

T. L. Saaty, 1965.[b]) It seems to me that it comes much closer to being finitary than your earlier proof.

With cordial greetings and best wishes, also from my wife,

Yours truly,

Kurt Gödel

---

65. Bernays to Gödel

8002 Zurich, 20 July 1968
Bodmerstrasse 11

Dear Mr. Gödel,

Many thanks for your letter of 16 May. I'm sorry to hear that you were sick in April and were still not well again even in May. I would like to hope that the medical treatment was successful and that by this time you are rid of the illness.

I very much welcome your proposal that you will send the revised philosophical introduction to your *Dialectica* paper as a contribution for the symposium. Perhaps it is not much trouble for you to write the contribution in German. (Possibly you first thought about the alteration to the earlier introduction in German anyway.)

We will also be very grateful for the list you offered of your published remarks on foundations.

As to the presentation of induction up to ϵ_0 for the new edition of the foundations book, I intend at any rate to give in the supplement[a]— which is to contain the presentation of Kalmár's consistency proof and that of Ackermann (from 1940)—a proof of that induction that is more closely tied to the 0-ω-figures than the proof in §5, which is based on the number-theoretic well-ordering $a \preccurlyeq b$. I thank you very much, moreover, for the reference to Kreisel's article in the *Lectures of modern mathematics*.

In der Hoffnung, dass Sie bald in der Lage sind, den Symposium Beitrag
und die revidierte Uebersetzung Ihrer Dialectica Abhandlung uns zu
schicken,[1] und mit herzlichen Grüssen an Sie und Ihre Frau

Ihr Paul Bernays

[1]Möchten Sie die Sendung an mich adressieren, da Herr Engeler in diesem Sommer nach der Schweiz kommt.

66. Gödel to Bernays

Princeton, 17./XII. 1968

Lieber Herr Bernays!

Entschuldigen Sie bitte, dass ich so lange nichts von mir hören liess.
Leider war mein Gesundheitszustand in den letzten Monaten ziemlich
schlecht. Erst seit kurzem fühle ich mich wieder etwas kräftiger. Ich
möchte Ihnen auch, zwar verspätet, aber darum nicht weniger herzlich,
meine besten Glückwünsche zur Erreichung des 80-ten Lebensjahres
übermitteln.

Was die Übersetzung meiner Dialectica-Arbeit betrifft, so habe ich die   A
Idee einer neuen philosophischen Einleitung wieder aufgegeben, da mir
die neue schliesslich ebensowenig gefiel als die alte. Ich habe statt dessen
ca. 1200 Worte in erklärenden Anmerkungen hinzugefügt, die dem Leser
bestimmt sehr erwünscht sein werden. Besondere Schwierigkeiten machte
⟨es⟩ mir, eine befriedigende Formulierung der zweiten Hälfte der Fussnote 1 auf p. 283 zu finden. Das dort gesagte ist zwar nicht mathematisch, aber doch erkenntnistheoretisch sehr wichtig. Es scheint aber sehr
2   schwierig (oder unmöglich) zu sein, es zu präzisieren u. | dabei doch in
vollem Umfange aufrechtzuerhalten. Ich beabsichtige Ihnen bald ausführlicher zu schreiben. Ich vermute, es ist schon zu spät, einen Beitrag für
das schriftliche Symposium einzusenden.

Mit besten Wünschen für die Weihnachtsfeiertage u. das kommende
Jahr u. herzlichen Grüssen, auch von meiner Frau

Ihr Kurt Gödel

In the hope that you are soon in the position to send[1] the symposium contribution and the revised translation of your *Dialectica* paper, and with cordial greetings to you and your wife,

Yours truly,

Paul Bernays

---

[1]Would you please address the mailing to me, since Mr. Engeler is coming to Switzerland this summer.

66. Gödel to Bernays

Princeton, 17 December 1968

Dear Mr. Bernays,

Please excuse that you haven't heard from me for such a long time. Unfortunately my state of health was rather bad in recent months. Only a short time ago did I feel somewhat stronger again. I would also like (belatedly, to be sure, but no less cordially on that account) to convey my most sincere congratulations to you on reaching the 80th year of your life.

As to the translation of my *Dialectica* paper, I've again given up the idea of a new philosophical introduction, since in the end I liked the new one just as little as the old. Instead I've appended about 1200 words of clarifying remarks, which to the reader will certainly be very welcome. I had particular difficulties in finding a satisfactory formulation of the second half of footnote 1 on p. 283. What is said there is very important, not mathematically, to be sure, but rather epistemologically. But it seems to be very difficult (or impossible) to make it precise and, in doing so, still maintain it to the full extent. I intend to write you in detail soon. I presume it is already too late to send a contribution for the written symposium.

With best wishes for the Christmas holidays and the coming year, and with cordial greetings, also from my wife,

Yours truly,

Kurt Gödel

67. Bernays to Gödel

8002 Zürich, 6. I. 1969
Bodmerstr. 11.

Lieber Herr Gödel!

Haben Sie vielen Dank für Ihren freundlichen Brief vom 17.XII. und
für Ihre Glückwünsche, die Sie mir noch zu meinem 80-ten Geburtstag
sandten.

Es tat mir sehr leid zu hören, dass Ihr Ergehen bis vor kurzem so
schlecht war. Hoffentlich hält die eingetretene Besserung weiter an.

Mit der Revision der Übersetzung Ihrer Dialectica-Abhandlung hatten
Sie ja noch recht viel zu tun. Doch da sich ja an diese Abhandlung so
2   vieles knüpft, so werden sich die | erklärenden Hinzufügungen jedenfalls
lohnen.—Sie können mir nun wohl schon bald die Abhandlung für die
Dialectica schicken.

Was das schriftliche Symposium betrifft, so sagte mir Herr Gonseth,
dass es auf zwei Dialectica-Hefte verteilt werden soll. Es ist daher noch
nicht zu spät, wenn Sie noch einen Beitrag einsenden wollen.

Ich lege diesen Zeilen die Fassung des Beweises für die Induktion bis
ϵ_0 bei, wie ich sie mir für die zweite Auflage des Grundlagenbuchs (zum
Anschluss an den Kalmár'schen Wf.-Beweis) zurechtgelegt habe; d. h. es
handelt sich um die schwache Form der Induktion, die besagt, dass jede
3   absteigende Folge nach endlich vielen Schritten zum | Abschluss kommt[a]
—Der erste Band von der 2-ten Auflage des Grundlagenbuchs ist kürzlich
erschienen. Ich will Ihnen ein Exemplar schicken.

Ihre Wünsche für das begonnene Jahr erwidere ich aufs beste! Seien
Sie und Ihre Frau herzlich gegrüsst

von Ihrem Paul Bernays

---

[a]The text Bernays included consists, literally, of the definition of the 0-ω-figures as
presented in *Hilbert and Bernays 1970* on p. 525 and the proof of the "Abstiegsendlichkeit" as given there on pp. 533–535.

67. Bernays to Gödel

8002 Zurich, 6 January 1969
Bodmerstrasse 11

Dear Mr. Gödel,

Many thanks for your friendly letter of 17 December and for your congratulatory wishes, which you sent me still for my 80th birthday.

I was very sorry to hear that your health was so bad up to a short time ago. Hopefully the improvement that took place has continued further.

You certainly still had quite a lot to do with revising the translation of your *Dialectica paper*. Yet since so much is tied to that paper, the clarifying additions will be worthwhile in any case.—Now you can probably send me the paper for *Dialectica* very soon.

As to the written symposium, Mr. Gonseth told me that it is supposed to be split into two issues of *Dialectica*. Therefore it is not yet too late, if you still want to send in a contribution.

With these lines I am enclosing the version of the proof for induction up to ϵ_0, as I laid it out for the second edition of the foundations book (in connection with Kalmár's consistency proof); i.e., it is about the weak form of induction, which says that every decreasing sequence comes to an end after finitely many steps.[a]—The first volume of the second edition of the foundations book has recently appeared. I will send you a copy.

I reciprocate most heartily your wishes for the year ahead! Cordial greetings to you and your wife

from yours truly, Paul Bernays

68a. Gödel to Bernays[a]

Princeton, /VII. 1969

Lieber Herr Bernays!

Vielen Dank für die Zusendung der Neuauflage des I. Band der Grundlagen der Mathematik u. für Ihren ~~sehr~~ *interessanten* Brief vom Jänner,
sowie auch für Ihre freundliche Karte aus Oberwolfach; u. bitte entschuldigen Sie die lange Schreibpause meinerseits. In der Beilage Ihres Briefes
haben Sie zweifellos den bisher überzeugendsten Beweis für den Ordi-
nalzahlcharakter von ϵ_0 gegeben. Sie verwenden ja nur konstante Funktionen vom zweiten u. variable vom ersten Niveau u. ein Rekursions-
schema das zweifellos finit ist. Da man Funktionen des ersten Niveaus
als freie Wahlfolgen interpretieren kann u. dieser Begriff offenbar ent-
scheidbar ist, so enthält ein Satz der Form: "Für alle freien Wahlfolgen
...." keine intuitionistische Implikation u. Sie haben daher die intuitionistische Logik vollkommen eliminiert. Wenn man die Wahlfolgen zur
finiten Mathematik rechnet ist Ihr Beweis sogar finit. Ich möchte diese
interessante Tatsache in einer (diesem Brief beiliegenden) Anmerkung
zu Fussn. 2, p. 286 meiner Dialectica Arbeit erwähnen. Bitte lassen Sie
mich wissen ob Sie mit dem Gesagten übereinstimmen, *insbes. ob Hilbert
die freien Wahlfolgen wirklich als nicht finit betrachtete.* Haben Sie etwas
dagegen wenn ich Ihren Beweis Kreisel oder Dana Scott zeige? Ich habe
jetzt endlich eine befriedigende Präzisierung des inder zweiten Hälfte
der Fussn. 1, p. 283 Gesagten gefunden. Ich glaube man kann wirklich
jede Zirkularität in der Definition der logischen Konstanten vermeiden.
Andrerseits bin ich jetzt stark im Zweifel ob das auf p. 280–281 über die
Grenzen des Finitismus Gesagte wirklich stimmt. Denn er scheint mir
jetzt, nach reiflicher Überlegung, dass die Wahlfolgen etwas Anschauliches
u. daher im Hilbertschen Sinn Finites sind, wenn auch Hilbert selbst
vielleicht anderer Meinung war.

[Text of Gödel's proposed remark to footnote 2 of his *Dialectica* paper:]

Another possibility is the admission of more intricate schemata of recursive definition. Bernays has given a most interesting proof (to be published in the 2nd ed. of "Grundlagen der Mathematik") of the finiteness
of descending sequences of ordinals $< \epsilon_0$, which uses only functional constants of level 2 and variables of level one, combined with a schema of
recursive definition which is doubtless finitary. Since functions of level 1

---

[a]Unsigned draft, bearing the shorthand annotation "nicht abgeschickt" ["not sent"].

68a.   Gödel to Bernays[a]

Princeton, July 1969

Dear Mr. Bernays,

Many thanks for sending the new edition of the first volume of *Grundlagen der Mathematik* and for your *interesting* letter of January, as well as for your friendly card from Oberwolfach; and please excuse the long lapse in correspondence on my part. In the enclosure with your letter you undoubtedly have given the most convincing proof to date of the ordinal-number character of ϵ_0. You employ only constant functions of the second level and variables of the first level and a recursion schema that is undoubtedly finitary. Since functions of the first level can be interpreted as free choice sequences and that concept is obviously decidable, a statement of the form "For all free choice sequences ...." contains no intuitionistic implication, and you have consequently completely eliminated the intuitionistic logic. If one reckons choice sequences to be finitary mathematics, your proof is even finitary. I would like to mention this interesting fact in a remark (enclosed with this letter) to footnote 2, p. 286 of my *Dialectica* paper. Please let me know whether you agree with what is said there, *in particular, whether Hilbert really considered the free choice sequences not to be finitary*. Do you have any objection to my showing your proof to Kreisel or Dana Scott?

I have now at last found a satisfactory way of making precise what I said in the second half of footnote 1, p. 283. I think every circularity in the definition of the logical constants really can be avoided. On the other hand I now strongly doubt whether what was said about the boundaries of finitism on pp. 280–281 is really right. For it now seems to me, after more careful consideration, that choice sequences are something concretely evident and therefore are finitary in Hilbert's sense, even if Hilbert himself was perhaps of another opinion.

〚Text of Gödel's proposed remark to footnote 2 of his *Dialectica* paper:〛

Another possibility is the admission of more intricate schemata of recursive definition. Bernays has given a most interesting proof (to be published in the 2nd ed. of *Grundlagen der Mathematik*) of the finiteness of descending sequences of ordinals $< \epsilon_0$, which uses only functional constants of level 2 and variables of level one, combined with a schema of recursive definition which is doubtless finitary. Since functions of level 1

may be interpreted as free choice sequences (which is a decidable concept) the use of intuitionistic logic (or of the concept of evidence referred to in note (k)[a]) is thus completely avoided. Hilbert did not regard choice sequences (or recursive functions of them) as finitary, but this position may be challenged on the basis of Hilbert's own point of view.

---

[a]See footnote bf on p. 72.

68b. Gödel to Bernays

Princeton, 25./VII. 1969.

Lieber Herr Bernays!

Ich danke herzlich für die Zusendung der Neuauflage des I. Bd. der Grundlagen der Mathematik u. für Ihren interessanten Brief vom Jänner, sowie auch für Ihre freundliche Karte aus Oberwolfach; u. bitte entschuldigen Sie die lange Schreibpause meinerseits.—Der Beweis für den Ordinalzahlcharakter von ϵ_0, den Sie in der Beilage Ihres Briefes geben, ist ausserordentlich elegant u. einfach. Man hat zunächst auch den Eindruck, dass er dem Finitismus näher kommt als die anderen Beweise. Bei näherer Überlegung scheint mir das aber sehr zweifelhaft. Die Eigenschaft "abstiegsendlich" enthält ja zwei Quantoren u. einer davon bezieht sich auf alle Zahlfolgen (was ⟨die⟩ wohl als Wahlfolgen zu interpretieren wären). Um die Quantoren zu eliminieren, müsste man jeder absteigenden Folge von Figuren $k+1$-ter Stufe eine ⟨oder mehrere⟩ k-ter Stufe so zuordnen, dass man aus dem Abbrechen der letzteren finit auf das Abbrechen der ersteren schliessen kann. Es scheint mir, dass man dazu eine verschränkte Rekursion brauchen würde. Verschränkte Rekursionen sind aber nicht finit im Hilbertschen Sinn (d. h. nicht anschaulich), obzwar wohl jeder Mathematiker sie ebenso überzeugend finden wird wie primit. rek. Def.[1] Oder glauben Sie das nicht? Dass man mit verschränkten Rekursionen (sogar ohne Wahlfolgen) bis ϵ_0 kommt, hat meines Wissens Tait in Math. Ann. 143 (1961) p. 236[a] bewiesen. | *Hilbert wollte Wahlfolgen wohl nicht zulassen?* Mir scheinen sie durchaus anschaulich zu sein, aber den Finitismus nicht wesentlich zu erweitern.—Ich habe jetzt endlich eine befriedigende Präzisierung der zweiten Hälfte der Fussn. 1, p. 283 meiner Dialectica Arbeit gefunden. Man kann wirklich

---

[1]Allerdings gilt das nur für verschränkte Rekursionen nach *kleinen* Ordinalz. (z. B. ω^n) u. diese werden wohl nicht hinreichen.

---

[a]*Tait 1961.*

may be interpreted as free choice sequences (which is a decidable concept), the use of intuitionistic logic (or of the concept of evidence referred to in note (k)[a]) is thus completely avoided. Hilbert did not regard choice sequences (or recursive functions of them) as finitary, but this position may be challenged on the basis of Hilbert's own point of view.

68b. Gödel to Bernays

Princeton, 25 July 1969

Dear Mr. Bernays,

I thank you cordially for sending the new edition of the first volume of *Grundlagen der Mathematik* and for your interesting letter of January, as well as for your friendly card from Oberwolfach; and please excuse the long lapse in correspondence on my part.

The proof for the ordinal number character of ϵ_0 that you give in the enclosure with your letter is extraordinarily elegant and simple. At first one also has the impression that it comes closer to finitism than the other proofs. But on closer reflection that seems very doubtful to me. The property of being "well-founded" contains two quantifiers after all, and one of them refers to all number sequences (which probably are to be interpreted as choice sequences). In order to eliminate the quantifiers, one would have to assign to each decreasing sequence of figures of level $k + 1$ one or several of level k, in such a way that from the termination of the latter one can finitistically infer the termination of the former. It seems to me that one would use a nested recursion for that. But nested recursions are not finitary in Hilbert's sense (i.e., not intuitive), although probably every mathematician will find them just as convincing as primitive recursive definitions.[1] Or don't you believe that? To my knowledge, Tait proved in *Mathematische Annalen* 143 (1961), p. 236,[a] that one gets up to ϵ_0 with nested recursions (even without choice sequences). *Hilbert, I presume, didn't want to permit choice sequences?* To me they seem to be quite concrete, but not to extend finitism in an essential way.

I've now at last found a satisfying way of making precise the second half of footnote 1, p. 283, of my *Dialectica* paper. One can actually avoid

---

[1] Of course, that holds only for nested recursions on *small* ordinal numbers (e.g. ω^n), and this probably will not suffice.

jede Zirkularität in der Def. der logischen Konstanten vermeiden. Ich
glaube, ich werde jetzt an der revidierten Übersetzung der Dialectica
Arbeit u. den hinzugefügten Anmerkungen kaum noch etwas ändern.—
Ich möchte schliesslich noch eine Frage über Hilbert's Arbeit "Über das
Unendliche"[b] stellen. Auf p. 181, unmittelbar vor Lemma I, spricht er
von einem allgem. metamath. Lemma. Auf Grund des vorher über die
Lösbarkeit aller math. Probleme Gesagten sollte man glauben, dass dieses Lemma so lautete: Man kann widerspruchsfrei annehmen, dass jede
Zahlenfolge rekursiv definierbar ist. War wirklich das gemeint? In diesem
Fall wäre die Struktur des intendierten Widerspruchsfreiheitsbew. der
Kont. Hyp. der meines Beweises tatsächlich *sehr* ähnlich. Man brauchte
dann nur Hilbert's transfinit rekursive Funktionen durch meine "konstruktiblen" zu ersetzen. Vielleicht sollte man das zu meiner Bemerkung,
die Jean v. Heijenoort in "From Frege to Gödel"[c] (Harvard Un. Press
1967) p. 368 publiziert hat, hinzufügen.

Ich hoffe, dass Ihr Gesundheitszustand zufriedenstellend ist, u. verbleibe mit herzlichen Grüssen, auch von meiner Frau

Ihr Kurt Gödel

3 | P.S. Soeben entnehme ich aus einem Brief von Kreisel, dass ein Schüler
von ihm (Ressayre) bewiesen hat, dass man mit verschränkten Rekursionen *nach* ω u. Wahlfolgen genau bis ϵ_0 kommt. Wenn das wirklich
stimmt, müsste es sich wohl auch aus einer Analyse Ihres Beweises ergeben u. würde dann zeigen, dass dieser doch dem Finitismus wesentlich
näher kommt als alle bisher bekannten Beweise, einschliesslich des Tait-
schen. Nach meiner Meinung müsste auf Grund davon der Gentzensche
Widerspruchsfreiheitsbew. für jeden Mathematiker ganz ebenso überzeugend sein wie ein primitiv rekursiver Beweis, wenn er auch nicht finit
im Hilbert-schen Sinn ist, weil der Hilbertsche Finitismus (durch die
Forderung der "Anschaulichkeit") eine ganz unnatürliche Grenze hat.
Würden Sie darin mit mir übereinstimmen?

Princeton, 27./VII.1969.

P.S. (zu meinem Brief vom 25./VII.). Es ist mir eingefallen, dass das
Ressayresche Resultat nicht stimmen kann. Verschränkte Rekursionen

---

[b] *Hilbert 1926.*

every circularity in the definition of the logical constants. I think I will now change hardly anything else in the revised translation of the *Dialectica* paper and in the added remarks.

Finally, I would still like to ask you a question about Hilbert's paper "Über das Unendliche".[b] On p. 181, directly before Lemma I, he speaks of a general metamathematical lemma. On the basis of what was said before about the solvability of all mathematical problems, one should think that this lemma runs thus: One can consistently assume that every number sequence is recursively definable. Was that really what was meant? In that case the structure of the intended proof of the consistency of the continuum hypothesis would really be *very* similar to my proof. One would then need only to replace Hilbert's transfinite recursive functions by my "constructible" ones. Perhaps that should be added to my remark Jean van Heijenoort published in "From Frege to Gödel" (Harvard U. Press 1967), p. 368.[c]

I hope that your state of health is satisfactory and remain, with cordial greetings, also from my wife,

Yours truly,

Kurt Gödel

P.S. Just now I learned from a letter of Kreisel that a student of his (Ressayre) has proved that with nested recursions *on* ω and choice sequences one gets precisely up to ϵ_0. If that is really correct, it would probably have to emerge also from an analysis of your proof and would then show that the latter indeed comes essentially closer to finitism than all proofs known hitherto, including Tait's. In my opinion, on the basis of [such an analysis] Gentzen's consistency proof would have to be exactly as convincing for any mathematician as a primitive recursive proof, even if it is not finitary in Hilbert's sense, because Hilbert's finitism (through the requirement of being "intuitive") has a quite unnatural boundary. Would you agree with me in that?

Princeton, 27 July 1969

P.S. (to my letter of 25 July). It has occurred to me that Ressayre's result cannot be correct. Nested recursions on ω can be reduced to

---

[c] *van Heijenoort 1967.*

nach ω kann man doch auf prim. rek. höheren Typs zurückführen. Also
könnte man ϵ_0 mit einigen wenigen Typen meines Systems T der Arbeit
in Dial. 12 erreichen, was offenbar unmöglich ist.

Herzliche Grüsse

Ihr Kurt Gödel

69. Bernays to Gödel

8002 Zürich, 7. Januar 1970
Bodmerstr. 11.

Lieber Herr Gödel!

Zuerst möchte ich Ihnen vielmals für Ihre freundlichen Neujahrsgrüsse
danken und Ihnen meine herzlichen Wünsche für Ihr Ergehen im be-
ginnenden Jahre aussprechen. Möchten insbesondere die verschiedenen
Umstände, von denen Sie schreiben, dass sie Ihre Arbeit sehr behinder-
ten, möglichst bald behoben sein!—
Hauptsächlich aber sollen diese Zeilen der Erwiderung Ihres inhaltrei-
chen Briefes dienen, den Sie mir im Spätsommer sandten. Sie entschul-
digten sich damals wegen einer längeren Schreib⟨e⟩pause. Wie viel mehr
muss ich mich heute entschuldigen, da ich Ihnen so lange nicht geantwortet habe! Die hauptsächliche Ursache der argen Verzögerung war die,
dass ich jedesmal, wenn ich mich anschickte, Ihren Brief zu erwidern, den
Eindruck hatte, ich sollte über die darin berührten Fragen erst noch etwas nachdenken.
Sie knüpften an die Ihnen von mir (in meiner damaligen Briefbeilage)
vorgelegte Fassung des Beweises für die Gültigkeit der Induktion bis ϵ_0,
im Sinne der "Abstiegsendlichkeit", an. Sie haben recht, dass auch bei
diesem Beweise der deutliche Abstand gegenüber den "finiten" Methoden
bestehen bleibt, indem Allheiten über Wahlfolgen genommen werden und
solche auch in Vordergliedern von Implikationen auftreten.
Sie meinten, dass die Elimination der Quantoren eventuell mittels verschränkter Rekursion gelinge, und bemerkten anknüpfend, dass ⟨William⟩
Tait in der Abhandlung Mathematische Annalen 143[a] bewiesen ~~habe~~
⟨hat⟩, dass man mit verschränkter Rekursion⟨en⟩ bis zu ϵ_0 kommt. Ich

---

[a] *Tait 1961.*

primitive recursion of higher type. Thus one could reach ϵ_0 with some few types of my system T of the paper in volume 12 of *Dialectica*, which is obviously impossible.

Cordial greetings,

Yours truly, Kurt Gödel

69. Bernays to Gödel

8002 Zurich, 7 January 1970
Bodmerstrasse 11

Dear Mr. Gödel,

First I would like to thank you for your friendly New Year's greeting and to express to you my cordial wishes for your health in the year ahead. In particular, may the various circumstances which you wrote were very much hindering your work be resolved as soon as possible!

But chiefly these lines should serve as the reply to your substantive letter you sent me in late summer. At that time you apologized for a long lapse in correspondence. How much more must I apologize today, since I've not answered you for so long! The principal cause of the bad delay was that every time I set out to reply to your letter I had the impression that I first ought to think a bit more about the questions mentioned in it.

You referred to the version of the proof for the validity of induction up to ϵ_0 in the sense of "well-foundedness" that I presented to you at that time (in an enclosure to my letter). You are correct that even in this proof the clear distance to "finitary" methods persists, in that universal generalizations are taken over choice sequences and occur also in antecedents of implications.

You believed that the elimination of the quantifiers may perhaps be accomplished by means of nested recursion, and remarked in that connection that William Tait proved in the paper [published in] Mathematische Annalen, [vol.] 143,[a] that one gets up to ϵ_0 with nested recursion. I've

2   habe in diese Tait'sche Abhandlung einen (allerdinges nur | unvollkom-
menen) Einblick getan. So viel ich sehe, handelt es sich dabei nicht um
verschränkte Rekursionen *in dem Sinne*, wie der Ausdruck in den "Grundlagen der Mathematik" gebraucht wird, sondern die behandelten "nested
recursions" sind eingeschachtelte Rekursionen, die nur in Bezug auf *ein*
Argument der zu definierenden Funktion fortschreiten (während für die
anderen Argumente Einsetzungen gemacht werden können). ~~Diese~~ ⟨Und
zwar⟩ sind ⟨es⟩ im allgemeinen nicht gewöhnliche Rekursionen in dem
Sinne, dass der Rückgang von $n + 1$ zu n stattfindet, sondern ordinale
Rekursionen, bei denen der Rückgang gemäss einer Wohlordnung der
Zahlenreihe erfolgt.

Nun scheint mir aber folgende Alternative zu bestehen: Entweder der
Typus der betreffenden Wohlordnung ist kleiner als ϵ_0; dann kann ⟨doch
wohl⟩ die Verwendung einer solchen Rekursion nicht die Überschreitung
des zahlentheoretischen Formalismus ⟨(Z_μ)⟩ liefern,[1] da ja die Induktion
bis zu einer kleineren Ordinalzahl als ϵ_0 noch in (Z_μ) beweisbar ist. Oder
jene Wohlordnung hat mindestens den Typus ϵ_0; dann wird bei der Anwendung der ordinalen Rekursion implizite die Abstiegsendlichkeit für ϵ_0
benutzt.

Was andererseits jene verschränkten Rekursionen betrifft, die in den
"Grundl. der Math." betrachtet werden, so haben Sie ja früher (zugleich
wie von Neumann) gezeigt, dass diese nicht über den zahlentheoretischen
Formalismus (Z_μ) hinausführen. Diese verschränkten Rekursionen (wel-
é   che ja von Rózsa Péter als "mehrfache Rekursionen" bezeichnet werden) erscheinen mir übrigens als im selben Sinne finit wie die primitiven
Rekursionen, d. h. wenn man sie auffasst als Angabe von Berechnungsverfahren, bei denen sich erkennen lässt, dass die durch das jeweilige
Verfahren bestimmte Funktion den Rekursionsgleichungen (für jedes Sy-
3   stem von Ziffernwerten der | Argumente) genügt. In der Tat kommt ja
die Berechnung des Wertes einer Funktion gemäss einer verschränkten
Rekursion bei gegebenen Ziffernwerten der Argumente auf die Anwendung mehrerer primitiver Rekursionen hinaus, deren Anzahl jeweils durch
ein Ziffernargument bestimmt ist.—

V   Die Art der Verwendung von Wahlfolgen, wie sie bei dem Beweis der
Abstiegsendlichkeit von ϵ_0 erfolgt, findet sich ähnlich auch bei dem *ursprünglichen* Gentzen'schen Widerspruchsfreiheitsbeweis, von dem ich
seinerzeit meinte, dass er das *fan theorem* implizite benutze. Zur Zeit
meines erstmaligen Aufenthaltes in Princeton war ⟨ja⟩ davon die Rede.
Das gab den Anlass zu der Opposition gegen den Beweis, aufgrund deren

---

[1] Vorausgesetzt wenigstens, dass für die in der Rekursion benutzte Funktion $\theta(\hat{x}, y)$
die Eigenschaft des ordinalen Absteigens $\theta(\hat{x}, y') < y'$ in Z_μ bewiesen werden kann.

attained an (admittedly only incomplete) insight into this paper of Tait's.
So far as I see, it is not about nested recursions *in the sense* that the expression is used in *Grundlagen der Mathematik*, but rather, the "nested recursions" dealt with are interlocked recursions that only progress with respect to *one* argument of the function to be defined (whereas for the other arguments substitutions can be made). And in fact they are in general not ordinary recursions, in the sense that they recur from $n + 1$ to n, but rather ordinal recursions, recurring in accordance with a well-ordering of the number sequence.

Now, however, the following alternative seems to me to exist: Either the [order-]type of the well-ordering under consideration is less than ϵ_0—then, I think you'll agree, the use of such a recursion cannot provide the means of going beyond the number-theoretic formalism (Z_μ),[1] since induction up to an ordinal less than ϵ_0 is of course still provable in (Z_μ)—or, [alternatively,] that well-ordering has [order-]type at least ϵ_0; then in the application of the ordinal recursion the well-foundedness of ϵ_0 is used implicitly.

On the other hand, as to those nested recursions that are considered in *Grundlagen der Mathematik*, you showed earlier (at the same time as von Neumann) that they do not go beyond the number-theoretic formalism (Z_μ). Those nested recursions (which by Rózsa Péter are called "multiple recursions") appear to me, moreover, as finitary in the same sense as the primitive recursions, i.e., if one views them as a description of computational procedures for which it can be seen that the function determined by the specific procedure satisfies the recursion equations (for each system of numerical values of the arguments). As a matter of fact the computation of the value of a function in accordance with a nested recursion with given numerical values of the arguments amounts to the application of several primitive recursions whose number is determined in each case by a numerical argument.

The way in which choice sequences are used in the proof of the well-foundedness of ϵ_0 is similar to that used in Gentzen's *original* consistency proof, which at the time I thought implicitly used the *fan theorem*. At the time of my first visit to Princeton there was indeed talk of that. That gave rise to opposition to the proof, on the basis of which Gentzen

---

[1] Assuming at least that for the function $\theta(\hat{x}, y)$ employed in the recursion the property of ordinal diminution, $\theta(\hat{x}, y') < y'$, can be proved in Z_μ.

Gentzen ihn damals zurückzog und durch denjenigen mit Hilfe der Induktion bis ϵ_0 ersetzte. Heute glaube ich nicht mehr, das der ursprüngliche Gentzen'sche Beweis das fan theorem erfordert. Ich habe im letzten Sommer zu der Konferenz in Buffalo eine (ein wenig vereinfachte) Darstellung jenes ursprünglichen Beweises ⟨(mit einer einleitenden historischen Bemerkung)⟩ eingesandt, die in den Kongressberichten erscheinen soll. Ausserdem ist jener ursprüngliche Beweis in die kürzlich erschienene Publikation der Gesammelten Gentzen'schen Abhandlung⟨en⟩[b] auf Englisch durch Manfred Szabo aufgenommen.—

Sie fragten, ob wohl Hilbert die Wahlfolgen nicht zulassen wollte. So viel ich weiss, hat Hilbert zu den Wahlfolgen überhaupt nie Stellung genommen.

Was Ihre Frage zu dem Lemma I in Hilberts Abhandlung über das Unendliche betrifft, so meinte in der Tat Hilbert mit diesem Lemma, das man ohne Widerspruch annehmen könne, dass bei Zulassung höherer Variablentypen jede zahlentheoretische Funktion sich rekursiv definieren lasse. (Dazu ist allerdings zu bemerken, dass die formale Einführung der Typenprädikate wohl nicht ohne Anwendung | von Quantoren gelingt.)—

Sie schrieben, dass Sie eine befriedigende Präzisierung der zweiten Hälfte der Fussnote 1 auf S. 283 Ihrer Dialectica-Arbeit gefunden hätten und dass Sie nun an der revidierten Übersetzung der Dialectica-Arbeit mit den hinzugefügten Anmerkungen kaum noch etwas ändern würden. Wäre es Ihnen daraufhin nicht möglich, die Übersetzung in ihrer nun vorliegenden Fassung an Herrn Gonseth (12, chemin du Muveran, CH-1012 Lausanne) für die Dialectica einzusenden? Herr Gonseth würde sich über den Empfang gewiss sehr freuen.

Mit herzlichen Grüssen auch an Ihre Frau,

Ihr Paul Bernays

---

[b] *Gentzen 1969.*

70. Bernays to Gödel

CH-8002 Zürich, 12. Juli 1970
Bodmerstr. 11

Lieber Herr Gödel!

Schon seit längerem habe ich vor, Ihnen zu schreiben. Entschuldigen Sie freundlichst, dass es erst heute geschieht. (Mit der Ausführung meiner Vorhaben komme ich leider nicht so voran, wie ich wünschte.)

then withdrew it and replaced it with the one by induction up to ϵ_0. Today I no longer think that Gentzen's original proof requires the fan theorem. Last summer I submitted to the conference in Buffalo a (somewhat simplified) presentation of that original proof (with an introductory historical remark) which is supposed to appear in the report of the conference. In addition, that original proof has been rendered in English by Manfred Szabo in the recently published collection of Gentzen's papers.[b]

You asked whether Hilbert didn't want to admit choice sequences. As far as I know, Hilbert never took a position on choice sequences at all.

As to your question about Lemma I in Hilbert's paper on the infinite, Hilbert actually did mean by that lemma that one can assume without contradiction that by admitting variables of higher type every number-theoretic function can be recursively defined. (On that point it is of course to be noted that the formal introduction of the type predicates presumably doesn't work without the use of quantifiers.)

You wrote that you had found a satisfying way of making precise the second half of footnote 1 on p. 283 of your *Dialectica* paper and that you would now change hardly anything else in the revised translation of the *Dialectica* paper with the added remarks. On the strength of that, would it not be possible to submit the translation, in its present version, to Mr. Gonseth (12 chemin du Muveran, CH-1012 Lausanne) for *Dialectica*? Mr. Gonseth would certainly be very pleased to receive it.

With cordial greetings, also to your wife,

Yours truly,

Paul Bernays

---

70. Bernays to Gödel

CH-8002 Zurich, 12 July 1970
Bodmerstrasse 11

Dear Mr. Gödel,

I had intended to write you for a long while. Excuse most kindly that it happened only today. (In carrying out my intention, I unfortunately did not progress as much as I desired.)

Es war mir sehr erfreulich, Ihre überarbeitete Fassung Ihrer ins Eng-
c lische übersetzten Dialektika-Abhandlung[a] sowie Ihre Note "Some remarks . . ."[b] zu empfangen, die Sie mir durch Herrn Dana Scott zugehen
liessen. Herr Gonseth freut sich natürlich auch sehr, dieses beides nun in
den Dialectica bringen zu können.

Beim Lesen Ihrer neuen Fassung habe ich mir wiederum die haupt-
sächlichen Verifikationen überlegt und dabei gesehen, wie förderlich Ihre
hinzugefügten Notes für diese Verifikationen sind. Überhaupt haben Sie
ja durch diese Notes Ihre Abhandlung sehr bereichert und dem Leser
auch zugänglicher gemacht.

Auf eine Einzelheit, die mir auffiel, wollte ich Sie aufmerksam machen.
Auf S.3 Ihres Manuskriptes, in dem Absatz beginnend mit den Worten
"The concept 'computable function of type t' is defined as follows" heisst
2 es (in Zeile 7-9 dieses Absatzes): "and yields | a computable function of
type t_0 as result, and for which, moreover, this general fact is intuitionistically demonstrable."

Hier könnte wohl der Leser stutzen, da doch Ihr Verfahren bezweckt,
den Begriff des intuitionistischen Beweises zu vermeiden. Es scheint mir
jedoch, dass Sie de facto hier diesen Begriff auch gar nicht brauchen und
dass es nur einer geeigneten Umformulierung bedarf, um dieses zum Ausdruck zu bringen.

Falls Sie dem zustimmen, so bitte ich Sie, mir anzugeben, wie Sie den
Passus geändert haben möchten.

Übrigens: Würde es Ihnen recht sein, dass man mir die Druckbogen
zur Durchsicht schickt, oder legen Sie Wert darauf, die Korrekturen
selbst auszuführen?

Es tat mir sehr leid zu hören, dass es Ihnen längere Zeit gar nicht gut
ging. Inzwischen schrieb mir Herr Dana Scott im Juni, dass Ihr Befinden
sich sehr gebessert hat. Hoffentlich hält diese Besserung weiter an!

Mit diesem Wunsche und mit herzlichen Grüssen, auch an Ihre
Frau,

Ihr Paul Bernays

---

[a] *1972*; cf. sections 3.1 and 3.3 of the introductory note.

It was very pleasing to me to receive your rewritten version[a] of your *Dialectica* paper translated into English, as well as your note "Some remarks ...",[b] which you had forwarded to me via Dana Scott. Mr. Gonseth is of course also very pleased to be able to publish both of these now in *Dialectica*.

In reading your new version I pondered anew the main verifications and saw in them how beneficial your added notes are for these verifications. In general, through these notes you have very much enriched your paper and also made [it] more accessible to the reader.

I wanted to call your attention to one particular detail that struck me. On p. 3 of your manuscript, in the paragraph beginning with the words "The concept 'computable function of type t' is defined as follows", it says (in lines 7–9 of that paragraph): "and yields a computable function of type t_0 as result, and for which, moreover, this general fact is intuitionistically demonstrable."

Here the reader could well be taken aback, since your procedure is surely intended to avoid the concept of intuitionistic proof. It seems to me, however, that in fact you do not need that concept here at all, and that only a suitable reformulation is needed in order to make that clear.

In case you agree, I ask you to let me know how you would like to have the passage changed.

Anyway: Would it be all right with you that the printer's galleys be sent to me for proofreading, or is it important to you to make the corrections yourself?

I was very sorry to hear that for a long time things have not been going at all well for you. In the meantime Mr. Dana Scott wrote me in June that your state of health has very much improved. Hopefully that improvement is continuing further!

With that wish and with cordial greetings, also to your wife,

Yours truly,

Paul Bernays

---

[b] *1972a.*

71. Gödel to Bernays

Princeton, 14./VII. 1970

Lieber Herr Bernays,

Professor Scott teilt mir mit, dass er schon vor längerer Zeit meine
Übersetzung der Dialectica-Arbeit[a] samt den neuen Fussnoten, sowie
auch die 3 separaten Bemerkungen[b] über Turing, Widerspruchsfreiheits-
beweise im selben System u. Kompliziertheit von Axiomensystemen an
Sie einschickte u. dass diese Dinge in den Dialectica erscheinen werden.
Darf ich fragen, welche Nummer der Dial. dafür in Aussicht genommen
ist u. wann ich Korrekturbogen erwarten kann? Da ich hoffe, die Korrekturen selbst lesen zu können, möchte ich die Fussnote über Scott durch
2   die folgende ersetzen: |
I wish to express my best thanks to Professor Dana Scott, who super-
vised the typing of these papers and carried on the correspondence with
the editors, while I was ill.

Herr Dill⟦er⟧ hat mir einen Brief über seine Arbeit (gemeinsam mit
Nahm) geschrieben. Ich verstehe nicht, was es heissen soll, dass in meinem Beweis der Formel $p \supset p \land p$ ein (nicht möglicher) Übergang zum
charakteristischen Term einer Formel nötig ist sei. Was nötig ist, ist
die Entscheidbarkeit von intensionalen Gleichungen zwischen Funktionen. Dagegen werden die Mathematiker wahrscheinlich Einwände er-
heben, weil die heutige Mathematik durchaus extensional ist u. daher
kl keine klaren Begriffe von Intensionen entwickelt wurden. Es ist aber
doch sicher, dass, zumindest im Rahmen einer bestimmten Sprache, vollkommen präzise Begriffe dieser Art definiert werden könnten.

Ich hoffe, dass es Ihnen gesundheitlich gut geht, u. verbleibe mit
herzlichem Gruss

Ihr Kurt Gödel

[At top of page 1:] P.S. Ihre Ansicht über die Fussnote (k),[c] u. eventuell
auch über anderes neu hinzugekommenes, würde mich sehr interessieren.

---

[a] *1972*; cf. sections 3.1 and 3.3 of the introductory note.

71. Gödel to Bernays

Princeton, 14 July 1970

Dear Mr. Bernays,

Professor Scott informs me that he already sent you a long time ago
my translation of the *Dialectica* paper[a] together with the new footnotes,
as well as the three separate remarks[b] about Turing, consistency proofs
in the same system, and the complexity of axiom systems, and that these
things will appear in *Dialectica*. May I ask which issue of *Dialectica* is
projected for that and when I can expect galley proofs? Since I hope to
be able to read the proof sheets myself, I would like the footnote about
Scott to be replaced by the following:

I wish to express my best thanks to Professor Dana Scott, who super-
vised the typing of these papers and carried on the correspondence with
the editors, while I was ill.

Mr. Dill⟦er⟧ wrote me a letter about his work (done jointly with
Nahm). I don't understand what it is supposed to mean that in my proof
of the formula $p \supset p \wedge p$ an (impossible) passage to the characteristic
term of a formula is necessary. What is necessary is the decidability of
intensional equations between functions. The mathematicians will probably raise objections against that, because contemporary mathematics
is thoroughly extensional and hence no clear notions of intensions have
been developed. But it is nevertheless certain that, at least within the
framework of a particular language, completely precise concepts of this
kind could be defined.

I hope that things are going well for you in terms of health, and I remain, with cordial greeting,

Yours truly,

Kurt Gödel

[At top of page 1:] P.S. Your opinion about footnote (k),[c] and perhaps
also about other newly added material, would interest me very much.

---

[b] *1972a.*
[c] See footnote bf on p. 72.

72. Bernays to Gödel

Ch-8002 Zürich, 12. IX. 1970
Bodmerstr. 11.

Lieber Herr Gödel!

Über den Empfang Ihres Briefes vom 14.VII. habe ich mich sehr gefreut. Nur tat es mir leid, dass Ihr Brief sich gerade mit meinem kurz zuvor an Sie abgesandten Brief kreuzte. Ich hoffe, dass Sie diesen bald danach erhalten haben.

Um Ihnen zunächst Ihre Frage bezüglich der Publikation Ihrer ⟨vorliegenden⟩ Abhandlung und Ihrer "remarks on the undecidability results" zu beantworten, so ist beabsichtigt, dass diese im Vol. 24 (1970) erscheint. In der Herausgabe der Bände von Dialectica war im vorigen Jahr eine Stockung eingetreten in Zusammenhang damit, dass ein Wechsel der Druckerei erfolgte. Jetzt ist man dabei, den Rückstand aufzuholen, und es erscheinen mehrere Hefte in kürzerem Abstand. Vor nicht langem ist Heft 2 von Vol 23 herausgekommen.

Die Fahnen-Korrektur von Ihrer Abhandlung (nebst den remarks) wurde zunächst an mich gesandt. Da diese ziemlich viele Druckfehler enthält, habe ich eine erste Durchsicht gemacht und um eine zweite Fahnenkorrektur gebeten.

Sobald ich diese bekomme—ich sollte sie eigentlich heute schon erhalten haben—, schicke ich sie Ihnen zu. |

Bei der Durchsicht habe ich nur die Druckfehler-Korrekturen in den Fahnen vermerkt. Doch habe ich noch ein paar andere Korrektur-Vorschläge, die ich Ihnen vorlegen möchte.

1. In der Note (c),[a] auf Fahne 4 heisst es (in Zeile 3 von unten bis 2 von unten) "added to Axiom 5 in the present version of the paper". Vielleicht ist es für den Leser deutlicher, wenn Sie etwa sagen "added, in the present version of the paper, to the rule of defining a function by a term".[1]

2. In der Note (i) steht, auf Zeile 5 dieser Note, "4. (Kreisel loc. cit. note b)". Soll es hier nicht anstatt "note b)" heissen "note (j)"?[a]

3. In der Note (j), drittletzte Zeile dieser Note, steht in Klammern "see note (h)". Hier soll es wohl heissen "see note (i)"?

4. In der Note (k), dritte Zeile dieser Note, heisst es "into certain trivial supplements". Soll hier nicht anstatt "into" stehen "up to"? |

[1]Axiom 5 entsteht doch erst durch die Hinzufügung; (in der ursprünglichen Version sind die Axiome von T *noch gar nicht numeriert*).

[a]See note bf on p. 72.

72. Bernays to Gödel

Ch-8002 Zurich, 12 September 1970
Bodmerstrasse 11

Dear Mr. Gödel,

I was very pleased by the receipt of your letter of 14 July. Only I was sorry that your letter crossed directly with mine, sent to you shortly before. I hope you received the latter soon thereafter.

First of all, to answer your question concerning the publication of your present paper and your "remarks on the undecidability results", it is intended that they appear in Vol. 24 (1970). The publication of the volumes of *Dialectica* was delayed during the past year in connection with a change of the printer. We are now working to catch up with the backlog, and several issues are appearing at shorter intervals. Not long ago issue 2 of Vol. 23 came out.

The proof sheets of your paper (together with the remarks) were first sent to me. Since they contained rather many printing errors, I did a first proofreading and asked for a second [set of] proofs.

As soon as I get them—I really should have received them already today—I shall send them to you.

When proofreading I noted only the typographical corrections on the galleys. Nevertheless, I still have a few other corrigenda to suggest that I would like to put before you.

1. In note (c)[a] on galley 4 it says (in lines 3 from the bottom up to 2 from the bottom) "added to Axiom 5 in the present version of the paper". Perhaps it would be clearer for the reader if you were to say "added, in the present version of the paper, to the rule of defining a function by a term".[1]

2. In note (i) one reads, on line 5 of that note, "4. (Kreisel loc. cit. note b)". Shouldn't it say "note (j)" here instead of "note b)"?[b]

3. In note (j), third from last line of that note, we have within parentheses "see note (h)". Here it should presumably say "see note (i)"?

4. In note (k), third line of that note, it says "into certain trivial supplements". Shouldn't "up to" replace "into"?

---

[1] Axiom 5 in fact results only from the addition. (In the original version the axioms of T *are not numbered at all*.)

---

[b] Gödel made vertical bars to the left of sections 2 and 3.

In meinem Brief vom Juli wies ich auf die Stelle in der Erklärung des Begriffes der computable function of type t (Fahne 2, Zeile 35), wo es heisst "this general fact is intuitionistically demonstrable", wo die Bezugnahme auf den Begriff des intuitionistischen Beweises doch nicht im Einklang zu stehen scheint mit dem, was Sie ausdrücklich in der Fussnote [5] sagen. Genügt es hier nicht, anstatt "is intuitionistically demonstrable" etwa zu sagen "follows directly from the definition of the function in question and those of the functions in the k-tuple"?—

Für die Fussnote, mit der Sie sich bei Herrn Dana Scott bedanken, habe ich nicht die Stelle gefunden, wo Sie sie anbringen wollen. Doch Sie werden diese Einschaltung ja selbst ausführen.—

Was den Herrn Dill[er] betrifft, so hatte er ja die neue Version Ihrer Dialectica-Abhandlung noch nicht gesehen. Die Schwierigkeit, die ihm bei der Behandlung der Formel $p \supset p \wedge p$ nach Ihrem Verfahren wohl begegnete, wird ja durch die jetzt von Ihnen hinzugefügte Definitionsregel für Funktionen (die Sie in der Note (c) noch genauer angeben) in befriedigender Weise behoben. (Die Anwendbarkeit dieses Definitions-Verfahrens beruht natürlich auf dem Umstand, dass bei Formeln | ohne gebundene Variablen (also bei den Formeln von T) das tertium non datur in konstruktivem Sinne gültig ist.)—

Für die Behandlung der vollständigen Induktion bei dem Widerspruchsfreiheits-Beweis haben Sie in der Note (c) (unter 1.) die hier auftretende verallgemeinerte Form der vollständigen Induktion angegeben. Wäre es nicht gut für den Leser, wenn Sie hier auch erwähnten, dass doch diese Form des Induktions-Schemas nach der Methode von[c] Skolem (vgl. "Grundlagen der Math. I", zweite Aufl.[d] S. 349) auf das gewöhnliche Induktions-Schema zurückführbar ist?

Schliesslich noch eine Frage (welche zwar nicht die gegenwärtige Publikation, aber doch ihr Thema betrifft). Gibt es nicht bei Ihrem Beweisverfahren die Möglichkeit, dass man, anstatt zu sagen: "eine Sequenz Q kann gefunden werden, sodass die Formel $A(Q(x), z, x)$ in T beweisbar ist", die Behauptung nimmt "eine Sequenz Q kann gefunden werden, sodass die Formel $A(Q(x), z, x)$ 'verifizierbar' ist, d. h. bei beliebiger Ersetzung der Variablen in ihr durch entsprechende Konstanten nach Ausrechnung stets den Wert "wahr" ergibt"? Die Beweismethode bleibt dabei ja im wesentlichen die gleiche, d. h. man zeigt die "Verifizierbarkeit" für die den Axiomen von H entsprechenden Formeln und ferner für die Schlussregeln von H die Übertragung der Verifizierbarkeit von den Formeln, die den Prämissen entsprechen, auf die Formel, die der Konklusion

---

[c] Here Gödel marked an "X" in the margin.

In my July letter I pointed to the place in the explanation of the concept of computable function of type t (galley 2, line 35) where it says "this general fact is intuitionistically demonstrable", where the reference to the concept of intuitionistic proof doesn't seem to be in accord with what you say expressly in footnote [5]. Wouldn't it suffice here, instead of "is intuitionistically demonstrable", to say something like "follows directly from the definition of the function in question and those of the functions in the k-tuple"?

I've not found the place where you want to put the footnote in which you express your thanks to Mr. Dana Scott. But you will certainly make that insertion yourself.

As to Mr. Dill[[er]], he had not yet seen the new version of your *Dialectica* paper. The difficulty that presumably arose for him in treating the formula $p \supset p \wedge p$ according to your procedure will now surely be resolved in a satisfactory way by the defining rule for functions you've now added (which you state still more precisely in note (c)). (The applicability of this definition procedure rests of course on the fact that in formulas without bound variables (hence in the formulas of T) the law of excluded middle is valid in the constructive sense.)

For the treatment of complete induction in the consistency proof you stated in note (c) (under 1.) the generalized form of complete induction that occurs here. Would it not be good for the reader if you also mentioned here that this form of the induction schema is nevertheless reducible to the usual induction schema by the method of Skolem[c] (cf. *Grundlagen der Mathematik I*, second edition,[d] p. 349)?

One final question (which, to be sure, doesn't concern the present publication, but does concern its theme). Doesn't the possibility exist for your proof procedure that, instead of saying "a sequence Q can be found such that the formula $A(Q(x), z, x)$ is provable in T", one makes the assertion "a sequence Q can be found such that the formula $A(Q(x), z, x)$ is 'verifiable', i.e., by arbitrary replacements of its variables through corresponding constants the value 'true' is always obtained after computing [[the values of all the functions occurring in A]]"? The proof method thereby remains essentially the same; i.e., one establishes the "verifiability" of the formulas that correspond to the axioms of H, and furthermore, for the rules of inference of H, one shows that verifiability is preserved [[in passing]] from the formulas that correspond to the premises to the formulas that correspond to the conclusion. In that way one has

---

[d] *Hilbert and Bernays 1968.*

5   entspricht. Dabei | hat man die Vereinfachung, dass die Definition des
    "Systems T" wegfällt. Ich möchte sogar vermuten, dass Sie diese Art des
    Widerspruchsfreiheits-Beweises zunächst im Sinn hatten und erst durch
    die metamathematische Analyse des Beweises auf die vorliegende Fassung
    geführt wurden.—
       Hoffentlich hat die im Frühjahr erfolgte Besserung Ihres Befindens sich
    weiter erhalten.
       Mir geht es verhältnismässig passabel; doch kann ich immer nur ein
    paar Stunden im Tage für Arbeiten (im weiteren Sinne des Wortes) ver-
    werten.

              Mit herzlichen Grüssen, auch an Ihre Frau,

                        Ihr Paul Bernays

       P.S. Leider hat der Abschluss dieses Briefes sich sehr verzögert.
       Nun ist die zweite Fahnenkorrektur eingetroffen. Wie ich sehe, waren
    noch nicht alle Druckfehler korrigiert. Ich habe nun die noch zu ergän-
    zenden Korrekturen in die zwei Exemplare der neuen Fahnen, die ich Ih-
    nen schicke, eingetragen.
       Es wäre sehr schön, wenn Sie mir bald das eine der Exemplare mit
    Ihren Korrekturen zurücksenden könnten.

73. Gödel to Bernays[a]

    bitte mit umbruch warten[[.]] habe noten ck[b] neu formuliert[[.]] sende
    korrekturen naechste Woche[[.]] goedel

---

[a]Telegram, sent 2 October 1970.

74. Gödel to Bernays

                                          Princeton, 22./XII. 1970

       Lieber Herr Bernays!

       Vielen herzlichen Dank für Ihre beiden freundlichen Briefe u. die Sepa-
    rata, u. vor allem für die Mühe, die Sie auf die Korrekturen meiner Ar-
    beit verwendet haben, sowie auch für die Richtigstellung einiger Versehen

the simplification that the definition of the "system T" can be omitted. I would even conjecture that you first had this sort of consistency proof in mind and were only led to the present version by the metamathematical analysis of the proof.

Hopefully the improvement in your health that occurred in the spring has continued further.

For me things are going tolerably well; yet I can always make use of only a few hours a day for working (in the wider sense of the word).

With cordial greetings, also to your wife,

Yours truly,

Paul Bernays

P.S. Unfortunately the completion of this letter was very much delayed.

The second ⟦set of⟧ galleys has now arrived. As I see, not all mistakes were corrected. I have entered the corrections that are still to be made in the two copies of the new galleys that I am sending you.

It would be very nice if you could *soon* send back one of the copies with your corrections.

73. Gödel to Bernays[a]

Please wait with the page layout. Have reformulated notes c and k.[b] Am sending corrections next week. goedel

---

[b]See footnote bf on p. 72.

74. Gödel to Bernays

Princeton, 22 December 1970

Dear Mr. Bernays,

Many cordial thanks for your two friendly letters and the offprints, and above all for the trouble you took with the proof sheets of my paper, as well as for correcting a few mistakes in the manuscript. Please excuse

im Manuskript. Entschuldigen Sie bitte, dass die Fertigstellung der neuen Note (k)[a] wesentlich mehr Zeit in Anspruch genommen hat als ich in Aussicht stellte. Die Note (k) ist jetzt 4mal so lang geworden. Ich glaube, dass die präzise Formalisierung des darin Gesagten keine Schwierigkeit mehr machen wird, aber die Sache ist wegen der unvermeidlichen "self re-flexivities" doch nicht ganz einfach. Ich habe auch an einer Reihe anderer Stellen kleinere Änderungen vorgenommen, von denen die meisten nicht unbedingt nötig, aber doch sehr erwünscht sind. Ich bedaure, dass ich bei der Drucklegung solche Schwierigkeiten bereiten muss u. bin gerne bereit, für die dadurch entstehenden Mehrkosten aufzukommen. Der Zeitpunkt des Erscheinens scheint mir weniger wichtig zu sein als die Textverbesserungen[1]. Ich werde aber die Korrekturen sehr bald nach

2   Neujahr einsenden. Weitere Änderungen will ich dann unter | keinen Umständen mehr vornehmen, schon deswegen weil ich mich jetzt ganz einer höchst wichtigen mengentheoretischen Frage widmen möchte. Natürlich will ich Sie in keiner Weise mit dem Lesen weiterer Korrekturen bemühen, da ich das ja jetzt selbst tun kann. Ich habe mich sehr darüber gefreut, dass Sie der Meinung sind, ich hätte durch die hinzugefügten Anmerkungen meine Arbeit sehr bereichert u. auch zugänglicher gemacht. Alles weitere schreibe ich gleichzeitig mit der Einsendung der Korrekturen.

Gesundheitlich geht es mir jetzt ganz passabel u. ich hoffe, dass auch Ihr Gesundheitszustand zufriedenstellend ist.

Mit besten Weih[n]achts- u. Neujahrswünschen, auch von meiner Frau, u. herzlichen Grüssen

Ihr Kurt Gödel

[1]Die Note (c) ist zum Teil unrichtig u. auch die Note (k) enthält ein kleines Versehen.

---

[a]See footnote bf on p. 72.

75. Bernays to Gödel

CH-8002 Zürich 31. XII. 1970.
Bodmerstr. 11.

Lieber Herr Gödel!

Haben Sie vielen Dank für Ihren freundlichen Brief vom 22. XII. Ich freute mich zu hören, dass Sie die Texte der Noten zu Ihrer Dialectica-Abhandlung nunmehr in eine Sie befriedigende Form gebracht, und sehe

that completing the new note (k)[a] has taken substantially more time than I projected. Note (k) has now become four times as long. I think that the precise formalization of what is said in it will create no further difficulty, but the matter is nevertheless not entirely simple, on account of the unavoidable "self reflexivities". I have also made small changes at a number of other places, most of which are not strictly necessary, but are still very desirable. I'm sorry that I have to cause such difficulties with the printing and am readily prepared to be responsible for the additional expenses resulting therefrom. The time of publication seems to me to be less important than the improvements to the text.[1]

I will however submit the proofs very soon after the New Year. Under no other circumstances do I then want to make further changes, if for no other reason than that I want to devote myself now entirely to a most important set-theoretic question. I certainly don't want to trouble you in any way with the reading of further proofs, since I can do that myself. I was very pleased that you think I have very much enriched my paper and also made it more accessible through the added comments. I'll write everything else when I submit the galleys.

In terms of health things are now going tolerably well, and I hope that your state of health is also satisfactory.

With best wishes for Christmas and the New Year, also from my wife, and with cordial greetings,

Yours truly,

Kurt Gödel

---

[1] Note (c) is in part incorrect, and note (k) also contains a small mistake.

---

75. Bernays to Gödel

CH-8002 Zurich 31 December 1970
Bodmerstrasse 11

Dear Mr. Gödel,

Many thanks for your friendly letter of 22 December. I'm pleased to hear that you have now brought the text of the notes to your *Dialectica* paper into a form that is satisfying to you, and I look forward with great

dem Empfang der korrigierten Fassung Ihrer Abhandlung mit viel Interesse entgegen.

Auch hat mich ⟨natürlich⟩ Ihre Mitteilung sehr interessiert, dass Sie zurzeit eine bestimmte mengentheoretische Frage verfolgen. So ist es 2 wohl | ein angemessener Neujahrswunsch, dass die Behandlung dieser Frage Sie zu ⟨weittragenden⟩ fruchtbaren Ergebnissen führen möge.

Möchte ferner Ihr gebessertes gesundheitliches Ergehen sich weiter festigen!

In herzlicher Erwiderung Ihrer freundlichen Neujahrswünsche grüsst Sie und Ihre Frau vielmals

Ihr Paul Bernays

P.S. Den Sonderdruck[a] betreffend den ursprünglichen Gentzenschen 3 Widerspruchsfreiheitsbeweis | haben Sie wohl inzwischen erhalten. Ein Exemplar der zweiten Auflage des Band II von d. "Grundl. der Math."[b] ist an Sie unterwegs.

---

[a] *Bernays 1970.*

76. Bernays to Gödel

CH-8002 Zürich, 16. März 1972
Bodmerstrasse 11

Lieber Herr Gödel,

Als Sie mir Ihre Neujahrsgrüsse sandten, fügten Sie hinzu, dass Sie hofften, mir bald wieder einmal ausführlich schreiben zu können. Gewiss hatten Sie dabei unter anderem die Korrekturen für die neue englische Fassung Ihrer Dialectica-Abhandlung im Sinn.

Nun hörte ich kürzlich von Herrn Kreisel, dass Sie bezüglich dieser noch immer gewisse Bedenken haben. Diese beziehen sich wohl auf die Definition des Begriffes der "computable function of type *t*" (Fahne 2) und die kurz darauf folgende Bemerkung: "and for which, moreover, this general fact is intuitionistically demonstrable".

Vielleicht kann ich hier ein wenig zur Verminderung der Bedenken beitragen, indem ich auf den Unterschied zweier Schwierigkeiten hinweise: einer solchen, welche die Formulierung betrifft und sich beheben lässt,

interest to the receipt of the corrected version of your paper.

Of course, your communication that you are presently pursuing a particular question in set theory also interested me very much. So it is perhaps an appropriate New Year's wish that the treatment of this question may lead you to far-reaching fruitful results.

In addition, may your improved state of health be further strengthened!

In cordial reciprocation of your kind New Year's wishes, many greetings to you and your wife.

Yours truly,

Paul Bernays

P.S. In the meantime you've probably received the offprint[a] concerning Gentzen's original consistency proof. A copy of the second edition of volume II of *Grundlagen der Mathematik*[b] is on the way to you.

---

[b] *Hilbert and Bernays 1970.*

76. Bernays to Gödel

CH-8002 Zurich, 16 March 1972
Bodmerstrasse 11

Dear Mr. Gödel,

When you sent me your New Year's greeting you added that you hoped to be able to write me once again soon in detail. By that you surely had in mind, among other things, the corrections for the new English version of your *Dialectica* paper.

Now I heard recently from Mr. Kreisel that with regard to that ⟦paper⟧ you still have certain doubts. Those are presumably related to the definition of the concept of "computable function of type t" (galley sheet 2) and the remark following shortly thereafter: "and for which, moreover, this general fact is intuitionistically demonstrable".

Here, perhaps, I can contribute a little to reduce the doubts by pointing to a distinction between two difficulties: one that concerns the formulation ⟦of the problem⟧ and may be resolved, and one that lies in the

und einer solchen, die in der Natur der Sache liegt. Die erste besteht ja
darin, dass Sie mit den Worten "intuitionistically demonstrable" scheinbar auf den Allgemeinbegriff des intuitionistischen Beweises rekurrieren,
dessen Vermeidung doch gerade in der Absicht Ihres Verfahrens liegt.
Tatsächlich wollen Sie ja hier nur daran erinnern, dass bei der betreffenden Feststellung nicht die vom Intuitionismus ausgeschlossenen Beweis-
methoden benutzt werden dürfen. Ein Missverständnis dieser Art kann ja
durch eine etwas andere Wendung oder mittels einer Fussnote vermieden
werden.

Die andere Schwierigkeit, die sich wohl nicht beheben lässt, ist diejenige, dass auch nach der Ersetzung des Begriffes "intuitionistischer
Beweis" durch den des "Funktionals" eine gewissen Imprädikativität
verbleibt. Eine solche findet sich aber, genau genommen, in der finiten
2   Betrachtung, wenn man sich klar | macht, dass allgemein eine primitiv-
rekursive Definition ein effektives Berechnungsverfahren liefert. Dabei
muss man doch die Voraussetzung benutzen, dass, wenn n eine konstruierbare Ziffer ist und ferner ein Prozess anschaulich beschrieben ist, der
aus gegebenen Ziffern wieder eine Ziffer liefert, dann die n-fache Iteration
dieses Prozesses ausführbar ist. Eine entsprechende Voraussetzung muss
⟨ja⟩ auch für die rekursiven Definitionen von Funktionalen angewandt
werden. Mit der Erforderlichkeit solcher Vorausetzungen und damit einer
gewissen Imprädikativität muss man sich vermutlich abfinden. Es ist
übrigens einer der Gedanken der Gonseth'schen Philosophie, dass unser
Denken und die Methode unseres theoretischen Forschens nicht restlos
prädikativ ist.

Man mag die Sachlage noch von einer anderen Seite beleuchten: Es
lässt sich eine reservierte Art des Intuitionismus von einer weitergehenden unterscheiden. Der reservierte Intuitionismus sieht nur davon ab,
beliebigen, noch unentschiedenen mathematischen Behauptungen einen
Wahrheitswert zu erteilen; er verwendet daher auch nicht die Deutung
der aussagenlogischen Operatoren als Wahrheitsfunktionen, schliesst sich
vielmehr enger an deren sprachlich geläufige Verwendung an. Der weiter-
gehende Intuitionismus involviert die Doktrin, dass die mathematischen
Aussagen von dem handeln, was wir mathematisch *tun* und dabei erhalten. Hiernach sind die mathematischen Beweise nicht einfach Mittel
des Erkennens, sondern das mathematische Beweisen gehört wesentlich
zum Thema der Mathematik, und es ist danach sachgemäss, den Allgemeinbegriff des mathematischen Beweises zu verwenden. Gewiss macht
auch die Hilbert'sche Metamathematik die mathematischen Beweise zum
Gegenstand, aber doch nur, nachdem sie diese durch die Formalisierung
gleichsam in die mathematische Gegenständlichkeit proj[[i]]eziert hat.

Mir scheint nun, dass Sie mit der Methode Ihrer Dialectica Abhandlung
im Rahmen des reservierten Intuitionismus verbleiben können, d. h. dass
der Begriff des Funktionals nicht se ⟨in der Weise⟩ wie der Allgemein-

nature of the problem. The first consists in the fact that, by the words "intuitionistically demonstrable" you are apparently referring to the general concept of intuitionistic proof, whose avoidance is nevertheless the very purpose of your procedure. In fact, in doing so you only wish to recall here that for the determination in question the methods of proof excluded by intuitionism must not be used. A misunderstanding of this kind can be avoided by a somewhat different locution or by means of a footnote.

The other difficulty, which may well not be resolvable, is that even after the replacement of the concept "intuitionistic proof" by that of "functional" a certain impredicativity remains. But, strictly speaking, such ⟦an impredicativity⟧ is to be found in the finitary approach, if one recognizes that in general a primitive recursive definition provides an effective computation procedure. In doing so one must still use the assumption that if n is a contructible numeral and if, furthermore, a process is intuitively described which, from given numerals, again provides a numeral, then the n-fold iteration of that process can be carried out. A corresponding assumption must also be employed for the recursive definitions of functionals. Presumably one must be resigned to the necessity of such assumptions, and thus of a certain impredicativity. It is, by the way, one of the ideas of Gonseth's philosophy that our thinking and the method of our theoretical inquiry is not entirely predicative.

In addition, the situation may be illuminated from another side: A reserved sort of intuitionism may be distinguished from a more far-reaching sort. The reserved intuitionism only refrains from assigning a truth value to arbitrary, still undecided mathematical assertions; therefore it also does not employ the interpretation of the propositional operators as truth functions, ⟦but⟧ is tied much more closely to their everyday linguistic usage. The more far-reaching intuitionism involves the doctrine that mathematical statements are concerned with what we *do* mathematically and what we obtain thereby. Accordingly, mathematical proofs are not simply means of discovery, but rather, mathematical proving is intrinsic to the theme of mathematics, and thus it is appropriate to use the general concept of mathematical proof. To be sure, Hilbert's metamathematics also takes mathematical proofs as ⟦its⟧ subject, but nevertheless, only after it has projected them by means of the formalization, as it were, into the mathematically objective realm.

Now it seems to me that with the method of your *Dialectica* paper you are able to remain within the framework of the reserved intuitionism, that is, that the concept of functional does not go beyond the latter in

begriff des Beweises über diesen hinausführt.

3 | Ich weiss nicht, ob diese Bemerkungen Ihnen etwas sagen, was Sie sich nicht schon selbst überlegt haben. Sollten sie Ihnen zu einem Entschluss über die Fassung des Textes der kritischen Stellen in Ihrer Abhandlung etwas beitragen, so würde es mich sehr freuen.

Mit herzlichen Grüssen, auch an Ihre Frau

Ihr Paul Bernays

77. Bernays to Gödel

CH-8002 Zürich, Bodmerstr. 11
21. VII. 1972

Lieber Herr Gödel!

Habe ich Ihnen einen Sonderdruck von meiner Abh. "Zum Symposium über die Grundlagen d. Mathematik"[a] geschickt? Ich bin nicht sicher, und so würde ich Ihnen dankbar sein, wenn Sie mir im Falle, dass Sie die Abh. nicht bekommen haben, es mitteilen wollten. Mit dem Wunsche, dass Ihr Ergehen befriedigend ist, u. mit herzlichen Grüssen, auch an Ihre Frau,

Ihr Paul Bernays

---

[a]*Bernays 1971.*

78. Bernays to Gödel

CH-8002 Zürich, Bodmerstr. 11.
28. Oktober 1972.
(Schon einige Tage zuvor angefangen!)

Lieber Herr Gödel!

Aus einem Brief von Herrn Kreisel erfahre ich, dass es Sie interessieren würde zu wissen, was ich für ein Herzmittel nehme, und so erlaube ich mir, Ihnen diesbezüglich ein paar Angaben zu machen. Die Auskunft ist

the way that the general concept of proof does.

I don't know whether these remarks tell you anything that you have not already considered yourself. Should they contribute something to [help] you to [reach] a decision about the formulation of the text of the critical points in your paper, I would be very pleased.

With cordial greetings, also to your wife,

Yours truly,

Paul Bernays

77. Bernays to Gödel

CH-8002 Zurich, Bodmerstrasse 11
21 July 1972

Dear Mr. Gödel,

Did I send you an offprint of my paper "Zum Symposium über die Grundlagen der Mathematik"?[a] I'm not sure, and so I would be grateful to you if you would inform me in case you did not get the paper. With the wish that your health is satisfactory, and with cordial greetings, also to your wife,

Yours truly,

Paul Bernays

---

78. Bernays to Gödel

CH-8002 Zurich, Bodmerstrasse 11
28 October 1972
(Already begun a few days before!)

Dear Mr. Gödel,

From a letter of Mr. Kreisel I understand that it would interest you to know what heart medicine I am taking, and so I [will] permit myself to inform you about some particulars in that regard. The information is

insofern einfach, als mein Gebrauch von Medikamenten seit einiger Zeit ziemlich stationär ist.

Als eigentliches Herzmittel nehme ich Digitoxin, davon aber nur alle zwei Tage eine Tablette, als Mittel für den Blutkreislauf habe ich Ronicol, wovon ich täglich zwei Dragées nehme. Über die beiden genannten Medikamente lege ich Ihnen die erläuternden Angaben bei, die man in den Verpackungen mitbekommt. Freilich können die darin stehenden Anweisungen für die *Dosierung nicht* generell verwendet werden. Darüber kann ja nur von Fall zu Fall entschieden werden, und man muss sich an die Anweisungen des Arztes halten. (Als generelle ärztliche Direktive erhielt ich die, dass man die Medikamente nicht bei leerem Magen nehmen soll.)—

2   Von meiner Betrachtung zu dem Dialectica-Symposium über die Grundlagen der Mathematik haben Sie doch einen Sonderdruck von mir bekommen? Falls nicht, so möchten Sie es mich | wissen lassen.

Die Beiträge zu diesem Symposium waren ja recht verschiedenartig; es dürfte schwer sein, daraus einen einheitlichen Gedankengang zu entnehmen. Doch das entspricht ja auch in gewissem Masse der Situation in der heutigen mathematischen Grundlagenforschung.

Leider sind Sie in dem Symposium nicht vertreten. Jedoch die Diskussion ist ja in keiner Weise abgeschlossen, und eine Stellungnahme von Ihnen wird natürlich jederzeit willkommen sein.

Hoffentlich ist Ihr Ergehen zurzeit befriedigend, insbesondere insoweit, dass Sie sich Ihren philosophischen und mathematischen Interessen widmen können.

Mit herzlichen Grüssen, auch an Ihre Frau,

Ihr Paul Bernays

79. Gödel to Bernays

Princeton, 26./XII. 1972.

Lieber Herr Bernays!

Ich danke Ihnen herzlich für Ihren freundlichen Brief vom 28./X. sowie die Karte von der Tagung in Oberwolfach. Ich habe mich sehr gefreut von Kreisel zu hören, dass es Ihnen gesundheitlich jetzt viel besser geht, u. vermutete, dass das auf ein neues Herzmittel zurückzuführen ist. Das Digitoxin ist aber, so viel ich weiss, schon sehr lange bekannt. Hier verwendet man viel das "Digoxin", das neueren Datums zu sein scheint.

simple insofar as my use of medications has been rather stable for some time now.

I take Digitoxin as the primary heart medicine, but I take only one tablet of it every other day; as a medicine for the circulation I have Ronicol, of which I take two pills a day. I am enclosing for you the explanatory statements about these two medications that come with the packaging. To be sure, the instructions given therein can *not* be used generally to ⟦determine⟧ the *dosage*. That can only be decided from case to case, and one must follow the physician's instructions. (As a general medical directive I was told that the medications should not be taken on an empty stomach.)

Have you yet received an offprint from me of my discussion at the *Dialectica* symposium on the foundations of mathematics? If not, please let me know.

The contributions to that symposium were quite varied; it might be hard to extract a unified train of thought from it. But that corresponds also, to a certain degree, to the situation in today's research on the foundations of mathematics.

Unfortunately you were not represented in the symposium. However, the discussion is in no way closed, and a comment from you will of course be welcome at any time.

Hopefully your health is satisfactory for the time being—in particular, to the extent that you can devote yourself to your philosophical and mathematical interests.

With cordial greetings, also to your wife,

Yours truly,

Paul Bernays

79. Gödel to Bernays

Princeton, 26 December 1972

Dear Mr. Bernays,

I thank you cordially for your friendly letter of 28 October as well as the postcard from the meeting in Oberwolfach. I was very pleased to hear from Kreisel that your health is now much better, and I assumed that that is to be attributed to a new heart medicine. Digitoxin has however, so far as I know, already been well known for a very long time. Here one frequently uses "Digoxin", which appears to be of more recent

Ein anderes viel verwendetes Herzmittel ist Valium, das erst seit ca. 10 Jahren im Gebrauch ist. Sonderbarerweise wird es meistens als Beruhigungsmittel angesehen, weil es diese Wirkung *auch* hat. Aber ich habe damit, zusammen mit Vitamin B12, eine Herzinsuffizienz $1\frac{1}{2}$ Jahre lang korrigiert. Man kann es auch abwechselnd mit Digoxin nehmen. Diese beiden scheinen auch bei sehr langem Gebrauch das Herz nicht zu schädigen.

Einen Sonderdruck Ihrer Betrachtungen zum Dialectica Symposium[a] habe ich leider nicht erhalten, sondern nur Ihre Bemerkungen beim Internat. Kongr. f. Phil 1968[b] u. | zu Gentzen's Widerspruch⟦s⟧freiheitsbeweis.[c] Ich danke Ihnen auch bestens für Ihren Brief über die Frage, ob der allgemeine intuition. Beweisbegriff für die intuition. Interpretation meines Systems T nötig ist (was meine Interpetation der logischen Operatoren erkenntnistheoretisch wertlos machen würde). Ich glaube, dass das *nicht* der Fall ist, sondern dass ein *viel* engerer ⟨u. im Prinzip⟩ entscheidbarer Beweisbarkeitsbegriff genügt, den ich in Note k[d] der Übersetzung meiner Dialectica⟨arbeit⟩ eingeführt u. "reduktive Beweisbarkeit" genannt habe. Aber das im einzelnen befriedigend durchzuführen, ist nicht ganz leicht, hauptsächlich wegen der *nicht eliminierbaren* Imprädikativität auch dieses engeren Beweisbegriffes, welche mit der von Ihnen erwähnten Imprädikativität des Funktionsbegriffes nahe zusammenhängt. Es ist zweifelhaft, ob die Durchführung die Mühe lohnen würde. Ich habe mich daher bis jetzt nicht dazu entschliessen können, obwohl die weitere ~~Möglichkeit~~ Verfolgung dieser Fragen vielleicht wesentlich zur Aufklärung der Grundlagen des Intuitionismus beitragen könnte.

Mit herzlichen Grüssen u. besten Wünschen für Ihre Gesundheit

Ihr Kurt Gödel

---

[a] *Bernays 1971.*
[b] *Bernays 1969.*

date. Another frequently used heart medicine is Valium, which has been in use for the last 10 years. Oddly, it is usually regarded as a tranquilizer, because it has that effect *too*. But for the past $1\frac{1}{2}$ years I have corrected a heart insufficiency with it, together with vitamin B12. One can also take it alternately with Digoxin. These two also seem not to damage the heart even when used for a long time.

Unfortunately I did not receive an offprint of your reflections on the *Dialectica* Symposium,[a] but only your remarks at the 1968 International Congress for Philosophy[b] and on Gentzen's consistency proof.[c] I also thank you very much for your letter about the question whether the general intuitionistic concept of proof is necessary for the intuitionistic interpretation of my system T (which would make my interpretation of the logical operators epistemologically worthless). I think that that is *not* the case, but rather that a *much* narrower and in principle *decidable* concept of proof suffices, which I introduced in note k[d] of the translation of my *Dialectica* paper and called "reductive provability". But to carry that through satisfactorily in detail is not all that easy, mainly on account of the *non-eliminable* impredicativity also of this narrower concept of proof, which is closely connected with the impredicativity of the concept of function that you mentioned. It is doubtful whether carrying it through would be worth the trouble. Up to now, therefore, I have not been able to make up my mind to do it, although the further pursuit of that question could perhaps contribute in an essential way to the clarification of the foundations of intuitionism.

With cordial greetings and best wishes for your health,

Yours truly,

Kurt Gödel

---

[c] *Bernays 1970.*

[d] See footnote bf on p. 72

80. Bernays to Gödel[a]

21. Februar 1973.

Lieber Herr Gödel!

Sie sollten schon längst eine Erwiderung auf Ihren freundlichen Brief von Ende 1972 haben. Und besonders auch muss ich mich sehr entschuldigen, dass ich Ihnen noch nicht für Ihre Neujahrswünsche gedankt habe. Hoffentlich hat das Jahr 1973 für Sie und Ihre Frau mit Gutem begonnen, und ich wünsche Ihnen für das Weitere einen möglichst befriedigenden Verlauf.

Die Verspätung meiner Antwort hatte ausser manchen äusseren Ursachen besonders noch diejenige, dass ich mehrmals, wenn ich mir Ihren Brief zur Beantwortung vornahm, ins Ueberlegen verfiel anstatt zu schreiben, da ich unschlüssig war, wie ich mich zu dem, was Sie mir bezüglich Ihrer Dialectica-Abhandlung schrieben, äussern solle.

Nun möchte ich Ihnen immerhin die folgende Erwägung zur Methodik Ihres Widerspruchsfreiheitsbeweises in Ihrer Dialectica-Abhandlung mitteilen. Sie verwenden ja für diesen Beweis den Allgemeinbegriff einer "berechenbaren Funktion vom Typus t". Dieser Begriff wird aber nur dazu gebraucht, um bestimmte Erzeugungsprinzipien für Funktionen als angemessene Bildungsverfahren zu erkennen. Wie Sie erwähnen, sind die so gewonnenen Funktionen diejenigen einer typenmässigen Erweiterung der rekursiven Zahlentheorie. (wobei nur endliche Typen verwendet werden). Dass durch jene Erzeugungsprozesse sämtliche berechenbaren Funktionen der betrachteten Typen erhalten werden, wird für Ihren Wf-Beweis nicht erfordert und ist auch wohl—wenn ich mich nicht täusche—nicht der Fall.

Ist aber einmal der Bereich der durch die Erzeugungsprozesse (die
a   sich ja formal präzise aufstellen lassen) gelieferten Funktionenbereich gewonnen, so brauchen Sie doch für die Interpretation der Formeln $(\exists y)(z)A(y, z, x)$ nur auf diesen Funktionenbereich und nicht ⟨mehr direkt⟩ auf den ursprünglichen Begriff der berechenbaren Funktion Bezug zu nehmen. Dieser schärfer abgegrenzte Funktionenbereich verhält sich ja zu demjenigen der überhaupt berechenbaren Funktionen analog wie der Bereich der in einem formalen System formalisierten Beweise zu dem der inhaltlichen Beweise.
2 | Die Ueberlegungen zur Etablierung dieses Funktionenbereiches lassen sich von der Durchführung des Wf-Beweises mittels Ihrer Methoden der Interpretation der zahlentheoretischen Formeln (durch Verwendung

---

[a]The text of this letter is taken from the copy retained by Bernays.

80. Bernays to Gödel[a]

21 February 1973

Dear Mr. Gödel,

You should long ago have had a reply to your friendly letter from the end of 1972. And in particular, I must apologize very much for not yet having thanked you for your New Year's wishes. Hopefully the year 1973 has begun well for you and your wife, and I wish that its further course may be as satisfying as possible for you.

The delay in my reply had, apart from some external causes, in particular this [one]: Several times when I resolved to answer your letter I lapsed into reflection instead of writing, since I was undecided how to express myself about what you wrote me on your *Dialectica* paper.

Anyway I would like to impart to you the following consideration on the methodology of your consistency proof in your *Dialectica* paper. For this proof you employ the general concept of a "computable function of type *t*". This concept is used, however, only in order to recognize specific generation principles for functions as being appropriate formation procedures. As you note, the functions so obtained are those of an extension of recursive number theory by types (in which only finite types are used). That all computable functions of the types in question are obtained by means of those generation processes is not required for your consistency proof and—if I'm not mistaken—is probably also not the case.

But once the domain of the functions obtained by the generation processes (which can be formulated in a formally precise way) has been secured, then—for the interpretation of the formulas $(\exists y)(z)A(y,z,x)$—you only need to refer to that function domain, and not any more directly to the original concept of computable function. This more sharply delimited function domain stands in relation to that of all computable functions analogously as the domain of the proofs formalized in a formal system relates to that of the contentual proofs.

The considerations for setting up this function domain may be detached from carrying out the consistency proof by means of your method of interpretation of the number-theoretic formulas (by using the

der Funktionale aus jenem Bereich) abtrennen, entsprechend wie man
ja bei den ⟨Wf⟩-Beweisen von Gentzen, Kalmár und Ackermann die
Ueberlegungen zur Begründung der ordinalen Induktion bis zu ϵ_0 von der
Durchführung des Wf-Beweises, der unter Benutzung dieser Induktion
erfolgt, absondern kann.

Gesonderte Betrachtungen des besagten Funktionenbereiches sind ja
auch, angeregt durch Ihre Dialectica-Abhandlung, durch Kreisel, Howard
und Schütte erfolgt.

Ich weiss nicht, ob diese Bemerkungen Ihnen irgend etwas bieten; vielleicht können sie aber doch ein wenig dazu beitragen, dass Sie sich eher
dazu entschliessen, der englischen Uebersetzung Ihrer Dialectica-Abhandlung,
mit den von Ihnen angebrachten bereichernden und erläuternden Zusätzen,
die endgültige Fassung zu geben.

Mit[b] leider sehr verspätetem Dank—Dank auch für Ihre freundlichen
Mitteilungen bezüglich der Medikamente—und mit herzlichen Grüssen,
Ihnen und Ihrer Frau,

> Ihr,   Paul Bernays

Meine Bemerkungen "Zum Symposium über die Grundlagen der Mathematik"[c] habe ich Ihnen zugesandt.

---

[b]The text from this point to the end of the letter was transcribed from Bernays'
Gabelsberger shorthand.

81. Gödel to Bernays

> 18./XII. 1973

Lieber Herr Bernays!

Herzlichen Dank für Ihren freundliche[n] Brief vom 21./II., sowie auch
für den Sonderdruck Ihrer Bemerkungen zum Symposium über die Grundlagen der Mathematik.

Ich hoffe, dass Ihr Gesundheitszustand zufriedenstellend ist.

Mit besten Grüssen, auch von meiner Frau,

> Ihr Kurt Gödel

functionals from that domain), just as, correspondingly, in the consistency proofs of Gentzen, Kalmár and Ackermann the considerations for grounding ordinal induction up to ϵ_0 can be separated from carrying out the consistency proof that uses that induction.

Separate considerations of the said function domain, stimulated by your *Dialectica* paper, have also come about through ⟦the efforts of⟧ Kreisel, Howard and Schütte.

I don't know whether these remarks offer anything to you; but perhaps they can contribute a little toward your sooner making up your mind to give the final form to the English translation of your *Dialectica* paper, with the enriching and illustrative additions you've brought to it.

With,[b] unfortunately, very belated thanks—thanks too for your friendly communication regarding the medications—and cordial greetings to you and your wife,

Yours truly, Paul Bernays

I've sent you my remarks "Zum Symposium über die Grundlagen der Mathematik".[c]

---

[c] *Bernays 1971.*

81. Gödel to Bernays

18 December 1973

Dear Mr. Bernays,

Cordial thanks for your friendly letter of 21 February, as well as for the offprint of your remarks on the symposium on the foundations of mathematics.

I hope that your state of health is satisfactory.

With best greetings, also from my wife,

Yours truly, Kurt Gödel

82. Bernays to Gödel[a]

8002 Zürich, 16. Dezember 1974

Lieber Herr Gödel

Vor einiger Zeit erhielt ich ein paar Nummern der brasilianischen Zeitschrift "Boletim de Analise e Logica Matematica"[1] zugesandt. Leider erst kürzlich bemerkte ich, dass in dem einen der Hefte ein Schreiben eingelegt war, aus dem hervorgeht, dass zwei dieser Hefte für Sie bestimmt sind.

Ich will Ihnen diese schicken, möchte aber zuvor gern wissen, ob nicht ein Versehen vorliegt und ob Sie vielleicht auch Exemplare der Zeitschrift bekommen haben. Die beiden genannten sind: ano 1, No 1, Dez. 1969 und ano 2, No 1, Dez. 70.

Hoffentlich geht es Ihnen befriedigend und können Sie sich Ihren philosophischen und mathematischen Anliegen im erwünschten Mässe widmen.

Ich benutze die Gelegenheit um Ihnen und Ihrer Frau alles Gute für das kommende Jahr zu wünschen. Seien Sie herzlich gegrüsst von

Ihrem Paul Bernays

[1][Written vertically in shorthand in the left margin:] In dieser Zeitschrift is auch eine umfangreiche Abhandlung von Jorge Emmanuel Ferreira Barbosa[b] ⟨(auf Portugiesisch)⟩ erschienen, welche beansprucht, die formalen Systeme der Mengenlehre auf konstruktivem Wege als widerspruchsfrei zu erweisen.[c]

---

[a]The text of this letter is taken from the copy retained by Bernays.

[b]*Barbosa 1973*. See the postscript to letter 84.

83. Gödel to Bernays

Princeton, 17./XII. 74

Lieber Herr Bernays!

Ich hoffe, dass Ihre Gesundheit zufriedenstellend ist. Mir geht es soweit ganz gut u. ich mache Fortschritte in meiner Arbeit über die wahre Mächtigkeit des Continuums.

Mit herzlichen Grüssen, auch von meiner Frau

Ihr Kurt Gödel

82. Bernays to Gödel[a]

8002 Zurich, 16 December 1974

Dear Mr. Gödel,

Some time ago a few issues of the Brazilian journal *Boletim de Analise e Logica Matematica* were sent to me.[1] Unfortunately I noticed only recently that in the one volume a note was inserted, which makes it clear that two of these issues are for you.

I will send them to you, but before doing so I would very much like to know whether a mistake is involved and whether you have perhaps also received copies of the journal. The two I mentioned are: ano 1, No. 1, December 1969 and ano 2, No. 1, December 1970.

Hopefully things are going satisfactorily for you and you are able to devote yourself to your philosophical and mathematical concerns to the extent you desire.

I take advantage of this opportunity to wish you and your wife all that is good for the coming year. [Please] accept cordial greetings from

Yours truly, Paul Bernays

[1][Written vertically in shorthand in the left margin:] In this journal a long paper by Jorge Emmanuel Ferreira Barbosa[b] (in Portuguese) also appeared, which claims to establish the consistency for the formal systems of set theory in a constructive way.[c]

---

[c]The footnote is transcribed from Bernays' Gabelsberger shorthand.

83. Gödel to Bernays

Princeton, 17 December 1974

Dear Mr. Bernays,

I hope that your state of health is satisfactory. So far things are going quite well for me and I am making progress in my work on the true power of the continuum.

With cordial greetings, also from my wife,

Yours truly, Kurt Gödel

84. Gödel to Bernays

Princeton, 12./I. 1975.

Lieber Herr Bernays!

Herzlichen Dank für Ihre freundlichen Briefe vom 16./XII u. 7./I.
sowie auch für die lieben Neujahrswünsche. Die beiden erwähnten Nummern des "Boletim de Analise e Logica Matematica" wurden mir zuge-
schickt, aber ich kann keine Zuschrift finden, dass diese beiden Nummern
"für mich bestimmt sind." Haben irgendwelche Abhandlungen darin
speziell mit meinen Resultaten zu tun?

Besten Dank auch für die Zusendung des Separatums über Popper.[a]
Hat nach Ihrer Meinung Popper etwas wesentliches zur Aufklärung der
Grundlagen der Mathematik beigetragen? Ich freute mich, dass Sie auf
p. 603 einen vorsichtig platonistischen Standpunkt vertreten. Mir scheint
ein Platonismus dieser Art (auch hinsich[t]lich der mathematischen Begriffe) eine Selbstverständlichkeit zu sein u. seine Ablehnung an Schwach-
sinn zu grenzen.

Mit herzlichen Grüssen

Ihr Kurt Gödel

| P.S. Die gesuchte Zuschrift hat sich in einem *andern* Heft des "Bo-
letim ... " vorgefunden, nämlich Dez. 1973 (Art 5,1). Hier ist eine Arbeit von Barbosa publiziert in der er die Widerspruchsfreiheit der klassischen mengentheor. Axiomensysteme auf ca. 350 p[[p]]. "konstruktiv"
beweist u. zwar auf Grund der Lorenzenschen Arbeiten. Nun halte ich
ja so etwas ~~wen~~ für nicht unmöglich, wenn man "konstruktiv" genug weit
fasst (was ich "imprädikativen" Konstruktivismus nenne, vgl. Myhill, Zs.
math. Log. u. Grundl. d. Math 19 (1973)[b] p[[p]]. 93–96). Aber ich glaube
nicht, dass (wenn das überhaupt geht) man dazu 350 p[[p]]. braucht, oder
dass Lorenzen's Arbeiten dafür verwendbar sind ausser wenn er seinen
Standpunkt stark geändert hat). Haben Sie eine bestimmte Meinung
über diese Sache? Wette ist gegenwärtig in Amerika u. möchte mich
sprechen. Ich habe nie die Zeit gehabt, mir irgend eine seiner Arbeiten
über Widersprüche in der klassischen Mathematik anzusehen. Ist darin
irgend etwas Vernünftiges enthalten??

---

[a] *Bernays 1974.*

84. Gödel to Bernays

Princeton, 12 January 1975

Dear Mr. Bernays,

Cordial thanks for your friendly letters of 16 December and 7 January
as well as for the dear New Year's wishes. The two issues of the *Boletim de Analise e Logica Matematica* you mentioned were sent to me, but
I can find no note that these two issues "are for me". Are some of the
papers therein especially concerned with my results?

Thanks very much too for sending the offprints about Popper.[a] In
your opinion, has Popper contributed anything essential to the clarification of the foundations of mathematics? I'm pleased that on p. 603 you
advocate a cautiously platonistic point of view. To me a platonism of
this kind (also with respect to mathematical concepts) seems to be obvious and its rejection to border on feeble-mindedness.

With cordial greetings,

Yours truly,

Kurt Gödel

P.S. The note sought appeared in *another* issue of the *Boletim* ...,
namely December 1973 (Art 5,1). Here a work of Barbosa is published in
which he proves the consistency of classical set-theoretic axiom systems
"constructively" in about 350 pp., and indeed, on the basis of the works
of Lorenzen. Now I believe that such a thing is not impossible, if one
views "constructive" in a sufficiently wide sense (what I call "impredicative" constructivism; cf. Myhill, *Zeitschrift für mathematische Logik und
Grundlagen der Mathematik* 19 (1973), pp. 93–96[b]). But I don't think (if
it works at all) that one needs 350 pp. for it, or that Lorenzen's papers
are useful for it (unless he has starkly changed his point of view). Do you
have a definite opinion about this matter? Wette is presently in America
and would like to speak to me. I've never had the time to look at a single
one of his papers on contradictions in classical mathematics. Is *anything
at all reasonable* contained in them?

---

[b] *Myhill 1973.*

85. Bernays to Gödel

8002 Zürich, 24. Januar 1975
Bodmerstr.11

Lieber Herr Gödel

Haben Sie vielen Dank für Ihren Brief vom 12. Januar.

Was die lange Abhandlung von Prof. Barbosa betrifft, so kann ich
kaum sagen, dass ich darüber eine bestimmte Meinung habe. In der An-
kündigung, die mir im August 1974 von der Universidade Federal Fluminense, und Ihnen wohl auch, zugesandt wurde, steht ja unter anderem,
dass eine Uebersetzung ins Englische in Vorbereitung ist. Wenn dann
eine solche erscheint, wird man ja eher beurteilen können, was es damit
für eine Bewandtnis hat.

Sie erkundigten sich, ob Karl Popper etwas Wesentliches zur Aufklärung der Grundlagen der Mathematik beigetragen habe. Jedenfalls hat
er sich mit Fragen der Logik und der Axiomatik eingehender befasst. Bei
der Logik ist sein Bestreben, möglichst Axiome und Grundregeln durch
Definitionen zu ersetzen. Die Art dieser Betrachtung finden Sie z. B. in
der Abhandlung "Functional Logic without Axioms or Primitive Rules
of Inference"[a] (Proceedings of the Koninklijke Nederlandsche Akademie Van Wetenschappen, Vol. L, No. 9, 1947, North-Holland Publishing
Company).[1] In dieser ist auch eine frühere Publikation angegeben. Mit
der Axiomatik hat sich Popper speziell für die Wahrscheinlichkeitstheorie
abgegeben, für die er verschiedene Axiomensysteme aufgestellt hat, wobei
er unter anderem darauf abzielte, ohne die Voraussetzung auszukom-
men, dass die der Wahrscheinlichkeit unterworfenen Gegenstände einer
Boole'schen Algebra genügen, während er andererseits die Theorie der
reellen Zahlen voraussetzt. Sie finden dies z. B. im Appendix der Abhandlung "The Propensity Interpretation of Probability"[b] (The British
Journal for the Philosophy of Science, Vol. X, No. 37, 1959). Eine etwas
frühere Abhandlung von Popper über Axiomatisierung der Wahrschein-
lichkeitsrechnung ist im gleichen Journal (Vol. VI, No. 21, 1955)[c]
erschienen.—Kürzlich hat Popper auch eine Abhandlung über ein Axiomensystem der Geometrie publiziert, welches eine Verbesserung eines
Axiomensystems von einem Herrn Roehle ist.

---

[1]Die hier ausgeführte Zurückführung von Axiomen auf Definitionen ist freilich insofern mehr scheinbar, als die Definitionen keine eigentlich expliziten Definitionen sind.
Im Abschnitt VIII der genannten Abhandlung bemerkt auch Popper selbst, dass man
zur Anwendung ~~des~~ seines Systems auf eine Objektsprache zu den Definitionen noch
entsprechende Existenzaxiome hinzufügen muss.

---

[a]*Popper 1947.*

85. Bernays to Gödel

8002 Zurich, 24 January 1975
Bodmerstrasse 11

Dear Mr. Gödel,

Many thanks for your letter of 12 January.

As to the long paper of Professor Barbosa, I can hardly say that I have a definite opinion about it. In the announcement that was sent to me, and probably also to you, in August 1974 by the Universidade Federal Fluminense it states among other things that a translation into English is in preparation. When such a thing appears, then one will be better able to judge what's going on with it.

You inquired whether Karl Popper has contributed anything essential to the clarification of the foundations of mathematics. In any case, he has concerned himself in more detail with questions of logic and axiomatics. In logic he has striven to replace axioms and primitive rules by definitions. You find this sort of consideration, e.g., in the paper "Functional logic without axioms or primitive rules of inference"[a] (*Proceedings of the Koninklijke Nederlandsche Akademie van Wetenschappen*, vol. L, no. 9, 1947, North-Holland Publishing Company).[1]

In this [paper] an earlier publication is also mentioned. In axiomatics Popper has particularly concerned himself with probability theory, for which he has set up various axiom systems with which, among other things, he aimed to get along without the assumption that the objects subject to probability [considerations] constitute a Boolean algebra, while, on the other hand, he assumes the theory of real numbers. You find this, e.g., in the appendix to the paper "The propensity interpretation of probability"[b] (*The British Journal for the Philosophy of Science*, vol. X, no. 37, 1959). A somewhat earlier paper by Popper on the axiomatization of probability theory[c] appeared in the same journal (vol. VI, no. 21, 1955). Recently Popper has also published a paper on an axiom system for geometry, which is an improvement of an axiom system of one Mr. Roehle.

---

[1] The reduction of axioms to definitions carried out here is, to be sure, more apparent [than real], insofar as the definitions are not really explicit definitions. In chapter VIII of the aforementioned paper even Popper himself remarks that in the application of his system to an object language corresponding existence axioms still have to be added to the definitions.

---

[b] *Popper 1959.*
[c] *Popper 1955.*

2 | Von Herrn Wette habe ich über seine verschiedenen Untersuchungen jeweils Mitteilungen empfangen. Ich glaube nicht, dass seine Behauptungen über Widersprüchlichkeiten mathematischer Theorien zutreffen; aber es ist mir nicht gelungen, seinen Ausführungen so weit zu folgen, dass ich einen bestimmten Fehler der Argumentation aufweisen kann. Dass ein solcher wohl vorliegt, dafür spricht wohl besonders der Umstand, dass Herr Wette dazu geführt wird, immer elementarere Systeme als widersprüchlich zu erklären: zuerst war es die Mengenlehre, dann die Analysis, sodann die intuitionistische Zahlentheorie, und neuerdings erklärt er sogar den Aussagenkalkul als widersprüchlich. Die Schwierigkeit der Feststellung eines Fehlers beruht insbesondere darauf, dass die Beweise an verschiedenen Stellen nur angedeutet sind. In dieser Hinsicht findet immerhin eine gewisse Konvergenz statt. So ist der Beweis, der kürzlich in der "International Logic Review" (No. 9, June, 1974)[d] erschienen ist, besser verfolgbar als die früheren.—Ich empfinde es als ein Desiderat, dass die Ursache der vermeintlichen Widersprüchlichkeiten aufgedeckt wird.

Meinem Eindruck nach ist Herr Wette ein Mensch mit starker Begabung, der jedoch durch seinen Ehrgeiz in eine falsche Linie gekommen ist. An sich ist er ein freundlicher Mensch, aber sehr verbittert darüber, dass man begreiflicherweise seine Arbeiten—(es sind sehr mühevolle Arbeiten!)—nicht anerkennt.

Mit herzlichen Grüssen, auch an Ihre Frau,

Ihr Paul Bernays

---

[d] *Wette 1974.*

I've always received information from Mr. Wette about his various investigations. I don't think that his claims about inconsistencies of mathematical theories are correct; but I haven't succeeded in following his exposition far enough so that I can point to a definite mistake in the argumentation. That there is such a mistake is particularly suggested by the circumstance that Mr. Wette is led to declare ever more elementary systems to be contradictory: first it was set theory, then analysis, then intuitionistic number theory, and recently he is even declaring the propositional calculus to be contradictory. The difficulty of detecting a mistake is due especially to the fact that the proofs at various points are only sketched. In this respect, at least, a certain convergence is taking place. Thus the proof that recently appeared in the *International Logic Review* (no. 9, June, 1974)[d] is easier to follow than the earlier ones. I view it as a desideratum that the cause of the alleged inconsistencies be exposed.

In my opinion Mr. Wette is a highly talented person who, however, has been led astray by his ambition. By nature he is a friendly person, but very embittered by the fact that, as is understandable, his papers— (they are very difficult papers!)—are not properly recognized.

With cordial greetings, also to your wife,

Yours truly,

Paul Bernays

Kenneth Blackwell

Kenneth Blackwell (b. 1943) was the archivist of the Bertrand Russell Archives at McMaster University in Ontario from its inception in 1968 until 1996, and has been editor of the journal *Russell* since 1971. In 1971 he wrote Gödel, seeking copies of correspondence and asking for Gödel's reaction to some remarks in Russell's autobiography. The passage in question, relating to the year 1944 (and written in 1952) is this:

> The last part of our time in America was spent at Princeton, where we had a little house on the shores of the lake. While in Princeton, I came to know Einstein fairly well. I used to go to his house once a week to discuss with him and Gödel and Pauli. These discussions were in some ways disappointing, for, although all three of them were Jews and exiles and, in intention, cosmopolitans, I found that they all had a German bias towards metaphysics, and in spite of our utmost endeavours we never arrived at common premises from which to argue. Gödel turned out to be an unadulterated Platonist, and apparently believed that an eternal 'not' was laid up in heaven, where virtuous logicians might hope to meet it hereafter. (*Russell 1968*, p. 341)[a]

Gödel made a very rough draft of a reply to Blackwell, but sent nothing. He states that he never corresponded with Russell. This is false, although just barely so: Gödel wrote to Russell once, on 28 September 1943, to urge Russell to write a reply to *Gödel 1944* for inclusion in *Schilpp 1944*.[b]

Gödel goes on to note, quite fairly, that Russell himself had held the Platonistic position mockingly ascribed to Gödel in the passage. In support of this, a citation from *Russell 1919* is indicated, but not given. A likely candidate is this: "Logic is concerned with the real world just as truly as zoology, though with its more abstract and general features" (p. 169), which Gödel quotes in *1944*, p. 127.[c] However, it should be noted that in that paper Gödel does not take Russell's Platonism to be "unadulterated." Immediately after the citation from Russell, he says that Russell's "realistic attitude...always was stronger in theory than

---

[a]Compare *Russell 1937*, p. ix: "not even the most ardent Platonist would suppose that the perfect 'or' is laid up in heaven."

[b]The letter appears in these *Works*, vol. V.

[c]In *Wang 1987*, p. 112, where the full text of Gödel's draft is printed, Hao Wang confidently takes this to be the quotation Gödel intended. He reports the sentence was Gödel's "favorite," and "corresponds pretty closely to his own view."

in practice," and later goes on to ascribe to Russell a "constructivistic attitude" as well (*1944*, pp. 127 and 137ff.).

Warren Goldfarb

1. Blackwell to Gödel[a]

September 22nd, 1971

Dear Professor Gödel,

Since Bertrand Russell sold his personal papers to this university, we have been trying to obtain copies of his correspondence and manuscripts located elsewhere. Do you have any letters from Russell? We should be very grateful for copies if you do. I enclose Russell's authorization for the provision of copies to McMaster.

I wonder if you have any comments on Russell's passage about you in his autobiography (Vol. II, p. 341).[b] He calls you there "an unadulterated Platonist".

With apologies for any trouble this request may cause you.

Yours truly

Kenneth Blackwell
Archivist

---

[a]On letterhead reading: McMaster University
                          Hamilton, Ontario, Canada
                          Mills Memorial Library
                          Bertrand Russell Archives
[b] *Russell 1968*.

2. Letter draft from Gödel to Blackwell

Dear —

Replying to your letter of _____ I would like to say that I never had any cor. with Bertr R.

As far as ~~the passage about me of his au~~ p. _____ of his autobiog is conc. I have to say *first* (for the sake of truth) that I am not a Jew

(even though I don't think ~~that~~ this question is of any importance.)
2.) ~~it~~ ⟨that the passage⟩ gives the wrong imp that I had many disc.
with Russell, which was by no means the case. ⟨(I remember only one)⟩
3.) Concerning my "una⟋dulterated" Plat ~~as conc.~~ it ~~was is~~ ⟨is no more
"unal." [[th]]an Russell's own ~~who~~ ⟨Plat.⟩⟩ in 1921 when in ⟨the⟩ Introd.[a]
~~was even~~ ⟨he said:⟩ [com the "not" with an animal when by negating]
"........." At that time evidently Russell had met the "not" even in *this*
world but later on ⟨under the infl of Wittg.⟩ he chose to overlook it.

Sinc.

---

[a] *Russell 1919.*

Herbert Bohnert

Herbert G. Bohnert

Herbert G. Bohnert (1918–1984) was professor of philosophy at Michigan State University. He had taken courses with Rudolf Carnap at the University of Chicago in the early 1940s, and subsequently became personally close with him. Throughout his career he defended Carnapian positions. In his letter to Gödel he asks about the influence of Carnap on *Gödel 1931* and that of Gödel on *Carnap 1934a*.

Bohnert's formulations of his questions evidence an imperfect knowledge of the historical situation. From 1927 until 1930, Carnap's principal project was a treatise on logic, which he called *Untersuchungen zur allgemeinen Axiomatik*, that is, "Investigations into general axiomatics." In this work, his approach was what might be called Russellian, rather than metamathematical. Notions such as "logical consequence", "consistency", and so on, were taken by him to be explicated internally to the logical system. (Thus, for example, that a sentence G was a logical consequence of a sentence F was expressed *inside* the system by the universally quantified conditional obtained from $F \supset G$ by generalizing on all the non-logical vocabulary. This is just Russell's notion of "formal implication.") As a student in 1928, Gödel probably attended Carnap's presentations to the Vienna Circle that were based on Part I of the treatise. Carnap subsequently circulated that part to several logicians and philosophers, including Gödel. Gödel's understanding of Carnap's project is reflected in the third paragraph of his dissertation *1929*, where he says that "if we replace the notion of logical consequence... by implication in Russell's sense," then the assertion of completeness of the logical system is provable in a few straightforward steps; and he credits Carnap with this observation (see these *Works*, vol. I, p. 63). What he does not note, charitably, is that at this time Carnap thought that logical consequence *was* formal implication, and so took the easily proved result to be a bona fide completeness result. Carnap first began to understand the necessity to define logical notions metamathematically in early 1930, as a result of lectures given by Alfred Tarski in Vienna and further discussions with Tarski. Consequently, he abandoned the project of the *Untersuchungen* shortly thereafter.[a]

In contrast, from the time of his first contribution *1929*, Gödel was in complete command of the distinction between metatheoretic notions

---

[a]In this paragraph, I draw heavily on *Awodey and Carus 2001*, which contains a fuller account of Carnap's *Untersuchungen*, its role in Carnap's general philosophical aims, and its eventual abandonment by Carnap. Before abandoning the project, Carnap published a report of some of its results in *1930a*.

and notions expressed within a logical system. Thus the method of
Gödel 1931 (which was devised in 1930) could certainly not have been
influenced by Carnap.[b]

In his reply to Bohnert, Gödel does not go into detail about this;
nor does he comment on Bohnert's apparent conflation of the arithmetization of syntax and the notion of formalizing a metalanguage. He
simply remarks that the method of *1931* was "in no way prompted by
the Vienna Circle."

On the question of Gödel's influence on *Carnap 1934a*, Gödel is modest to the point of distortion. He says that Carnap was not "to any
considerable extent" influenced by him in that work, and that arithmetization was "*not* essential for the primary aims of 'Logical Syntax.'"
Carnap's account suggests more than this:

> My way of thinking was influenced chiefly by the investigations of Hilbert
> and Tarski in metamathematics... I often talked with Gödel about these
> problems. In August 1930 he explained to me his new method of corre-
> lating numbers with signs and expressions. Thus a theory of the forms of
> expressions could be formulated with the help of concepts of arithmetic.
> (*Carnap 1963*, p. 53)

Arithmetization was important to Carnap, because the ability to frame
syntax within a clearly unobjectionable arithmetical language answered
the doubts, stemming from Wittgenstein and shared by some members
of the Vienna Circle, that the logical structure of language could properly be described at all.

Much of *Logical syntax of language* is a response to the challenge
Carnap took to be posed by Gödelian incompleteness. Incompleteness
shows that the notion of mathematical truth cannot be captured by notions based on formal derivability. Carnap sought definitions that would
capture it, for the formal languages he considered, while using only the
resources he considered "syntactic" (a word he used in a wider sense
than it currently has). In this project Gödel was of technical assistance
as well: he showed Carnap that the strategy of his original draft was
incorrect, and pointed him in the direction of the appropriate definition.
(See letters 3–6 of the Gödel–Carnap correspondence in this volume.)
In manifold ways, *Logical syntax* is inconceivable without Gödel.

Gödel's disclaimer to Bohnert of any significant influence on Carnap
was clearly motivated by a desire to distance himself from Carnap's
philosophical stance, and that of the Vienna Circle generally. Gödel
disagreed profoundly with the philosophical gloss Carnap put on the

[b]The question of Carnap's impact on Gödel more generally is considered in the
introductory note to the Carnap correspondence in this volume.

technical work of *Logical syntax*, namely, that mathematical truth was essentially an artifact of language, and did not pertain to a reality of any sort. He did not publish his criticisms of these views during his lifetime.[c] Perhaps for that reason, he was at pains in the early 1970s to take opportunities to depict, and sometimes to elaborate, how much he differed from the views associated with logical positivism.[d]

Warren Goldfarb

[c]The most focused criticisms are published as *1953/9* in these *Works*, vol. III. Additional demurrals can be found in *1951* and *1961/?* in that volume.

[d]In addition to the material in *Wang 1974* that Gödel mentions to Bohnert, see also the reply to Burke Grandjean in this volume, and letter 17 in the Menger correspondence in volume V.

1. Bohnert to Gödel[a]

July 3, 1974

Dear Professor Gödel:

I have a question about conversation(s) between you and Carnap in 1930 and/or before. In writing an article on the development of Carnap's logicism I have been trying to account for the metalinguistic procedures in his *Logical Syntax*.[b] In his Autobiography in Schilpp[c] he remarks, p. 30, that he was awakened to the power of metalinguistic methods by Tarski's visit to Vienna in February 1930. Then on the same page he remarks that he had often discussed the problem of speaking about language with you. Later, p. 53, he more specifically remarks that in August 1930 you explained your method of arithmetization and your major result. A few sentences later he says that in January 1931 he wrote down the core ideas of *Logical Syntax* after thinking about these problems for several *years* (apparently stemming from disagreement with Wittgenstein on the possibility of self-referring language [note sentence in the middle of the same page]). Furthermore, in September of 1930 he spoke at the Königsberg symposium with the avowed purpose (p. 46) of effecting some

[a]On letterhead reading: Michigan State University
Department of Philosophy
East Lansing, Michigan, 48824.

[b]*Carnap 1937*.

[c]*Carnap 1963*.

rapprochement with formalism. I find these dates a bit hard to reconcile. I recall asking him, many years ago, about the striking similarity between many of his methods in *Logical Syntax* and those of your 1931 paper, expecting him to say that he had been inspired to *Logical Syntax* by your paper. He said then that he had had conversations with you but said little further about them—aside from giving you full credit for arithmetization and your results (in reply to my specific question on that matter). Now, questions arise anew in my mind. Nowhere in earlier formalist literature do I find such a complete formalization of the metalanguage as I do in your 1931 paper. Were you influenced by formalism toward that approach, by Carnap, or only by the needs of your proof? Did Carnap speak at all of a fully formalized syntax language in any of those conversations?

I recall, many years ago, being entranced when you animatedly described the thoughts that led you to your theorem (at Paul Oppenheim's house in 1952—or 1953). I forgot, of course, the foregoing at the time.

Even a brief line or two on any of the above matters would be appreciated.

Sincerely yours,

Herbert G. Bohnert
Professor

2. Gödel to Bohnert

Professor Herbert G. Bohnert
Department of Philosophy
Michigan State University
East Lansing, Michigan, 48824

September 17, 1974

Dear Professor Bohnert:

I have your letter of July 3. Without going into any details about specific dates or subjects of my discussions with Carnap (which, incidentally, were not very numerous) I can generally say this:

I don't think Carnap, to any considerable extent, was influenced by me in his "Logical Syntax of Language," (except, of course, in the passages where he explicitly refers to my work). The existence of a completely precise syntax and metalanguage was clear already from the work of Frege, von Neumann, Ackermann, and others.

The new idea which I brought in (and which is *not* essential for the primary aims of "Logical Syntax") was the expression of syntax in the same language. This idea was in no way prompted by the Vienna Circle. On the contrary there was a tendency there to consider such a procedure to be an *inadmissible* (i.e., *faulty*) commingling of things that have to be kept separated. Rather this idea arose as a necessary means for carrying through certain purely mathematical schemes of proof.

Moreover, I can say that I *never* agreed with the view that mathematics is syntax of language and that my work rather is based on an *opposition* to this view, as you can see from some letters and remarks of mine that were published recently in the book "From Mathematics to Philosophy" by Hao Wang[a] (cnf. in particular pp. 8–12 and the other passages indicated in the preface).

I owe a great deal to the Circle of Vienna. But it is solely the introduction to the problems and their literature.

Sincerely yours,

Kurt Gödel

---

[a] *Wang 1974*.

William Boone

William Boone

William Werner Boone (1920–1983) was a logician whose research centered on algebraic decision problems, particularly in group theory. He received his doctorate from Princeton University in 1952, and in 1954–1956 he was at the Institute for Advanced Study, where he got to know Gödel. During this period, he was occupied with the word problem for groups, and in 1956 he achieved his goal of constructing a finitely presented group whose word problem is undecidable (*Boone 1957* and *1957a*). The same result was obtained independently, by a different construction, slightly earlier, by P. S. Novikov (*Novikov 1955*). Consequently, the recursive undecidability of the word problem for groups is often called the Boone–Novikov theorem. Over the next few years, Boone worked at improving and refining his proof; the end product was the long paper *Boone 1959*. In 1958 Boone joined the mathematics department of the University of Illinois, becoming professor in 1960. He remained on the faculty there until his death. Boone had a second two-year stint at the Institute for Advanced Study in 1964–1966. He spent some of that time completing *Boone 1966* and *1966a*, in which he showed that every recursively enumerable degree of unsolvability can be represented by a word problem for a semigroup and for a group.

Boone and Gödel became close during Boone's first stay at the Institute. Gödel soon was quite interested in Boone's research, and discussed it with him in detail. Boone wrote Gödel several times thanking him for his interest, once even with specific examples of how their discussions benefitted Boone's work. Gödel clearly thought very highly of Boone, and liked him personally very much. He was most supportive of Boone's research, and wrote numerous strong letters of recommendation for Boone for fellowships, grants and academic positions. Moreover, Gödel refereed *Boone 1959* for the *Annals of mathematics*, writing in his report, "The paper contains a very ingenious proof for the algorithmic unsolvability of the Word Problem for Groups. ...Some parts of the proof of the main result could be simplified which would shorten the paper by a few pages, but for the most part the argument is very elegant and not too complex."[a] Boone remained in contact with Gödel by letter and later by telephone, nearly until Gödel's death. On 1 February 1978, Boone wrote a condolence note to Adele Gödel:

[a]Document 010186 in the Gödel *Nachlaß*.

I hope you know, dear Mrs. Gödel, how much Professor Gödel did to help me with my work. It was he who was willing to check my work and say it was correct. He made many suggestions as to what I should work on, too. But what was the most wonderful aspect of our relationship for me, is that he became my friend, always willing to talk to me about mathematics or the world in general.

Certainly he was the most brilliant person I have ever known. I treasured every moment to speak with him.

My wife, Eileen, and I extend to you our deepest sympathy. What can one say? There won't be another like him. Requiescat in pace.

Again with sympathy, Bill Boone[b]

The two letters from Gödel to Boone reproduced below express views about mathematical issues that do not appear elsewhere in Gödel's writing or correspondence. On 26 October 1956 Boone wrote Gödel from Oslo, reporting that he was searching for any errors in his unsolvability proof. In contrast, he felt that the task of verifying Novikov's proof was "gigantic", and he hoped Gödel did not think he should do this before publishing his own. In his reply (letter 1), Gödel advised Boone to bring his proof "close to formalization", by dividing it into many small lemmas, and added that this method often leads to simplifications or generalizations. No doubt the fact that Boone's proof was a long combinatorial argument, with little intuitive content in its parts, was a factor in Gödel's advice. Gödel's own practice in this regard was variable; of his writings perhaps *Gödel 1940* shows the most affinity to this expository approach. Letter 2 was a response to a letter from Boone of 15 October 1961, asking for Gödel's support as a reference, in his application to the National Science Foundation for a travel grant to the International Congress of Mathematicians in Stockholm in 1962, and adding, "Recently, too—partly as a result of refereeing—I've read a great deal on Hilbert's Tenth Problem. In a general way may I ask you: Do you think I should launch myself (as the French say) into that field?" Hilbert's Tenth Problem is that of deciding whether an arbitrary diophantine equation has a solution in the integers. Boone wrote this just after Martin Davis, Hilary Putnam and Julia Robinson had shown that the analogous problem for exponential diophantine equations was recursively unsolvable (*Davis, Putnam and Robinson 1961*, which may well have been the paper Boone had refereed). In response to Boone's question, Gödel cautioned him that a solution of the problem might require "a great deal of number theory". This turned out not to be the case: the undecidability of Hilbert's Tenth problem was shown by using only elementary facts about the Fibonacci series (*Matiyasevich 1970*).

[b]Document 010264 in the Gödel *Nachlaß*.

But Gödel's response did not evince the overarching pessimism about the possibility of settling Hilbert's Tenth Problem that was shared by many logicians at that time.[c]

Warren Goldfarb

A complete calendar of the correspondence with Boone appears on pp. 557–558 of this volume.

[c]See, for example, *Kreisel 1962b*.

1. Gödel to Boone

Princeton, Dec. 4, 1956.

Dr. Boone:

I was very pleased to hear that you are well and that your work is progressing so satisfactorily. I would suggest that you recheck carefully also the results published in the Proc. Dutch Acad.,[a] which, according to my recollection, have not been checked by others so carefully as your later work (except in so far as they agree with your dissertation[b]). I am making this suggestion, not because I am distrustful of your exactitude, but because I believe that in case of an ⟨really⟩ important theorem every precaution should be taken (in your own interest).

Incidentally, the best method, in my opinion, of checking a complicated proof is to bring it close to formalization by interpolating so many auxiliary theorems that the proof of each becomes trivial on the basis of the preceding ones. This procedure frequently also leads to simplifications or generalizations. Of course, eventually you will have to study also Novikoff's proof,[c] in view of the fact that you are supposed to write a book on the subject and also in view of your subsequent work.

The manner of publication which you suggest seems alright to me. Also I believe it is a very good idea to extend your stay in Europe by the visits you are planning. As to your suggestion to stay in en England for the next year I agree with Prof. Montgomery.

---

[a]Presumably *Boone 1954, 1954a, 1955* and *1955a*.

[b]*Boone 1952*.

[c]*Novikov 1955*.

With all good wishes and kind regards to Mrs. Boone.

Sincerely yours,

Kurt Gödel

2. Gödel to Boone

November 8, 1961

Dr. William Boone
Department of Mathematics
University of Illinois
Urbana, Illinois

Dear Dr. Boone:

I am terribly sorry that due to the fact that the International Congress is still so far away I have overlooked the fact that you needed a reply to your request before November 1. Of course I will be glad to support your application for a travel grant to the Congress. I will be glad to write a letter also without your having mentioned my name, although I believe *it will not* be necessary for you in order to obtain the grant.

Thank you for notifying me of Britton's work and sending me the seminar notes. I hardly think I shall have a chance to read them in the foreseeable future, but I might once come back to these things. I did not receive Britton's paper for refereeing. I suppose it is generally known by now that I am working on entirely different matters.

As far as your going into Hilbert's Tenth Problem is concerned, I think I told you once that I am under the impression a great deal of number theory will be required for its solution. So, unless you have a genuine taste for number theory (irrespective of applications to the foundations) or unless you have some definite idea of how to go about it without using much of number theory, I would not think it is the right field of work for you. In the second case, it might at any rate be worthwhile to spend some time in order to see whether this idea leads to anything.

With best wishes to you and your family, I am

Sincerely yours,

Kurt Gödel

Georg Brutian

The Soviet academician Georg[a] Abelovich Brutian was born in Sevkar, Armenia, in 1926. A graduate of Yerevan State University, he received his Ph.D. and Sc.D. degrees from Moscow State University. Since 1971 he has been Professor of Philosophy at Yerevan State University. He has served as President of the Armenian Philosophical Academy since 1987 and Vice President of the National Academy of Sciences of the Republic of Armenia since 1994. His writings comprise 53 books and nearly 200 articles in logic, methodology, epistemology, philosophy of language and the history of philosophy.

In mid-November 1969, as part of a two-month sojourn in the United States sponsored by the International Research and Exchanges Board, Brutian spent a week in Princeton where, as planned in his project itinerary, he secured an interview with Gödel.

According to Brutian[b] their long conversation was dominated by Gödel's questions rather than his own. In particular, perhaps on the basis of Brutian's paper *1968*, Gödel inquired about the relation between contradictions in logic and Hegel's theory of dialectical contradictions and asked Brutian about the views of Soviet philosophers and logicians on such matters. Late in the meeting Brutian asked Gödel for an evaluation of the philosophical significance of the incompleteness theorems, but Gödel said that he was then too tired to continue the conversation. He offered instead to send Brutian a written response, which he did in the letter reproduced below. Brutian subsequently wrote back, raising a few further questions, but Gödel did not reply.

John W. Dawson, Jr.

[a]Anglicized by Gödel as 'George'.
[b]Personal communication to J. Dawson, 10 January 1998.

1. Gödel to Brutian

Professor George A. Brutian
International Research and Exchanges Board
110 East 49th Street
New York, NY 10022

December 10, 1969

Dear Professor Brutian:

Here is *one* formulation of the philosophical meaning of my result, which I have given once in answer to an inquiry:

The few immediately evident axioms from which *all* of contemporary mathematics can be derived do not suffice for answering all Diophantine yes or no questions of a certain well-defined simple kind.[1]

Rather, for answering all these questions, infinitely many new axioms are necessary, whose truth can (if at all) be apprehended only by constantly renewed appeals to a mathematical intuition, which is actualized in the course of the development of mathematics.[a] Such an intuition appears, e.g., in the axioms of infinity of set theory.

There are other formulations, which ought to be added, in order to make the situation completely clear. Perhaps I can send them to you at some later date through the International Research and Exchanges Board.

Yours sincerely,

Kurt Gödel

[1]See: M. Davis, The Undecidable,[b] New York 1965, p. 73, last but one paragraph.

---

[a]An insertion indicated here and written at the bottom of the page reads "(and phil. of math)", followed by an arrow pointing to "nicht abgeschickt" ("not sent") in shorthand. Thus presumably this was added only to Gödel's retained copy, from which the text as printed here was taken.

[b]*Davis 1965.*

J. Richard Büchi

Julius Richard Büchi was born in Porto Allegre, Brazil, on 31 January 1924 to Swiss parents and as a citizen of Zell, Switzerland.[a] He grew up in Switzerland and in 1948 received a doctoral degree in mathematics from the Eidgenössische Technische Hochschule in Zurich; his thesis supervisor was Paul Bernays. After graduation he moved almost immediately to the United States and had a number of academic appointments, among others at the University of Michigan, Ann Arbor. In 1963 he became Professor of Mathematics and Computer Science at Purdue University and retained that position until his death in 1984. Büchi did important work in mathematical logic and, relatedly, theoretical computer science. Dirk Siefkes states in his *1985* that Büchi is "probably best known for using finite automata as combinatorial devices to obtain strong results on decidability and definability in monadic second-order theories and extending the method to infinite combinatorial tools".[b] Büchi's papers were collected by Mac Lane and Siefkes in *Büchi 1990*; his posthumously published book *Finite automata, their algebras and grammars* was edited by Siefkes.

Two letters were exchanged between Gödel and Büchi in November 1957; they throw some additional light on (Gödel's views of) Herbrand's role in the development of the notion of recursiveness. In his Princeton lectures *1934* Gödel presented a general notion of recursiveness, where the individual functions arise simply as unique solutions of systems of equations:

> If ϕ denotes an unknown function, and $\psi_1, \ldots, \psi_k$ are known functions, and if the ψ's and ϕ are substituted in one another in the most general fashions and certain pairs of resulting expressions are equated, then, if the resulting set of functional equations has one and only one solution for ϕ, ϕ is a recursive function.[c]

Gödel asserted in his lectures that Herbrand had suggested this definition to him in a private communication. Büchi calls this notion *recursive (1)*. It is to be contrasted with the concept *general recursive* that Gödel obtained from it by restricting the form of equations and by

---

[a]More detailed information concerning Büchi's life and work can be found in *S. Büchi 1990, Lipshitz, Siefkes and Young 1984* and *Siefkes 1985*. At the time of writing his letter to Gödel, Büchi was visiting the department of mathematics at the University of Notre Dame. As to the logical and historical issues raised by Büchi, cf. the (introductory note to the) correspondence between Gödel and Herbrand.

[b]*Siefkes 1985*, p. 7.

[c]*Gödel 1934*, p. 26; these *Works*, vol. I, p. 368.

specifying elementary replacement rules to be used in calculating the
value of functions. Büchi reports that he did not find the definition
of *recursive (1)* in any of Herbrand's papers; he did find, however, in
Herbrand's *1931* "a definition which comes much closer to your definition
⟦of general recursive⟧ of 1934." The definition Büchi alludes to allows
the introduction of functions f_i of n_i arguments into Herbrand's system
of arithmetic together with hypotheses (i.e., defining equations) such
that, as Herbrand requires there,

(a)   The hypotheses contain no apparent variables;
(b)   Considered intuitionistically, they ⟦the hypotheses⟧ make the actual
      computation of the $f_i(x_i, \ldots, x_{ni})$ possible for every given set of
      numbers, and it is possible to prove intuitionistically that we obtain
      a well-determined result.[d]

In a footnote to the first occurrence of "intuitionistically" Herbrand
explains that this expression means, "when they ⟦the hypotheses⟧ are
translated into ordinary language, considered as a property of integers
and not as mere symbols."

Büchi asks two questions concerning the notion *recursive (1)*: (a)
Was the definition actually suggested by Herbrand or did Gödel refer
to *1931*? and (b) Is it known that this notion is much weaker than
general recursive? As to the substantive mathematical issue underlying question (b) Büchi had obtained results, in particular, that there
are *recursive (1)* predicates that are not general recursive, indeed not
even arithmetical. Gödel refers to *Kalmár 1955* for an affirmative answer to (b).[e] Concerning the historical question Gödel reasserts that
Herbrand communicated the definition of *recursive (1)* to him in a letter. The definition in Herbrand's *1931*, Gödel says, "means nothing
else but *demonstrably recursive (1)*", where the demonstrations have to
be intuitionistic. The actual computation is not, according to Gödel,
to proceed according to formal rules, but rather by "any kind of intuitionistic reasoning". "Therefore," he continues, "it is a priori possible
that also the non-recursive functions which you mention in your letter
might be recursive in this sense." For a further discussion of Gödel's
analysis of Herbrand's notion(s) and Jean van Heijenoort's elaborations
of this analysis, see the introductory note to the correspondence with
Herbrand, these *Works*, vol. V, and the literature mentioned there.

Wilfried Sieg

---

[d] *Herbrand 1931*, pp. 290–291.
[e] The theorem established by Kalmár (on p. 94) is slightly weaker than Büchi's
and states that there is a system of equations with a unique solution ϕ such that ϕ
is arithmetically definable but not general recursive.

1. Büchi to Gödel

November 5, 1957

Dear Professor Gödel:

I would appreciate it very much to know the answers to a question concerning the history of the notion of recursiveness.

In the notes to your lectures at Princeton in 1934 it is stated that Herbrand has suggested the following definition,

(1) A function f is recursive if it is the unique solution of a system $E[g_1, \ldots, g_n, f]$ of equations.

I have not been able to find this in any of Herbrand's papers. On the other hand in Crelles Journal 1932[a] he states a definition which comes much closer to your definition of 1934. I would like to know,

a. Was (1) really ever suggested by Herbrand, or was the reference in your 1934 lectures to Herbrand's paper in the Crelle Journal?

b. Is it known that recursive (1) is much weaker than recursive.

I would like to know the answers because I have obtained the following result,

I. If f is arithmetical then f is recursive (1).
   And more generally,

II. If the predicate F on natural numbers is the unique solution of a 1st order formula $A[G_1, \ldots, G_n, F]$, then F is recursive (1).

Because A in II may be taken to be any inductive definition $A[+, \cdot, F]$ of F, this implies

III. There are predicates which are recursive (1) and not arithmetical.

I would be very glad for any information concerning these matters.

Sincerely yours,

J. Richard Büchi

---

[a] *Herbrand 1931.*

2. Gödel to Büchi

November 26, 1957

J. Richard Büchi
Department of Mathematics
University of Notre Dame
Notre Dame, Indiana

Dear Mr. Büchi:

The definition of recursive (1) to which you refer was communicated to me by Herbrand in a letter. The definition given in Crelle's journal, vol. 166, means nothing else but "demonstrably recursive (1)", where "demonstrably" refers to intuitionistic mathematics.[1] For "faire effectivement le calcul" evidently is not intended to mean: "to compute formally", but rather: "to compute by any kind of intuitionistic reasoning". Therefore it is a priori possible that also the non-recursive functions which you mention in your letter might be recursive in this sense. Even now the assertion that every intuitionistically computable function is recursively computable is not considered to be an intuitionistic axiom or theorem by Heyting.

That the concept of "recursive (1)" is wider than "recursive" has been proved by Kalmár in Zeitsch. f. math. Log. u. Grundl. d. Math. vol. 1, p. 93.[a] I understand that some student of Kleene's has stronger results, which, however, have not been published yet.

Sincerely yours,

Kurt Gödel

---

[1] Of course it is quite possible that Herbrand chose these two definitions only because he believed that "recursive (1)" is equivalent with "recursive" and that "intuitionistically computable" implies "recursively computable".

---

[a] *Kalmár 1955.*

Rudolf Carnap

Rudolf Carnap (1891–1970) was the most important philosopher of the twentieth-century movement known as logical positivism or logical empiricism.[a] He arrived at the University of Vienna as a *Privatdozent* in 1926 and immediately became active in the Vienna Circle, a group of philosophers and mathematicians that met regularly to discuss questions in "scientific philosophy". In June and July of 1928, Carnap gave presentations to the Circle on mathematical logic. Gödel, then in his fourth year of university studies, probably attended those meetings. Later that summer, Gödel read parts of *Principia mathematica*,[b] and in the fall of 1928, Gödel heard Carnap's course of lectures on logic at the University. Indeed, in 1975, in answer to an inquiry about the important influences on his becoming interested in problems of completeness, Gödel listed *Hilbert and Ackermann 1928*, where he found the problem for his doctoral dissertation, and Carnap's lectures.[c] Carnap circulated to Gödel the manuscript from which his lectures were drawn, the *Untersuchungen zur allgemeinen Axiomatik*; some remarks in the opening paragraphs of Gödel's dissertation are references to it.[d]

Over the next few years, Carnap and Gödel held discussions from time to time on questions in logic and the foundations of mathematics.[e] Carnap was among the first of those whom Gödel told of his incompleteness discovery, in a conversation on 26 August 1930, and he was in the audience at Gödel's first public announcements of his result during a conference in Königsberg on 5–7 September 1930.[f]

In the summer of 1931, Carnap moved to Prague to take up a professorship at the German University. Much of the correspondence below took place during Carnap's first years there, when he was principally occupied with writing his *Logical syntax of language* (*Carnap 1934a*, English edition *1937*). The idea for the book had come to him, he later

---

[a]Carnap's life, intellectual aims and work are nicely portrayed in his intellectual autobiography *Carnap 1963*.

[b]See the letter to Herbert Feigl of 24 September 1928 in this volume.

[c]See the correspondence with Burke Grandjean in this volume.

[d]See the introductory note to the correspondence with Herbert Bohnert in this volume. Carnap's manuscript was recently published (*Carnap 2000*).

[e]Several sets of discussion notes are preserved in Carnap's *Nachlaß*.

[f]See these *Works*, vol. III, pp. 13–15.

Rudolf Carnap

recounted, when

> the whole theory of language structure and its possible applications in philosophy came to me like a vision during a sleepless night in January 1931, when I was ill. On the following day, still in bed with a fever, I wrote down my ideas on forty-four pages under the title 'Attempt at a metalogic'. [*Carnap 1963*, p. 53]

He characterized his major technical end thus:

> One of my aims was to make the metalanguage more precise, so that an exact conceptual system for metalogic could be constructed in it. Whereas Hilbert intended his metamathematics only for the special purpose of proving the consistency of a mathematical system formulated in the object language, I aimed at the construction of a general theory of linguistic forms. [*ibid.*]

As part of his "general theory of linguistic forms", Carnap hoped to give a precise form to his central idea in the philosophy of logic and mathematics, namely, that logical and mathematical truths were not descriptions of some reality, but rather only artifacts of language, true merely by dint of the way the representational structure is set up. Incompleteness presents a challenge here: for it shows that one cannot think of a language as specified by an axiom system and deduction rules, if the specification is to yield all mathematical truths. To preserve his central idea Carnap would need a stronger notion of what is constitutive of a language. This impelled Carnap in the direction of giving a definition of truth for the purely mathematical sentences of a language.

Carnap first wrote Gödel about his book on New Year's Day 1932, asking whether Gödel would be interested in reading the first part of his manuscript, at that time titled "Metalogic". (In that letter, not reproduced here, Carnap also asked Gödel's help in converting assets he had in Vienna into Czechoslovakian currency; apparently he had been having some trouble about this, "on account of the currency regulations".) Gödel presumably replied with interest, for in February Carnap wrote to say that Gödel would be receiving the manuscript shortly (letter 1). Carnap continued with remarks on other current work in foundations of mathematics: a request for copies of the correspondence between Zermelo and Gödel,[g] some comments on a recent book of the German philosopher Hugo Dingler and expressions of puzzlement about articles and a manuscript of the Polish logician Leon Chwistek. The tone of the letter is friendly, and includes a mild jest about Gödel's practice of handwriting his letters rather than typing them, as well as encouragement for Gödel to write up the sequel to *1931* that Gödel had intended.

---

[g]This correspondence, dating from the fall of 1931, appears in these *Works*, vol. V.

In his reply (letter 2), Gödel says nothing about Carnap's manuscript, but does request that Carnap send him the second part, if it is done, and he makes some remarks on Chwistek's work.

The second part of Carnap's manuscript was called "Semantics". In September 1932 Gödel wrote Carnap with several important comments on it (letter 3). In the manuscript Carnap had attempted to give a truth-definition for the purely mathematical sentences of a language that incorporated the theory of types, that is, in Carnap's terminology, a definition of "analytic" for this language. The definition proceeded by induction on the complexity of formulas. Earlier, Carnap had seen that, for a first-order arithmetical language, one can stipulate that a sentence $(\forall x)\phi(x)$ is true if and only if each instance $\phi(\bar{n})$ is true. Similarly Carnap wanted to define the truth, e.g., of a second-order universal quantification $(\forall F)\phi(F)$ as amounting to the truth of each instance $\phi(P)$, where P is any second-order predicate in the language. Gödel points out that this procedure cannot work. It fails due to the impredicativity of the quantification: the truth of some of the instances $\phi(P)$ can depend on the truth of $(\forall F)\phi(F)$, which is what is meant to be defined, and so the definition is circular. More generally, Gödel says, higher-order languages cannot be understood "semantically", by which he apparently means construed as limiting the range of higher-order quantifiers to predicates or sets definable in the language (that is, as it might be put today, construed as treating the quantifiers substitutionally). For on such a construal one gets essentially the ramified theory of types, a predicative theory that does not yield the classical laws of, say, the real numbers. Gödel also says that he wants to include a definition of truth for this mathematical language in the planned sequel to *Gödel 1931*.

Gödel also raises an objection to what was apparently a very detailed exposition of the incompleteness proof in Carnap's manuscript. He descries a circularity in the definitions of a certain number and a certain formula, and concludes that the existence of the desired objects cannot be secured by definitions, but "probably" needs some kind of "diagonal inference". In *Logical syntax*, no proof with the level of intricacy indicated in Gödel's comment survives. Rather, Carnap presents the first incompleteness theorem by first formulating a general diagonal lemma (or fixed point lemma), and then constructing the Gödel sentence by an application of this lemma to the predicate that expresses non-provability. In fact, Carnap was the first to separate out these two elements of the proof and to formulate the general diagonal lemma, in the way that is now standard. It is tempting to speculate that it was Gödel's comment that stimulated Carnap to see the proof in this way.

The next two letters show Carnap trying to work out Gödel's suggestion about the truth-definition. He finds difficulty in the need to quantify over all sets in the definition, in particular concerning what metalanguage one would use to do so. Two issues seem to be of concern: first, that the metalanguage be able to capture the needed generality, and second that the metalanguage be a semantic one, that is, be about language in some sense. In letter 4, of 25 September 1932, he asks, "Can you define the concept 'set' within a definite formalized semantics?" In letter 5, written two days later, he sees what is needed is a metalanguage in which the universal quantifier can express the required generality, even though not every item in the range of the quantifiers can be defined in the language. Moreover, the generality needed is over what he calls "valuations" of sentences, and so is semantic in the sense he sought.

Traces of these considerations appear in *Logical syntax*. (In the book Carnap uses the term "syntactical" rather than "semantic".) He credits Gödel with the observation that the definition of "analytic" cannot proceed by substitution, because more than definable properties are at issue. After giving the definition, he notes that it

> must not be limited to the syntactical properties which are definable in S [the metalanguage] but must refer to all syntactical properties whatsoever. But do we not by this means arrive at a Platonic absolutism of ideas, that is, at the conception that the totality of all properties...is something which subsists in itself, independent of all construction and definition? From our point of view, this metaphysical conception...is definitely excluded. We have here absolutely nothing to do with the metaphysical question as to whether properties exist in themselves...The question must rather be put as follows: can the phrase "for all properties"...be formulated in the symbolic syntax-language S? This question may be answered in the affirmative. The formulation is effected by the help of a universal operator with a [predicate variable], i.e., by means of '$(F)(...)$', for example.[h]

In late November 1932, Gödel replied to Carnap's September letters (letter 6), explaining his delay by the amount of work he had to do on "a report about research on foundations". (This is a joint project he had agreed to undertake with Arend Heyting, but for which he wound up not producing anything.[i]) He also admits that the planned sequel to

---

[h] *Carnap 1937*, p. 114. The definition of "analytic" and related material did not appear in the original German edition of *Logical syntax*, due to space constraints. It was published separately as *Carnap 1935*, and then included in the English edition.

[i] See the introductory note to the correspondence with Heyting in these *Works*, vol. V.

Gödel 1931 "exists only in the realm of ideas". As for the truth-
definition, he tells Carnap that it appears from letter 5 that Carnap
has understood his suggestions, and then goes on to make two further
comments. First, the definition of truth for the entire type-theoretic
language will require variables of type ω, that is, variables that range
over objects of arbitrarily high finite type. Second,

> the interest of this definition does not lie in a clarification of the con-
> cept "analytic", since one employs in it the concepts "arbitrary sets",
> etc., which are just as problematic. Rather I formulate it only for the
> following reason: with its help one can show that undecidable sentences
> become decidable in systems which ascend farther in the sequence of
> types.

These remarks substantiate footnote 48a of *Gödel 1931* (these *Works*,
vol. I, p. 181), where Gödel says that the addition of type ω to the type-
theoretic system would make decidable the undecidable sentences he
constructs. This suggests that Gödel saw how to frame a truth-definition
for the type-theoretic language as early as 1931, that is, before Tarski
began to publish his results on the topic. It also shows a difference
in attitude from Tarski, as Tarski *did* think that the truth-definition
provided clarification of the concept of truth as applied to formalized
languages. Moreover, Gödel's assertion that the concept of "arbitrary
set" is problematic, in this context, does seem to indicate that he might
not at this time have held quite the strong Platonistic view which he
defended in subsequent years.[j]

Gödel does not comment on whether the truth-definition supports
Carnap's "syntactical" view of the nature of logical and mathematical
truth. Later on Gödel strongly criticized that view: *Gödel *1953/9*
(these *Works*, vol. III, pp. 324–362) contains what Gödel thought to be
conclusive arguments against it. He dismisses the relevance of the truth-
definition to the view, for, he says,

> it will have to be required of the *rules of syntax* that they be "finitary", i.e.,
> that they must not contain phrases such as: "If there exists an infinite
> set of expressions with a certain property"... [*Gödel 1953/9-III*, §9]

and, in a footnote to this paragraph, explicitly criticizes the remarks
cited above from *Logical syntax*, p. 114. Gödel never submitted *1953/9*,
so Carnap did not have the opportunity to react to this point.

---

[j]See also the introductory note to *Gödel *1933o*, these *Works*, vol. III, pp. 39–40.
On Gödel and Tarski regarding the notion of truth, see also *Feferman 1984a*, pp.
557ff. (reprinted in *Feferman 1998*, pp. 160ff.)

After 1932, the correspondence between Gödel and Carnap became more sporadic. On 9 October 1933 Carnap wrote Gödel to enlist his aid in amassing letters of support for Carnap's application for a Rockefeller Foundation fellowship to support a trip to the United States. Gödel at that point was already in Princeton, and Carnap asked, among other things about life in America, "What does one do in Princeton in the summer?" (This letter is not reproduced here.) Carnap did not obtain the fellowship, and was still in Prague the following fall, when he wrote Gödel again (letter 7). *Logical syntax* had been published, and Carnap asked whether Gödel had received a copy, and also asked about his experiences in America. He reports on the lectures he gave at the University of London: these are simplified reports of his research program, published as *Carnap 1935c*.

By this time Carnap clearly wished to leave Prague. As he later put it:

> With the beginning of the Hitler regime in Germany in 1933, the political atmosphere, even in Austria and Czechoslovakia, became more and more intolerable. ...Therefore I initiated efforts to come to America, at least for a time. [*Carnap 1963*, p. 34]

Carnap arrived in the United States in the summer of 1935, and became professor at the University of Chicago a year later. During the spring semester of 1939, Gödel taught at the University of Notre Dame, which is fairly close to Chicago. In a letter of 4 February 1939 (not reproduced here), Carnap invited Gödel to visit him and even to attend the seminar of Bertrand Russell, who was at the University of Chicago for the academic year, and was giving a seminar on material that eventually became *Russell 1940*. It appears that Gödel did visit Carnap in Chicago; and during the 1940s they had at least two meetings. Closer contact was renewed in 1952–1954, when Carnap was in residence at the Institute for Advanced Study from 1952 to 1954. Gödel's last letter to Carnap is a note of condolence on the death of Carnap's wife Ina in 1964.

Letters 2–6 of this correspondence have been previously published, in German with a translation into French, in *Heinzmann and Proust 1988*.[k]

Warren Goldfarb

The translation is by John Dawson and Warren Goldfarb. A complete calendar of the correspondence with Carnap appears on p. 559 of this volume.

[k]I am most grateful to André Carus, John Dawson, Solomon Feferman and Charles Parsons for helpful comments and suggestions.

1. Carnap to Gödel

Prag XVII, den 23. Februar 1932.
N. Motol, Pod Homolkou

Lieber Herr Gödel!

In der nächsten Zeit wird Ihnen Hempel den bisher geschriebenen ersten Teil meiner "Metalogik" zuschicken. Sie können das Ms ruhig einige Wochen behalten; wenn Sie es fertig gelesen haben, bitte es nicht zurückzuschicken, sondern mich zu benachrichtigen. Für kritische Bemerkungen wäre ich Ihnen dann sehr dankbar.

Ich habe im Math.∧Kränzchen über Ihre Arbeit referiert und zwei Wochen vorher über Hilbert, zur Vorbereitung. Ihre Methoden und Ihre Ergebnisse fanden hier starkes Interesse. Es würde mich interessieren, was Zermelo dazu bemerkt hat. Könnten Sie mir vielleicht seine Briefe (und vielleicht die Durchschläge Ihrer Briefe an ihn, falls Sie sich inzwischen zu zivilisierten Methoden der wissenschaftlichen Korrespondenz bekehrt haben) zuschicken?

Dinglers Bemerkungen über die Wiener sind allerdings sehr töricht.[a] Im Uebrigen halte ich aber sein Buch für im Ganzen vernünftig. Es fördert natürlich nicht die eigentliche Grundlagenfrage der Mathematik, aber macht doch ganz nette Ueberlegungen zur Entstehung der Symbolik als Handlung.

Chwistek hat mir seinen neuen Aufsatz[b] geschickt. Auch den alten[c] habe ich mir nochmal angesehen. Er hat nämlich vor einiger Zeit ein neues Ms für die "Erkenntnis"[d] geschickt. Ich muss leider gestehen, dass ich alle 3 Aufsätze nicht kapiere. Falls Sie die "zweite Mitteilung" verstanden haben, oder wenigstens die beiden ersten Seiten, wäre ich Ihnen für einige Erläuterungen | sehr dankbar. Z. B.: Wie sieht die Funktion aus, die keine Wahrheitsfunktion ist? Inwiefern ist die Wittgensteinsche Wahrheitstheorie "mit der Grundidee der Semantik im Widerspruch"? Welche seiner Buchstaben gehören der ersten Sprache an, welche der Metasprache? Uebrigens habe ich auch seine Arbeit "Une méthode ..."[e] gelesen, die hat mir aber auch nicht geholfen.

---

[a] *Dingler 1931.* Chapter I, part III, §3 discusses views of logic common to the Vienna Circle, particularly those of Wittgenstein and Carnap. Chapter I, part II, §1 is entitled "Calculi as Actions" [*Kalkule als Handlungen*].

[b] *Chwistek 1932.*

1. Carnap to Gödel

Prag XVII, 23 February 1932
N. Motol, Pod Homolkou

Dear Mr. Gödel,

In the coming days Hempel will send you the first part of my "Meta-logic", which is what I've written so far. Feel free to keep the manuscript for a few weeks. When you have finished reading it, please do not send it back, but rather let me know. I would be very grateful to you for critical comments.

I reported on your work in the mathematics circle, and two weeks earlier, in preparation for that, on Hilbert. Your methods and your results met with keen interest here. I would be interested to know what Zermelo said about it. Could you perhaps send me his letters (and perhaps the carbon copies of your letters to him, in case in the meantime you have become a convert to civilized methods of scientific correspondence)?

Dingler's remarks about the Viennese are certainly very foolish.[a] However, on the whole I regard his book as reasonable. Of course, it doesn't help to advance the real foundational questions of mathematics, but it does contain some very nice reflections on the creation of a symbolism as an action.

Chwistek sent me his new paper.[b] I have also looked at the old one once again.[c] Some time ago, you know, he sent me a new manuscript for "Erkenntnis".[d] Unfortunately I must confess that I didn't grasp any of the three papers. In case you understood the "second communication", or at least the first two pages of it, I would be very grateful to you for some elucidations. For example: What does the function that is not a truth function look like? To what extent is Wittgenstein's theory of truth "in contradiction to the basic idea of semantics"? Which of his letters belong to the first [object] language and which to the metalanguage? Moreover I have also read his work "A method...",[e] which, however, also did not help me.

---

[c] *Chwistek 1929.*

[d] Presumably, a version of what was later published as *Chwistek 1933.*

[e] *Chwistek 1930.*

Wann erscheint der zweite Teil Ihrer Arbeit? Hoffentlich veranlasst der Umstand, dass auch Leute wie Zermelo (und sicher viele andere) Ihre Arbeit nicht verstanden haben, Sie dazu, uns allen im zweiten Teil die Sache durch ausführlichere und verständlichere Darstellung ~~die Sache~~ leichter zu machen.

Mit den besten Grüssen

Ihr R. Carnap

2. Gödel to Carnap

Wien 11./IV. 1932

Lieber Herr Carnap!

Anbei sende ich Ihnen das Separatum und Manuskript von Chwistek mit bestem Dank zurück. Eine Erläuterung ~~dar~~ über ⟨das letztere⟩ kann ich leider nicht schreiben, denn das meiste darin (und gerade das wichtigste z. B. Typenlehre u. reelle Zahlen) ist in so vager Form behandelt, daß man ohne Kenntnis der Originalarbeiten daraus nichts entnehmen kann. Das einzige, was etwas ausführlicher dargestellt ist, nämlich die "elementare" Semantik, finde ich ganz verständlich. Die Symbole *, c und . muß man, glaube ich, als für sich allein sinnlose *Teilzeichen* auffassen, welche erst in gewissen Verbindungen eine Bedeutung bekommen, ebenso wie etwa Indices an Variablen oder der i-Punkt der gewöhnlichen Schrift. Ich würde es für sehr wünschenswert halten wenn Chwistek auch den weiteren Ausbau seines *formalen Systems* für die "Erkenntnis" in
2 analoger Weise skiz|zieren würde, wie die elementare Semantik, womöglich mit erläuternden Beispielen, damit die Sache doch irgendwo in genießbarer Form steht.

Ihr Manuskript über Metalogik habe ich an Kaufmann geschickt. Falls der II. Teil schon fertiggestellt ist, wäre ich Ihnen für dessen Zusendung sehr dankbar.

Mit herzlichen Grüßen

Ihr Kurt Gödel

When will the second part of your work appear? Hopefully the fact that even people like Zermelo (and certainly many others) have not understood your work will induce you to make the matter in the second part easier for us all, by means of a more detailed and more comprehensible presentation.

With best wishes,

Yours truly,

R. Carnap

2. Gödel to Carnap

Vienna, 11 April 1932

Dear Mr. Carnap,

With this letter I am sending back to you Chwistek's offprint and manuscript, with best thanks. Unfortunately I cannot write an elucidation about the latter, for most of what is in it (and precisely what is most important, e.g., type theory and real numbers) is treated in such a vague form that one can get nothing at all out of it without knowledge of the original works. The only thing that is presented in somewhat more detail, namely the "elementary" semantics, I do find fully intelligible. I think one must view the symbols ∗, c and . as *sign-parts*, meaningless in themselves, which only acquire a significance in certain combinations, just like, say, indices on variables or the dot over the i in ordinary writing. I would regard it as very desirable if Chwistek would also sketch for "Erkenntnis" the further development of his *formal systems* in a way analogous to that for the elementary semantics, if possible with illustrative examples, so that the material is available somewhere in digestible form.

I have sent your manuscript on metalogic to Kaufmann. In case the second part is already finished, I would be very grateful to you for sending it.

Cordially,

Kurt Gödel

3. Gödel to Carnap

Wien 11./IX. 1932.

Lieber Herr Carnap!

Besten Dank für Ihre Karte aus Burgstein. Ich war diesen Sommer nur für kurze Zeit fort (eine Woche am Semmering). Das M.S. "Semantik" habe ich Ende August an Behmann geschickt. Leider bin ich nicht dazu gekommen, es bis in alle Details zu lesen u. muß mich daher in meinen Bemerkungen auf die Hauptpunkte beschränken.

Zunächst glaube ich, daß die Df. von "analytisch" für die erweiterte Sprache u. daher auch der W-Beweis fehlerhaft sind. Sie geben Regeln an durch welche die Frage, ob eine Formel analytisch ist, zurückgeführt wird auf dieselbe Frage für andere Formeln. Dadurch würde ein Begriff "analyt." nur dann definiert sein, wenn dieses Verfahren schließlich immer zu Formeln führen würde, von denen anderweitig feststeht, ob sie analyt. sind (z. B. $0 = 0$). Dies ist aber, wie mir scheint, | nicht immer der Fall. Denn nehmen wir z. B. die Formel $(F)\ F(0) \lor \overline{F(0)}$; um festzustellen, ob sie analyt. ist, muß man dies für alle Formeln der Gestalt $P(0) \lor \overline{P(0)}$ tun. Zu diesem Zweck muß man jedes konstante Prädikat P durch sein Definiens ersetzen; in diesem können aber wieder gebundene Prädik.-Variable vorkommen u.s.w., so daß man auf einen regressus in inf. kommt. Am deutlichsten wird dies dadurch, daß u. U. *dieselbe* Formel immer wiederkehren kann. Setzt man z. B. in der Formel $(F).F(0) \lor 0 = 0$ für F das konst. Präd $(F)F(x)$ ⟨ein⟩ und bringt auf die Normalform, so erhält man wieder die ursprüngliche Formel. Dieser Fehler läßt sich m. E. nur dadurch vermeiden, daß man als Laufbereich der Funktionsvariablen nicht die Präd. einer bestimmten Sprache, sondern alle Mengen u. Relationen überhaupt ansieht.[1] Ich werde auf Grund dieses Gedankens im II. Teil[a] meiner Arbeit eine Df. für "wahr" geben u. bin der Meinung, daß sich | die Sache anders nicht machen läßt u. daß man den höheren Funktionenkalkül *nicht* semantisch auffassen kann. D. h. man kann natürlich auf semantischer Basis einen höheren Funkt.-K. aufbauen, es sind dann aber gerade die Gesetze, welche man für die klass. Theorie der reellen Zahlen braucht, nicht erfüllt, weil man notwendig zur verzweigten Typentheorie (ohne Red. Ax.) geführt wird.

[1] Dies involviert nicht etwa einen platonistischen Standpunkt, denn ich behaupte nur, daß sich diese Df. ⟨für "analyt"⟩ innerhalb einer bestimmten Sprache, in der man die Begriffe "Menge" u. "Rel" schon hat, durchführen läßt.

[a] The planned sequel to *Gödel 1931*, which never appeared.

3. Gödel to Carnap

Vienna, 11 September 1932

Dear Mr. Carnap,

Thank you very much for your card from Burgstein. This summer I was away only for a short time (one week at Semmering). I sent the manuscript "Semantics" to Behmann at the end of August. Unfortunately I have not gotten down to reading it in all its details and must therefore restrict myself in my remarks to the principal points.

First of all, I think that the definition of "analytic" for the extended language, and therefore the W-proof as well, is faulty. You specify rules by means of which the question of whether a formula is analytic is traced back to the same question for other formulas. Accordingly, then, a concept "analytic" would only be defined if this procedure were always to lead finally to formulas for which it was settled by other means whether they are analytic (for example, $0 = 0$). But it seems to me that this is not always the case. For let us take, e.g., the formula $(F)\, F(0) \vee \overline{F(0)}$; in order to establish whether it is analytic, one must do that for all formulas of the form $P(0) \vee \overline{P(0)}$. Towards that end, one must replace each constant predicate P by its definiens; but in the latter, bound predicate variables could again occur, and so on, so that one runs into an infinite regress. This becomes most evident in that, under certain circumstances, *the same* formula can always recur. For example, if in the formula $(F).F(0) \vee 0 = 0$ one substitutes the constant predicate $(F)F(x)$ for F and brings it into normal form, one again obtains the original formula. In my judgment, this error may only be avoided by regarding the domain of the function variables not as the predicates of a definite language, but rather as all sets and relations whatever.[1] On the basis of this idea, in the second part[a] of my work I will give a definition for "truth", and I am of the opinion that the matter may not be done otherwise, and that one can *not* view the higher functional calculus semantically. That is, one can of course build up a higher functional calculus on a semantic basis, but then just those laws that one needs for the classical theory of the real numbers are not satisfied, because one is led of necessity to ramified type theory (without the axiom of reducibility).

I also have an objection to the descriptive rendering of the undecidability proof on pages 308ff. The number e and the formula G_6 seem to me to be defined circularly. For e is defined as the "sequence number of the set of the atomic formulas which express that the formula G_6 stands

---

[1]This doesn't necessarily involve a Platonistic standpoint, for I assert only that this definition (for "analytic") be carried out within a definite language in which one already has the concepts "set" and "relation".

Einen Einwand habe ich auch gegen die deskriptive Nachbildung des Unentsch.-Beweises p. 308ff. Die Zahl e und die Formel G_6 scheinen mir zirkelhaft definiert zu sein. Denn e ist definiert als "~~Meng~~ Reihenzahl der Menge der Atomformeln, welche aussagen, daß an den Stellen a bis $a + 16$ die Formel G_6 steht." (p. 309). In der Df. von e kommt also G_6 vor, während umgekehrt in der Df. von G_6 e vorkommt. Damit ist natürlich nicht gesagt, daß eine Zahl e und eine Formel G_6 mit den durch die beiden Df. ausgesprochenen Eigenschaften nicht ~~geben~~ ⟨existieren⟩ kann. Aber ihre Existenz müßte eigens bewiesen werden, wozu man

4   wahrscheinlich wieder irgend eine Art von | Diagonalschluß brauchen wird.

Nicht ganz klar war mir, wozu Sie 2 Arten von Funktionsvariablen (a^n u. p^n) brauchen. Man muß doch wohl mit einer auskommen können.

Ich möchte jetzt noch eine Frage anschließen. Ist Ihnen vielleicht bekannt, ob Frege außer dem Nachwort zu den "Grundgesetzen"[a] noch etwas über die Antinomien publiziert hat? Soviel ich weiß, hat er doch das Jahr 1903 ziemlich lange überlebt, u. es wäre doch merkwürdig, wenn er darauf nicht mehr zurückgekommen wäre. Sind Ihnen überhaupt Publikationen nach dem Erscheinen der Grundgesetze bekannt?

Haben Sie schon Zermelos unsinnige Kritik an meiner Arbeit im letzten Bd. des Jahresberichtes 2. Abt. p. 87[b] gelesen?

Es hat mir sehr leid getan, daß Ihre Sommerreise Sie nicht über Wien geführt hat. Hoffentlich kommen Sie nächstens doch wieder einmal her.

Mit den besten Grüßen

Ihr ergebener Kurt Gödel

---

[a] *Frege 1903.*

4. Carnap to Gödel

Prag XVII, den 25. Sept. 1932.
N. Motol, Pod Homolkou

Lieber Herr Gödel!

Haben Sie besten Dank für Ihren Brief. Ich bin Ihnen sehr dankbar, dass Sie mich auf die Fehlerhaftigkeit der Def. für "analytisch" aufmerksam gemacht haben. Ich habe inzwischen versucht, eine bessere Def. aufzustellen, etwa in der Richtung Ihrer Ande⟨u⟩tungen (falls ich diese richtig verstanden habe). Aber da treten gewisse Schwierigkeiten auf, die ich noch nicht überwinden kann. Jedenfalls sehe ich ein, dass man nicht zu den Formeln der klass. Math. über reelle Zahlen gelangt, wenn man

in places a through $a + 16$" (p. 309). Thus G_6 occurs in the definition of e, while conversely e occurs in the definition of G_6. That is not to say, of course, that a number e and a formula G_6 with the properties expressed by the two definitions can not exist. But their existence would have to be proved expressly, for which one will probably again need some kind of diagonal inference.

It was not quite clear to me why you need two kinds of function variables (a^n and p^n). Surely one ought to be able to get by with one.

I would now like to add one more question. Do you perhaps know whether Frege published anything about the antinomies apart from the afterword to the "Grundgesetze"?[a] So far as I know, he lived a rather long time after 1903, and it would really be curious if he had not come back to it again. Are any publications at all after the appearance of the Grundgesetze known to you?

Have you already read Zermelo's nonsensical criticism of my paper in the last volume of the Jahresbericht, 2nd part, p. 87?[b]

I was very sorry that your summer travels did not lead you to Vienna. Hopefully you will indeed come here again soon.

With best wishes,

Yours truly,

Kurt Gödel

---

[b] *Zermelo 1932.*

4. Carnap to Gödel

Prag XVII, 25 September 1932
N. Motol, Pod Homolkou

Dear Mr. Gödel,

Thanks very much for your letter. I am very grateful to you for having made me aware of the faultiness of the definition of "analytic". In the meantime I have attempted to work out a better definition, somewhat in the direction of your suggestions (if I have correctly understood them). But in this certain difficulties arise that I still cannot overcome. In any case, I realize that one will not arrive at the formulas of classical mathematics about real numbers if the universal quantifier with predicate

den Alloperator ~~in~~ mit Prädikatsvariableℓ (od.$_\wedge$Funktorvar.) nur auf die   n ⊔
in einem bestimmten begrenzten System definierbaren Prädikate bezieht.
Sie sagen: man muss ihn auf "~~A~~alle Mengen" beziehen; aber was heisst
das? Ich habe so versucht: unter einer Prädikatsbewertung vers⟨t⟩ehe ich
eine Regel, die jedem Zahlausdruck (z. B. "0''''") bezw. jedem n-tupel von
solchen entweder "0 = 0" oder "0 ≠ 0" zuordnet; eine Funktorbewer-
tung ist eine Regel, die jedem n-tupel von Zahlausdr. einen Zahlausdr.
zuordnet. Mit Hilfe dieses Begriffs wird "Auswertung" definiert: in der
betr. Formel wird anstelle von "F(...)" die æ zugeordnete Formel gesetzt
und anstelle von "f(...)" der zugeordnete Ausdruck. "[F](...)" ist analytisch, wenn "..." inbezug auf jede Bewertung für F analytisch ist. Im
Ganzen wird die Def. ziemlich kompliziert, lässt sich aber machen. Nun
aber das Bedenken. Will man wirklich das Erstrebte mit der Def. erreichen, so darf man für die Regel, die ja$_\wedge$in der Sprache der Semantik   ⊔
formuliert werden muss, nicht eine begrenzte Sprache der Sem. nehmen,
sondern: die Regel darf mit beliebigen semantischen Begriffen gebildet
werden. (Denn sonst bleiben ja wieder gewisse Zahlmengen draussen,
die man zwar plausibel machen, aber nicht im System erfas⟦s⟧en kann).
Aber ist das nicht bedenklich? Es scheint mir nicht bedenklich, wenn
~~die Def für~~ "analytisch in der Sprache S" nicht definiert werden kann
in einer Sem., die in S formalisiert ist, sondern nur in einer Sem., die
in einer weiteren Sprache S$_2$ formalisiert ist. Aber mit einem Begriff zu
operieren, für den es überhaupt keine Sprache gibt, in der er ⟨streng⟩
definiert werden kann, ist doch wohl ziemlich bedenklich.

Aber vielleicht sehen Sie einen Weg, den Begriff in einer bestimmten
Sprache zu definieren? Oder wie ist Ihre Bemerkung zu verstehen: "...
dass sich diese Def. für "~~A~~anal." innerhalb einer bestimmten Sprache, in
der man die Begriffe "Menge" und "Rel." schon hat, durchführen lässt"?
Können Sie den Begriff "Menge" innerhalb einer bestimmten, formalisierten Semantik definieren?

Eine weitere Schwierigkeit betriff⟨t⟩ die höherstufigen Prädikate; ein
solches müsste dargestellt werden durch eine Zuordnung zu Bewertungen
niederer Stufe (oder durch eine semantische Menge von Mengen usw. von
Ausdrücken).

2 | Werden Sie vielleicht bald schon einen ersten Entwurf Ihrer Arbeit
niedergeschrieben haben, sodass ich sie lesen und verwerten könnte? Das
wäre mir ausserordentlich wertvoll. Oder können Sie mir wenigstens
brieflich jetzt schon Näheres angeben über die Def. von "Menge" und
"anal."? Für den Zusammenhang meines Buches ist der Begriff "anal."
~~d~~sehr wichtig, wei~~l~~l ich die übrigen Begriffe (besonders "Gehalt" usw.)   l
daran anknüpfen will. Falls Ihre Def. für meine Zwecke brauchbar ist
(was doch sicher zu vermuten ist), wäre es für mich am praktischsten, sie
von Ihnen zu übernehmen (selbstverständl. unter Hinweis auf Ihre Urhe-
berschaft), anstatt erst selbst eine zu suchen, die sich dann womöglich

variables (or function variables) ranges only over the predicates definable in a definite delimited system. You say: it must range over "all sets"; but what does that mean? My attempt goes as follows: by a predicate valuation I understand a rule that assigns to each numerical expression (for example, "$0''''$"), or, respectively, to each n-tuple of such, either "$0 = 0$" or "$0 \neq 0$"; a function valuation is a rule that assigns a numerical expression to each n-tuple of numerical expressions. With the aid of this concept, "evaluation" is defined: in the formula under consideration the assigned formula is put in place of "$F(\ldots)$" and the assigned expression in place of "$f(\ldots)$". "$[F](\ldots)$" is analytic if "$\ldots$" is analytic with respect to each valuation of F. In its entirety the definition is rather complicated, but it can be done. But now the objection. If the definition is to achieve what was striven for, then for the rule—which must of course be formulated in the language of semantics—one may take not a restricted semantic language; rather the rule may be constructed with arbitrary semantic notions. (For otherwise certain sets of numbers would again remain outside, sets that are indeed plausible, but that cannot be comprehended within the system.) But is that not questionable? It seems to me not questionable if "analytic in the language S" cannot be defined in a semantics that is formalised in S, but only in a semantics that is formalised in a more extended language S_2. But to operate with a concept for which there is no language at all in which it can be rigorously defined is certainly rather questionable.

But perhaps you see a way to define the concept in a definite language. Or how is your remark to be understood: "$\ldots$ that this definition for 'analytic' may be carried out within a definite language in which one already has the concepts 'set' and 'relation'"? Can you define the concept "set" within a definite formalized semantics?

A further difficulty concerns the higher-type predicates; such things must be represented by assignments to valuations of lower type (or by a semantic set of sets, and so on, of expressions).

Will you perhaps soon be writing up a first draft of your paper, so that I could read it and make use of it? That would be extraordinarily valuable to me. Or can you at least tell me now by letter further details about the definition of "set" and "analytic"? For the purposes of my book, the concept "analytic" is very important, because I want to tie the usual concepts (especially "content", and so on) to it. In case your definition is usable for my purposes (which is surely to be expected), it would be most practical for me to adopt it from you (of course, with mention of your authorship), instead of first seeking one myself, which subsequently

nachträgl. als schlechter erweist als Ihre. Jedenfalls wäre es für mich
höchst wertvoll, zu sehen, in welchem Sinne Ihre Def. überhaupt zu
einem formalisierbaren Begriff führt.

Zur Terminologie: der Terminus "wahr" scheint mir sehr unzweck-
mässig; jedenfalls wäre sein Gebrauch nicht im Einklang mit dem allg.
Sprachgebrauch. Denn hiernach heisst der Satz "Wien hat so und so viel
Einwohner" doch wahr, während die von Ihnen beabsichtigte Def. doch
wohl a⟨u⟩f ihn nicht zutrifft. Man müsste also doch wohl "logisch-wahr"
sagen, oder "logisch-gültig" oder "tautologisch" oder "analytisch"; und
von diesen Ausdrücken scheint mir der letzte der zweckmässigste.

Ja, Zermelos Bemerkungen im Jahresber. über Sie habe ich mit Erstaunen gelesen; trotz seines Scharfsinns greift er manchmal so erstaun-
lich daneben.

Frege hat, soviel ich weiss, nichts mehr über die Antinomien geschrieben, obwohl er erst 1925 gestorben ist. Er hat noch veröffenlicht: Der
Gedanke; eine log. Untersuchung. In: Beiträge z.∧Pheilos. d.∧dt. Idealismus (das sind die späteren "Bl.∧f.∧dt. Philos.") I, 1918.[a] Ferner in
ders. ZS etwas später, ich glaube über die Negation.[b] Eben finde ich
in Jörgensen, Treatise,[c] I 174 Anm. 84 einen Hinweis auf eine Note von
Frege in: Quart. Journ. of Math. 43, p. 251, 1912;[d] diese Note dürfte
viell. wichtiges enthalten. Lewis, Survey 393,[e] nennt noch: Translated
portions, Monist 25(1915)–27(1917);[f] viell. sind das aber nur übersetzte
Stücke aus den "Grundgesetzen".[g] Sonst ist mir nichts bekannt.

Was macht man in Wien? Wird Hahn das Seminar über Grundl. fortsetzen oder Schlick den Zirkel wieder aufmachen? Wann wird Ihre Arbeit
fertig und wann erscheint sie? Denken Sie ja daran, ausführlicher und
leichter verständlich zu schreiben!

Mit den besten Grüssen

Ihr R. Carnap

---

[a] *Frege 1918.*
[b] *Frege 1918a.*
[c] *Jørgensen 1931.*

might prove to be worse than yours. In any case it would be of utmost value to me to see in what sense your definition really leads to a formalizable concept.

As to terminology: The term "true" seems to me very unsuitable; in any case, its usage would not be in accord with general linguistic usage. For according to the latter, the sentence "Vienna has so and so many inhabitants" is of course true, whereas the definition proposed by you surely does not apply to it. Thus one would surely have to say "logically true" or "logically valid" or "tautological" or "analytic"; and of those expressions the last seems to me the most suitable.

Yes, I read Zermelo's remarks about you in the Jahresbericht with astonishment. Despite his powers of discernment, he sometimes astonishingly fails to grasp things.

So far as I know, Frege wrote nothing more about the antinomies, although he did not die until 1925. He published in addition: "Der Gedanke; eine logische Untersuchung", in Beiträge zur Philosophie des deutschen Idealismus (which was later called "Blätter für deutsche Philosophie"), volume I, 1918;[a] and something more, somewhat later in the same journal, I think about negation.[b] Just now I found in Jörgensen, Treatise, volume I, p. 174, remark 84,[c] a reference to a note by Frege in the Quarterly Journal of Mathematics 43, p. 251, 1912;[d] that note may perhaps contain something important. Lewis, Survey, p. 393,[e] mentions in addition: Translated portions, Monist 25(1915)–27(1917);[f] but perhaps those are only translated pieces from the "Grundgesetze".[g] Nothing else is known to me.

What is going on in Vienna? Will Hahn continue the seminar on foundations or Schlick again start up the Circle? When will your paper be finished and when will it appear? Think about writing in more detail and in a way that is easier to understand!

With best wishes,

Yours truly,

R. Carnap

---

[d] Remarks written by Frege to Philip E. B. Jourdain, and published in *Jourdain 1912*.
[e] *Lewis 1918*.
[f] *Frege 1915, 1916* and *1917*.
[g] *Frege 1893* and *1903*.

5. Carnap to Gödel

Prag XVII, den 27. Sept. 1932.
N. Motol, Pod Homolkou

Lieber Herr Gödel!

Vorgestern schrieb ich Ihnen über eine Schwierigkeit, ~~die ich~~ bei der
Def. für "analytisch". Gestern fand ich die Lösung: die in der Def. vorkommende Wendung "für jede Bewertung ..." kann doch in einer in be-
stimmter Sprache formulierten Sem. ausgedrückt werden, nämlich durch
"[F](...)", da ja eine Bewertung ein sem. Prädikat ist. Dies ist möglich,
obwohl in der betr. Sem. nicht alle möglichen Bewertungen, d. h. Prädikate, definiert werden können. Bei der Anwendung auf Prädikate und
Funktoren höherer Stufe und gemischten Typus wird die Sache allerdings
doch recht kompliziert.

Ich schreibs Ihnen gleich, damit Sie sich keine unnötige Mühe bei der
Beantwortung mehr zu machen brauchen. Trotzdem würde mich natürlich Ihre Auffassung hierzu und eine Andeutung über den von Ihnen
eingeschlagenen Weg lebhaft interessieren.

Noch eine Frage: kann man ohne Einbusse an definierbaren Präd. u.
Funktoren auf rekursive Def. verzichten, wenn man in der Sprache Präd.-
und Funktorvariable in Operatoren zulässt?

Mit den besten Grüssen

Ihr R. Carnap

6. Gödel to Carnap

Wien 28./XI. 1932.

Lieber Herr Carnap!

Besten Dank für Ihre beiden Briefen und die Karte. Ich habe jetzt leider mit dem Bericht über die Grundlagenforschung viel zu tun gehabt
u. bin daher bis heute nicht dazu gekommen, Ihnen zu schreiben. Auch
der II. Teil meiner Arbeit existiert erst im Reiche der Ideen. Meine Andeutungen über die Df. von "analytisch" haben Sie, wie ich aus Ihren
zweiten Brief entnehme, ganz in meinem Sinn verstanden. Um die Sache
allgemein d. h. für Funktionen beliebigen endlichen Typs durchführen zu
können, braucht man eine Variable nächst höheren Typs (Typus ω),

5. Carnap to Gödel

Prag XVII, 27 September 1932
N. Motol, Pod Homolkou

Dear Mr. Gödel,

The day before yesterday I wrote you about a difficulty in the definition for "analytic". Yesterday I found the solution: The locution "for every valuation ..." that occurs in the definition can still be expressed in a semantics formulated in a definite language, namely by "$[F](...)$", since a valuation is of course a semantic predicate. This is possible even though in the semantics under consideration not all possible valuations, that is, predicates, can be defined. To be sure, in application to predicates and functions of higher types and mixed types the matter becomes quite complicated indeed.

I am writing you at once so that you need take no more unnecessary trouble with the reply. Nevertheless your opinion on this matter and an indication of the course you adopted would interest me keenly.

One more question: If in the language one allows predicate- and function-variables in operators, can one do without recursive definitions, without loss to definable predicates and functions?

With best wishes,

Yours truly,

R. Carnap

6. Gödel to Carnap

Vienna, 28 November 1932

Dear Mr. Carnap,

Thank you very much for your two letters and the postcard. Unfortunately I have had much work to do recently with a report about research on foundations, and thus did not get around to writing to you until today. Also, the second part of my paper exists only in the realm of ideas. As I gather from your second letter, you have understood my suggestions about the definition of "analytic" entirely as I meant them. In order to be able to carry out the matter in general, that is, for functions of arbitrary finite type, one needs a variable of the next higher type (type ω),

d. h. eine solche, die *alle* ⟨endlichen⟩ Typen durchläuft, was a priori vor-
auszusehen ist, da man eben niemals "anal." in demselben System definieren kann, sonst würden sich Widersprüche ergeben. Ich glaube übrigens, das Interesse dieser Df. liegt nicht in einer Klärung des Begriffs
"analyt.", weil man in ihr die ebenso problematischen Begriffe "beliebige
2　Menge" etc. verwendet, sondern ich stelle sie | nur aus dem Grunde auf,
weil man mit ihrer Hilfe zeigen kann, daß die unentsch. Sätze in Systemen, welche in der Reihe der Typen weiter hinaufsteigen, entscheid-
bar werden. Übrigens hat Tarski, wie Sie vielleicht wissen, eine ähnliche
Df. für "anal." in einer in poln. Sprache erscheinenden Abhandlung[a]
gegeben, worüber er im Anzeiger der Wiener Akad. d. Wiss. 1932 №
2[b] berichtet.

　　Die Frage, die Sie am Schluß Ihres Briefes vom 27./IX. stellen, ist mit
ja zu beantworten, wenn das Red. Ax. oder ein Äquivalent (z. B. Ihre
Df.-Regel für Funktionen) im System enthalten ist.

　　Wie hat Ihnen die Reise nach Schweden gefallen? Haben Sie Skolem
kennengelernt? Hat Chwistek sein Manuskript für die Erkenntnis abge-
ändert? Wenn ja, würde ich es gerne lesen. Der Schlickzirkel wurde zwar
wieder aufgenommen, aber die Diskussion mit Waismann wird nicht fortgesetzt, sondern Referate über Ramsey u. a. gehalten. Wie ich hörte,
beabsichtigen Sie Mitte Dezember nach Wien zu kommen. Ich würde
mich dann sehr freuen, ausführlich über die logischen Fragen mit Ihnen
zu sprechen, was brieflich schwer möglich ist.

　　Mit den besten Grüßen

　　　　　　　　　　Ihr Kurt Gödel

---

[a] *Tarski 1933a.*

7. Carnap to Gödel

　　　　　　　　　　　　Prag XVII, den 7. November 1934.
　　　　　　　　　　　　Pod Homolkou 146

　　Lieber Herr Gödel,

　　ich habe lange nichts von Ihnen gehört, weiss nicht einmal, ob Sie in
Wien oder wieder in Amerika sind. Haben Sie meine Syntax bekommen?
Es würde mich interessieren, zu hören, wie Ihre Meinung über Ihre Zeit
in Amerika ist, und über die Leute dort. Ich war jetzt einige Wochen in

i.e., one which runs through *all* finite types. This could be foreseen a priori, since one can never define "analytic" in the same system–otherwise contradictions will result. I believe moreover that the interest of this definition does not lie in a clarification of the concept "analytic", since one employs in it the concepts "arbitrary sets", etc., which are just as problematic. Rather I formulate it only for the following reason: with its help one can show that undecidable sentences become decidable in systems which ascend farther in the sequence of types. By the way, as you perhaps know, Tarski has given a similar definition for "analytic" in a paper to appear in the Polish language,[a] about which he reported in the Anzeiger der Wiener Akademie 1932, no. 2.[b]

The question you pose at the end of your letter of 27 September is to be answered affirmatively if the axiom of reducibility or an equivalent (e.g., your definition rule for functions) is contained within the system.

How did you like the trip to Sweden? Did you become acquainted with Skolem? Has Chwistek revised his manuscript for Erkenntnis? If so, I would like to read it. The Schlick Circle has indeed resumed, but the discussion with Waismann is not being continued; rather, reports about Ramsey and others are being delivered. I heard that you intend to come to Vienna in mid-December. I would then very much enjoy talking in detail with you about logical questions, which is difficult to do by letter.

With best wishes,

Yours truly,

Kurt Gödel

---

[b] *Tarski 1932.*

7. Carnap to Gödel

Prag XVII, 7 November 1934
Pod Homolkou 146

Dear Mr. Gödel,

I've heard nothing from you in a long time, and don't even know whether you are in Vienna or again in America. Did you receive my Syntax? I would be interested to hear your opinion of your time in America and of the people there. Just now I was in London for a few weeks and

London, und bin recht begfriedigt zurückgekommen. Ich habe dort auf
Einladung der Universität Vorträge[a] gehalten. Die eigentlichen Fach-
philosophen sind ja mit Ausnahme von Prof. Stebbing gar nicht an un-
sern Dingen interessiert, aber ich hatte einige anregende Gespäche, mit
andern, zumeist jünger[e]n Wissenschaftlern.

Church hat mir ⟨jetzt⟩ 2 Sonderdrucke geschickt, die Sie wohl auch
bekommen haben werden: 1) A Set of Postulates (Second Paper).[b] Da
schreibt er S. 843 oben: Gödels Ueberlegungen seien auf sein System
nicht anwendbar. Das erscheint mir sehr zweifelhaft.— 2) "The Richard
Paradox".[c] Das scheint mir ziemlicher Unsinn zu sein. Er redet ernstlich
von einer ~~Menge~~ abzählbaren Menge, die eine nicht abzählbare Teilmenge
hat.

Lassen Sie von sich hören!

   Mit besten Grüssen

                        Ihr R. Carnap

Eben sehe ich, dass Sie mir über Church (1) schon im Februar geschrieben haben.

---

[a]Published as *Carnap 1935c*.

8. Gödel to Carnap

                                      Princeton, 1./VI. 1964.

   Lieber Herr Carnap!

Wir haben mit grösstem Bedauern von dem Tode Ihrer lieben Frau
Ina erfahren. Besonders meine Frau ist durch die Nachricht ganz nieder-
geschlagen. Wir haben die Verstorbene bei ihrem Aufenthalt in Princeton lieben und schätzen gelernt. Sie wird uns immer eine liebe Erinnerung bleiben. Bitte nehmen Sie unser aufrichtig gefühltes Beileid entgegen.

   Mit den herzlichsten Grüssen

                        Ihr Kurt Gödel

came back quite pleased. I gave lectures there, at the invitation of the University.[a] The professional philosophers proper, with the exception of Professor Stebbing, are of course not at all interested in our things, but I had some stimulating conversations with others, mostly younger scientists.

Church has now sent me two offprints that you have probably also received: 1) A Set of Postulates (Second Paper).[b] There, at the top of page 843 he writes: Gödel's considerations do not seem applicable to his system. That appears very doubtful to me.—2) "The Richard Paradox".[c] This seems to me rather nonsensical. He speaks seriously of a countable set that has an uncountable subset.

Let me hear from you!

> With best wishes,
>
> Yours truly,
>
> R. Carnap

Just now I see that you already wrote me about Church (1) in February.

---

[b] *Church 1933.*
[c] *Church 1934.*

8. Gödel to Carnap

Princeton, 1 June 1964

Dear Mr. Carnap,

We have learned with greatest sorrow of the death of your beloved wife Ina. My wife especially was deeply affected by the report. We learned to love and esteem the departed during her stay in Princeton. She will always remain a beloved memory to us. Please accept our sincerely felt sympathy.

> With most heartfelt greetings,
>
> Yours truly,
>
> Kurt Gödel

Alonzo Church

Alonzo Church

Alonzo Church (1903–1995) will be well known to readers of these *Works* for his important contributions to logic and for his founding and long-time editing of the Reviews section of *The journal of symbolic logic.*[a] During the whole period within which he corresponded with Gödel, he was on the faculty of the department of mathematics at Princeton University.

The correspondence of Church and Gödel consists of four separate exchanges, in 1932, 1946, 1965, and 1966. None of them deals directly with what was probably the most important connection of their work, in the early development of the analysis of computability and the concept of recursive function.[b]

Church's early work in logic provided the occasion for the first exchange. He developed a comprehensive system of logic based on the concept of function. One of Church's aims was that the system should be comprehensive enough to develop mathematics. This work appeared in *Church 1932* and *1933*. Not long afterward the system proved to be inconsistent,[c] but Church extracted the λ-calculus from it and, in particular, was able to define the concept of λ-definability, one of the coextensive concepts that came to be identified with effective computability.

Prompted by a visit to Vienna by Oswald Veblen,[d] Gödel wrote to Church on 17 June 1932 (letter 1) with two sharp questions about *Church 1932*. The first question is how "absolute existence propositions" such as the axiom of infinity can be proved in Church's system. The second is whether, if the system is consistent, it is possible to inter-

---

[a] A brief account of Church's career is given in *Enderton 1995*. *Barendregt 1997*, *Sieg 1997*, *Anderson 1998* and *Enderton 1998* survey Church's work on the λ-calculus, the analysis of computability, philosophy and intensional logic, and the Reviews, respectively. An edition of Church's collected papers is in preparation.

[b] On this subject see *Davis 1982* and *Sieg 1997*.

[c] *Kleene and Rosser 1935*. Gödel already shows awareness of this result in his review *1934e* of *Church 1933*. It was obtained in the spring of 1934, when Gödel was at the Institute for Advanced Study. (See *Dawson 1997*, p. 99.) Ironically, Gödel's mention of this then unpublished result would have violated a policy Church enforced for the reviews in *The journal of symbolic logic*, that a review should not be the occasion for announcing a new result.

[d] Oswald Veblen (1880–1960) had been Church's teacher and was at the time about to become one of the founding professors at the Institute for Advanced Study. He played a role in arranging both Gödel's visits to the Institute in the 1930s and his coming there permanently in 1940; see these *Works*, vol. I, p. 8 and p. 11, and *Dawson 1997*, pp. 95–97. Evidently the impression Gödel made during Veblen's visit to Vienna in 1932 led to his wish to bring Gödel to the Institute for its first year; see *Dawson 1997*, pp. 82–83.

pret it in type theory or set theory and whether its consistency could be made plausible in some other way.

Church replied at length on 27 July (letter 2) in his inimitable handwriting. He sketches how to prove in his system the Dedekind–Peano axioms for arithmetic but concedes that he has not worked out the further step of defining "infinite class" and showing that the class of natural numbers is infinite.

Gödel's second question subsequently became moot because of the inconsistency of Church's system.[e] But Church's reply is still of interest. He says he "cannot see" that a consistency proof relative to the consistency of *Principia mathematica* would be of much value "because the freedom from contradiction of *Principia Mathematica* is itself doubtful, even improbable." The evidence for the consistency of PM is empirical, in that many consequences have been derived, and in time evidence of the same kind for the consistency of his system might be obtained. He says he has made unsuccessful attempts to give a proof-theoretic consistency proof.

Finally, Church remarks that he has been unable to see how Gödel's second incompleteness theorem applies to his system. "Possibly your argument can be modified so as to apply to my system, but I have not been able to find such a modification of your argument." He was of course right for the wrong reason, since the theorem applies to *consistent* systems. But apart from that, in 1932 it was of course not clear exactly on what premises the second incompleteness theorem rested; this was not made explicit in print until *Hilbert and Bernays 1939*. Church seems also to have underestimated the generality of the *first* incompleteness theorem. At the time he wrote,

> His [Gödel's] argument makes use of the relation of implication U between propositions in a way which would not be permissible under the system of this paper, and there is no obvious way of modifying the argument so as to make it apply to the system of this paper.[f]

Many years later he wrote,

---

[e] In fact the system of *Church 1932* is shown inconsistent already on pp. 839–840 of *Church 1933*, which modifies the system in an unsuccessful attempt to avoid paradox.

[f] *Church 1933*, p. 843. This remark seems directly addressed to the second theorem, but if correct it would surely apply to the first. Cf. the doubting reference to it in Carnap's letter to Gödel of 7 November 1934 in this volume. But by that time Church had already been convinced by Gödel of the greater generality of the theorem, on the basis of the considerations presented in *Gödel 1934*, §6. See *Sieg 1997*, pp. 161–162.

I was one of those who thought that the Gödel incompleteness theorem might be found to depend on peculiarities of type theory (or, as I might later have added, of set theory) in a way that would show his results to have less universal significance than he was claiming for them. There was a historical reason for this, and that is that even before the Gödel results were published I was working on a project for a radically different formulation of logic which would (as I saw it at the time) escape some of the unfortunate restrictiveness of type theory.[g]

Perhaps it was in some way fitting that the full generality of the first incompleteness theorem, which Gödel stressed frequently in later years, was attained by means of the analysis of computability to which Church contributed.

The second exchange between Gödel and Church, in 1946, not included here, concerned the organization of the session on Mathematical Logic at the conference on Problems of Mathematics, which took place 17–19 December 1946 as part of the Princeton Bicentennial. The lead lecture was given by Alfred Tarski,[h] and one of the shorter following talks was *Gödel 1946*.

The third exchange arose through the reviewing by *The journal of symbolic logic* of *Gödel 1962a*, a translation of *Gödel 1931* with an introduction by the British philosopher of science R. B. Braithwaite, both of which had been prepared without consulting Gödel.[i] It apparently begins with a letter of Gödel to Church of 26 January 1965 which has not been located. It appears that Church may have sent or shown Gödel the review of the translation submitted by Stefan Bauer-Mengelberg, since on 29 January he writes, "I will see that this first paragraph of Dr. Bauer-Mengelberg's review is replaced by something more suitable."[j] The published review *Bauer-Mengelberg 1965* consists of very detailed criticisms, and its overall effect is devastating.[k] Church

[g]Letter to John W. Dawson, Jr., 25 July 1983, quoted in full in *Sieg 1997*, pp. 177–178.

[h]*See Sinaceur 2000*, which contains a transcript of Tarski's talk together with other documents and much information concerning this session.

[i]Cf. *Dawson 1997*, pp. 215–216.

[j]The first paragraph of *Bauer-Mengelberg 1965* reads, "This translation was published by translator and editor on their own initiative, and Professor Gödel did not see the material before publication" (p. 359). It is not known what statement this replaced.

[k]Bauer-Mengelberg was at the time a specialist on translation of German texts in logic and foundations; he did a number of the translations in *van Heijenoort 1967* and some in volume I of these *Works*. It is possible that either he or Jean van Heijenoort sent the review to Gödel.

and Gödel agreed that the translation was deficient. Gödel, however, reacting to Church's remark in the same letter that Braithwaite's introduction is "probably considerably better", is somewhat more charitable about the translation but goes into some detail about errors in the Introduction (letter 3, 2 March 1965).[1] The remark in the second paragraph expresses a concern that Gödel expressed frequently during this period; cf. the Postscriptum to *Gödel 1934*.

In the event the Introduction was reviewed by Church himself (*Church 1965*). The review begins with generalized praise, but by far the greater part of the review consists of criticism. Gödel's criticism in the third paragraph of his letter is reflected in the review, but that in the second paragraph is not. Church is also critical of Braithwaite's account of Hilbert's program, as Gödel is in the draft but not in the letter as sent.

The fourth and final exchange between Gödel and Church, a little over a year later, arose because Paul Cohen was to receive a Fields Medal at the International Congress of Mathematicians in Moscow in August 1966. His celebrated independence results in set theory were a prominent reason for the award of the medal.[m] Custom called for a short talk on the work of the recipient, and Church accepted an invitation to give this talk.[n] He wrote to Gödel on 27 June 1966 (letter 4) saying that the talk would confine itself to the independence results and inviting Gödel's suggestions. It is not known whether they had any further discussion of the matter before Church wrote again on 4 August saying that the first draft of his talk was not ready but that he would get it to Gödel so that he could see it before his own departure for Moscow on 14 August. He brings up the issue that dominates the rest of the correspondence: correcting a claim of Mostowski (*1964*, p. 124) that Gödel had anticipated Cohen's independence results already in 1938.[o]

---

[1]A draft (document no. 010331 in Gödel's papers) goes into yet more detail.

[m]See *Cohen 1963* and *1964*; cf. the letters of Gödel to Cohen in this volume. Note c of the introductory note to that correspondence describes the Fields Medals.

[n]Since very little of his research had been in set theory, Church seems an odd choice to give this lecture. The question arises whether Gödel was asked and declined. No letter inviting him is to be found in his papers. It is still possible that he might have been approached informally by one of the Princeton or IAS mathematicians.

[o]See also the introductory note to the correspondence with Wolfgang Rautenberg in volume V of these *Works*. Church does not state the claim of Mostowski he is correcting or give a reference, but I know of no other place where he claims independence results for Gödel than in the article quoted in Rautenberg's letter to Gödel. In particular, no such claim is made in *Mostowski 1965*.

In his letter of 4 August, Church mentions the possibility of correcting Mostowski's statement in a review of his paper in *The journal of symbolic logic*. However, *Mostowski 1964* was never reviewed or even listed.

Church says he will "correct Mostowski implicitly" by including "a statement that you had at some time in the 1940's a proof of the independence of the axiom of choice, confined to type theory."

Without having received Church's draft, Gödel replied on 10 August (letter 5) and attached a statement saying he had obtained in 1942 "a partial result, namely the independence of the axiom of choice in type theory" but that he had "never worked this proof out in detail".

Whether Gödel saw a draft of the talk before Church's departure is not clear. With a letter of 12 September, not included here, Church enclosed what he described as "an amended copy of my address about Paul Cohen, embodying all the last minute amendments I made before presenting the address at the Congress." However, he does not attribute any amendments to comments by Gödel. He describes a paragraph on pp. 6–7 as "in effect a quotation from you", but it seems most likely that it quotes the statement accompanying letter 5. Neither the version Church sent at this point nor any other version of the paper has been located in Gödel's papers.

Gödel responded on the 29th (letter 6). With regard to Gödel's comment on the philosophical significance of Cohen's result, Church had said that he had not intended to argue a philosophical position in any detail "but mainly just to put the question as a means of arousing interest"; however, the reader may guess that his own position is the "intermediate position" referred to in the last two sentences.[P]

About his own work on independence problems, Gödel had in the meantime looked at his notes from that time and was now prepared to claim only that he had a proof of the independence of the axiom of constructibility from type theory with choice "which, he believed, could be extended to an independence proof of the axiom of choice." Church adopted the revision Gödel proposed with a very slight modification (see editorial note d to Gödel's letter). From Church's subsequent letter of 3 October, we learn that they discussed the matter on the telephone.

In the following year Gödel revisited the question of his earlier work on independence problems. The German logician Wolfgang Rautenberg inquired with him about the same claim of Mostowski that had concerned Church. His reply to Rautenberg of 30 June 1967, published in volume V of these *Works*, has a slight difference from his final position in the exchange with Church, in that he says he was "in possession of" a proof of the independence of the axiom of choice in type theory; however, he again says he can only reconstruct "without difficulty" the proof of the independence of the axiom of constructibility.

[P]Presumably the same as in the published version, *Church 1968*, p. 18.

The reader of the later correspondence of Gödel and Church will be struck by its formal tone, although they had by 1966 known each other for 33 years, for many of which they had lived in the same town. Very probably that reflects the general habits of both men and does not indicate any coolness between them.

Charles Parsons[q]

A complete calendar of the correspondence with Church appears on p. 560 of this volume. The translation of letter 1 is by Charles Parsons, revised using suggestions of John W. Dawson, Jr.

[q]I am indebted to Tyler Burge, John Dawson, Solomon Feferman and Wilfried Sieg for helpful comments.

1. Gödel to Church

Wien 17/VI. 1932

Sehr geehrter Herr Church!

In einem Gespräch mit Prof. Veblen über Ihre vor kurzem in den Ann. of Math. 33 erschienene Arbeit,[a] haben sich die beiden folgenden Fragen ergeben, von denen Prof. Veblen meinte, es wäre vielleicht nützlich Ihnen davon Mitteilung zu machen:

1. Wie ist es in Ihrem System möglich absolute Existenzsätze z. B. das Unendlichkeitsaxiom zu beweisen?
2. Falls das System widerspruchsfrei ist, wird es dann nicht möglich sein, die Grundbegriffe in einem System mit Typentheorie bzw. im Axiomensystem der Mengenlehre zu interpretieren, und kann man überhaupt auf einem andern Wege als durch eine solche Interpretation die Widerspruchsfreiheit plausibel machen?

Ich übersende Ihnen gleichzeitig einige Separata meiner Arbeiten.

Ihr ergebener   Kurt Gödel

---

[a] *Church 1932.*

[English translation follows:]

1. Gödel to Church

Vienna, 17 June 1932

Dear Mr. Church,

In a conversation with Professor Veblen about your paper that appeared recently in the Annals of Math. 33,[a] the following two questions came up. Professor Veblen thought it might be useful to communicate them to you:

1. How is it possible to prove absolute existence propositions in your system, for example the axiom of infinity?
2. In case the system is consistent, won't it then be possible to interpret the fundamental concepts in a system with type theory, or in the axiom system of set theory, and can one make the consistency plausible at all in another way than through such an interpretation?

I am sending you at the same time some offprints of my papers.

Sincerely, Kurt Gödel

---

2. Church to Gödel

Princeton, N.J., July 27, 1932

Dear Dr. Gödel,

I have your letter of June 17 concerning my recent paper in the Annals of Mathematics.

In reply to your first question, whether the axiom of infinity is provable as a consequence of the set of postulates proposed in this paper, let me say this. If we make the definitions;

$$1 \to \lambda f \lambda x \,.\, f(x)$$
$$S \to \lambda \rho \lambda f \lambda x \,.\, f(\rho(f, x))$$
$$N \to \lambda y \,.\, [\varphi(1) \,.\, \varphi(x) \supset_x \varphi(S(x))] \supset_\varphi \varphi(y)$$

then the five axioms of Peano are provable in the following forms:

1. $N(1)$
2. $N(x) \supset_x N(S(x))$

3. $N(x) \supset_x \sim \,.\, S(x) = 1$
4. $[S(x) = S(y)] \supset_{xy} \,.\, x = y$
5. $[\varphi(1) \,.\, \varphi(x) \supset_x \varphi(S(x))] \supset_\varphi \,.\, N(y) \supset_y \varphi(y)$

Equality is here to be defined, essentially as in *Principia Mathematica*:

$$x = y \to \varphi(x) \supset_\varphi \varphi(y)$$

The proofs of propositions 1, 2, and 5 are almost immediate. The proofs of 3 and 4 are | somewhat longer, but I have seen the proofs, as they were worked out by a student here at Princeton,[a] and I have checked them through in detail, so that I am quite sure that the thing can be done.

As a matter of fact there are many ways of defining 1, S, and N so that axioms of Peano shall be provable, but this particular way has the advantage of providing a convenient notation for the nth power of a function, and therefore a convenient notation for definitions by induction.

You will grant that with the axioms of Peano proved, the essential part of the axiom of infinity is already present. It remains to provide a definition of what it is to be an infinite class, and to present a formal proof that the class of positive integers, $K(N)$, is infinite. This part of it I have not worked out, but I think it probable that it can be done.

Now as to the freedom from contradiction of the system, I cannot see that a proof of freedom from contradiction which began by assuming the freedom from contradiction of (say) *Principia Mathematica* would be of much value, because the freedom from contradiction of *Principia Mathematica* is itself | doubtful, even improbable.

In fact, the only evidence for the freedom from contradiction of *Principia Mathematica* is the empirical evidence arising from the fact that the system has been in use for some time, many of its consequences have been drawn, and no one has found a contradiction. If my system be really free from contradiction, then an equal amount of work in deriving its consequences should provide an equal weight of empirical evidence for its freedom from contradiction.

This is perhaps all that should be said.

But it remains barely possible that a proof of freedom from contradiction for my system can be found somewhat along the lines suggested by Hilbert. I have, in fact, made several unsuccessful attempts to do this.

Dr. von Neumann called my attention last fall to your paper entitled "Über formal unentscheidbare Sätze der Principia Mathematica".[b] I have

---

[a]This student was undoubtedly S. C. Kleene; cf. *Kleene 1935.*
[b] *1931.*

been unable to see, however, that your conclusions in §4 of this paper apply to my system. Possibly your argument can be modified so as to make it apply to my system, but I have not been able to find such a modification of your argument.

Sincerely yours,

Alonzo Church

3. Gödel to Church

March 2, 1965

Professor Alonzo Church
Fine Hall
Princeton University
Princeton, New Jersey

Dear Professor Church:

I thank you very much for your letter of January 29. It seems to me that the translation of my paper is considerably better than the introduction. It is true that it contains some errors and is rather poor in several other passages. But the errors are few and everything could easily be set right.

The introduction, however, contains a considerable number of wrong or misleading statements, even on important matters. Moreover it entirely disregards the subsequent developments in connection with the concept of computable function. Thereby it withholds from the philosophically interested reader what would interest him most, namely that my theorems can now be proved rigourously for *all* formal systems (containing number theory).

Even what I proved in 1931 is stated clearly (or more or less clearly) only at the end of the introduction (on p. 32).[1] From the rest of the introduction the reader gets the wrong impression that my theorems apply only to "arithmetical systems" (see p. 1), and that also P is only "the arithmetical part" of PM (see p. 6 and other passages), while in fact it contains all the mathematics of today, except for a few papers dealing with large cardinals.

---

[1] "proclaim" of course should be replaced by something like "made it highly probable."

What I mentioned are by no means the only serious shortcomings of the introduction.

Yours sincerely,

Kurt Gödel

4. Church to Gödel

Princeton University, June 27, 1966

Dear Prof. Gödel,

One of four Fields prizes to be given at the forthcoming International Congress in Moscow goes to Paul Cohen. Because of the number of prizes, talks describing the work of each recipient are to be limited to fifteen minutes. I have agreed to give the fifteen-minute talk in reference to Cohen. — From Prof. de Rham,[a] who is in charge of the matter, I obtained permission to confine the discussion, in view of limitation of time, to the continuum problem, the independence of the axiom of choice and of the continuum hypothesis, and matters immediately related. I also obtained from de Rham permission to consult you about the content of the talk, but he asks me to ask you to treat the information about prize recipients confidential until the date of presentation of the prizes.

I have no intention of making you responsible for anything I say, and this should not be necessary even if there are remarks that I ought to credit to you. But I would greatly appreciate any suggestions you have to make, or wishes you would like to express, about the content of the talk.

A secondary point is to ask whether you have a copy of Dana Scott's notes bearing on the result presented in his recent lecture,[b] and whether

---

[a]Georges de Rham, of the University of Lausanne, was chairman of the Fields Medals Committee for the International Congress of Mathematicians in Moscow in 1966.

[b]We have not been able to determine precisely what was the subject of Scott's notes or of the lecture he had evidently given at Princeton. In a communication of 21 June 2000, Professor Scott stated that he did not have a clear recollection but that they may well have concerned the method of Boolean-valued models for independence proofs, perhaps what went into the expository paper *Scott 1967*. If so, the result was very likely the existence of a Boolean-valued model of set theory (or of the third- or higher-order theory of the reals, as in *Scott 1967*) in which the continuum hypothesis fails. Comment on the relation of forcing and Boolean-valued models occurs on pp. 108–109 of Scott's paper.

I could borrow them briefly. I | don't mean to use the occasion of an en-
comium for Paul Cohen to announce a new (and subsequent) result by
Scott, as this would be inappropriate. But Scott had a very neat characterization of Cohen's method, which I no longer remember exactly, and
which—although Scott intended it in order to relate Cohen's method to
his own—I might nevertheless be able to make some use of. At least I
would like to look at it if you have the notes available.

The notes can be sent by Institute mail if you have them available.
Let me know how quickly you need them back and I will be governed accordingly. I am sure I can if necessary have a few pages Xeroxed quickly
in order to return the notes in two or three days. I will be out of town
briefly, from July 3 to 8; if it appears that you are currently using the
notes I will get them back to you before then, but otherwise may hold
them a little longer.

Sincerely yours,

Alonzo Church

5. Gödel to Church

August 10, 1966

Professor Alonzo Church
Fine Hall
Princeton University
Princeton, N. J.

Dear Professor Church:

Thank you very much for your letter of August 4.

I am enclosing herewith a brief statement correcting Mostowski's claims.
I'll be very much interested in seeing a draft of your Moscow talk. Perhaps it would be best, as soon as you are finished with it, to let me know
by phone where I can pick up a copy.

Scott's notes, unfortunately, are not mine, but Kreisel's. For this reason I would like to have them back before you leave (no matter when).

Very sincerely yours,

Kurt Gödel

It is not true that, for many years, I have been in the possession of a
proof of the independence of the continuum hypothesis and of the axiom
of choice in set theory. I did obtain (in 1942) a partial result, namely
the independence of the axiom of choice in type theory. However, due
to the fact that my interests, soon after that, veered to entirely different subjects, I never worked this proof out in detail and I never returned
to these questions. What I had at that time was far from sufficient for
proving the independence of the continuum hypothesis from the axiom of
choice.

K. Gödel, August 1966.

6. Gödel to Church

September 29, 1966

Professor Alonzo Church
Fine Hall

Dear Professor Church:

Thank you very much for sending me the manuscript of your talk.
You know that I disagree about the philosophical consequences of Cohen's result. In particular I don't think realists need expect any perma-
nent ramifications (see bottom of p. 8)[a] as long as they are guided, in the
choice of the axioms, by mathematical intuition and by other criteria of
rationality.

I have explained these things in more detail on p. 270 ff. of the 2nd
edition of my paper "What is Cantor's Continuum Problem?",[b] which
was published in PHILOSOPHY OF MATHEMATICS, edited by P. Benacerraf and H. Putnam in 1964.[c]

As far as your mention of my result of 1942 is concerned, I find, on
looking up my old notes, that, because of their sketchiness, I cannot, off
hand, reconstruct the independence proof of the axiom of choice, but
only of $V = L$ (in type theory including the axiom of choice). So I can
vouch only for the correctness of the latter proof.

---

[a]What Gödel refers to probably corresponds to the last three paragraphs of *Church
1968*, on p. 18.

[b]I. e., *Gödel 1964*.

[c]*Benacerraf and Putnam 1964*.

If you wish to say something with which I am in complete agreement, I would suggest that you replace the part of your talk beginning with the inception of line 4 from below on p. 6 and closing with the end of the paragraph, by the following:

"of the axiom of constructibility in type theory, which, he believed, could be extended to an independence proof of the axiom of choice. But, due to a shifting of his interests toward philosophy, he soon afterwards ceased to work in this area, without having settled its main problems. The partial result mentioned was never worked out in full detail or put in form for publication."[d]

I think this should suffice for the purpose of your talk. In addition to the shifting of my field of work, there were various other motives for my not publishing the partial result, e.g., that I disliked the version of the proof I had found, that the result is of little philosophical interest, that I hoped to return to these questions sometime later and to obtain something more satisfactory (while as a matter of fact I more and more lost contact with my former work), etc. At any rate, the fact that I had not | proved the corresponding set theoretical result was of little importance   2 in this connection because I believed (and I think rightly) that I could extend my proof to this case without serious difficulty. But these details would have no interest for the readers of your talk.

If you would like to add something to the formulation I suggested I'll be glad to discuss it with you on the phone or in person.

Sincerely yours,

Kurt Gödel

---

[d]These words occur almost verbatim in *Church 1968*, p. 17. The only change worthy of mention is that Church replaced "which, he believed ..." by a sentence break and "According to his own statement (in a private communication), he believed ...".

Paul J. Cohen

The most famous result in the metamathematics of set theory, after Gödel's proof in 1938 of the consistency with the Zermelo–Fraenkel axioms of the axiom of choice and the continuum hypothesis, was the proof of their independence, obtained by Paul J. Cohen in 1963. In terms of the usual abbreviations, Cohen showed that AC is independent of ZF and that CH is independent of ZFC (i.e., ZF + AC), in both cases assuming ZF is consistent.[a] The major part of the correspondence between Gödel and Cohen is devoted to the details of his work and its publication. The latter part of the correspondence turns to other questions in set theory and its metamathematics. Cohen did not permit his side of the correspondence to be published here; however, there are a number of items from him in the Gödel *Nachlaß* to whose contents it will be necessary to refer for background.

Paul Joseph Cohen was born in New Jersey in 1934. He received a Ph.D. in mathematics at the University of Chicago in 1958 for his work on topics in the theory of uniqueness of trigonometric series, under the direction of Antoni Zygmund.[b] After holding short-term positions at the University of Rochester (prior to obtaining his Ph.D.) and Massachusetts Institute of Technology (directly following it), and visiting the Institute for Advanced Study for the years 1959–1961, he joined the faculty of the mathematics department at Stanford University, where he has remained ever since. Cohen received the prestigious Bôcher Prize from the American Mathematical Society in 1964 for his work, "On a conjecture of Littlewood and idempotent measures" (*Cohen 1960*); principally for his work in set theory he received the even more prestigious Fields Medal at the International Congress of Mathematicians held in

---

[a]*Fraenkel 1922* had established the independence of AC from ZFU, i.e. the system ZF modified so as to allow "urelements" or atoms other than the empty set. Mostowski (*1939*) extended Fraenkel's methods to obtain the independence of various forms of AC from each other over ZFU. However, these methods did not apply directly to ZF nor did they touch the independence of CH from AC.

[b]For personal information regarding Cohen see the interview with him in *Albers, Alexanderson and Reid 1990*, pp. 43–58.

Moscow in 1966.[c] In his presentation of the award, Alonzo Church (*1968*) explained the background to and importance of Cohen's outstanding achievement. In the course of that, he mentioned that a partial, though unpublished, step toward the results had previously been taken by Gödel in 1942 when he found a proof of the independence of the axiom of constructibility from the theory of types that looked as though it could be extended to a proof of the independence from that system of the axiom of choice.[d] For his part, Gödel lauded Cohen's work in his 1966 Postscript to *1964* (pp. 269–270) as "no doubt... the greatest advance in the foundations of set theory since its axiomatization."

The story of how Cohen arrived at his results is recounted in *Moore 1988*; the parts relevant to the correspondence are recapitulated here. Known for his broad interests in mathematics, especially in analysis and number theory, and his talents as a "problem-solver", Cohen sought out major unsolved problems. He had a long-standing largely self-developed interest in logic, and after he came to Stanford he discussed some of the big open problems with the resident logicians, especially Georg Kreisel and the undersigned. At first he proposed to establish the consistency of analysis, and even conducted a seminar on this work, but abandoned his approach when it failed to give anything essentially beyond arithmetic. In 1962 Cohen then turned to set theory, where his first result was to obtain a minimal standard model M of ZF, where M is of the form L_α with α countable (*Cohen 1963a*); unbeknownst to him, this had previously been established by Shepherdson (*1953*). The existence of the minimal model showed that there could be no inner model construction to settle the leading independence questions. At the end of 1962 Cohen moved on to the question of the independence of AC from ZF. Gödel's proof of the consistency of AC and GCH (generalized continuum hypothesis)

---

[c]The Fields Medal has long been considered the most important prize in mathematics, since there is no Nobel Prize in that subject. It was established by a bequest in the will of the Canadian mathematician John Charles Fields, which stipulated that up to four awards are to be made every four years at a meeting of the International Congress of Mathematicians to mathematicians in recognition of their existing work and the prospect of future achievement; to meet the latter wishes, the Fields Medal is awarded only to mathematicians under the age of 40. The first award was made in 1936. In more recent years, other somewhat more munificent prizes have been established for mathematicians, but the Fields Medal still confers special distinction on the awardee.

[d]See, in this respect, the correspondence between Gödel and Church in this volume as well as between Gödel and Rautenberg in volume V. The former shows that the particular statement made in *Church 1968* follows quite closely a draft by Gödel himself. The introductory notes in these volumes to this pair of correspondences calls attention to the difference between what Gödel wrote Church in 1966 and what he wrote Rautenberg in 1967. Cf. also *Moore 1988*, pp. 149–151.

with ZF had proceeded by showing that they are both consequences of the axiom of constructibility, V = L, and that ZF + (V = L) can be interpreted in ZF. Cohen thus took as his first, easier challenge, the problem of showing the independence of V = L from ZF. This he tried to do by adjoining a new subset S of ω to the minimal model L_α in such a way that the collection of sets constructible relative to S up to α—in symbols $L_\alpha(S)$—forms a model of ZF; but then S would not be in the constructible sets interpreted in that model, so $L_\alpha(S)$ would satisfy (V $\neq$ L). Since α is countable there are many S that do not belong to L_α, but for S with special properties (for example, that it codes a well-ordering of order type α), $L_\alpha(S)$ is *not* a model of ZF. Cohen's key idea, arrived at early in the spring of 1963, was to produce S which are *generic*, in the sense that each property of S in the language of $L_\alpha(S)$ is *forced* by a finite amount of information about S. Once he obtained an appropriate definition of the forcing relation, Cohen was able to show that $L_\alpha(S)$ with S generic *is* a model of ZF and in this way to establish the consistency of ZF + (V $\neq$ L) relative to that of ZF; in fact, both AC and GCH hold in that model. From there, by using modified notions of forcing and one or more generic sets, he was able swiftly to move on to prove the independence of AC from ZF and of CH from ZFC; for this last, he adjoined to L_α a generic set coding a sequence of $\aleph_2$ (in the sense of the minimal model) subsets of ω, thus making $2^{\aleph_0} = \aleph_2$ in the resulting model.

Soon after obtaining these results, Cohen prepared a draft paper explaining his arguments, and gave a public lecture on them at Stanford in April 1963. When apprised of the work, many logicians—and especially those on the scene at Stanford and Berkeley—began to examine the work in detail. Though the general plan seemed sound, due to the novelty of the approach and the complexity of its elaboration some began to raise questions about points that were considered troublesome and would need more attention. Cohen felt that only Gödel's imprimatur on his proof would satisfy the critics. He thus wrote Gödel a brief letter on 24 April, announcing his results, and requesting a personal meeting to discuss his arguments.[e] A few days later, Cohen left for Princeton where he was to give another lecture on his work prior to talking to Gödel. A phone call from Dana Scott raising an objection to one point in that lecture increased Cohen's distress about questions concerning his work, and he thus wrote Gödel from New York on 6 May saying that only he, as the pre-eminent figure in the field, could give the "stamp of approval" which he so fervently desired. In the same letter, Cohen wrote that such objections put him under great nervous strain, and he hoped that Gödel

---

[e]Prior to this, Gödel had been informed of Cohen's work by Kreisel on 15 April.

would have a chance to study the manuscript thoroughly and that they could go over it together. Finally, he asked whether Gödel would consider communicating a description of the work to the *Proceedings of the National Academy of Sciences (PNAS)* in his capacity as a member of the Academy.[f] This was agreed to, and Cohen started to work in earnest on a submission for that purpose when he returned to Stanford later in May.

The first item we have from Gödel to Cohen in Gödel's *Nachlaß* is dated 5 June 1963. It is a handwritten, very messy draft, with many crossouts, and it is not clear whether a letter based on that was actually sent. The following has been extracted from it.

> Let me repeat that it is really a delight to read your proof of the ind[ependence] of the cont[inuum] hyp[othesis]. I think that in all essential respects you have given the best possible proof & this does not happen frequently. Reading your proof had a similarly pleasant effect on me as seeing a really good play.

Cohen wrote Gödel on 14 June saying that he was working away, but slowly, on his note for the *PNAS*, and that he had cleared up some points. In addition, he outlined a proof of the consistency of $2^{\aleph_0} = \aleph_\tau$ for τ not excluded by König's theorem (*1905*), according to which the cofinality of τ must be uncountable. In his response of 20 June (letter 1), Gödel said that he found the proof "too sketchy to be convincing", and tried to discourage Cohen from improving his results while writing up his work, as that would unduly delay the publication. Another point to which Gödel responded with interest was Cohen's suggestion that there could be models of ZF in which the set of constructible reals is countable. Finally, Gödel was skeptical about a conjecture that Cohen had made, according to which the existence of a definable well-ordering of V implies V = L.

Cohen lectured on his work on Independence Day, the Fourth of July, at the Theory of Models Symposium held at the University of California, Berkeley, 25 June–11 July 1963 (cf. *Cohen 1965*). The first further applications of the method of forcing and generic sets were also reported at the same symposium. Besides extending the method to arithmetic and analysis, in *Feferman 1965* the undersigned used the method to establish the consistency with ZFC + GCH of the scheme that no definition in the language of set theory establishes a well-ordering of the continuum, a result to which Cohen also referred in his letter of 14 June. And in the same proceedings, Robert Solovay (*1965*) announced the result conjectured by Cohen, under the title "$2^{\aleph_0}$ can be anything it ought to

---

[f]The letter is excerpted in *Moore 1988*, p. 157.

be", as well as some generalizations.[g] It was not long before there was a flood of further independence results by the method of forcing, and systematization of the arguments in one respect or another; among specific results *Lévy 1963* announced the existence of a model of ZF in which the constructible sets are countable, as conjectured by Cohen, and *Feferman and Lévy 1963* announced consistency with ZF of the statement that the set of reals is a countable union of countable sets, thus improving a result of Cohen's (as he noted in a letter to Gödel of 5 September).[h] Meanwhile, Cohen and Gödel were corresponding about the details of the presentation of Cohen's work to the *PNAS*, concerning which Gödel suggested the title, "The independence of the continuum hypothesis", and about which he had many specific suggestions to make. This part of the correspondence (not published here) stretched through August. In the process, Cohen became increasingly frustrated by what he thought were picky points and the consequent delay in publication. Exasperated, on 5 September he wrote Gödel granting him absolute freedom to make any changes he thought necessary, without needing to consult him. Gödel responded on 18 September that he had confined himself to those changes that he thought were "strongly indicated"; and as to Cohen's invitation for him to take over final responsibility, "please, write me explicitly whether you want me to send in the manuscript in this form, since it is an incorrect procedure to submit a manuscript with changes (even very minor ones), unless the author agrees to the individual changes." Cohen quickly agreed to all the suggestions, and the manuscript was submitted to the *PNAS* on 27 September 1963; it appeared as *Cohen 1963* and *1964*.

On 21 January 1964, Gödel responded negatively to a dinner invitation for 30 January from a group called the Research Corporation, in which he, Paul Cohen and perhaps others were to be conferred a special award. He wrote that "in order to preclude any misinterpretation of my absence I would like to say that in my opinion Paul J. Cohen's work in set theory is one of the milestones in the development of this field and of great importance also for the foundations of mathematics generally." Letter 2 from Gödel to Cohen, of 22 January 1964, begins by referring to this dinner, and goes on to try to attract his interest in the "Pantachie Problem" concerning scales of functions. Cohen did not take this up; Gödel's own pursuit of related problems was to be the subject matter of **1970a,b* and *c*. The letter concludes with a promise to write Cohen

---

[g]A far-reaching generalization was obtained by William B. Easton in his 1964 dissertation; cf. *Easton 1970*.

[h]Cf. *Moore 1988* for some of the subseqent history of the applications of forcing.

at some other time about another problem "of considerable interest, but not in connection with set-theoretical, but rather with foundational questions"; there is no subsequent evidence as to what he had in mind here.

In his letter to Gödel of 5 September 1963 referred to above, Cohen had also said that he thought he was close to proving the independence of the existence of Lebesgue non-measurable sets from ZF plus the countable axiom of choice. This was considered the next major independence result to be obtained in set theory. However, Cohen did not succeed in carrying through his ideas. Instead, the conjectured result (strengthened with the countable axiom of choice replaced by the axiom of dependent choices) was obtained within the year by Solovay, who lectured on his solution of the measure problem at a meeting in Bristol, England in July 1964. The second part of letter 3 from Gödel to Cohen of 11 January 1965 refers to this work. Solovay had considered submitting a note on it to Gödel for publication in the *PNAS*, but—as he stated in an interview with Gregory Moore in 1981 (cf. *Moore 1988*, p. 163)—Gödel kept asking him how he knew that the proof was correct. In letter 3, Gödel asked Cohen to confirm its correctness, and Cohen responded on the 18th of January to say that he had heard an oral presentation by Solovay, but could not see "how the main difficulty was overcome" (thus repaying in kind the suspicions others had originally held concerning his work), though he thought he could "rapidly" convince himself if presented with a written account. Solovay's faith in the correctness of his proof was confirmed instead through study of it by his peers, Silver, Kunen and Martin.[i] In any event, a note was never submitted to *PNAS*; instead, Solovay announced his result on the measure problem in an abstract (*1965b*), and subsequently published the work in full in the paper *Solovay 1970*. Incidentally, the consistency of "all sets are Lebesgue measurable" was established there only relative to the consistency of ZF plus the existence of an inaccessible cardinal; the necessity of that additional assumption was demonstrated in *Shelah 1984*.

The first part of letter 3, on the other hand, responds to a letter from Cohen of 20 November 1964, in which he sketched a generalization of forcing using partial orderings. Cohen seems not to have carried this any further; a generalization centering on partial orderings was presented by Joseph Shoenfield under the rubric "unramified forcing" to the Institute on Axiomatic Set Theory held at UCLA in the summer of 1967 (*Shoenfield 1971*). As Shoenfield explains in his introduction,

---

[i]Personal communication, 23 September 2000.

that approach owes much to the simplifications and generalizations of forcing developed by various set theorists subsequent to Cohen's work, but he is vague about just what he drew from these sources and the "folk literature". In fact, Solovay had realized as early as 1963 that with each countable transitive model M of ZF and partially ordered set $(P, <)$ in M is associated an extension $M[G]$, whose properties are completely determined by $(P, <)$ and relative to which G is "generic".[j] That subsequently became one standard way to present independence results via the method of forcing (cf. *Kunen 1980*). Another speculative remark in Cohen's November letter to which Gödel was responding in letter 3 is Cohen's belief expressed there that he can formulate a principle that can be recognized as an axiom of infinity and that implies the existence of measurable cardinals. For Gödel, that harked back to the ideas about axioms of infinity that he had floated in his remarks (*1946*) at the Princeton Bicentennial conference.

There is no letter from Cohen in Gödel's *Nachlaß* to which letter 4 (13 August 1965) appears to be a response. It deals with some of the same issues broached before, including justification of axioms of infinity by means of strong reflection principles, of the sort already formulated in *Lévy 1960*. The correspondence died off after that except for a few isolated items. On 30 November 1966, Cohen wrote Gödel in his capacity as one of the organizers[k] of the UCLA Set Theory Institute being planned for the summer of 1967, with a fervent invitation for him to attend the conference, at least for some part of it. Cohen spoke of Gödel as the man whose name is most associated with the subject of set theory and of logic in general, and that his presence would make this a memorable occasion, especially for younger mathematicians who had never had a chance to meet him. In a postscript, Cohen mentioned the appearance of the book (*Cohen 1966*) in which his work on independence results in set theory was exposited for a broad mathematical audience, a copy of which, of course, he had arranged to have sent to Gödel. In his response, Gödel thanked him kindly for the book, the praise, and the invitation. In declining the last, he cited his poor health and the direction of his own thought on set theory, which was more philosophical than mathematical. But in a postscript, Gödel mentioned for the first time in their correspondence (though it was already in print in *1947*) his belief that there is an objective question as to "the true power of

---

[j] *Ibid.* Solovay added that he had lectured on this approach in his course at the University of California at Berkeley in 1965. He also remarked that Cohen, in his letter to Gödel of 20 November 1964, seemed to miss the point that the properties of $M[G]$ should depend only on $(P, <)$.

[k] The other organizers were Abraham Robinson and Dana Scott.

the continuum", thus (implicitly) one that is left unsettled by the consistency and independence results. And once more he pointed to the "very neglected subject" of "orders of growth", i.e. scales of functions, as an important tool in determining that.

In May 1969, after a long hiatus in the correspondence, Cohen wrote Gödel asking him to join as a signatory on a telegram on behalf of a Soviet student who was in some kind of difficulty. In a long, messy draft of a response in the *Nachlaß*, Gödel tries to explain why he is begging off; we do not know if this was ever sent.

Solomon Feferman[1]

A complete calendar of the correspondence with Cohen appears on pp. 561–562 of this volume.

[1]I am indebted to Aki Kanamori, John W. Dawson, Jr., Charles Parsons, Wilfried Sieg and Robert M. Solovay for helpful comments on a draft of this note.

1. Gödel to Cohen

Princeton, June 20, 1963

Dear Professor Cohen:

A theorem to the effect that $2^{\aleph_\alpha} = \aleph_\tau$ is consistent for any τ permissible by Cantor's and Koenig's inequalities would be very interesting. However, what you write about its proof is too sketchy to be convincing. The "collapsing" (by my method) in this case is complicated by the fact that the collapsed set must also be closed for C, which may bring in new a_δ. The problem is substantially the same as that which Hajnal tries to settle by his Theorem 1 (p. 133 of Zs. f. math. Log. u. Gr. d. Math. vol. 2).[1] But he seems to need the regularity of $\aleph_\tau$.

That the constructible subsets of an enumerable set ⟨with any structure⟩ are enumerable is an old conjecture of mine. A consistency proof therefore would be very welcome for me.

But if I were in your place, I [would] not now try to improve the results, since this is an infinite process. I said that one month delay in

[1]The ful[l] proofs have been published in Act. math. Ac. Sc. Hung. 12, 1961, p. 321.[a]

---

[a]*Hajnal 1956* and *1961*.

the publication would not make much difference. On the other hand I think you should not delay it much longer. I don't see that any changes are necessary in part 3 of your paper except those I mentioned to you, namely the very easy one in connection with Lemma 16 and the one in connection with the final paragraph of your paper. The other results you could announce in a second paper a little later. Of course you could also make a much briefer announcement of your results than I suggested. At any rate I would suggest that you mention when and where you first presented your results in a lecture.

I hope you are not under some nervous strain which hampers you in your ~~your~~ work. You have just achieved the most important progress in set theory since its axiomatization. So you have every reason to be in high spirits.

With all good wishes, and hoping to hear from you soon.

I am sincerely yours

〚no signature 〛

Please give my regards to Professor Kreisel.

P.S. I don't think that the exist. of a definable well-ordering of V implies V = L.

2. Gödel to Cohen

January 22, 1964

Dr. Paul J. Cohen
Department of Mathematics
Stanford University
Stanford, California

Dear Dr. Cohen:

I am very sorry I cannot attend the dinner on January 30. I hope that, in view of the letter of which I enclose a copy herewith, you won't take it amiss if I am absent.

I have promised to write you about some important problems in connection with your method. Here is one of them: Once the continuum hypothesis is dropped the key problem concerning the structure of the continuum, in my opinion, is what Hausdorff calls the "Pantachie Problem",[1]

---

[1] In German the problem is frequently called "Problem der Wachstumsordnungen". Perhaps there exists some standard English expression for it, too.

i.e., the question of whether there exists a set of sequences of integers of power $\aleph_1$ which for any given sequence of integers contains one majoring it from a certain point on. Hausdorff evidently was trying to solve this problem affirmatively (see Abh. Saechs. Ges. Wiss. 31, 1909[a] and Ber. Saechs. Ges. Wiss. 59, 1907[b]). I was always suspecting that, in contrast to the continuum hypothesis, this proposition is correct and perhaps even demonstrable from the axioms of set theory. Moreover I have a feeling that, if your method does not yield a proof of independence here, it may lead to a proof of this proposition. At any rate it should be possible to prove the compatibility of the "Pantachie Hypothese" with $2^{\aleph_0} > \aleph_1$.

There is another problem which also is of considerable interest, but not in connection with set theoretical, but rather with foundational questions. I shall write you about it some other time.

Sincerely yours,

Kurt Gödel

---

[a] *Hausdorff 1909.*
[b] *Hausdorff 1907.*

3. Gödel to Cohen

Princeton, Jan. 11, 1965.

Dear Professor Cohen,

Many thanks for your letter. I am very glad to hear that you are having a pleasant time in Spain. The development of the concept of forcing which you sketch is very interesting and, I am sure, very fruitful. It was clear from the beginning that your∧method permits a great deal of generalization. This is exactly why your work is so important. It would be quite interesting if the problem of measurable cardinals could be solved in the manner you suggest. That it can be follows from the hypothesis that every number[-]theoretic question is decidable by some axiom of infinity, which is a rather plausible assumption.

I shall be glad to submit an abstract of Solovay's work on the measure problem to the Proc. Nat. Ac.,∧as he requested. However, I understand

from him that the proof is extremely involved. Therefore, checking it myself may cause considerable delay. I presume that Solovay's work, in substance, is known to you. Hence I should appreciate it if you would confirm to me the correctness of the arguments, at least to the extent that
you are sure the theorems which he asserts can be proved substantially
in this manner. One of his theorems says that the existence of a unique
association of fundamental sequences to the ordinals of the II class is independent from the denumerable choice axiom. This result seems even
more important to me than the independence of the existence of non-
measurable sets.

With best wishes

Yours sincerely

⟦no signature⟧

4. Gödel to Cohen

August 13, 1965

Dr. Paul J. Cohen
Department of Mathematics
Stanford University
Stanford, California

Dear Professor Cohen:

The passages in which Cantor states the continuum hypothesis are
on pp. 132, 244, 257 in his "Gesammelte Abhandlungen",[a] edited by
A. Fraenkel[b] in 1932. Also what is said on p. 462 of this book and in
Act. Math. 50 (1927),[c] p. 16ff. will probably interest you in this connection.

Hausdorff's proof of the existence of Ω^* gaps in the orders of growth is
given in Abh. Sächs. Ges. Wiss., Math. Phys. Kl. 31 (1909),[d] p. 320ff.

When we spoke about the power set axiom, what I had in mind, but
could not recall at the moment, was that $2^{\aleph_0} \leq \aleph_{\tau+1}$ can be proved for
your model without once more using the concept of forcing.

---

[a] *Cantor 1932.*

[b] Edited not by Fraenkel, but by Zermelo.

[c] *Schoenflies 1927.*

[d] *Hausdorff 1909.*

As far as the axiom of the existence of inaccessibles is concerned I think I slightly overstated my view. I would not say that its evidence is due *solely* to the analogy with the integers. But I do believe that a clear analogy argument[1] is much more convincing than the quasi-constructivistic argument in which we imagine ourselves to be able somehow to reach the inaccessible number. On the other hand, Levy's principle might be considered more convincing than analogy.

I'll be very much interested to see the notes of the lectures you gave in Harvard last year.

Sincerely yours,

Kurt Gödel

[1]such as, e.g., the one obtained if an inaccessible α is defined by the fact that sums and products of fewer than α cardinals $< \alpha$ are $< \alpha$.

5. Gödel to Cohen

April 27, 1967

Dr. Paul J. Cohen
Department of Mathematics
Stanford University
Stanford, California 94305

Dear Professor Cohen:

I have received the copy of your book on set theory and your invitation to attend the summer conference on set theory. I would like to thank you for both, and especially for the high regard for my work which you, who have played such a decisive role in the recent developments, evidence by setting such great value on my taking part in the conference.

I am extremely sorry I have to disappoint you. Even disregarding the fact that, on account of my poor health, I have not made any long trips for many years, there is another reason which makes it impossible for me to attend the conference. It is the fact that, for many years, my own thinking has moved along lines entirely different from those of the conference and even of a, perhaps envisaged, philosophical section of it. Namely, I have been trying first to settle the most general philosophical and epistemological questions and then to apply the results to science. On the other hand I have not yet advanced far enough to make such applications. For this reason I have not participated actively in the recent

most interesting developments and am, at the present moment, not in a position to participate in their continuation.

I wish to the conference the best of success which, in view of the considerable number of highly gifted mathematicians interested in the subject, I am sure, will not be wanting.

Yours sincerely,

Kurt Gödel

P.S. I may add that practically the only point where I have taken issue with the recent developments is the question of the true power of the continuum. You know that my views on this point are rather unorthodox. It would therefore be of little use to present them without sufficient objective reasons.

The only point about which there should be general agreement is that the orders of growth (to which I referred as an important tool in this context) are an interesting and very neglected subject. In my opinion it would be much more illuminating to draw consequences from the highly a   plausible axioms about them thén from the continuum hypothesis.

K. G.

Burton Dreben

Burton Dreben

In the 1950s–1970s, Burton Dreben (1927–1999), professor of philosophy at Harvard University, was investigating Herbrand's Theorem and its applications to proof theory and decision problems.[a] In 1962, Peter Andrews, Ståle Aanderaa and Dreben discovered an error in Herbrand's thesis (*1930*), in the argument for a lemma crucial to Herbrand's proof of the Theorem, and developed counterexamples that showed the lemma was in fact false. When Gödel learned of these findings, he told Alonzo Church that he had been aware of errors in Herbrand. Hence Dreben wrote to inquire (letter 1), and enclosed abstracts announcing the falsity of the lemma as well as a way to correct Herbrand's argument. The latter is far from trivial: it requires a rather complex argument that yields a weakened form of the crucial lemma; this weakened lemma then enables a proof of the Theorem along the lines that Herbrand envisaged. (The correction, improved from Dreben's original version, was first published in *Dreben and Denton 1966.*) Gödel did not reply to Dreben's inquiry, nor to a subsequent letter of 3 May 1963, with which Dreben enclosed a preprint of *Dreben, Andrews and Aanderaa 1963.* Apparently Gödel did mention the error in Herbrand to Jean van Heijenoort in a conversation they had in late September, 1963, for in a letter to Gödel of 14 October 1963 van Heijenoort wrote:

> I mentioned to Wang and Dreben what you had said about the proof of Herbrand's main lemma during our conversation. They expressed the hope that you would permit us to publish in the Source Book [*van Heijenoort 1967*] what you wrote about the proof. We had planned that the introductory note on Herbrand, written by Dreben, would contain his corrections...; but, of course, if your earlier corrections were made available, this would enhance the historical value of the book.

How much Gödel had said about his "earlier corrections" is unknown. Gödel did not respond to the suggestion of publishing them, and it appears van Heijenoort did not bring up the matter again. As we now know, in the early 1940s Gödel had worked on Herbrand's thesis, had seen the fallacy in Herbrand's argument (but did not find counterexamples that showed Herbrand's original lemma to be false) and had devised a correction. Moreover, his statement of the weakened crucial lemma and his argument for it were in all essentials the same as that in

---

[a]For a brief account of Dreben's life and work, emphasizing his contributions to logic, see *Goldfarb 2003.*

Dreben and Denton 1966. This work of Gödel's remained unknown until the late 1980s.[b]

Dreben's second letter to Gödel concerns a decision problem. In 1933, Gödel had shown that any prenex formula of pure quantification theory having a quantifier prefix of the form $\forall\forall\exists\ldots\exists$ is satisfiable over a finite universe, if satisfiable at all. At the end of the paper, he remarked that this result "can also be proved, by the same method, for formulas that contain the identity sign" (*Gödel 1933i*, p. 443). Dreben had been working on a monograph on solvable classes of the decision problem for quantification theory since the early 1950s (a work that eventually appeared as *Dreben and Goldfarb 1979*); in the mid-1960s he tried to verify Gödel's claim and could not see how to do so. Stål Aanderaa, then a graduate student at Harvard, formulated counterexamples to the obvious ways of proceeding; one of these was sent to Gödel. Gödel replied promptly (letter 3), but what he said was unhelpful. In a postscript to a letter to Hao Wang of 7 December 1967, Gödel expressed his continued belief that his proof can be extended. Wang, in a letter of 23 April 1968, raised further difficulties.[c] Clearly the matter weighed on Gödel; for in early 1970 he asked Dana Scott to write to Dreben and to Hao Wang, outlining a strategy for extending the proof.[d] The strategy outlined in Scott's letter does not succeed: see these *Works*, vol. I, pp. 229–230. Both Wang and Dreben wrote Gödel to express their doubts and communicate further counterexamples. In fact, Gödel was completely mistaken: once identity is allowed in the formulas, the class becomes undecidable; indeed, even the class of formulas with the simplest prefix of the Gödel type, $\forall\forall\exists$, becomes undecidable (*Goldfarb 1984, 1984a*).

On an unrelated issue, Dreben wrote in late 1969 to ask Gödel for substantiation of a remark in *Gödel 1944* regarding the relation of an idea in Russell to an (unspecified) consistency proof for arithmetic (letter 4). This letter was unanswered, although in the *Nachlaß* it was found with its envelope inscribed "Reply." Dreben wrote again, repeating the question, on 14 December 1971; the later inquiry also went unanswered. An answer came to light in the *Nachlaß*. At the end of his notes for his 1941 lectures in Princeton on what subsequently became known as the *Dialectica* interpretation,[e] Gödel remarks:

---

[b]Gödel's work on Herbrand's thesis was found in the *Nachlaß*, in an *Arbeitsheft* written in Gabelsberger shorthand. It was transcribed by Cheryl Dawson in 1988, which enabled the author of this note to publish a full account of Gödel's correction in *Goldfarb 1993*.

[c]The relevant parts of the Wang correspondence appear in these *Works*, vol. V.

[d]Scott's letter, dated 3 April 1970, appears in these *Works*, vol. V, Appendix A.

[e]See these *Works*, vol. III, p. 186. The citation comes from *Nachlaß* document 040408, p. 87.

Finally, I wish to remark that this whole scheme of defining the logical notions has a certain relationship to what Russell intended in the §9 of the *Principia Mathematica.* Namely it is chiefly the question of defining the meaning of the logical operations for expressions involving quantifiers provided that their meaning for no-quantifier expressions is given.

Thus the consistency proof he refers to in *1944* was his own, which remained unpublished for fourteen more years.

Warren Goldfarb

1. Dreben to Gödel[a]

March 6, 1963

Dear Professor Gödel:

Mr. Peter Andrews has just told me that after he had mentioned in a letter to you that there was a crucial difficulty in Herbrand's thesis you informed Professor Church, over the telephone, that you had known of errors in Herbrand's work. Since Mr. Andrews and I are preparing a paper on this topic, we would very much appreciate your telling us whether among the errors of which you have been aware is the one described in the enclosed abstract entitled *Errors in Herbrand.*[b]

Sincerely yours,

Burton S. Dreben

Enclosures: *Errors in Herbrand*
                Corrections to Herbrand[c]

---

[a]All of Dreben's letters are on letterhead of Harvard University, Department of Philosophy.

[b]Typescript of *Dreben, Andrews and Aanderaa 1963a.*

[c]Typescript of *Dreben 1963.*

2. Dreben to Gödel

May 24, 1966

Dear Professor Gödel:

At the end of your article ZUM ENTSCHEIDUNGSPROBLEM DES LOGISCHEN FUNKTIONENKALKÜLS[a] you remark that Theorem I also holds for formulas that contain the identity sign. However, I have been unable to extend your proof of Theorem I to cover such formulas, and I would be most grateful to you if you would tell me how your argument is to be modified so as to apply to the following example, given to me by Mr. Stål Aanderaa.

$$(x_1)(x_2)(Ey)[(Fyy \equiv \overline{Fx_1x_1}) \,\&\, ((Fx_1x_1 \,\&\, Fx_2x_2) \supset x_1 = x_2)$$
$$\&\, ((Fyy \,\&\, y \neq x_2 \,\&\, (Fyx_1 \equiv Fyx_2)) \supset x_1 = x_2)$$
$$\&\, (Fx_1x_1 \supset (y \neq x_2 \,\&\, (Fyx_1 \equiv x_1 \neq x_2)))]$$

Yours sincerely,
Burton Dreben

---

[a] *1933i.*

3. Gödel to Dreben[a]

July 19, 1966

Dear Professor Dreben:

I am sorry I don't have any notes about the exact procedure for proving Th. 1 of my paper in Mon. H. Math. Phys. vol. 40 in case the formula contains "=". However, I remember that the idea was to formulate the auxiliary concepts and the lemmas under the assumption that in addition to the relations F_i an equivalence relation leaving the F_i invariant is given. No difficulty arose in carrying this through.

I am sorry I cannot be more helpful at the present moment.

Sincerely yours,

Kurt Gödel

---

[a] Text taken from retained copy.

4. Dreben to Gödel

December 30, 1969

Dear Professor Gödel:

On page 143 of your article "Russell's Mathematical Logic"[a] you say that Russell's definition of the truth functions as applied to propositions containing quantifiers "proved its fecundity in a consistency proof for arithmetic". I would very much appreciate your telling me which consistency proof you were referring to here.

Sincerely yours,

Burton Dreben

---

[a] *1944*.

5. Dreben to Gödel

April 15, 1970

Dear Professor Gödel:

I am extremely sorry to hear that at present you are not well, and I very much hope that your health will soon improve.

I am grateful to you for having Dana Scott write to me about your procedure for extending your solvable case to cover equality. The reduction to what Scott called in his letter "sharp formulas" presents no problem, but I do not see how to apply your original method to such formulas. A series of examples will bring out some of the difficulties that arise. (I owe these examples to Dr. Stål Aanderaa, and my comments on them result from discussions with Aanderaa and Mr. Warren Goldfarb.)

Consider the *satisfiability* question for formulas in the form

$$(A) \qquad \forall x \forall y \exists z (x \neq y \supset (x \neq z \,\&\, y \neq z \,\&\, Hxyz)),$$

where $Hxyz$ does not contain "=" (hence all formulas in this form are essentially sharp), and let $B(x, y, z)$ be the matrix of (A).

Now let F_1 be (A) with $Hxyz$ as $(\neg Gx \vee \neg Gy) \,\&\, (Gz \equiv (\neg Gx \,\&\, \neg Gy))$. Then F_1 is satisfiable in all domains of cardinality ≥ 3. However, there

exist for F_1 no sets $\mathcal{P}$ and $\mathcal{Q}$ of tables of orders 1 and 2 that fulfill conditions (I) and (II) of your paper (p. 435). For, if M is any model that satisfies F_1, then there must be an individual $\underline{a}$ in the domain of M such that $\varphi^M(\underline{a})$ is true, where φ^M is the function interpreting G. But then the set $\mathcal{P}$ must contain the table $[M|\underline{a}]$. Hence, if we set $R = [M|\underline{a}]$ and $S = [M|\underline{a}]$, from condition (I) there must be a table T in $\mathcal{Q}$ such that $[M|\underline{a}] = [T|1] = [T|2]$. But this table T *cannot* be extended to a table T' of order 3 such that $B_{T'}(1,2,3)$ is | true. Hence (II)-2 fails. Thus, when applied to sharp formulas conditions (I) and (II) need not be necessary for satisfiability.

2

Moreover, if we were to take condition (I) to require that R and S always be distinct tables, then a difficulty still remains. For although, when so read, conditions (I) and (II) are now necessary for satisfiability, they no longer assure it. Let F_2 be (A) with $Hxyz$ as the conjunction of (1), (2), and (3), and let F_3 be (A) with $Hxyz$ as conjunction of (1), (2), (3), and (4):

(1) $\neg Kxx \lor \neg Kyy$
(2) $Kzz \equiv (\neg Kxx \,\&\, \neg Kyy)$
(3) $Kyz \equiv \neg Kzy$
(4) $Kxx \lor Kyy \lor (Kxy \equiv Kyx)$.

F_2 is satisfiable in all domains of cardinality ≥ 3, but F_3 is not satisfiable in any domain of cardinality ≥ 2. Nevertheless, there exist a set $\mathcal{P}$ and a set $\mathcal{Q}$ that fulfill, under the new reading, conditions (I) and (II) for both F_2 and F_3.

The discussion of F_1 shows that if we simply drop from F_1 the clauses containing "$=$", the resulting formula is not satisfiable. And the following formula F_4 shows that another problem arises if we construe "$=$" as an arbitrary dyadic letter E, even if we were to add to the matrix of F_4 expressible natural properties of equality, such as Exx and $Exy \equiv Eyx$. (I see no way of expressing transitivity.) Let F_4 be (A) with $Hxyz$ the conjunction of the following four clauses:

(1) $Kzz \equiv (\neg Kxx \,\&\, \neg Kyy)$
(2) $\neg Kxx \lor \neg Kyy$
(3) $Kxx \lor Kyy \lor (Kzx \equiv \neg Kzy)$
(4) $Kxx \supset (Kyz \,\&\, \neg Kzy)$.

F_4 is not satisfiable because (1), (2), (3) require that there be no more than three objects, but (1) and (4) require at least four objects. On the

3

other hand, replacement of "$=$" by E in | F_4 (with the additional clauses mentioned above) yields a formula satisfiable in domains of cardinality ≥ 5. Thus there is no obvious way to eliminate "$=$" from the matrix of sharp formulas.

The difficulties raised by each of the above examples turn on a basic difference between sharp formulas and pure Gödel formulas. There exist sharp formulas all of whose satisfying models M contain "distinguished" individuals, that is, individuals $\underline{a}$ such that if the table $[M|\underline{a}]$ is identical with the table $[M|\underline{b}]$, for any individual $\underline{b}$, then $\underline{a} = \underline{b}$.

I am sending copies of this letter to Scott and Wang.

Yours sincerely,

Burton Dreben

Herbert Feigl

Herbert Feigl

Herbert Feigl (1902–1988) was a fellow student and friend of Gödel's at the University of Vienna in the 1920s. He was a member of the Vienna Circle, and one of its most influential representatives in the United States. He joined the philosophy department at the University of Iowa, Iowa City in 1933, and moved to the philosophy department at the University of Minnesota, Minneapolis in 1940. He founded the Center for Philosophy of Science at the latter in 1953, and remained active in the department and center until his retirement in 1971. He is probably best known today for his defense of a form of logical empiricism, and his defense of the thesis that mental events are contingently identical with physical events.[a]

In the letter, Gödel coaches his friend on the standard "hydrodynamic" interpretation of the basic operations of vector calculus (divergence and curl), and the interpretation it yields, derivatively, of Maxwell's equations. It was written in response to an inquiry from Feigl (no longer extant), who, apparently, was looking for a non-technical paraphrase of the equations for inclusion in a short book he was writing (*Feigl 1929*).[b] The letter is of interest, and has a certain charm, because it shows that Gödel was willing to take the time to write at length to a friend to help him understand some relatively elementary ideas in mathematical physics. In a similar position, many people would simply refer the friend to a textbook.

Of interest, as well, is Gödel's report in the final paragraph that he has read a section of *Principia mathematica* (for the first time), and is less enthusiastic about it than he had expected given the work's reputation.[c]

David B. Malament

The translation is by John Dawson, revised in consultation with David Malament.

---

[a]Feigl's life and work is reviewed in an essay by C. Wade Savage that is available online (http://www.mcps.umn.edu/fbiofr.htm); see *Savage 1988*.

[b]One finds such a paraphrase on pages 62–63. It is very much in the spirit of Gödel's remarks, but does not follow them exactly.

[c]Gödel would later explain his dissatisfaction in *Gödel 1944*: " It is to be regretted that this first comprehensive and thorough-going presentation of a mathematical logic and the derivation of mathematics from it [is] so greatly lacking in formal precision in the foundations (contained in *1–*21 of *Principia*) that it presents in this respect a considerable step backwards as compared with Frege."

1. Gödel to Feigl

Brünn 24./IX. $\overline{28}$

Lieber Feigl!

Deine Karte aus St. Georgen habe ich erhalten und mich gefreut, daß Dein Büchlein[a] solche Fortschritte gemacht hat. Hoffentlich war auch Natkin[b] im Sommer nicht allzu faul, denn ich bin schon riesig neugierig, was der Verl. W. u. H.[c] dazu meinen wird. Was Deine Bitte betrifft, muß ich Dir leider mitteilen, daß Du mir fälschlich imputierst, Dir jemals eine *brauchbare* populäre Formulierung der M. Gl. gesagt zu haben. Ich habe Dir nur einmal die Ehrenhaftsche, als abschreckendes Beispiel erzählt. Diese besteht in Folgendem: E. sagt statt rot "Wirbel" (div verdeutscht er komischer Weise nicht.) und spricht dann einfach die einzelnen Gleichungen in Worten aus. Also z. B. Wirbel der magn Kraft is prop. der zeitlichen Änderung der elektrischen Kraft (in Leitern = Summe aus Stromvektor und dieser Änderung) u.s.w. Die Hauptsache wäre jetzt natürlich eine Definition von rot und div. Dabei beschränkt er sich aber auf folgendes:

Kraftlinienverlauf für positive div

1.) (div ist ja ein Skalar)   ↖ ↗
                              ← · →
                              ↙ ↘

2.) für negative div   ↓
                      → · ←
                      ↗ ↖

3.) für rot senkrecht aufwärts vom Papier

---

[a] *Feigl 1929.*

[b] Marcel Natkin, a fellow student and close friend of Gödel and Feigl, attended meetings of the Vienna Circle with them. He later became a photographer in Paris.

1. Gödel to Feigl

Brünn 24 September 1928

Dear Feigl!

I received your card from St. Georgen and am pleased that your booklet[a] has made such progress. Hopefully, Natkin[b] was also not too lazy during the summer, for I am mighty curious what the publisher W. u. H.[c] will think of it. As concerns your request, I must regrettably inform you that you falsely impute to me that I ever conveyed to you a useful popular formulation of Maxwell's equations. I only once told you, as a cautionary example, that of [Felix] Ehrenhaft. That consists of the following: Instead of 'rot' [curl] Ehrenhaft says 'whirl' (oddly, he doesn't render 'div' into German) and then simply expresses the individual equations in words. Thus, for example, the whirl of the magnetic force [field] is proportional to the temporal [rate of] change of the electric force [field] (in conductors it equals the sum of the current vector and this change), and so on. The main issue would now of course be a definition of 'rot' and 'div'. But in that regard he restricts himself as follows:

Configuration of the lines of force for positive div

1.) (div is of course a scalar) $\begin{array}{ccc} \nwarrow & \uparrow & \nearrow \\ \leftarrow & \cdot & \rightarrow \\ \swarrow & \downarrow & \searrow \end{array}$

2.) for negative div $\begin{array}{c} \downarrow \\ \rightarrow \cdot \leftarrow \\ \nearrow \quad \nwarrow \end{array}$

3.) for rot outwardly perpendicular to the paper

[c]Perhaps a reference to the series "Wissenschaft und Hypothese" (Science and Hypothesis), published by Teubner in Leipzig.

Natürlich sind das nur Beispiele und der Kraftlinienverlauf kann für jeden
2 | der 3 Fälle auch ganz anders sein.

Doch gibt es eine sehr anschauliche hydrodynamische Definition für
rot und div, die ich Dir empfehlen möchte. Ich schreibe sie Dir für den
Fall, daß Du sie noch nicht kennen solltest. Man deutet das Kraftfeld als
stationäre Strömung einer inkompressiblen Flüssigkeit.[1] Im allgemeinen
wird es aber keine physikalisch mögliche solche Strömung darstellen; denn
dazu müßte noch die Bedingung erfüllt sein, daß die in einen Volumteil in
der Zeiteinheit einströmende Flüssigkeit = der aus[s]trömenden ist. Um
die Deutung trotzdem möglich zu machen muß man annehmen, daß an
jeder Stelle des Raumes Flüssigkeit entstehen bzw. verschwinden kann.
Dann ist div = der pro Zeiteinh. an der betr. Stelle entstehenden (neg.
verschwindenden) Flüssigkeitsmenge)[2]

Zur Erklärung der rot denkt man sich an der betr. Stelle ein so kleines
Flüssigkeitsteilchen abgegrenzt, daß man es während einer sehr kurzen
Zeit als starr betrachten kann. Dieses Teilchen hat dann im allg. 1. eine
fortschreitende 2. eine drehende Bew. ~~so~~ der Vektor rot hat dann die
Richtung der Drehungsachse und die Größe der ⟨doppelten⟩ Winkelgeschwindigke

Diese ganze Geschichte wird Dir ja wahrscheinlich bekannt sein. Für
eine populäre Darstellung scheint mir das sehr geeignet. Die Maxw. Gl.
3 selbst würde ich an Deiner | Stelle für den Fall $\epsilon = 1$ angeben, wo dann,
der Unterschied zwischen Kraft und Verschiebung ($\mathfrak{E}, \mathfrak{D}$ bzw. $\mathfrak{H}, \mathfrak{B}$) wegfällt.
Sie stellen dann bloß eine Beziehung zwischen der Feldgrößen $\mathfrak{E}, \mathfrak{D}$ und
den sie verursachenden Ladungsgrößen ρ, i = [Strömvektor=Bewegungszustand
der Ladung] dar

$$\text{rot } \mathfrak{E} = -\frac{1}{c}\frac{\partial \mathfrak{H}}{\partial t} \qquad\qquad \text{div } \mathfrak{E} = 4\pi\rho$$

$$\text{rot } \mathfrak{H} = \frac{1}{c}\left[\frac{\partial \mathfrak{E}}{\partial t} + 4\pi i\right] \qquad \text{div } \mathfrak{H} = 0$$

Übrigens kann man auch der obigen Ehrenhaftschen Darstellung einen
exakteren Sinn geben. Es entspricht nämlich jedem Punkt eines Feldes
ein Feld von der Art 1.) oder 2.) je nachdem ob div $\lessgtr 0$ und ein Feld
3.) wobei die Ebenen der Kreise immer senkrecht sind zum Vektor der

---

[1]D. h. man faßt das vorgelegte Kraftfeld als Geschwindigkeitsfeld einer Strömung
auf.[d]

[2]Genau müßte man sagen der in einem kleinen Volumteil entstehenden Menge/Volumteil.

---

[d]Footnote 1 is taken from the last page and begins "Auf Blatt 1.)..." ("On sheet
1.)...").

Of course, these are only examples, and the lines of force for each of the three cases can also be entirely different.

Yet there is a very intuitive hydrodynamic definition for rot and div that I would like to recommend to you. I'll write it down for you, in case you're not yet familiar with it. One interprets the force field as the stationary flow of an incompressible fluid.[1] In general, however, it will not represent any physically possible such flow; for that, the condition that the fluid flow into a volume element in unit time equal the flow out would have to be satisfied in addition. Even so, in order to make the interpretation possible one must assume that at each point of the space fluid can emerge or disappear. Then div equals the amount of fluid emerging (or, if negative, disappearing) per unit time at the point under consideration.[2]

To clarify the rot one imagines demarcated at the point under consideration a fluid particle so small that one can regard it as fixed during a very short time. In general this particle then has first a translational [and] second a rotational motion. The vector rot then has the direction of the rotation axis and magnitude twice the angular velocity.

This whole story is probably familiar to you. For a popular presentation it seems to me very suitable. Were I in your place, I would state Maxwell's equations themselves for the case $\epsilon = 1$, when the distinction between field strength and displacement ($\mathfrak{E}, \mathfrak{D}$ or, respectively, $\mathfrak{H}, \mathfrak{B}$) disappears. They then represent merely a relation between the field strengths $\mathfrak{E}, \mathfrak{D}$ and the quantities of charge ρ, i that produce them [the current vector = the state of motion of the charge]:

$$\operatorname{rot} \mathfrak{E} = -\frac{1}{c}\frac{\partial \mathfrak{H}}{\partial t} \qquad\qquad \operatorname{div} \mathfrak{E} = 4\pi\rho$$

$$\operatorname{rot} \mathfrak{H} = \frac{1}{c}\left[\frac{\partial \mathfrak{E}}{\partial t} + 4\pi i\right] \qquad \operatorname{div} \mathfrak{H} = 0$$

Moreover, one can also give the aforementioned representation of Ehrenhaft a more exact meaning. Namely, to each point of a field there corresponds a field of type 1.) or 2.), according to whether div is > 0 or < 0, and a field 3.) in which the planes of the circles are always perpendicular to the vector rot at the point under consideration; and by superposition

---

[1] That is, one conceives the aforesaid force field as the velocity field of a flow.

[2] More precisely, one would have to say: the amount/volume element emerging in a small volume element.

rot an der betr. Stelle; und durch Superposition aller dieser Felder erhält
man wieder das ursprüngliche (selbstverständlich läuft die Superposition auf eine Integr. hinaus da ja $\aleph$ Summanden vorhanden sind). Doch
scheint mir diese Interpretation viel komplizierter zu sein als die hydrodynamische und außerdem gibt sie ja keine direkte Definition von rot u.
div sondern nur eine (allerdings sehr wichtige) Eigenschaft, nämlich die
Zerlegbarkeit des ursprünglichen Feldes in Teilfelder die durch div und
rot an den verschiedenen Raumstellen bestimmt sind—Wie ich sehe hat
sich mein Brief etwas in die Länge gezogen, aber ich glaube, es wird Dir
4   ja zu viel lieber als zu wenig sein, und außerdem besitze | ich in diesen
Dingen eine angeborene Gründlichkeit. Du wirst ja in Deinem Buch die
Maxwellschen Gl. wahrscheinlich nur kurz erwähnen wollen und deshalb
die ganz im Anfang gegebene Erklärung als kürzeste bevorzugen. Ich
weiß allerdings nicht ob jemand davon mehr hat als wenn man ihm sagt:
"Die M. Gl. setzen die Feldgrößen miteinander in Beziehung".

—Hätte gern gehört, wie Du ansonsten den Sommer verbracht hast.
Du warst doch zuerst mit Deinen Eltern in Salzkammergut? Von Natkin
habe ich eine Karte aus Zoppot[e] bekommen. Ich selbst war die ganze
Zeit in Brünn und habe unter anderem einen Teil der Principia mathematica[f] gelesen, von denen ich aber weniger begeistert war, als ich es dem
Ruf nach erwartet hatte. Nach Wien komme ich Anfang Oktober. Also
auf baldiges Wiedersehen und herzliche Grüße

Dein

Kurt Gödel

---

[e]Polish town, northwest of Gdańsk.

of all those fields one again obtains the original one (the superposition is of course tantamount to an integration, since $\aleph$ summands occur). But this interpretation seems to me to be much more complicated than the hydrodynamic one, and furthermore it gives no direct definition of rot and div, but rather only a property [of them] (admittedly a very important one), namely the decomposability of the original field into subfields that are determined by div and rot at the various points of space.—As I see, my letter has become somewhat extended in length, but I believe it will be too much rather than too little for you, and besides, in these things I am possessed of an inborn thoroughness. In your book you will probably want to mention Maxwell's equations only briefly, and therefore will prefer as the shortest explanation that given right at the beginning. Indeed, I don't know whether anyone will get more out of it than if you [just] say: "Maxwell's equations relate the field strengths to one another."

—I would like to have heard how else you spent your summer. At first you were with your parents in the Salzkammergut, weren't you? I received a card from Natkin from Sopot.[e] I myself was in Brno the whole time and among other things read a part of Principia mathematica,[f] about which, however, I was less enthusiastic than I had expected from its reputation. I am coming to Vienna the beginning of October. So, until we soon see each other again, heartfelt greetings.

Yours,

Kurt Gödel

---

[f]Probably the second edition, *Whitehead and Russell 1925*.

Paul Finsler

Paul Finsler

The Swiss mathematician Paul Finsler (1894–1970) is best known for his work in differential geometry. In his doctoral dissertation, written at Göttingen in 1918 under the direction of Constantin Carathéodory, he generalized the notion of Riemannian manifold to what are now called Finsler spaces. In 1922 he was appointed as a lecturer at the University of Cologne, and five years later he became an associate professor for applied mathematics at the University of Zürich, where he remained for the rest of his academic career. Besides geometry, he also published papers in number theory, probability theory and set theory; a bibliography is given in *Gross 1971*.

Finsler's correspondence with Gödel was prompted by his desire to claim priority for having established the existence of undecidable propositions—a claim based on his paper *1926*, in which he adapted Richard's paradox to exhibit a false but "formally" undecidable proposition. In contrast to Gödel, however, Finsler made no attempt (and saw no need) to give a precise specification of the syntax of his "formal" language; and since metamathematical considerations are meaningless in the absence of restrictions on linguistic resources, Gödel's results appeared to him as mere specializations of those he had already obtained. He could not understand, therefore, why Gödel's paper had attracted widespread notice while his had not.

In his letter to Yossef Balas (see page 9 of this volume) Gödel attested that he was unaware of Finsler's paper at the time he wrote his own *1931*. But when Finsler brought the paper to his attention, he immediately recognized its shortcomings: While avoiding direct confrontation, he responded forthrightly to Finsler's challenge, politely but firmly pointing out the essential differences between Finsler's procedures and his own. He noted in particular that the antidiagonal construction Finsler had employed would, if applied to a "truly formal" system, define a sequence that lay outside the system itself.

Finsler reacted angrily to Gödel's criticisms, and his subsequent papers, particularly *1944*, reveal a persistent inability to understand the issues involved. Like Gödel he was a staunch Platonist; unlike Gödel, however, he failed to appreciate the value of formalization and never grasped the fundamental distinction between use and mention that is central to the modern conception of logic. As a result, his proposed axiomatization of set theory was rejected by other logicians as ill-defined

or inconsistent, a rebuff that wounded him and probably retarded his professional advancement. (He was not promoted to full professor until 1944.)

For a recent reappraisal of Finsler's foundational views and set-theoretic work see *Breger 1992*. A compilation of all of his set-theoretical papers, accompanied by unreservedly laudatory commentary and a brief

1. Finsler to Gödel

Zürich, den 11. März 1933
Schmelzbergstr. 27

Sehr geehrter Herr Dr. Gödel!

Am letzten Zürcher Kongress und sonst schon hörte ich von Ihren Sätzen über die Entscheidbarkeit in formalen Systemen. Es würde mich interessieren, Ihre Arbeiten, die ich hier nur flüchtig durchsehen konnte, noch näher kennen zu lernen; besonders Herr Prof. Hahn riet mir, mich mit Ihnen in Verbindung zu setzen.

Soweit ich sehe, handelt es sich im Prinzip um etwas Aehnliches, wie ich es in meiner Arbeit über "Formale Beweise und die Entscheidbarkeit"[a] (Math. Zeitschr, 25, 1926) behandelt habe, und worüber ich auch schon im Anschluss an einen Düsseldorfer Vortrag mit Herrn Prof. Hahn korrespondiert hatte. Dabei legen Sie jedoch einen engeren und deshalb schärferen Formalismus zugrunde, während ich, um den Beweis kürzer führen und nur das Wesentliche hervorheben zu können, einen allgemeinen Formalismus angenommen habe. Es ist natürlich von Wert, den Gedanken auch in einem speziellen Formalismus wirklich durchzuführen, doch hatte ich diese Mühe gescheut, da mir das Ergebnis doch schon festzustehen schien und ich deshalb für die Formalismen selbst nicht genügend Interesse aufbringen konnte. Auch scheint mir mein Resultat in der Anwendung auf die Hilbertsche Theorie noch etwas weiter zu gehen. Es würde mich freuen, Ihre Ansicht über diese Dinge noch näher kennen zu lernen.

Als Drucksache sende ich Ihnen einige Separate.

Hochachtungsvoll

P. Finsler

---

[a] *Finsler 1926.*

chronology of Finsler's life, has also appeared recently in English translation (*Booth and Ziegler 1996*).

For another putative anticipation of Gödel's incompleteness discovery see *Post 1994*, as well as Post's correspondence with Gödel in volume V.

John W. Dawson, Jr.

1. Finsler to Gödel

Zürich, 11 March 1933
Schmelzbergstr. 27

Dear Dr. Gödel:

At the last Zürich Congress, and elsewhere even before that, I heard of your theorems about decidability in formal systems. I would be interested to become still better acquainted with your papers, which I could look over only fleetingly here; in particular, Professor Hahn advised that I contact you.

So far as I see, in principle it is a question of something similar to what I treated in my paper on "Formal proofs and decidability"[a] and about which I had also already corresponded with Professor Hahn following a lecture in Düsseldorf. In that regard, however, you take as a basis a narrower and therefore more precise formalism, whereas I, in order to be able to shorten the proof and emphasize only the essentials, assumed a general formalism. It is of course also worthwhile actually to carry out the ideas in a special formalism; nevertheless, I had avoided that trouble, since the result seemed to me already to have been established and I therefore could not muster sufficient interest in the formalism itself. Moreover it seemed to me that my result went somewhat further in its application to Hilbert's theory. I would be happy to become better acquainted with your view on these matters.

I am sending along a few offprints as printed matter.

Respectfully,

P. Finsler

2. Gödel to Finsler[a]

Wien, 25./III. 1933

Sehr geehrter Herr Professor!

Ich danke Ihnen bestens für die Übersendung Ihrer Separata u. schicke mit gleicher Post Separata meiner Arbeiten an Sie ab. Zu Ihrem Aufsatz "Formale Beweise und die Entscheidbarkeit" habe ich folgendes zu bemerken:

Das System $\mathfrak{S}$ mit dem Sie operieren ist überhaupt nicht definiert, denn[b] bekanntlich wird die Frage, was ein "logisch einwandfreier Beweis" ist,[1] von verschiedener Mathematikern verschieden beantwortet. Versucht man aber (und das ist das Entscheidende) Ihren Beweis auf ein wirkliches formales System P anzuwenden, so erweist er sich als falsch. Die von Ihnen (p. 681 oben) definierte Antidiagonalfolge und daher auch der formal unentscheidbare Satz ist nämlich *niemals* in demselben formalen System P enthalten, von dem man ausgegangen ist. Das ist leicht zu beweisen und kann auch folgendermaßen plausibel gemacht werden: Würde die im System P vorhandenen | formalen Mittel ausreichen, die genannte Folge zu definieren, so würden sie offenbar auch ausreichen im folgende ganz ähnlich gebildete Dualfolge zu definieren: An allen ungeraden Stellen steht 0 und an $2n$-ter Stelle steht eine Ziffer, die verschieden ist von der $2n$-ten Ziffer der n-ten Folge aus der Reihe derjenigen Dualfolgen, für welche ein Beweis existiert, daß sie unendlich viele 0 enthalten. Die Annahme, daß diese Folge in P liegt führt aber auf einen Widerspruch, da ~~un[?]bar~~ unmittelbar aus ihrer Definition ⟨formal⟩ gezeigt werden kann, daß sie unendlich viele 0 enthält.—Ihre Begründung, daß ~~der~~ ⟨die genannte Folge und daraus⟩ von Ihnen konstruierte Satz formal darstellbar ~~ist~~ sind (p. 681, Zeile 8 v. unten) enthält 2 Fehler[c] 1. Daraus, daß kein Widerspruch vorliegt, folgt nicht, daß keiner existiert.[2] 2. Selbst wenn

---

[1] vgl p. 680, Zeile 12 von unten, Ihrer Arbeit

[2] Tatsächlich existiert ein solcher Widerspruch, doch würde es zu weit führen, dies auseinanderzusetzen.

---

[a] Finsler's copy of this letter has not been found. The texts reproduced here are those of two fragmentary drafts preserved in Gödel's *Nachlaß*. This is the longer, apparently first, of the two drafts. (See fn. b.)

[b] From here to the bottom of the page Gödel wrote some apparently intended revisions without crossing out what is reproduced here. For the most part, these would have resulted in the shorter draft following this letter. See the textual notes for further description.

[c] Written in shorthand above the line: Das sich auf ein formales System überhaupt nicht anwendet, da die Frage, ob ~~irgendein Satz [?] [?] oder eine Folge~~ irgendein Gegenstand in einem formalen System darstellbar ist, nicht davon abhängt, ob ohne *Wider-*

2. Gödel to Finsler[a]

Vienna, 25 March 1933

Dear Professor!

Thank you very much for forwarding your offprints. In the same mail ⟦with this⟧ I am dispatching offprints of my papers to you. As to your essay "Formal proofs and decidability", I have the following remarks to make:

The system ⑤ with which you operate is not really defined at all, for as is well known, the question of what a "logically unobjectionable[b] proof" is[1] is answered differently by different mathematicians. But (and that is the deciding ⟦criterion⟧) if one tries to apply your proof to a truly formal system P, it turns out to be wrong. Namely, the antidiagonal sequence you defined (top of p. 681) and therefore also the formally undecidable statement is *never* contained in the same formal system P ⟦as that⟧ from which one starts out. That is easy to prove and can also be made plausible as follows: Were the formal means available in the system P sufficient to define the designated sequence, it would obviously also be sufficient to define the following quite similarly formed dual sequence: At all the odd positions there stands a 0, and at the $2n$th position stands a number that differs from the $2n$th number of the nth sequence in the series of those dual sequences which can be proved to contain infinitely many zeroes. But the assumption that this sequence lies in P leads to a contradiction, since directly from its definition it can be formally shown that it contains infinitely many zeroes. Your justification that the designated sequences you constructed are formally representable contains two mistakes:[c] 1. From the fact that no contradiction presents itself it does not follow that none exists.[2] 2. Even if one knew that the sequence

---

[1]Cf. your paper, p. 680, line 12 from the bottom.

[2]Such a contradiction actually exists, but to explain that would lead too far ⟦afield⟧.

---

spruch recht ist, sondern ob es eine Kombination ~~der~~ von Zeichen des Systems gibt, das dieses Gegenstand bezeichnet.—Abgesehen davon kann man daraus, daß kein Widerspruch vorliegt, nicht schließen, daß keiner vorhanden ist—Tatsächlich führt ja auch die Annahme... ("That doesn't apply to a formal system at all, since the question whether some object is representable in a formal system does not depend on whether ⟦the system⟧ is correct without *contradiction*, but rather on whether there is a combination of signs of the system that denotes that object.—Apart from that, one cannot conclude from the fact that no contradiction presents itself that none exists.—The assumption also actually leads...")

man wüßte, daß die ⟨Annahme, die⟩ betrachtete Folge ⟨liege in P,⟩ zu keinem Widerspruch ~~in~~ ~~P~~ führt, so könnte man noch nicht schließen, daß sie in P enthalten sein muß.—Da der von Ihnen konstruierte Satz in dem System P, von dem man ausgegangen ist, überhaupt nicht vorkommt, kann er natürlich auch nicht in P. [The draft breaks off here]

3. Gödel to Finsler[a]

Wien 25./III. 1933

Sehr geehrter Herr Professor!

Ich danke Ihnen bestens für die Zusendung Ihrer Separata und sende mit gleicher Post Sonderdrücke meiner Arbeiten an Sie ab. Zu Ihrem Aufsatz "Formale Beweise und die Entscheidbarkeit" möchte ich folgendes bemerken:

Das System $\mathfrak{S}$, mit dem Sie operieren ist überhaupt nicht definiert, denn Sie verwenden zu seiner Definition den Begriff des "logisch einwandfreien Beweises"[b] der ohne nähere Präzisierung der Willkür den weitesten Spielraum läßt. Versucht man aber (und das ist der springende Punkt) Ihren Gedankengang auf ein wirkliches formales System anzuwenden, so erweist er sich als falsch. Die von Ihnen p. 681 oben definierte Antidiagonalfolge und daher auch der unentscheidbare Satz ist nämlich *niemals* in demselben formalen System P darstellbar, von dem man ausgeht. Der Beweis dafür ist nicht schwer, es würde aber doch zu weit führen ihn hier mitzuteilen. Die Begründung, die Sie | dafür geben, daß die genannte Folge u. der von Ihnen konstruierte unentscheidbare Satz formal darstellbar sind (p. 681, Zeile 8 v. unten) läßt sich auf ein formales System überhaupt nicht anwenden, da die Frage, ⟨ob⟩ irgend etwas in einem bestimmten formalen System darstellbar ist, nichts mit dem Vorhandensein eines Widerspruchs zu tun hat. Außerdem kann man daraus, daß ein Widerspruch nicht vorliegt, nicht schließen, daß keiner existiert. Tatsächlich führt ja auch die Annahme, daß die genannte Antidiagonalfolge im System P enthalten ist, auf einen Widerspruch, wenn dies auch nicht sofort ersichtlich ist. [End of draft]

2

---

[a] The shorter, presumably revised draft.

[b] "vgl. p. 680, Zeile 12 v. unten Ihrer Arbeit" ("Cf. your paper, p. 680, line 12 from

considered leads to no contradiction, one could still not conclude that it must be contained in P.—Since the proposition you constructed doesn't actually occur in the system P from which you started out, in P it also cannot ⟦The draft breaks off here.⟧

3. Gödel to Finsler[a]

Vienna, 25 March 1933

Dear Professor!

Thank you very much for sending your offprints. In the same mail ⟦with this⟧ I am sending offprints of my papers to you. As to your essay "Formal proofs and decidability", I would like to make the following remarks:

The system 𝔖 with which you operate is not really defined at all, for in its definition you employ the notion of "logically unobjectionable proof",[b] which without more precise restriction on its arbitrariness allows the widest scope for interpretation. But (and that is the salient point) if one tries to apply your reasoning to a truly formal system P, it turns out to be wrong. Namely, the antidiagonal sequence you define at the top of p. 681, and therefore also the undecidable proposition, is never representable in the same formal system P from which one starts out. The proof of that is not difficult, but it would lead too far ⟦afield⟧ to impart it here. The justification that you give that the designated sequence and the proposition you constructed are formally representable (p. 681, line 8 from the bottom) cannot be applied to a formal system at all, since the question of whether something is representable in a specified formal system has nothing to do with the existence of a contradiction. In addition, from the fact that no contradiction presents itself one cannot conclude that none exists. Actually, the assumption that the designated antidiagonal sequence is contained in the system P also leads to a contradiction, though that is not immediately evident. ⟦End of draft⟧

---

the bottom.") was written at the bottom of the page. By comparison with the preceding longer draft, we assume that this was to be a footnote flagged at this point.

4. Finsler to Gödel

Zürich, den 19. Juni 1933
Schmelzbergstr. 27

Sehr geehrter Herr Dr. Gödel!

Ich danke Ihnen für die zugesandten Separata.

Auf die Bemerkungen zu meiner Arbeit möchte ich folgendes erwidern: Wenn man über ein System $\mathfrak{S}$ Aussagen machen will, so ist es durchaus nicht notwendig, dass dieses System scharf definiert vorgelegt ist; es genügt, wenn man *es als gegeben annehmen kann* und nur einige Eigenschaften desselben kennt, aus denen sich die gewünschten Folgerungen ziehen lassen. Dies ist aber hier der Fall, und die angebliche "Vagheit" des Systems ist daher kein Einwand. Mit grösserem Recht könnte ich Ihre Arbeit (Ueber formal unentscheidbare Sätze[a] ...) wenigstens solange als "vollkommen unbrauchbar" bezeichnen, als Sie nicht für die darin verwendeten Axiome, insbesondere die Axiome Peanos, die absolute Widerspruchsfreiheit nachweisen können. Uebrigens scheint mir auch die "Eineindeutigkeit" Ihrer Abbildung (Seite 179, Zeile 6) reichlich vage zu sein; wie ist denn die Abzählung der Zeichen (Seite 176, Anmerkung 17) festgelegt?

Eine mathematische oder logische Untersuchung braucht auch keineswegs in einer Rechnung oder einem bestimmten Schema zu bestehen. Auch Sie bedienen sich für viele Ueberlegungen der wörtlichen Ausdrucksweise, ohne zu behaupten, dass sie deshalb nur vage oder ungültig seien. Wie könnten Sie sonst zeigen, dass der formal unentscheidbare Satz doch richtig ist? Wie können Sie einen bestimmten Formalismus einführen, ohne Worte zu gebrauchen?

Wie schliessen Sie ferner, dass die von mir definierte Antidiagonalfolge *niemals* in demselben System formal darstellbar ist, von dem man ausgeht, während doch das System $\mathfrak{S}$ ein Gegenbeispiel liefert? (Oder wie ist ein "wirkliches" formales System allgemein definiert?) Dass man für sehr enge Systeme andere Beispiele wählen muss, ist verständlich, ändert aber am ganzen Sachverhalt gar nichts. Einen *prinzipiellen* Unterschied zwischen Ihrem Beispiel und dem meinigen kann ich tatsächlich nicht entdecken. Dass ein solcher nicht vorhanden ist, geht auch aus Ihrer *Anmerkung 14 (Seite 175)* hervor.

Die Tatsache, dass es Sätze gibt, die auf Grund eines gegebenen Formalismus logisch eindeutig entschieden sind, die aber in dem System

---

[a] *1931.*

4. Finsler to Gödel

Zurich, 19 June 1933
Schmelzbergstr. 27

Dear Dr. Gödel:

Thank you for the offprints you sent.

I would like to reply as follows to the remarks on my work: If one wants to make assertions about a system 𝔖, it is not at all necessary that that system be presented in a sharply defined way; it is enough if one *can accept it as given* and if one knows only a few of its properties, from which the desired consequences can be drawn. But this is the case here, and the alleged "vagueness" of the system is therefore no objection. I could with greater justice characterize your paper ("On formally undecidable propositions...".[a]) at least as much as "completely unusable", since you cannot demonstrate the absolute consistency of the axioms employed in it, in particular the Peano axioms. Moreover, the "unicity" of your mapping (page 179, line 6) also seems to me to be abundantly vague; for how is the enumeration of the symbols (page 176, note 17) established?

A mathematical or logical investigation also need in no way consist of a computation or a specified schema. You too, for many considerations, make use of verbal means of expression, without asserting that they are therefore only vague or invalid. Otherwise how can you show that the formally undecidable proposition is in fact correct? How can you introduce a specified formalism without using words?

Furthermore, how do you conclude that the antidiagonal sequence I defined is *never* formally representable in the same system from which one starts out, whereas in fact the system 𝔖 furnishes a counterexample? (Or how is a "truly" formal system defined in general?) That for very narrow systems one must choose other proofs is understandable, but that does not at all alter the situation as a whole. I really cannot discover a distinction *in principle* between your proof and mine. That such a thing does not exist also follows from your note 14 (page 175).

The fact that there are propositions which are decided in a logically unambigous way on the basis of a given formalism but which "do not

selbst "überhaupt nicht vorkommen", ist übrigens für die Beurteilung der Tragweite und der logischen Widerspruchsfreiheit eines Systems ebenfalls von wesentlichster Bedeutung. Dass Ihr Beweis "intuitionistisch einwandfrei" ist, mag für manche Mathematiker einen Vorteil bedeuten. Der Fundamentalsatz der Algebra existiert aber auch nicht erst dann, wenn er intuitionistisch einwandfrei bewiesen ist.

Sie bemerken in Ihrer Arbeit, dass der Widerspruchslosigkeitsbeweis gewisser Systeme nicht in diesen Systemen selbst formalisiert werden kann, lassen aber die Möglichkeit zu, dass dies in übergeordneten Systemen geschehen kann, wobei aber doch die Widerspruchsfreiheit dieser Systeme auch wieder in Frage steht. Dass man nun in solcher Weise allein mit finiten Mitteln überhaupt nicht zum Ziel kommen kann, das zeigen die allgemeinen Betrachtungen, wie ich sie angegeben habe.

Hochachtend,

P. Finsler

occur at all" in the system itself is moreover of the most essential significance as well for judging the scope and logical consistency of a system.

That your proof is "intuitionistically unobjectionable" may signify an advantage for many mathematicians. But the fundamental theorem of algebra also exists, not just when it is proved in an intuitionistically unobjectionable way.

You observe in your paper that the proof of the consistency of certain systems cannot be formalized in those systems themselves, but you admit the possibility that that can be done in systems laid on top of [[them]], in which, however, the consistency of the latter systems once again stands in question. That one thus cannot at all arrive at the goal in such a way, solely by finitary means—that shows the general considerations, as I have referred to them.

Yours truly,

P. Finsler

Wilson Follett

Wilson Follett was one of the editors at New York University Press during the negotiations between Gödel, Ernest Nagel and NYU Press concerning the publication of one or both of *1934* and a translation of *1931* as an appendix to the planned *Nagel and Newman 1958*. See the correspondence with Allan Angoff (this volume, p. 1) and for a full account of the background, the introductory note to the correspondence with Ernest Nagel, vol. V, p. 135.

1. Gödel to Follett

Princeton, Aug. 29, 1957

tt Dear Mr. Folley:

Unfortunately your letter has reached me with considerable delay. I am sorry to say that the plan of the book as you now suggest it is very poor and that, therefore, I cannot give the permission to reprint my paper of 1931[a] in this setting and under these conditions. First, that paper is not well suited to serve, by itself, as an appendix to a book which is destined for the general public. My lectures[b] would be much better for this purpose. Moreover they contain some of the advances made since 1931. Second, as I have stated several times already, neither my paper nor my lectures ought to be republished without mentioning certain advances made since. Failure to do so would amount to withholding from the reader some very important facts concerning the scope and meaning of my results. I̶t̶ *It goes without saying that I cannot leave it to the judgment of others what should be added.*

All difficulties that have arisen could easily have been avoided if the authors had discussed with me the plan of the book beforehand. Under the present circumstances, since apparently no agreement between the authors and myself can be reached, it seems best to me to abandon the idea of reprinting some of my work in an appendix.

---

[a] *Gödel 1931.*
[b] *Gödel 1934.*

Very truly yours,

Kurt Gödel

P.S. I am sorry that my insistence ~~to~~ ⟨on⟩ see⟨ing⟩ the text of the book
has created ill-feeling. I understand from friends that it is usual for the
authors in such a case to send a copy of the manuscript on their own
2   initiative.[c] A note as | you suggest it would not release me from *all* re-
sponsibility for the book. Moreover such a note would be inappropriate,
if I cooperate to the extent of writing a few pages of additions. That,
however is absolutely necessary, because ~~otherwise a mis~~ ⟨failure to do so
would be tantamount to a mis⟩representation of the scientific situation of
today ~~would result~~.

---

[c] "A note as" occurs in handwriting at this point on the first side; the P.S. continues typewritten on the back. We here follow the copy of the letter in the Ernest
Nagel papers.

2. Follett to Gödel[a]

September 26, 1957

My dear Dr. Gödel:

Your letter of August 29, received September 3, could not be answered
until after consultation with the Messrs. Newman and Nagel. If I had
had any idea how long it might take to hear from them, I should have
sent an *ad interim* acknowledgment of your letter on September 3, the
day on which I dispatched copies of it to Mr. Newman and Mr. Nagel.
Their answer was delayed for various reasons, including vacation ab-
sences, and I had no word from them until yesterday, September 25.

Their answer, the only one possible in the circumstances, is that they
relinquish with great regret the plan to print a careful translation of
your 1931 paper in the appendix of their book. The book will therefore
be brought out without the inclusion of any original material by you,
whether in German or in English.

---

[a] On letterhead of New York University Press, Washington Square, New York 3,
Editorial Department.

For myself only and without prompting from any source, I wish to add that the position assumed by you seems to me a deeply shocking contravention of the generally recognized canons of free inquiry. In a long life of frequent participation in such negotiations over permissions to reprint or to translate, I have never before encountered or even heard of a scholar who demanded, as a precondition of quoting him verbatim, a power of censorship over other scholars' interpretation of his words. These coauthors proposed to reproduce a document of cardinal importance that you printed and signed. They proposed to satisfy you as to the accuracy and completeness of their version. They proposed to pay fairly and even liberally for your consent, though the material in question is evidently not even under copyright. They proposed to state unequivocally that their reproduction of the material in no way committed you to approval of their interpretation of it. It seems to me astounding and incomprehensible that any scholar, in these circumstances, should not permit and even welcome the use of his article, secure in the knowledge that the public prints are | open to him for any correction, amplification, or disagreement 2 that he thinks it desirable to state. If learned men and original thinkers generally were to operate on this strange new principle of refusing to be quoted without the right to dictate the construction put upon what is quoted, what effect on learning and on original thought could their attitude possibly have, other than the effect of a disastrous cripplement?

Yours sincerely,

Wilson Follett

Leonard Goddard

Leonard Goddard (b. 1925) was a senior lecturer in philosophy at the University of New England in New South Wales, Australia when he wrote to Gödel in 1965. Later he taught at St. Andrews University in Scotland, and from 1977 until his retirement in 1989 he was professor and chair of the philosophy department at the University of Melbourne. His letter inquires about the connection between Bertrand Russell's vicious circle principle and the theory of types, and in particular Russell's apparent claim in *Whitehead and Russell 1910*, p. 48, that the simple theory of types is a consequence of the vicious circle principle. In his draft reply, which was never sent, Gödel asserts that, contrary to Russell's claim, there is no route from the vicious circle principle to the theory of types, because the vicious circle principle at best supports a cumulative hierarchy. Gödel thus sharply distinguishes the theory of types from any theory that allows a cumulative hierarchy, taking it to be central to type theory that a propositional function can have arguments of only one type. In taking this to be so fundamental, Gödel differs from W. V. Quine, who in *1969*, §38, suggests that a move from a strict (simple) theory of types to a cumulative theory is just a technically straightforward matter of simplification. For Gödel the sharp distinction is grounded in the philosophical rationale for type theory, especially the role of Russell's notion of "ranges of significance". This emphasizes the importance for Gödel of the considerations he canvassed in *1944*, pp. 147–149 (these *Works*, vol. II, pp. 136–138).

Warren Goldfarb

1. Goddard to Gödel

The University of New England
Armidale, N.S.W., Australia
Department of Philosophy
16. February, 1965

Dear Professor Gödel,

I hope you will forgive my troubling you, but I should be very grateful if you could spare the time to give me your opinion on the following matter.

I have recently completed a book,[a] one half of which is devoted to a discussion of Russell's type-theory. In the course of this I say that *Russell supposed that both the theory of orders and the theory of simple types arose from the vicious circle principle*. However, the referee has said that this is a mistake and he refers to your article in the Schilpp volume[b] in support of this.

Now I certainly do not want to deny that the theory of simple types is a different theory from a theory of orders, but it seems to me that Russell did not fully realize this and that partly this confusion arose just because he believed them both to be derivable from the VCP. The crucial passage where I find this view stated most clearly occurs in *PM*[c] Introduction to 1st edn., Chap II, V, "The Hierarchy of Functions and Propositions". The first eight lines of this seem to me to state the view that the theory of simple types arises from the VCP, & is a summary of earlier arguments to the effect that a propositional function involves the totality of its argument-values & hence ⟨by VCP⟩ cannot be among its own argument-values—which I take it establishes a type, not an order, difference between e.g. φ and x etc. Then the passage continues: "But the hierarchy which has to be established is not so simple as might at first appear:" & the remainder of the paragraph states the view that it has to be supplemented by the theory of orders and that this theory too arises from the VCP.

Of course, even if it is true that Russell believed that both theories arise from the VCP, the question still remains whether he was justified in believing this & whether there is in fact any connection between the VCP and the theory of simple types. Here again, however, I find difficulty in seeing that there is no connection. For if "whatever involves all of a collection cannot be one of the collection", then a class, which by definition involves all of a collection, cannot | be one of the collection, i.e. cannot be a member of itself. Thus the VCP seems to justify the hierarchical distinction between a class and its members, but this distinction is a type difference not an order difference.

In general it seems to me that since Russell proposed the VCP as a means of eliminating the paradoxes, and since some of the paradoxes are removed by simple type theory alone, there has to be a direct connection between VCP and simple type-theory.

As I say, I do not want to deny that there are two different theories involved, & especially I do not want to deny that the theory of orders is

---

[a]Never published (personal communication to J. Dawson, 14 August 2000).

[b] *Gödel 1944*.

[c] *Whitehead and Russell 1910*.

constructive while the theory of types is not, but this seems to me to be due to other reasons than the fact that the VCP relates to one and not the other.

I should be most glad to know whether you disagree with these conclusions or not, because the referee has interpreted your article as establishing: (1) that Russell did not take VCP as the source of both theories and (2) that in fact there is no connection at all between VCP and the theory of types.

Yours sincerely,

Leonard Goddard

2. Gödel to Goddard[a]

Dear Mr. Goddard,

It is true that Russell in the first sentence of Ch. II, sect V of the Introduction ⟨to PM⟩ claims that the simple theory of types follows also from the vicious circle principle. But this is incorrect and no such derivation is found in previous sections. It seems to me that Russell does not take to⟦o⟧ seriously the considerations preliminary to setting up the formal system (probably because one does not have a precise logic at this stage or for this purpose.)

It is correct that something which might perhaps be called a weakened form of the simple theory of types does follow from the vicious circle principle + the principle that a function presupposes ~~its~~ ⟨the totality of it/s possible⟩ arguments. It is ⟨substantially⟩ the theory of "ranks" of set theory, ⟨namely a theory⟩ which differs ~~from~~ | from the simple theory   2
of types ~~1. by considering proposition sentences with faulty type combinations as wrong (instead of meaningless) 2.~~ by admitting ~~iterated~~ mixtures of preceding types ~~the~~ and transfinite types. ~~Since however the negation of item 1. is quite essential for the theory of types, it is doubt- ful that the name is appropriate~~ Note, however, ⟨that⟩ mixtures of types flatly contradict the main idea of the ⟨simple⟩ theory of types, as it is commonly understood. ~~Many logicians, I would expect, will find the name in~~

Sincerely,

⟦No signature⟧

---

[a] Apparently unsent draft.

Marianne Gödel

Marianne Gödel

Shortly after Gödel's death his wife Adele reportedly destroyed all of his mother's letters to him, so that of their correspondence only his letters to her survive. Those that are extant extend from September 1945 until Marianne Gödel's death in July 1966. Numbering 245 in all, they are preserved in the Wiener Stadt- und Landesbibliothek.

Unsurprisingly, the letters are concerned primarily with family matters and daily affairs, and as such are an important biographical resource. Otherwise, however, they are of little scholarly interest, since Marianne Gödel, though a cultured woman for her time, was not one with whom her son was wont to discuss intellectual issues. The five letters excerpted here, in which Gödel expounds upon his religious views, are a singular exception.

As an infant, Gödel was baptized in the Lutheran church in Brno, and as a youth he attended Protestant schools there. But his parents were not churchgoers, and in the first of the letters excerpted below he disparages the religious instruction he experienced in school. At the same time he shared his mother's disdain for the Catholic Church and its influence in Austrian political affairs.[a]

Nevertheless, unlike his brother Rudolf, Gödel was not unmoved by religious concerns. On the contrary, his library included many books and tracts devoted to various religious sects; among his notebooks are two devoted to theology;[b] and in a shorthand manuscript found in his *Nachlaß*[c] he remarked that "Die Religionen sind zum größten Teil schlecht, aber nicht die Religion." ("Religions are for the most part bad, but not religion itself.") In his letter to his mother of 12 September 1961 (excerpted below) he decried the efforts of "contemporary philosophers", 90 per cent of whom, he contended, "see their principal task to be that

---

[a]In his letter to her of 18 March 1961 (not reproduced here) he contrasted religious attitudes in America with those in Europe; there, he declared, religion was not a matter of conviction, but of manifestly harmful Catholic political parties. ("In Europa macht sich die Religion auch noch stark bemerkbar, aber nicht als Überzeugung, sondern in Form katholischer politischer Parteien, was offenbar schädlich ist.")

[b]Numbered 1 and 3, they are contained in *Nachlaß* folders 03/107 and 03/108. Volume 2 is missing.

[c]Item #060186, folder 06/15.

of beating religion out of men's heads"; and in his posthumously published essay *1961/?*, written about the same time as the last four of the
letters included here, he spoke out, unusually polemically, against the
"leftward" drift of philosophical world-views "away from metaphysics
(or religion)".

In his (unsent) reply to the Grandjean questionnaire, p. 448, item
13 in this volume, Gödel described his own religious beliefs as "theistic rather than pantheistic, following Leibniz rather than Spinoza"; but
his innate caution prevented him from giving public expression to those
views. He refrained from publishing his formalization (*1970*) of Leibniz's ontological argument (so he told his friend Oskar Morgenstern)
because he feared that if he did so it would be thought that he "actually believed in God", whereas, he claimed, he had carried out that
work only as a "logical investigation".[d] Two years later he declined a request to submit a theological paper to a journal, on the grounds that he
had "not yet worked out ... matters sufficiently" and that, especially in
theology, "publication of half-finished work would do more harm than
good."[e] No wonder, then, that he discussed his views concerning an
afterlife only in his letters to his mother.

---

[d]Quoted by Morgenstern in his diary entry of 29 August 1970 (Morgenstern
Papers, Box 15, Special Collections Department, Duke University Library, Durham,
North Carolina).

[e]Gödel to Paul A. Lee, 7 September 1972 (*Nachlaß* item 011416.4, folder 01/104).

---

1. Gödel to Marianne Gödel

Princeton, 27./II. 1950.

Liebste Mama!

* * *

... Was Du über *Einsteins Biographie* schreibst, stimmt, so viel mir be-
2 |kannt, soweit, dass der Anblick eines *Kompasses* als Kind zuerst das In-
teresse für Physik ⟨in ihm⟩ erweckte u. dass durch den *Religionsunter-*

Mindful of his mother's lack of training in philosophy, Gödel wrote to her informally and somewhat apologetically. His remarks are ostensibly speculative ("If the world is rationally organized and has a sense ...") and will likely strike more sophisticated readers as philosophically naïve. They combine certain elements of traditional Christian belief (man's fallibility in this world and his perfectibility in the next) with a version of the anthropic principle ("if God had created beings in our place that had nothing to learn ... we wouldn't then exist at all"). Above all, they reflect Gödel's abiding "rationalistic optimism" (to borrow Hao Wang's term): his belief that "the world is not at all chaotic and capricious", but that "everything has a cause" (a tenet on which—contrary to the doctrines of quantum mechanics—he claimed "the whole of science rests"); his faith that in the next world "we will perceive everything of importance with unerring certainty"; and his suggestion that "our incredible ignorance about ourselves" might be remedied if only "we could ... look deeply enough within ourselves with scientific methods of self-examination"—the sort of thing he suggested in *1961/?* might be achieved through cultivation of Husserl's ideas.

The views expressed in these letters must not be judged according to the standards of academic philosophical discourse. Nonetheless, they are consistent with Gödel's personality and with the overall world-view that he repeatedly expressed in his scientific writings.

John W. Dawson, Jr.

1. Gödel to Marianne Gödel

Princeton, 27 February 1950

Dearest Mama,

* * *

... What you write about *Einstein's biography* is correct, so far as I know, to the extent that the sight of a *compass* as a child first awakened in him the interest in physics and that ⟨for him⟩ the basis of the endeavor to seek

richt ⟨bei ihm⟩ der Grund zu dem Bestreben gelegt wurde, nach einer
einheitlichen Theorie für die ganze Welt zu suchen. Das muss allerdings
ein ⟨sehr⟩ guter u. interessanter Religionsunterricht gewesen sein. Denn
bei einem solchem, wie wir ihn hatten, wäre das wohl kaum möglich gewesen. Was Du über das Vergehen der Zeit in Deinem Bericht über das
Schankalsche Buch[a] schreibst hängt sehr nahe mit dem Thema meines
Aufsatzes im Einstein-Band[b] zusammen. Mit der Traurigkeit hast Du
recht: Wenn es eine vollständig hoffnungslose Traurigkeit gäbe, so wäre
nichts Schönes mehr in ihr. Aber ich glaube eine solche kann es vernünftiger
Weise gar nicht geben. Denn wir verstehen weder warum diese Welt existiert, noch warum sie gerade so beschaffen ist wie sie ist, noch warum
wir in ihr sind, noch | warum wir gerade in diese u. keine anderen äusseren
Verhältnisse hineingeboren wurden. Warum also sollen wir uns einbilden,
gerade das eine ganz bestimmt zu wissen, dass es keine andere Welt gibt
u. dass wir nie in einer anderen waren noch sein werden?

3

* * *

Mit tausend Bussis

immer Dein Kurt

---

[a]The editors have been unable to identify this book.

2. Gödel to Marianne Gödel

Princeton, 23./VII. 1961.

Liebste Mama!

* * *

Du stellst in Deinem letzten Brief die schwerwiegende Frage, ob ich an
ein Wiedersehen glaube. Darüber kann ich nur folgendes sagen: Wenn
die Welt vernünftig eingerichtet ist u. einen Sinn hat, dann muss es das
geben. Denn was sollte es für einen Sinn haben ein~~e~~ Wesen (den Menschen) hervorzubringen, der ein so weites Feld von Möglichkeiten der
eigenen Entwicklung u. der Beziehungen zu andern hat, u. ihn dann nicht
einmalen 1/1000 davon erreichen zu lassen. Das wäre ungefähr so, als

a unified theory for the whole world was laid through *religious instruc-tion.* That must certainly have been a very good and interesting [kind of] religious instruction. For with such a one as we had that would hardly have been possible. What you write about the passage of time in your re-port on *Schankal*'s book[a] is very closely connected with the theme of my essay in the Einstein volume.[b] You are right about sadness: If there were a completely hopeless sadness, there would no more be anything beauti-ful in it. But I think that from a rational point of view there cannot be such a thing at all. For we understand neither why this world exists, nor why it is constituted just as it is, nor why we are in it, nor why we were born in just these and no other external circumstances. Why then should we fancy that we know precisely the one thing for sure, that there is no other world and that we never were nor ever will be in another?

* * *

With a thousand kisses,

ever yours, Kurt

---

[b] *Gödel 1949a* or *1955*.

2. Gödel to Marianne Gödel

Princeton, 23 July 1961

Dearest Mama,

* * *

In your last letter you pose the weighty question whether I believe we shall see each other again [in a hereafter]. About that I can only say the following: If the world is rationally organized and has a sense, then that must be so. For what sense would it make to bring forth a being (man) who has such a wide range of possibilities of individual development and of relations to others and then allow him to achieve not one in a thou-sand of those? That would be much as if someone laid the foundation for

5   wenn jemand mit grösster | Mühe u. Kostenaufwand den Grund für ein
    Haus legen u. dann alles wieder verkommen lassen würde. Hat man aber
    einen Grund anzunehmen, *dass* die Welt vernünftig eingerichtet ist? Ich
    glaube ja. Denn sie ist durchaus nicht chaotisch u. willkürlich, sondern es
    herrscht, wie die Wissenschaft zeigt, in allem die grösste Regelmässigkeit
    u. Ordnung. Ordnung ist aber eine Form der Vernünftigkeit. Wie ein an-
    deres Leben zu denken ist? Darüber gibt es natürlich nur Vermutungen.
    Aber es ist interessant, dass gerade die moderne Wissenschaft Stützen
    dafür liefert. Denn sie zeigt, dass diese unsere Welt mit allen Sternen u.
    Planeten, die darin sind, einen Anfang gehabt hat u. aller Wahrschein-
    lichkeit nach, ein Ende haben wird.[1] Warum soll es aber dann nur diese
6   eine Welt gebe u. da wir uns in dieser Welt eines Tages vorgefun|den ha-
    ben, ohne zu wissen wieso u. wohin, so kann dasselbe auf dieselbe Weise
    auch im einer andern sich wiederholen. Die Wissenschaft bestätigt jeden-
    falls den im letzten Buch der Bibel vorausgesagten Weltuntergang u. lässt
    Raum für das, was dann folgt: "Und Gott schuf einen neuen Himmel u.
    eine neue Erde". Natürlich kann man fragen: Wozu diese Verdoppelung,
    wenn die Welt vernünftig eingerichtet ist? Aber auch darauf gibt es sehr
    gute Antworten. So, jetzt habe ich Dir also einen philosophischen Vor-
    trag gehalten u. hoffe Du hast ihn verständlich gefunden.

            Mit tausend Bussis u. herzlichen Grüssen an Rudi

                        immer Dein Kurt

                            * * *

[1]D.h., buchstäblich zu "nichts" werden wird.

3. Gödel to Marianne Gödel

                                            Princeton, 14./VIII. 1961

        Liebste Mama!

                            * * *

    Wenn Du schreibst, Du betest die Schöpfung an, so meinst Du wahr-
scheinlich, dass die Welt überall schön ist, wo der Mensch nicht hinkommt
etc. Aber gerade das könnte des Rätsels Lösung enthalten, warum es

a house with the greatest trouble and expense and then let everything go to ruin again. But do we have reason to assume that the world is rationally organized? I think so. For the world is not at all chaotic and capricious, but rather, as science shows, the greatest regularity and order prevails in all things; [and] order is but a form of rationality.

How is another life to be imagined? About that there are of course only conjectures. But it is interesting that modern science is the very thing that provides support for them. For it shows that this world of ours, with all the stars and planets that are in it, had a beginning and, in all probability, will have an end.[1] But why then should there be only this one world? And since we one day found ourselves in this world, without knowing how [we got here] and whither [we are going], the same thing can be repeated in the same way in another [world] too. In any case, science confirms the end of the world prophesied in the last book of the Bible and allows room for what follows next: "And God created a new heaven and a new earth". Of course, one can ask: Why this duplication, when the world is rationally organized? But to that also there are very good answers. So now I've delivered a philosophical lecture to you, and I hope you've found it understandable.

With a thousand kisses and hearty greetings to Rudi,

ever yours, Kurt

* * *

[1] That is, will literally become "nothing".

3. Gödel to Marianne Gödel

Princeton, 14 August 1961

Dearest Mama,

* * *

When you write that you worship the creation, you probably mean that the world is beautiful everywhere that man has not come along, etc. But precisely that could contain the solution of the riddle why there are

zwei Welten gibt. Tiere u. Pflanzen, im Gegensatz zum Menschen, haben
nur in geringem Masse die Fähigkeit zu lernen, ⟨leblose Dinge überhaupt
keine.⟩ Nur der Mensch kann durch Lernen zu einer besseren Existenz
kommen, d. h. seinem Leben mehr Sinn geben. Eine, u. oft die einzige,
4   Methode, zu | lernen, besteht aber darin, es erst falsch zu machen. Und
das geschieht ja in dieser Welt wirklich in ausreichendem Masse. Jetzt
kann man natürlich fragen: Warum hat Gott den Menschen nicht so erschaffen, dass er gleich vom Anfang an alles richtig macht? Aber dass
uns diese Frage berechtigt vorkommt, könnte sehr leicht seinen einzigen
Grund in der unglaublichen Unwissenheit haben, in der wir uns über
uns selbst heute noch befinden. Wir wissen ja nicht nur nicht, woher u.
warum wir da sind, sondern auch nicht *was* wir sind (nämlich im Wesen
u. von innen gesehen). Wenn man aber einmal mit wissenschaftlichen
Methoden der Selbstbetrachtung genug tief in sich hineinblicken könnte,
um diese Frage zu beantworten, so würde sich doch wahrscheinlich her-
ausstellen, dass jeder von uns ein etwas mit ganz bestimmten Eigenschaften ist. D. h. jeder könnte dann von sich sagen: Unter allen möglichen
Wesen bin "ich" gerade diese so u. so beschaffene Verbindung von Ei-
5   genschaften. | Wenn dann aber zu diesen Eigenschaften gehört, dass wir
nicht alles gleich richtig machen, sondern vieles erst auf Grund von Erfahrung, so folgt, dass, wenn Gott an unserer Stelle Wesen erschaffen
hätte, die nichts zu lernen brauchen,[1] diese Wesen eben nicht *wir* wären.
D. h. wir würden dann überhaupt nicht existieren. Nach der üblichen
Auffassung wäre die Frage "Was bin ich" dahin zu beantworten, dass
ich ein etwas bin das ⟨von sich aus überhaupt⟩ keine Eigenschaften hat,
so etwas ähnliches wie ein Kleiderstock, an den man beliebige Kleider
anhängen kann. Man könnte über all diese Dinge natürlich noch sehr
viel sagen. Ich glaube in der Religion, wenn auch nicht in den Kirchen,
liegt viel mehr Vernunft als man gewöhnlich glaubt, aber wir[2] werden
von frühester Jugend an zum *Vorurteil dagegen* erzogen durch die Schule,
den schlechten Religionsunterricht, durch Bücher u. Erlebnisse.

Mit tausend Bussis immer Dein Kurt

---

[1] Es ist natürlich anzunehmen dass solche (oder nahezu solche) Wesen auch ir-
gendwo existieren oder existieren werden.

[2] d. h. die Mittelschichte der Menschheit, der wir angehören, oder zumindest die
meisten Menschen in dieser Schichte.

two worlds. Animals and plants, in contrast to man, have the capacity to learn only to a meager extent, lifeless things not at all. Only man can arrive at a better existence through learning, that is, his life has more meaning. One, and often the only, method of learning consists, however, in first making mistakes. And indeed, in this world that actually happens in abundant measure. Now one can of course ask: Why didn't God create man so that he does everything right immediately from the beginning? But that this question appears justified to us could very easily have its own basis in the incredible state of ignorance about ourselves that we find ourselves in even today. We not only don't know where we came from and why we are here, we also don't know what we are (that is to say, in essence and seen from within). But if we could once look deeply enough within ourselves with scientific methods of self-examination in order to answer this question, it would probably turn out that each of us is a thing with quite definite characteristics. That is, each one could then say of himself: Among all possible beings "I" am precisely this such-and-such constituted combination of characteristics. But then if one of those characteristics is that we do not do everything right immediately, but many times only on the basis of experience, it follows that if God had created in our place beings that had nothing to learn,[1] we just wouldn't be those beings. That is, we wouldn't then exist at all. According to the usual conception, the question "What am I?" is supposed to be answered, in effect, [by saying] that I am a thing that, in and of itself, has no characteristics at all, something similar to a clothes rod on which one can hang any clothes whatever. One could of course say a lot more about all these things. I think in religion, if not also in the churches, lies much more reason than is usually thought, but from earliest youth on we[2] are brought up with *prejudice against* it through the schools, the bad religious instruction, through books and experiences.

With a thousand kisses, ever yours, Kurt

---

[1] It is of course assumed that such (or nearly such) beings also exist somewhere, or will exist.

[2] That is, the middle class of people, to which we belong, or at least most people in this class.

4. Gödel to Marianne Gödel

Princeton, 12./IX. 1961

Liebste Mama!

* * *

Dass es Dir Schwierigkeiten gemacht hat, den "theologischen" Teil
meines letzten Briefes zu verstehen, das ist ja ganz selbstverständlich u.
4 hat nichts mit Deinem Alter zu tun. Ich habe mich ja | sehr kurz gefasst
u. manche ziemlich tiefliegenden philosophischen Fragen berührt. Auf
den ersten Blick sieht ja diese ganze Anschauung, die ich Dir auseinander-
setzte, höchst unwahrscheinlich aus. Aber ich glaube, wenn man genauer
darüber nachdenkt, so stellt sie sich als durchaus möglich u. vernünftig
heraus. Insbesondere muss man sich vorstellen, dass das "Lernen" zum
grossen Teil erst in der nächsten Welt stattfinden wird, nämlich dadurch,
dass wir uns an unsere Erlebnisse in dieser Welt erinnern u. diese erst
dann wirklich verstehen werden, so dass unsere hiesigen Erlebnisse sozusagen
nur das Rohmaterial für das Lernen sind. Denn was könnte z. B. ein
Krebskranker *hier* aus seinen Schmerzen lernen? Es ist aber durchaus
denkbar, dass ihm in der zweiten Welt klar werden wird, durch welche
Fehler seinerseits (nicht in hygienischer, sondern vielleicht in ganz an-
5 derer Hinsicht) diese Krankheit verursacht wurde u. dass | er dadurch
nicht nur diesen Zusammenhang mit seiner Krankheit, sondern zugleich
andere ähnliche Zusammenhänge verstehen lernt. Natürlich setzt das
voraus, dass es viele Zusammenhänge gibt, von denen sich die heutige
Wissenschaft u. Schulweisheit nichts träumen lässt. Aber davon bin ich
auch unabhängig von jeder Theologie überzeugt. Hat doch sogar der
Atheist Schopenhauer einen Artikel über die "scheinbare Absichtlichkeit
im Schicksal des Einzelnen"[a] geschrieben. Wenn man einwendet, es sei
unmöglich, dass wir uns in einer andern Welt an die Erlebnisse in dieser erinnern, so ist das ganz unberechtigt, denn wir könnten ja in der
andern Welt schon mit diesen latenten Erinnerungen geboren werden.
Ausserdem muss man natürlich annehmen, dass unser Verstand dort
6 wesentlich besser sein wird als hier, so dass wir alles Wichtige | mit der-
selben untrüglichen Sicherheit erkennen werden, wie $2 \times 2 = 4$,[1] wo eine

---

[1]Sonst könnten wir ja z. B. gar nicht wissen, ob wir in der andern Welt nicht auch
sterben werden.

---

[a]*Schopenhauer 1850.*

4. Gödel to Marianne Gödel

Princeton, 12 September 1961

Dearest Mama,

* * *

That you had difficulties in understanding the "theological" part of
my last letter is quite natural and has nothing to do with your age. I expressed myself very succinctly and touched upon many rather profound
philosophical questions. At first glance that whole view that I set forth
to you seems most improbable. But I think that if one reflects upon it
more closely it turns out to be entirely possible and rational. One must
in particular imagine that the "learning" will in large part take place
only in the next world, namely in this way, that we will recall our experiences in this world and only then really understand them, so that our
experiences here are, so to speak, only the raw material for the learning.
For what, for example, could a cancer victim learn here from his pains?
It is entirely conceivable, however, that in the second world it will be
clear to him through what mistakes of his (not in hygienic matters, but
perhaps in quite other respects) that sickness was caused, and that he
thereby learns to understand not only that connection with his illness,
but at the same time other similar connections. That assumes, of course,
that there are many connections of which modern science and academic
wisdom cannot dream at all. But of that I am also convinced independently of every theology. After all, even the atheist Schopenhauer wrote
an article about the "apparent purposefulness in the fate of individuals".[a]
If someone protests that it is not possible that we recall in another world
the experiences in this one, that is quite unjustified, because we could be
born in the other world already with those latent recollections. Besides,
one must of course assume that our understanding there will be better
than it is here, so that we will perceive everything of importance with the
same unerring certainty as that $2 \times 2 = 4$,[1] where delusion is objectively

---

[1] Otherwise, for example, we couldn't know at all whether we won't also die in the
other world.

Täuschung objektiv ausgeschlossen ist. So können wir dann auch absolut
sicher sein, alles wirklich erlebt zu haben, woran wir uns erinnern. Aber
ich fürchte ich komme wieder etwas zu viel in die Philosophie hinein. Ich
weiss nicht, ob man die letzten 10 Zeilen überhaupt verstehen kann, ohne
Philosophie studiert zu haben. N. B. hilft auch das heutige Philosophie-
studium nicht viel zum Verständnis solcher Fragen, da ja 90% der heutigen Philosphen ihre Hauptaufgabe darin sehen, den Menschen die Religion aus dem Kopf zu schlagen, u. dadurch im selben Sinne wirken wie
die schlechten Kirchen.

   Mit tausend Bussis

                                             immer Dein Kurt

Viele herzliche Grüsse an Rudi

5. Gödel to Marianne Gödel

                                             Princeton, 6./X. 1961

        Liebste Mama!

                            * * *

   Die religiösen Ansichten ⟨über⟩ die ich Dir schrieb, haben nicht mit
Okkultismus zu tun. Der religiöse Okkultismus besteht darin, in spiri-
tistischen Sitzungen den Geist des Apostels Paulus oder den Erzengel
Michael ⟨etc.⟩ zu zitieren u. von ihnen Auskünfte über religiöse Fragen
einzuholen. Was ich Dir schrieb, ist ja nichts als eine anschauliche Darstellung u. so zu sagen eine "Adaptierung" Anpassung an unsere heutiger
Denk⟨weise⟩ von gewissen theologischen Lehren, die seit 2000 Jahre gepredigt
werden, allerdings mit vielem Unsinn gemischt. Wenn man liest, was so
im Laufe der Zeit in den verschiedenen Kirchen als Dogma behauptet
wurde ⟨u. noch wird,⟩ muss man sich freilich wundern. Z. B. hat nach
katholischem Dogma der allgütige Gott die meisten Menschen ausschliesslich
zu dem Zweck geschaffen, um sie für alle Ewigkeit in die Hölle zu schicken,
nämlich alle ausser den guten Katholiken, die ja auch von den Katholiken
5   nur ein Bruch|teil sind. Ich glaube nicht, dass die Anwendung des Ver-
standes in irgend einem Gebiet etwas Ungesundes ist (wie Du andeutest).

excluded. We can then also be absolutely sure really to have experienced everything that we recall. But I fear I am again entering somewhat too much into philosophy. I don't know whether the last ten lines can be understood at all without one's having studied philosophy. N. B.: The contemporary study of philosophy also doesn't help much for understanding such questions, since 90% of contemporary philosophers see their principal task to be that of beating religion out of men's heads, and in that way have the same effect as the bad churches.

* * *

With a thousand kisses,

ever yours, Kurt

Many hearty greetings to Rudi

5. Gödel to Marianne Gödel

Princeton, 6 October 1961

Dearest Mama,

* * *

The religious views that I wrote you about have nothing to do with occultism. Religious occultism consists in conjuring up the ghost of the Apostle Paul, or the Archangel Michael, etc., in spiritualistic séances and seeking information about religious questions from them. What I wrote you is nothing but a concrete representation and an adaptation to our present way of thinking of certain theological teachings that have been preached for two thousand years, though mixed with much nonsense. If one reads what in the course of time has been, and still is, asserted as dogma in the various churches, one must be truly amazed. For example, according to Catholic dogma the all-benevolent God created most people—namely, all except the good Catholics, which even of the Catholics are only a fraction—expressly for the purpose of sending them to Hell for all eternity. I don't think that the application of the understanding in any domain is something unsound (as you intimate). It

Es wäre auch ganz unberechtigt, zu sagen, dass man gerade in diesem Gebiet mit dem Verstande nichts ausrichten kann. Denn wer hätte vor 3000 Jahren geglaubt, dass man von den fernsten Sternen wird bestimmen können, wie gross, wie schwer, wie heiss u. wie weit entfernt sie sind u. dass viele von ihnen 100mal grösser sind als die Sonne. Oder wer hätte geglaubt, dass man Fernsehapparate bauen wird? Als vor 2500 Jahren zuerst die Lehre aufgestellt wurde, dass die Körper aus Atomen bestehen, muss das damals ebenso phantastisch u. unbegründet ausgesehen haben als heute vielen die religiösen Lehren erscheinen. Denn es war damals buchstäblich keine einzige Beobachtungstatsache bekannt, welche zur Aufstellung der Atomtheorie hätte veranlassen können, sondern das geschah aus rein philosophischen Gründen. Nichts desto weniger hat sich diese Theorie heute glänzend bestätigt u. ist die Grundlage für einen sehr

6 grossen Teil der | modernen Wissenschaft geworden.—Man ist natürlich heute weit davon entfernt, das theologische Weltbild wissenschaftlich begründen zu können, aber ich glaube, schon heute dürfte es möglich sein rein verstandesmässig (ohne sich auf den Glauben u. irgend eine Religion zu stützen) einzusehen, dass die theologische Weltanschauung mit allen bekannten Tatsachen (einschliesslich den Zuständen, die auf unserer Erde herrschen) durchaus vereinbar ist. Das hat schon vor 250 Jahren der berühmte Philosoph u. Mathematiker ⟨Leibniz⟩ zu tun versucht, u. das ist es auch, was ich in meinen letzten Briefen versucht habe. Was ich theologische Weltanschauung nenne, ist die Vorstellung, dass die Welt u. alles in ihr Sinn u. Vernunft hat, u. zwar einen guten u. zweifellosen Sinn. Daraus folgt unmittelbar, dass unser Erdendasein, da es an sich höchstens einen sehr zweifelhaften Sinn hat, nur Mittel zum Zweck für eine andere Existenz sein kann. Die Vorstellung, dass alles in der Welt einen Sinn hat, ist übrigens genau analog dem Prinzip, dass alles eine Ursache hat, worauf die ganze Wissenschaft beruht.

Mit tausend Bussis

immer Dein Kurt

* * *

would be quite unjustified to say that precisely in this domain one can achieve nothing with the understanding. For who would have thought, three thousand years ago, that one would be able to determine of the most distant stars how large, how heavy, how hot and how far away they are, and that many of them are a hundred times larger than the sun. Or who would have believed that someone would build television sets? When, 2500 years ago, the doctrine was first enunciated that bodies consist of atoms, that must at that time have seemed just as fantastic and unfounded as many of the religious doctrines appear today. For at that time there was literally not a single known observational fact that could have led to the enunciation of the atomic theory; rather, that happened on purely philosophical grounds. Nevertheless, that theory has today been brilliantly confirmed and has become the basis for a very large part of modern science.—Of course, today we are far from being able to justify the theological world view scientifically, but I think already today it may be possible purely rationally (without the support of faith and any sort of religion) to apprehend that the theological world view is thoroughly compatible with all known facts (including the conditions that prevail on our earth). Two hundred fifty years ago the famous philosopher and mathematician Leibniz already tried to do that, and that is also what I have attempted in my last letters. What I call the theological world-view is the idea that the world and everything in it has meaning and reason [to it], and in fact a good and indubitable meaning. From that it follows directly that our earthly existence, since it in itself has a very doubtful meaning, can only be a means toward the goal of another existence. The idea that everything in the world has meaning is, after all, precisely analogous to the principle that everything has a cause, on which the whole of science rests.

With a thousand kisses,

ever yours, Kurt

* * *

Burke D. Grandjean

Prior to his death, only the scantiest information about Gödel was available in published sources; interest focussed on his work rather than his life, and Gödel's personal remoteness tended to deter biographical inquiries. In 1974, however, that dearth of information prompted Burke D. Grandjean, then a doctoral student in sociology at the University of Texas, to send Gödel a questionnaire dealing with his youth and intellectual development.

Unknown to Grandjean, the questionnaire reached Gödel shortly after his recovery from a serious prostate crisis and only a short time before the beginning of his final physical and mental decline.[a] No wonder, then, that Gödel did not at first respond.

But Grandjean persisted. He sent a number of reminder letters and, on 24 February 1975, spoke with Gödel by phone. Another reminder letter followed in May, after which Gödel finally composed a typewritten reply (his letter of 19 August 1975). Characteristically, however, he held back from sending it.

The two spoke again by telephone on 24 September, and Grandjean, who had by then accepted a one-year appointment at the Russell Sage Foundation in New York, sent yet another reminder letter on 20 October. It seemed to him that Gödel "was obviously pleased" at being asked about his early life and intellectual influences but "was ultimately unwilling to entrust his recollections of those matters" to him[b]—a view that may explain Gödel's action on 2 December 1975, when he telephoned Grandjean once more to request further information on *his* background and research interests.

The following month Gödel finally sent off the letter he had written five months before, dutifully crossing out the word "August" and replacing it by "Jan". Along with it he enclosed a page of handwritten addenda, not found in his *Nachlaß* nor previously published, which has been incorporated into the text printed here.

Gödel never did return the questionnaire, but he did fill it out: It was found among his papers after his death, together with an undated draft of a letter to Grandjean containing some further amplificatory remarks.

[a]For details, see *Dawson 1997*, pp. 245–253.

[b]Personal communication to J. Dawson, 27 March 1997.

Both items are reproduced below. Together with the letter of 19 August, they constitute Gödel's only personal commentary on his own early life; all three first appeared in print in *Wang 1987*.

John W. Dawson, Jr.

1. Grandjean to Gödel

University of Texas at Austin
Austin, Texas 78712
July 16, 1974

Dear Professor Gödel:

I am presently engaged in research concerning the social and intellectual atmosphere in central Europe during the first third of the twentieth century. Your work in the 1930's, in particular your seminal paper "On Formally Undecidable Propositions ...,"[a] constitutes an important facet of this research. However, treatments of your personal and intellectual background in the sources available to me are inadequate for my purposes. In fact, the only biographical information I have about you comes from a 1952 article in the *New Yorker*,[b] or from brief sketches in *Who's Who* and in a volume commemorating your sixtieth birthday (edited by Bulloff, Holyoke, and Hahn, 1969).[c]

I am therefore writing this letter to ask if you will assist me in two ways:

(1) If you are aware of a more complete biographical source which I may have overlooked, would you please send the reference to me?

(2) I would greatly appreciate it if you could find the time to answer some or all of the specific questions on the enclosed sheet[d] and mail your responses to me. You should of course feel entirely justified in ignoring any question which, for whatever reason, you would prefer not to answer. Furthermore, I currently have no plans to publish

---

[a] *1931*.

[b] Titled "Inexhaustible", the article appeared in the *New Yorker*, 23 April 1952, pp. 13–15.

[c] *Bulloff, Holyoke and Hahn 1969*.

[d] See item 4 below.

the results of my research, but should I decide to do so, you have my assurance that none of the information you supply to me would be used without your permission.

Please let me know whether or not you can be of assistance to me in either of these ways. Thank you very much.

Sincerely,

Burke D. Grandjean

2. Gödel to Grandjean

August 19, 1975[a]

Dear Mr. Grandjean:

Replying to your inquiries I would like to say first that I *don't* consider my work a "facet of the intellectual atmosphere of the early 20th century," but rather the opposite.[1] It is true that my interest in the foundations of mathematics was aroused by the "Vienna Circle," but the philosophical consequences of my results, as well as the heuristic principles leading to them, are anything but positivistic or empiristic. See what I say in Hao Wang's recent book "From Mathematics to Philosophy"[b] in the passages cited in the Preface. See also my paper "What is Cantor's Continuum Problem?"[c] in "Philosophy of Mathematics" edited by Benacerraf and Putnam in 1964,[d] in particular pp. 262–265 and pp. 270–272.

---

[1]This is demonstrably true at least for the heuristic principles which led to my results ⟨which are Platonistic⟩ (see the passsages of my writings noted in this letter ⟨in particular p⟦p⟧. 8–13 of W̶ Hao Wang's book) ⟨But exactly in Positivism *"leading to verifiable consequences"* is the most important criterion of truth. ~~Moreover I believe I have not shown (as is usually said) that mathematical reason is limited [but, on the contrary, that it is "boundless" (see the same passages)].~~ So my work ~~fits with positivism or empiricism only in a certain interpretation while in my own interpretation it~~ points toward an entirely different world view.

---

[a]A handwritten correction to the copy received by Grandjean had 'August' crossed out and 'Jan' written above. Overwriting on the '1975' may have been intended to alter it to '1976'.

[b] *Wang 1974.*

[c] *1964.*

[d] *Benacerraf and Putnam 1964.*

I was a conceptual and mathematical realist since about 1925. I never held the view that mathematics is syntax of language. Rather this view, understood in any reasonable sense, can be *disproved* by my results.

I am enclosing a brief biographical sketch. I never studied in Brunn.[2] My interest in mathematics and philosophy started in high school at the age of about 15. None of the scholars mentioned under 5 in your questionnaire, except perhaps Kant, had any direct influence on the development of my interests.[3]

I am of German-Austrian descent, but my family and I were always friendly disposed toward the Jews. My family was upper middle class. My father was engaged in the manufacturing industry.

Sincerely yours,

Kurt Gödel

P.S. Thank you for the material you sent me.

[2] Except for high school.
[3] Also the interest of the Vienna Circle in the foundations of mathematics began long before Wittgenstein's Tractatus.[d]

.

---

[d] *Wittgenstein 1922.*

3. Grandjean to Gödel

Russell Sage Foundation
230 Park Avenue
New York, NY 10017
December 2, 1975

Dear Professor Gödel:

Thank you for your telephone call this morning. I am happy to give you some additional details about myself and my research on European intellectual currents in the first third of this century.

Enclosed is a copy of my *curriculum vitae,* summarizing my background and academic experience. As I mentioned on the telephone, I am currently devoting most of my attention to completing my Ph.D. dissertation, a study of the attitudes of faculty members in three different types of educational organizations: nursing schools, high schools, and

university departments of sociology. This research, like my previous articles on social stratification in Italy and on cross-national differences in systems of education, relies primarily on statistical analysis of large-scale survey data. (Copies of these publications are enclosed.[a])

⊔   For some ∧ time, however, I have been pursuing a very different type of research in connection with my long-term interest in European intellectual history. This, of course, is the study about which I have been in communication with you. My major goal for the research is simply to learn as much as I can about a topic which fascinates me—the social and intellectual climate in central Europe during an extremely creative and productive period (approximately 1900–1935). Toward this end I have been reading works by and about outstanding scholars of the time. Eventually I plan to distill what I have learned into an article or short monograph, and if I am satisfied with the results, submit it for publication. I have not yet settled on a single central theme for any such article, but my interest has come to focus on a group of scholars, yourself included, who are often referred to as affiliated with the Vienna Circle. In particular, I believe that some of these scholars' ideas are in fact very different from the positivism characteristic of the Vienna Circle. For rather distinct reasons, Wittgenstein's works, your own, and those of a few other scholars seem to diverge in important respects from the works of the Viennese positivists. My reading of some of your work leads me to this conclusion, but the biographical material available to me is inadequate for a thorough understanding of your intellectual and philosophical position. As a result, I began writing you last summer in an effort to learn more about your background and ideas. Available sources do a much better job of supplying information about some of your contemporaries than they do about yourself.

| William Johnston's *The Austrian Mind* (1972)[b] is a masterful survey of the lives and work of a large number of highly creative individuals of the time, as is *Wittgenstein's Vienna* (1973),[c] by Allan Janik and Stephen Toulmin. The latter book, in fact, argues that Wittgenstein's ideas are considerably less positivistic than those who have discussed him in connection with the Vienna Circle have indicated. Yet neither these nor other sources of which I am aware treat your own work or philosophical position in sufficient detail for me to make informed comparisons of your ideas with Carnap's, say, or Wittgenstein's, or those of any number of other scholars of the period. For this reason I sincerely hope that

    2

---

[a]No such copies were found in Gödel's *Nachlaß*.

[b] *Johnston 1972.*

[c] *Janik and Toulmin 1973.*

you will see fit to provide me with as much of the information I have requested as you think is appropriate. As I have previously assured you, I
will give you absolute veto power over the publication in any form of any
of a̶n̶y̶ o̶f̶ the information you supply me.

If I can tell you anything further about myself or my research, please
write or telephone and I shall be pleased to do so. We might even set
up a personal meeting to get to know each other better; I could be in
Princeton at your convenience for such a meeting. If you wish, you may
write or call the individuals in the list of references on my vita, or
Dr. Hugh F. Cline, president of the Russell Sage Foundation. Professor
Gideon Sjoberg at the University of Texas has been advising me in my
research on this topic, so you might also contact him about me.

I look forward to hearing from you again and, if you see fit, to receiving your answers to my written questions. Thank you.

Sincerely,

Burke D. Grandjean

4. Grandjean's questionnaire

Intellectual and Biographical Information: Kurt Gödel

1. Please sketch for me your educational background. My research
   indicates that your primary schooling was at Brünn (now Brno,
   Czechoslovakia), that you took a degree in engineering at a Brünn
   technical college, and that in 1924 you began your studies in mathematics at the University of Vienna, where you received the Ph.D.
   degree in 1930. How complete and accurate is this account? [Gödel
   crossed out "that you took a degree ... technical college" and wrote
   "wrong" beside it in the margin. After "studies" an arrow leads to
   "first in physics 1 or 2 years".]

2. (a) As well as you can recall, when did your interest in mathematics
   begin? [Gödel: 14 years of age]
   (b) Are there any influences you would single out as especially important in the development of your interests (for example, a certain
   schoolteacher, a particular book or author; parents' interests, etc.)?
   [Gödel: Introduction to calculus in "Göschen's collection"[a]]

---

[a]Presumably *Junker 1919*, the second volume of which was found among Gödel's
books.

3. When did you first study *Principia Mathematica*? ⟦Gödel: 1929⟧

4. (a) As well as you can recall, when did you become interested in the problem, central to your 1930 and 1931 papers, of the completeness of logic and mathematics? ⟦Gödel: 1928⟧
   (b) Are there any influences you would single out as especially important in this regard? ⟦Gödel: Hilbert Ackermann: Introduction to math Logic,[b] Carnap: Lectures of math logic⟧

5. When, if at all, did you first study any of the works of the following: (a) Ludwig Boltzmann; (b) Jan Brouwer; (c) Paul Finsler; (d) Immanuel Kant; (e) Karl Kraus; (f) Fritz Mauthner; (g) Jules Richard; (h) Ludwig Wittgenstein ⟦Gödel: (a) phil paper 1960 (a) 1940 (b) 1940 (c) 1932 (alt two papers (d) 1922 (e) 1960 (f,g) never (h) 1927 (never thoroughly)⟧

6. How much importance, if any, do you attribute to each of the scholars in the above list, in the development of your interests? ⟦Gödel: Only Kant was imp.⟧

7. Your name is frequently mentioned in connection with the "Vienna Circle" (also known as the "Ernst Mach Society", and later, the "Society for Unified Science"), which began to meet in 1924 and included Rudolf Carnap, Hans Hahn, Otto Neurath, Moritz Schlick, and others.
   (a) When did your association with this group begin? ⟦Gödel: 1926⟧
   (b) How close was this association? ⟦Gödel: 1926–28 frequ. disc. with ⟨report on my paper⟩ some ⟨younger⟩ members[1]⟧
   (c) Do you regard this group or some of its members as influential in the development of your intellectual interests? ⟦Gödel: Yes, this group aroused my int. in the foundations but my views about them differed ⟨subst⟩ *fund.* from theirs.⟧

8. Your philosophical leanings have been described by some as "mathematical realism," whereby mathematical concepts ⟦written above by Gödel: "and sets"⟧ and theorems are regarded as describing *objects* of some kind.
   (a) How accurate is this characterization? ⟦Gödel: correct⟧
   (b) In particular, how well does it describe your point of view *in the 1920's and early 1930's*, as compared with your later position? ⟦Gödel: was my pos. since 1925 before ⟦attending?⟧⟧

---

[1]but mostly ⟨very often⟩ I took the non-pos. ~~approach~~ pos. ~~(even though I agreed at some point with posit.~~ Attendance at sessions

---

[b]Presumably Gödel refers here to *Hilbert and Ackermann 1928*.

(c) Are there any influences to which you attribute special significance in the development of your philosophy? [Gödel: ~~Inspired
infl.~~ by ⟨introduct.⟩ phil lectures of Gomp. & math by Phil Furtw.]

9. Is there a published source, either written by yourself or by another author, which you regard as a particularly apt statement of
your philosophical point of view? (Or perhaps you would like to
elaborate on it briefly.) [Gödel: Book by Wang,[c] Paper on Cont.
Probl.,[d] on Russell[e]]

10. (a) When and where were your parents born?
(b) What was their native language? [Gödel: German]

11. (a) What was your father's occupation during your childhood?
[Gödel: Manager of a cloth factory]
(b) What was your mother's occupation (if employed)? [Gödel: not
employed]

12. What was the highest level of education attained by each parent?
[Gödel: Father technical school, some classes ~~in school~~ equi. with
coll. courses]

2 | 13. (a) What was your parents' religion? [Gödel: Father Old Cath.,
Mother Lutheran]
(b) What is your religion? [Gödel: Baptized as Lutheran (but not
member of any rel. cong.) My belief is theistic not pantheistic
⟨(following Leibniz rather than Spinoza)⟩.]

14. (a) How many other children did your parents have? [Gödel: 1]
(b) In what year was each born? [Gödel: 1902]
(c) What is the highest level of education attained by each of your
siblings (approximately)? [Gödel: Doctor of medicine (X-ray
spec.)]

15. (a) How would [you] characterize your family's financial situation
during your childhood: quite poor, poor, average, above average, or
wealthy? [Gödel: close to wealthy]
(b) To what extent was your family affected by World War I and
the post-war inflation? [Gödel: not much affected]

16. (a) To what extent were you and your family aware of and/or affected by the Czech nationalist movement for independence from

---

[c] *Wang 1974.*
[d] *1947.*
[e] *1944.*

the Austro-Hungarian Empire? ⟦Gödel: Emigrated to Austria
⟨1929⟩ (after my father's death)⟧
(b) To what extent were you and your family aware of and/or affected by anti-Semitism in the period from 1910 to 1935? ⟦Gödel:
not affected⟧

17. Before you enrolled at the University of Vienna, how much contact
did you and your family have with the cultural and intellectual life
in Vienna (for example, through visits, communication with friends
or relative, newspapers, etc.)? ⟦Gödel: little contact except through
newspaper "Neue freie Presse"⟧

Your time and attention to these questions are much appreciated, Professor Gödel. Thank you very much.

Sincerely,

Burke D. Grandjean

5. Gödel to Grandjean[a]

Dear Sir

I am sorry I ~~am not well enou~~ have no time for an interview. Moreover I have not been so well recently.

Here are some answers to your questions (excluding those that are too
~~private~~ personal or too complicated to answer). In part they are given
on the form you sent me.

2a   14 years of age
2b   current textbooks in calculus
3   1929
4a   1928
4b   Hilbert–Ackermann Introduction[b]
     Lectures by R. Carnap
     (a) ca 1960 (phil. writings) (b) 1940
     (c) only 2 papers 1932 and later
     (d) 1922 (e) ca 1960 (f)(g) not all

---

[a] Undated draft.
[b] See footnote b to the questionnaire.

(h) ca 1927 only very superficially only Kant & only as to my
general philo. views.

As to 5 I ~~don't~~ ⟦?⟧ would like to say that only Kant had some infl. on
2   my | phil thinking in gen. ⟨& that I got acqu with him about 1922⟩, that
I knew Wittgen. very superficially, that I read only two papers by Finsler
(~~after 1927~~ in or after 32) finally that the greatest phil. infl. ~~was~~ on me
~~was that of~~ came from Leibniz which I studied ⟨about⟩ 1943–46.

7a   1926
(b) 1926–28 reg. att. of meetings & frequent disc. with some ⟨of
the younger⟩ members after 1930 the assoc. became more ⟦and⟧
more loose. ⟨Generally speaking⟩ I only agreed with some of
their tenets. ⟨E. g.⟩ I never believed that math is syntax of lang.
In the course of time I moved farther & farther away from their
views. (c) Yes, they aroused my interest in the found. of math.

8   (a) accurate
(b) held this view since 1925
(c) Heinrich Gomp. Prof. of Ph of Vienna

Marvin Jay Greenberg

Marvin Jay Greenberg (1935–) is Professor of Mathematics, Emeritus, at the University of California, Santa Cruz. His research areas of specialization include algebraic geometry and algebraic topology. Greenberg wrote Gödel on 5 September 1973 in connection with an undergraduate textbook on the development and history of Euclidean and non-Euclidean geometries (*1974*) that he was then in the process of completing. His text emphasizes the revolutionary change in consciousness resulting from the discovery of another geometry besides Euclid's which is equally consistent. Chapter 8 is devoted to the philosophical implications of that discovery, surveying various points of view about geometry and mathematics more generally. It concludes with topics for essays in which students are asked to comment on provocative quotes from famous mathematicians and physicists. For essay topic 10, Greenberg asked for permission to quote the following passage from "What is Cantor's continuum problem?" (*Gödel 1947* and *1964*): [a]

> I don't see any reason why we should have less confidence in this kind of perception, i.e., in mathematical intuition, than in sense perception, which induces us to build up physical theories and to expect that future sense perceptions will agree with them, and, moreover, to believe that a question not decidable now has meaning and may be decided in the future. The set-theoretical paradoxes are hardly any more troublesome for mathematics than deceptions of the senses are for physics. ... Evidently the "given" underlying mathematics is closely related to the abstract elements contained in our empirical ideas. It by no means follows, however, that the data of this second kind, because they cannot be associated with actions of certain things upon our sense organs, are something purely subjective, as Kant asserted. Rather, they, too, may represent an aspect of objective reality, but as opposed to the sensations, their presence in us may be due to another kind of relationship between ourselves and reality. (*1964*, pp. 271–272).

Less than a month later, Gödel replied (letter 1) saying that he had no objection to Greenberg's using the quotation, as long as it was given more fully by adding the sentence which opened the first of the two paragraphs from which it was taken, namely:

> But, despite their remoteness from sense experience, we do have something like a perception also of the objects of set theory, as is seen from the fact that the axioms force themselves upon us as being true.

---

[a]For the reader's convenience, this and the further quotations below from *1964* that were in question are given in the order they were taken up in the exchange.

Then, in a footnote to the letter, Gödel made the additional request that
the reference be given in full, "because you omit important parts of my
exposition", and because the quotation occurred only in the supplement
to *1947* provided for its 1964 re-publication. The extensive part that
had been omitted in the ellipsis by Greenberg was as follows:

> That new mathematical intuitions leading to a decision of such problems
> as Cantor's continuum hypothesis are perfectly possible was pointed out
> earlier.
>
> It should be noted that mathematical intuition need not be conceived
> of as a faculty giving an *immediate* knowledge of the objects concerned.
> Rather it seems that, as in the case of physical experience, we *form* our
> ideas also of those objects on the basis of something else which *is* imme-
> diately given. Only this something else here is *not*, or not primarily, the
> sensations. That something besides the sensations actually is immedi-
> ately given follows (independently of mathematics) from the fact that
> even our ideas referring to physical objects contain constituents qualita-
> tively different from sensations or mere combinations of sensations, e.g.,
> the idea of object itself, whereas, on the other hand, by our thinking we
> cannot create any qualitatively new elements, but only reproduce and
> combine those that are given.

Besides this, in granting permission, Gödel also said that it was "absolutely necessary" for Greenberg to include a *new* paragraph in which
set-theoretical intuition is distinguished from geometrical intuition, and
in which, further, the mathematical and physical aspects of geometrical
intuition are distinguished. The formulation of that paragraph is found
here in letter 1.

Greenberg of course acceded fully to Gödel's requests, but was in-
trigued by the new paragraph. Pursuing this in letter 2, he said, "[i]t
seems to me that your remarks which I quoted *do* apply to Non-Euclidean
Geometry, not just to set theory..." and offered as evidence the intuitions that Bolyai and Lobachevsky had for non-Euclidean geometry
prior to the discovery of Euclidean models for them, and the intuitions
that Riemann had concerning his geometries. Gödel did not reply to this
letter, but composed an unsent draft (letter 3), in which he said that
intuition is not absolutely necessary in order to develop a mathematical theory, and besides, one might develop a non-Euclidean intuition by
combining the elements of Euclidean intuition in a different way, with
sufficient practice. He then mentioned some problems with that kind of
intuition, but then the draft broke off. One may speculate that perhaps
he thought those problems a serious objection to the points made in the
first part of the draft, or perhaps he only wished to formulate the matter more clearly before sending the letter, and then got distracted. This
correspondence is of particular interest because it gives some insight into
Gödel's thoughts on the question of the analogy between non-Euclidean
geometries and undecidability in set theory. Traditionally, Euclidean
geometry was thought to describe an objective reality, and at least in
the time of Kant, was thought to be known a priori to be true, rather
than empirically true. Developments in physics showed that if it is true

a priori, its truth is of a mathematical entity (Euclidean space) rather than a physical entity. Gödel thought that set theory was true about an objective, external reality. His general incompleteness results and Cohen's later, more specific independence results, invited comparisons and analogies with the development of non-Euclidean geometry. Is truth in set theory analogous to truth in geometry, where one has a choice of axiom systems, none of which correspond exactly to physical reality? Is there something objective that either Euclidean or non-Euclidean intuition or set-theoretical intuition is about? While no conclusive answers are to be found in this correspondence, the various considerations and questions are brought into sharper focus. No consensus exists among mathematicians on these philosophical issues, and the fact that the correspondence ends in an unsent draft invites the guess that Gödel too was unsettled concerning them.[b]

Michael Beeson[c]

[b]One difference between geometry and set theory, not brought out in the correspondence, is that in first-order logic both Euclidean and non-Euclidean geometries are complete theories, while set theory is essentially incomplete and incompletable. Another difference was emphasized by Kreisel (*1971a*): "CH is (provably) not independent of the full (second order) version of Zermelo's axioms.... In contrast, ...the parallel axiom is independent of the remaining geometric axioms even if the continuity principle is treated as a second-order axiom." (op. cit., pp. 195–196). Of course, this difference is significant only if the "full" semantics of second-order logic is granted as meaningful. In any case, it is not clear that these differences have any bearing on the philosophical issues, other than to show that the analogy between the independence results, if there is one, cannot be perfect.

[c]I am very grateful to Solomon Feferman and Charles Parsons for their attention to and substantial improvements of a draft of these introductory remarks.

1. Gödel to Greenberg

October 2, 1973

Dear Professor Greenberg:

I have no objection to the quotation mentioned in your letter of September 5, *provided* you add the following:

Gödel in this passage speaks ⟨primarily⟩ of *set theoretical intuition.* As far as geometrical intuition is concerned the following, according to Gödel, would have to be added: "Geometrical
intuition, strictly speaking, is not mathematical, but rather a
priori physical, intuition. In its purely mathematical aspect our
Euclidean space intuition is perfectly correct, namely it represents correctly a certain structure existing in the realm of mathematical objects. Even physically it is correct 'in the small'."

This addition is absolutely necessary in view of the fact that your
book deals with geometry, and that, moreover, in your quotation, you
omit the first sentence of the paragraph in question. See Benacerraf-
Putnam, Philosophy of Mathematics, Prentice[-]Hall, 1964,[a] p. 271.[1]

Sincerely yours,

Kurt Gödel

[1] I also have to request that you give *this reference in full,* because you omit important
parts of my exposition and, moreover, the passage you quote does not occur in my
original paper, but only in the supplement to the second edition.[b]

---

[a] *Benacerraf and Putnam 1964.*
[b] Gödel here refers to the supplement to the reprinting of *1947* in *Benacerraf and
Putnam 1964.*

2. Greenberg to Gödel

October 15, 1973

Dear Professor Gödel,

Thank you for your letter of October 2 giving me permission to quote
you in my forthcoming geometry text. I have given the reference in full
and added your remark about geometrical intuition as you requested.
Since you have taken the trouble of sending me this interesting remark,
may I comment on it?

First you say "geometrical intuition, strictly speaking, is not mathematical, but rather a priori physical, intuition." Then you make a statement about "our Euclidean space intuition". Are you here equating "geometrical intuition" with "Euclidean space intuition"? If so, perhaps

I should emphasize that my book will be entitled EUCLIDEAN AND NON-EUCLIDEAN GEOMETRIES.[a] What about the intuition J. Bolyai and Lobachevsky had for Hyperbolic Geometry before it was shown that there were Euclidean models? What about Riemann's intuition? It seems to me that your remarks which I quoted *do* apply to Non-Euclidean Geometry, not just to set theory (as you qualified them in your letter to me). Am I mistaken?

Yours sincerely,

Marvin Jay Greenberg

---

[a] *Greenberg 1974.*

3. Gödel to Greenberg[a]

Dear Mr. Greenberg

~~I have not studied never heard of a non-Euclid. space intu~~ ⟨I am not sufficiently well acqu. with⟩ the original papers by Lob[achevsky], Bol[yai] & Riemann ~~sufficiently~~ to ~~say~~ know whether ⟨they claim to have developed a non-Eucl. space *intuition*⟩. ⟨Note ⟨that⟩ ~~an abstract theory need not be accompanied by an *int.* of it's objects~~ for developing a theory from given ax. an *int* of it/s objects is quite unnecessary⟩. If you can quote to me statements by them or bring reports ~~about them~~ to this effect I'll be very much interested. ~~Helmholtz, as far, tried to develop such a~~ It is not imp. that out of ~~the elements of~~ the Eucl. ⟨space⟩ int. (which we all have) a non-Eucl. ⟨space⟩ int. ~~by a sufficient practice~~ ⟨could be developed⟩ by cons. its el. differently & by ~~practice~~ sufficient practice, ~~But~~ ⟨even though⟩ in order to do that one would ⟨have to solve e. g.⟩ ~~among other things have to solve~~ the problem of imagining two ~~assympt~~ lines assympt⟨otically⟩ ~~to each o~~ approx. each other and being at the same time ~~concave to~~ everywhere concave toward each other | (which *trivially* exist in hyperb. geom.). Also we would have ~~Sincerely yours~~ to ~~imagine~~ picture to ourselves a closed ⟨& finite⟩ three-dim. space.

2

---

[a] Draft of an answer to the previous letter.

Gotthard Günther

Gotthard Günther

Gotthard Günther (1900–1984) began an academic career in Germany and published a book on Hegel's logic (*Günther 1933*), based on his 1932 dissertation. He left Germany in 1937 and after brief stays in Italy and South Africa came to the United States in 1940. In the 1940s he taught for a time at Colby College, but he did not have a regular academic position again until 1961, at the end of his correspondence with Gödel, which begins in 1953. During that correspondence he lived in Richmond, Virginia, where his wife seems to have been employed. He had research grants and earned money as a flying instructor and by freelance writing. During the correspondence his relation with the University of Hamburg began, where he obtained a visiting position for the winter semester of 1955–1956. In 1961 he became a research professor in the department of electrical engineering at the University of Illinois in Urbana. He was given the title of Professor Emeritus at Hamburg, and in 1971, after his retirement from Illinois, he moved permanently to Hamburg (after further visits) and continued to give lectures there until the 1982–1983 academic year.

Günther's body of writing is considerable, but it is unlikely to be known to most readers of Gödel's works. His original philosophical background was Hegelian, and he continued to see philosophy from that point of view, though he was also influenced by Leibniz and by twentieth-century German figures. Moreover, although he lived for 30 years in the United States, even during that period his philosophical writing was mostly in German.[a] A project that he pursued for many years, which is one of the themes of his correspondence with Gödel, was how formal logic ought to be revised to accommodate what he took to be insights about the nature of thought and its relation to reality from the German idealist tradition. He also became interested in and wrote about "cybernetics". Norbert Wiener, who publicized the term, characterized cybernetics as the science of "control and communication, in the animal and the machine."[b] Its concerns derived from engineering and theoretical biology, but what seems to have most interested Günther was the idea of artificial intelligence. He was one of the earlier thinkers to write

---

[a]There is a bibliography of Günther's published writings in *Günther 1980*, pp. 305–310. *Günther 1975* is partly autobiographical.

[b]In the title of *Wiener 1948*.

from a philosophical point of view on that subject.[c] He was thus a very unusual intellectual figure for his time, a Hegelian philosopher with an interest in modern logic and involvement in what later came to be called computer science.

The occasion for Günther's correspondence with Gödel was an inquiry Gödel received from the American Committee for Emigré Scholars, Writers and Artists, which wished to support an application by Günther for a grant from the Bollingen Foundation. Günther followed this up by writing to Gödel on 2 August 1953 and sending him several papers. Shortly thereafter he took the liberty of using Gödel's name as a reference. Gödel wrote what was evidently a supportive letter, of which handwritten drafts survive in his papers.[d] On 12 December 1953 Günther wrote to inform Gödel that he had received a three-year grant and to thank him for his support. The more substantive correspondence began the following spring. On 29 April 1954 Günther wrote expressing some views and raising some questions about the law of excluded middle. The exchange continued for several years, with Günther, however, writing more and longer letters than Gödel. On 17 September 1956 they met in Gödel's office and had a morning of discussion, apparently largely of Günther's work.[e] Gödel's last known letter to Günther was in January 1959, but Günther continued to write to Gödel through 1959 and 1960. There is no evidence known to us that Gödel sent any further replies.

1. Günther on metaphysics and logic

Before we describe the course of the correspondence, it is necessary to say something about Günther's point of view at the time and about the philosophical project, aspects of which he lays out in his letters, in

---

[c] *Günther 1952* is his most direct discussion of whether machines can be conscious. It seems to be presupposed in *Günther 1957a*, the clearest and most accessible presentation of the ideas Günther discussed with Gödel, as Gödel himself seems to have thought (letter 13, but see note aa below). Since *Günther 1963* incorporates the latter and contains a reprint of the former, it is the best introduction to his ideas.

[d] The Bollingen Foundation acknowledged receipt of the letter, which was dated October 24, 1953 (letter of Ernest Brooks, Secretary of the Foundation, to Gödel, 26 October, 1953). Correspondence of Gödel with others cited in this note is from the Gödel papers, filed under Günther.

[e] In letter 9, 20 September 1956, Günther thanks Gödel for such a discussion the previous Monday. The date can be inferred from the fact that 20 September was a Thursday. I know of no evidence that they met on any other occasion.

whose service he was making inquiries with Gödel. The conviction with which he began and which animated his whole involvement with logic is that the philosophical insights of German idealism from Kant through Hegel required a revision of logic. This view was already expressed in the published version of his dissertation.[f] Probably by the time he left Germany he was convinced that modern mathematical logic was relevant to the project of such a revision,[g] but he was never convinced that in its classical version it was what was required. He considered Hegel's logic an effort in this direction, although he was aware that it was not a logic in the sense in which the systems constructed in modern logic are logics. He thought Hegel's logic a grand failure, but he remained interested in the project of expressing it in more formal terms. That logic should be intimately related to metaphysics was a lifelong conviction of his and no doubt a point of agreement with Hegel. Moreover, to express his view of metaphysics he constantly uses as basic categories thought or consciousness, its relation to objects, and self-consciousness. In this respect the idealist tradition determines how he describes even non-idealist philosophy. Both the perspective from idealism and the metaphysical conception of logic are epitomized in his remark, "A logic is the metaphysical self-definition of a subject" (*1957*, p. 29).

Günther describes his project repeatedly as that of constructing a "non-Aristotelian logic." By "Aristotelian" he means what we would call classical, so that most of the vast extension of logic that has taken place since the mid-nineteenth century still counts as Aristotelian.[h] However, what is decisive for him is a certain metaphysical interpretation of the foundations of logic. His own preferred means for carrying out the project of a non-Aristotelian logic is many-valued logic. That is not itself non-Aristotelian, but it offers the technical means of carrying out

---

[f] *Günther 1933*; cf. *Günther 1978*.

[g] See *Günther 1940* and *1957*, of which the latter appears to have been drafted in 1935; see *Günther 1980*, p. 305 n. Unlike other publications of Günther in the 1950s dealing with logic, it makes no mention of three-valued logic. Günther sent Gödel the published version without having mentioned the paper earlier in the correspondence; that would suggest he was not working on it during their exchange.

[h] "We can leave out of account the 'logistic' criticism [of Aristotelian logic] beginning with Leibniz, because it represents no philosophical critique of the metaphysics of this logic but a generalization and extension (functional calculus) of the classical ways of proceeding" (*1957*, p. 5 n.). He does go on to say that his own work would not be possible without modern logic, and already in *1940* he argued that philosophers interested in "transcendental logic" needed to pay attention to modern logic.

the idea of a non-Aristotelian logic. Or so Günther thought in the 1950s at the time of his exchange with Gödel.[i]

Günther's conception of classical logic is intimately bound up with a view of the relation of logic and metaphysics. He sees the metaphysical tradition from Plato and Aristotle at least through pre-Kantian modern philosophy as a unity and classical logic as obtaining its rationale and metaphysical foundation from that tradition. He describes as "axioms" of the classical tradition in logic the principles of identity, contradiction, and excluded middle (*1957*, p. 5). Whenever he discusses the foundations of logic, however, the framework is that of a theory of thought, in particular of a subject thinking about an object, or being conscious of an object.

What is most distinctive in Günther's point of view comes out in his remarks about the principle of identity. Repeatedly he says that the metaphysical tradition is based on the presupposition of the identity of thought and being. He also took this as a metaphysical presupposition of classical logic. As he said in his letter of 23 May 1954 to Gödel, "Classical logic presupposes the metaphysical identity of thought and being" (letter 3, p. 17). The "original phenomenon" of thought is expressed by "I think something." The relation of "I" and "something" is characterized as identity. What this comes to is that the object, the "something" is identical with itself, but also that the ego is in the end identical with its object.

However, without some further elaboration we do not do justice to Günther's thought. What he took the metaphysical meaning of the principle of identity to be is first of all that thought aspires to complete objectivity:

> While consciousness in the judgment "I think something" determines its definitive and final subject matter, that is the "something", simply as "identity", it claims implicitly that all thinking that is possible at all intends as its definitive metaphysical goal the objective In-itself, identical with itself. The transcendent essence of all self-identical In-itself, however, is Being. Consequently Being is the only, original, and last metaphysical subject matter of reflecting consciousness (*1957*, p. 8).

The object of thought can preserve its identity through the different perspectives from which it is experienced and thought about. Thus

---

[i] *Günther 1958* sets forth what was to be the basic logical construction for the second volume of the work of which *1959* was the first. He states (*1978a*, p. xxii) that the calculus was not able to bear the philosophical weight it was meant to carry. He apparently came to this conclusion through exchanges with cyberneticists early in his time at the University of Illinois. No attempt is made here to follow Günther's thought after his exchange with Gödel.

that it is identical with itself means not simply what is expressed by the logical truth that everything is identical with itself, but, one might say, that the concept of identity has a non-trivial application, so that the same object can be presented to consciousness under different circumstances and in different ways, which we might call modes of presentation.

A clear conclusion from the identity of thought and Being that Günther draws in his principal systematic work, *Idee und Grundriß einer nicht-aristotelischen Logik*,[j] is that what is objective or what is true will be perfectly intersubjective (*1978a*, p. 11). If a subject has a "true concept" of an object, that will in principle be communicable to any other subject. But Günther sees this as implying that the division of subjectivity in general into individual subjects is only "provisional and apparent." I think what he means by this is that an intrinsic goal of the thinking of individuals is to converge on some absolute thinking, in which each one's thoughts would represent the world as it is and thus not in any way differently from those of any other. This train of thought leads to the idea of an absolute subject, in effect God, in whom the relativity of consciousness to a perspective would be overcome. One of the challenges to this way of thinking is the thought, prominent in post-Hegelian philosophy, that actual subjectivity is formed by history (*1978a*, p. 10).

To get a fuller idea of Günther's conception of classical metaphysical thinking, we have to see a little of what he thought overthrew it. One development he mentions but does not give a prominent place in his philosophical analysis is the gradual divorce of fundamental science from philosophy. It is because modern logic arose primarily within mathematics, and thus on the scientific side of this divide, that Günther thought it did not break with the classical logical tradition on the metaphysical level.

What is central to Günther's story is German idealism beginning with Kant but especially as embodied in Fichte, Schelling and Hegel. His picture of philosophy before Kant clearly owes much to idealism; for example he views the classical tradition as in Kantian terms transcendental realist. However, if one thinks of pre-Kantian philosophy as transcendental or metaphysical realism, one will see the main revolutionary element in German idealism as Kant's Copernican turn and its further development by his successors. That is for Günther only part of

---

[j] *Günther 1959*. This work was Günther's principal project during his exchange with Gödel, and its Preface contains a generous acknowledgment to him (*1978a*, p. xxi). Quotations from this work are from the second edition, *Günther 1978a*. Translations of Günther's published writings are my own, of his letters to Gödel by Thomas Teufel and me.

the truth. What is central, though related, is the role self-consciousness plays in the thought of these philosophers. It is attention to subjectivity that was lacking in earlier thought. To Gödel he writes the following:

> In the classical tradition the subject of thought and the process of reflection do not count at all. The goal of thought is to grasp the sense of *absolutely objective being*. And truth means absolute agreement of thought with the absolutely objective object. That is, all categories of logic must, if they are to be true, be absolutely objectively definable. Everything "subjective" is quite simply to be eliminated (letter 3, p. 3).

Now, whatever one might think of this as a characterization of ancient and medieval thought, it is a commonplace that subjectivity becomes a philosophical theme with Descartes. Günther has much more to say about what is new in Kant and his successors; the central concept that he uses to describe it is "reflection". It is not easy to say what he means by it. A basic meaning is certainly "self-consciousness". More generally, it is represented as a feature of thought about objects when the conception of the objects takes into account the subject's thought about them. In this way it comes also to cover what I myself would call semantic reflection, that is, the passage from the straightforward use of words "taken at face value" to discourse in which they are mentioned and something is said about their reference, truth or meaning.

Some helpful explanations are found in *Günther 1957* (already quoted above). In this paper Günther distinguishes stages of reflection, which are fundamental stances of consciousness to the world and itself, which, however, also have some relation to stages in the history of thought. The stages are described as "R-levels", levels of reflection. The idea seems to have some inspiration from the theory of types (*1957*, p. 5 n. 1), but also from Hegel's *Phenomenology of Spirit*.

What he calls the 0th R-level is one at which there is no self-consciousness at all; consciousness simply reflects the world:

> This elementary state or 0th R-level of consciousness mirrors wholly immediately the objectually closed connection of being and produces, with the naturalness of an optical camera a simple image of the objective world (*1957*, p. 19).[k]

The first R-level is, as one might expect, that at which the 0th level becomes an object. Günther describes it as arising from an "existential contradiction" in the 0th level, because at that level the subject realizes itself as *negation*, as mere *object*, that is as *non*-subject. But at the first level it is able to distinguish being and consciousness. It discovers that

---

[k]This stage is identified with the "sense-certainty" of Hegel's *Phenomenology*.

being is mirrored in consciousness, but nothing else is so reflected. So consciousness is in reality being.

This, according to Günther, is the position of Aristotle and therefore of the origins of logic. But it is an unstable position:

> For Aristotle consciousness knows of itself and of its experienced opposition to being, but it relativizes this opposition and rescinds it by means of a reflection. The self-consciousness of man begins its history with a denial of itself (*1957*, p. 20).

A second R-level, which is conscious of the first level and its limitations, seems to arise at the end of the Middle Ages. Although he agrees with the historians of logic that nothing came of it for formal logic, he seems to attribute some awareness of the limitations of Aristotelian logic, which as belonging to the first level can only be about the 0th level, to those who rebelled against medieval logic during the Renaissance. But no new logical system was possible at the time.

Every R-level, according to Günther, can itself be object of a further reflection. Thus the iteration involved is infinite. Günther draws the further conclusion that it cannot at this point be characterized axiomatically, as a new logic would require, "because it is in no way possible to arrive at final, most general propositions about this open subject and to define it as self-consciousness (thus as a closed whole)" (*1957*, p. 23). Günther had stumbled on a kind of paradox, which is familiar from reflection on the theory of types. If "reflection", whatever that is, is iterated an arbitrary finite number of times, one arrives only at a particular stage of reflection from which it is possible to go a step further. Therefore general propositions about all stages can't be formulated.

Günther formulates the problem as one about self-consciousness. At the second level consciousness is related to a relation of consciousness and objects, but self-consciousness is not part of what one is conscious of. It seems that at each level from the first on one has self-consciousness which is, however, limited to its own consciousness at lower levels and so does not really take in *itself*. Günther argues that this leads to a predicament in which metaphysics is impossible.

In the theory of types, eventually a way was found of so formulating the theory that the iteration of progression to higher types can be iterated into the transfinite, and of course with variables of type ω one can make statements about entities of arbitrary finite types. This does not abolish the dilemma, however, because the progression to higher types can be carried still further. The understanding of the language of set theory that comes most naturally takes the quantifiers as ranging over absolutely all sets, and thus as encompassing all types in the simple type hierarchy. That is not the end of the story, however, since that interpretation is not beyond question, and one still has to cope with

another form of ascent, the semantic ascent that, in a perfectly classical setting, parallels ascent in Russell's ramified hierarchy.[1]

Günther sees the achievement of idealism as at least posing the task of developing a conception of self-consciousness that would have the required closure property, so that self-consciousness would be truly consciousness of *itself* and not just a representation of itself as an object. In developing his own scheme he does not stay with the idea that there will be an nth R-level for every n, and on into the transfinite. That there is a higher standpoint is, according to Günther, shown by the fact that it is possible to conceive such an iteration and make judgments about it, whether or not one holds that there is a higher standpoint.

> With this thought of the infinite iterability of the reflected consciousness, we have already elevated ourselves above the infinite series of reflections proceeding from the second R-level and made it the "object" of a reflection that by definition cannot itself belong to this sequence. The content of this new reflection is thus the idea of the totality of the infinite sequence of iterations (and not itself an iteration on which others can follow) (*1957*, p. 27).

Günther concludes that we have a true third R-level that is not just a term in the sequence of iterations of reflection, in which the subject reflects on itself and thus "defines the ego as total self-reflection". It might seem that Günther denies Hume's famous point and thinks that one can directly capture oneself in one's thinking. If that were so, the elaborate story about levels of reflection would be unnecessary. On the contrary, following Hegel, Günther maintains that the compulsion of consciousness toward objectivation can itself be "reflected" and thus seen as a feature of consciousness. It is that that makes "total reflection" possible.

In other writings Günther uses a scheme that he finds in Hegel, which finds three different levels of reflection: "Reflexion-in-anderes", the simple thought of an object by a subject; "Reflexion-in-sich", the thought by a subject of a subject that, however, plays the role of object; and

---

[1]If, as I have suggested, "reflection" as Günther understands it includes semantic reflection, then complete closure, that is a theory that would express its own semantics without any remainder, is impossible. "Total reflection" is not so clearly defined that one can say definitely whether it implies semantic closure in this sense. An affirmative answer is suggested in a passage in *Günther 1957a* (*1963*, pp. 78–79). It is somewhat difficult to interpret because it is not always clear when he is rendering Hegel and when speaking for himself. Although the I/R/D three-valued logic in a way renders total reflection, it describes "a thinking that is not and cannot be thought by anyone" (p. 79). To capture the subject of this thought, one must ascend to a four-valued logic, and indefinite further ascent can be forced. Thus Günther writes in letter 11 that "total reflection is not three-valued but indefinitely n-valued, where $n > 2$ always holds". In the end he seems not to claim to escape an ascent like the ascent of types.

"Reflexion-in-sich der Reflexion-in-sich-und-anderes",[m] which seems at first sight to be just the thought by a subject of a subject's thinking of an object, where both the subject thought of and the object are objects of the first's subject's thought.[n] There is a rough correspondence between these and the 0th, 1st and 2nd R-levels, but Günther uses these concepts for different purposes, in particular explaining the truth-values of his three-valued logic, as we shall see shortly.

One of the other purposes is the "deduction" of the notion of the Thou (*Du*). Given the fundamental role of consciousness in his thinking, and its distinctness from "being", it is not surprising that other minds should be an ontological category in their own right. The Thou is a subject, thus with the same reflective closure we have been considering. That taking the Thou seriously should lead to a fundamental change follows from Günther's view of earlier philosophy, because the identity of thought and being tended in the end to abolish the differences between subjects.

Günther expresses the view that takes the Thou seriously in the form of two "metaphysical theorems":

I. Being and Thought are only partially identical.

II. The object has one metaphysical root, the subject has two (*1978a*, p. 85).

This view leads, according to Günther, to a questioning of the law of excluded middle. In some way he identifies the rejection of a "third" between subject and object and the rejection of a third truth-value, or an alternative to being true and being false. The "home" of the law of excluded middle is a classical subject-object schema in which one or more subjects relate to an object. But once one accepts theorem I, two subjects will not be equivalent in the way the tradition had it. If two subjects think of an object, one will be able to think of the other as thinking of the object. That is what Günther calls "Reflexion-in-sich der Reflexion-in-sich und anderes", schematized as $S^s \rightarrow (S^o \rightarrow O^s)$. This suggests an asymmetry that is, however, not the final point of view because it leaves out the fact that the other subject is also capable of thinking of me. We have, in Husserl's well-known phrase, transcendental intersubjectivity.[o]

---

[m] One might translate these roughly as "reflection into other", "reflection into self" and "reflection into self of reflection into other".

[n] *Günther 1978a*, p. 98; cf. *1953*, p. 48, and *1958*, pp. 390–391.

[o] The symmetry is illustrated by Günther's diagram in *1978a*, p. 98. He also remarks that Hegel "defines total reflection as 'Reflexion-in-sich der Reflexion-in-sich-und-Anderes'" (*1958*, p. 379). I take Günther's analysis of the Thou as implying that he regards this characterization as inadequate.

Günther argues that from this point of view one must distinguish two negations, one of which is expressed in the statement that the subject is not its object, another in the statement that the subject (as ego) is not the Thou. Günther seems to be driven toward many-valued logic by the fact that he doesn't consider an alternative to a truth-functional interpretation of propositional logic, and at least a third truth-value is needed in order to make the distinction between the two negations.

How, then, does he interpret his "truth"-values? At this point he does something that is from a logician's point of view crazy, because the values seem not to be *truth*-values at all. He uses the designations I, R and D, which he reads as "irreflexive", "reflexive" and "double-reflexive". In other words, they represent stages of reflection coming out of the analysis we have discussed. He even says in one place that all the values are "true" (*1953*, p. 48). The concepts of truth and falsity should "disappear without remainder" from the sort of logic he is constructing because they exclude a genuine third.[P] It seems that he has simply changed the subject, as a result of taking the relation of the I to what is not I as the paradigm of all negation.

Günther is, however, a somewhat slippier target, and I don't think I have grasped his thought at this point. He says that our thought is in a way necessarily two-valued. What the three-valued logic does is allow for the fact that two-valued thought can occur at different levels of reflection. How he conceives this is not at all clear to me. But he does say something about how it works in propositional logic. He singles out pairs of values and notes that one might treat that pair as truth and falsity, and certain functions might behave like, say, conjunction when just these two values are considered, perhaps behaving differently when the third value is taken into account. He saw the fact that two-valued structures can occur in different places in a three- or more-valued system as analogous to the place-value feature of Arabic or binary notation for numbers:

> A many-valued logic is now nothing but a system that allows us to give to our single "actual" logic different place-values in the system of consciousness of such a kind that each place-value is connected with a different semantic meaning of the two-valued calculus that thus repeats itself. Such a many-valued system allows us thus to read off the structural interrelation of the different two-valued stages of consciousness.[q]

This remark would suggest that the two truth-values retain their status as genuine truth-values and that the values of the many-valued system

[P] *1953*, p. 47. Cf. the remarks about truth and falsity in letter 3, pp. 8–9, and letter 8, pp. 1–2.

[q] *1958*, p. 393.

have a quite different role. It is not clear how this would be reconciled with Günther's claim that his constructions constitute a genuine revision of logic.

2. The correspondence

Günther wrote Gödel on 29 April 1954 with questions about the law of excluded middle, prompted by an allusion in *Menger 1933* to Gödel's translations of classical logic and arithmetic into intuitionistic. Although Menger gives a reference to *Gödel 1933e*, Günther says he cannot obtain it and struggles with the issues as best he can. He concerns himself with the difference between a version of the law of excluded middle in propositional logic and the quantificational schema $\neg\forall x F x \to \exists x \neg F x$, which he regards as a formulation of the law although it is better described as a consequence of it. His comments on the latter are not very clear, but he seems to hold that its truth in a given case presupposes that an object is given that satisfies $\neg F x$ if anything does.

Gödel's reply of 15 May to this rather confused letter is a gem, an introductory lesson on the relation of classical to intuitionistic logic and mathematics. In the context of setting forth some of the basic technical facts, he agrees with Günther that the basis of the difference of intuitionist and classical logic is that they apply different conceptions of being. He adopts Günther's term 'aufweisbare Existenz' ('exhibitable existence') for the conception at work in intuitionism. "But the core of the intuitionistic objections lies surely in the fact that it is shown that for exhibitable existence certain statements of classical mathematics are unproved and others are even demonstrably false."

Gödel notes a point of philosophical agreement with Günther, that "the application of the method and results of mathematics [in particular, no doubt, mathematical logic] should not be limited to positivistic philosophy." Günther had remarked that he had been trying, thus far unsuccessfully, to persuade his fellow metaphysicians to pay attention to modern logic.

In his reply of 23 May 1954 Günther does more than before to set forth his basic ideas about the connection of metaphysics and logic. Some of the general ideas discussed above, in particular the connection Günther sees between classical logic and a certain metaphysical tradition, are sketched, and he introduces the idea of three-valued logic, which in his writings of this time is the main technical device for carrying out his logical program. Some of the discussion (following up the earlier exchange on intuitionism) is about different conceptions of being, and early in the letter Günther poses a challenge to Gödel. He remarks on

the aim stated in *Gödel 1944* "to set up a consistent theory of classes and concepts as objectively existing entities" (p. 152) and asks in terms of which of the different conceptions of being the "objectively existing entities" should be understood.

One might object to Günther's point of view in many ways. Gödel's response (letter 4, 30 June) is limited in what it takes on. He first denies that Günther's philosophical claims contradict his own results "although my results make impossible certain forms of a subjectivistic interpretation of mathematics and in general speak strongly against every such interpretation." He remarks on the relevance of undecidability theorems for the law of excluded middle.

Gödel agrees in general terms with Günther's idealistic way of reading intuitionism. But in a very striking remark, he simultaneously agrees about the general importance of some basic ideas from idealism and gives a ringing affirmation of his own realism:

> The reflection on the subject treated in idealistic philosophy (that is, your second theme of thought), the distinction of levels of reflection, etc., seem to me interesting and correct. I even consider it entirely possible that this is "the" way to the correct metaphysics. However, I cannot go along with the denial of the objective meaning of thought that is connected with it, [although] it is really quite independent of it. I do not believe that any Kantian or positivistic argument or the antinomies of set theory or quantum mechanics has proved that the concept of objective being (no matter whether for things or abstract entities) is senseless or contradictory. When I say that one can (or should) develop a theory of classes as objectively existing entities, I do indeed mean by that existence in the sense of ontological metaphysics (pp. 3–4).

Whether idealist philosophers deny "the objective meaning of thought" is a highly controversial matter. I don't think Gödel has really assimilated Günther's conception of "ontological metaphysics", which on his view is an expression of the identity of thought and being that he associates with the classical tradition. Gödel may be affirming a form of transcendental realism, but the very fact that he thinks it can coexist with what he agrees to be insights of idealism implies that it is not exactly ontological metaphysics as Günther understood it. The matter was pursued only to a limited degree in the correspondence. But it illustrates a puzzle about Gödel's philosophy that is not confined to this exchange. Gödel is reported by Hao Wang to have much later described his philosophy as idealistic (*Wang 1996*, remark 0.2.2, p. 8), and after 1959 he was much attracted to the thought of Husserl, who described his position as transcendental idealism. Yet Gödel, in writing about mathematics and in a few places (including the above) about physics,

expresses strong realistic convictions. He gives little explanation of how realism and idealism can coexist.[r]

Gödel declines to go into Günther's attempts in three-valued logic; he doesn't find a more detailed explanation of the connection of the truth-tables with Günther's philosophical ideas. But I will postpone until §3 a discussion of Gödel's reaction to Günther's own logical ideas.

Günther wrote two letters (2 October 1954 and 19 June 1955) before Gödel replied again. The issues he raises again concern the sense in which objects of different kinds are independent of thought. He mentions again some realistic remarks in *Gödel 1944* and quite reasonably queries the sense in which objects are said to be independent of thought (or of something else belonging to "us", e.g. "our definitions and constructions" (*Gödel 1944*, p. 134)). What concerns him is that different objects might be independent in different senses, although according to the first letter "the previous theory of thought from Aristotle to the present knows only *one* concept of logical (thought-independent) *object!*" This could be questioned, but it might be more relevant to object that classical formal logic does not distinguish different concepts of object or of thought-independence simply because the relevant differences are not in the province of formal logic; that need not make logic inapplicable across these differences. Perhaps intuitionistic logic does in its intended application call for a different conception of object, but it does not follow that classical logic is not applicable to a number of different conceptions.

Somewhat revealing is a remark Günther makes in letter 5 about the ontological status of space:

> If there is only one logically comprehensible form of existence, then the absence of everything "physical" is, as Plato thinks, mere nothing. If, however, we assume two forms of objectivity, then "empty" space is also a genuine "objectual" (*gegenständliches*) object of thought (p. 7).

But he seems to think that two-valued logic forces one to take the first view and thus to hold that space is "*objectively* considered nothing", which he claims to have been the purport of Kant's view that space is a form of intuition.

It's not surprising that in the next letter Günther gives some exposition of three-valued logic. But the context is a different line of thought,

---

[r]Of course this is not to say that they cannot. In particular, Husserl has been interpreted so that his position is at least compatible with "common-sense realism", and many philosophers, including perhaps Husserl, have undertaken to overcome the opposition between realism and idealism or some other opposite of realism. In his introductory note to *1961/?* (these *Works*, vol. III, pp. 364–373), Dagfinn Føllesdal evidently does not see an opposition between Husserl's "idealism" and Gödel's realism concerning mathematics.

the distinction between reflection on objects thought of as independent of thought and reflection on reflection itself. He claims that these two types of reflection require fundamentally different concepts of object and the difference consists in whether the object is altered by the act of thought directed at it. He maintains that this obtains when one thinks about one's own subjective act of thought.

Gödel finally replied on 10 August 1955 (letter 7). He directs his criticism mainly at the formulation in Günther's second letter; even in intuitionistic logic, which he concedes is "a reflection on thought", objects of thought are as little altered by acts of thought directed at them as the objects of physics. He agrees, however, that objects of the second type of reflection are totally different from objects of the first. This letter also contains Gödel's most extended comment on Günther's ideas about many-valued logic; see §3 below.

There followed a series of letters of Günther to which we have no replies by Gödel. Only the first (letter 8, 18 September 1955) represents an attempt to respond substantively to Gödel. On 22 June 1956, after returning to Richmond from his visit to Hamburg and a few months' stay in Chicago,[s] Günther asked for Gödel's support in his application for a renewal of his grant from the Bollingen Foundation and sent him a manuscript which appears to have been a preprint of *Günther 1958*, his most extended exposition of the ideas on many-valued logic in relation to metaphysics. On 17 September they met for a morning in Gödel's office, and evidently they discussed this paper.[t] Gödel wrote to the Foundation on 20 October to express his "wholehearted support."[u] He says that the recent paper "Die aristotelische Logik des Seins und die nicht-aristotelische Logik der Reflexion" (i.e., *Günther 1958*) "fully

---

[s]See *Günther 1963*, p. 13.

[t]See above, in particular note e. In a card sent from Hamburg on 20 December 1955, Günther had said he would return to the United States in April or May and would like to drop in on Gödel at his office in Princeton. No other correspondence discusses a meeting; the actual arrangement could well have been made by telephone. That they discussed a version of *Günther 1958* is shown by Gödel's letter to Günther of 4 April 1957.

[u]A typed draft of this letter, with corrections in Gödel's hand, is in Gödel's papers (document no. 010767). Quotations are from this draft, whose text is in all probability very close to that of the letter as sent. That 20 October was the date of the actual letter is confirmed by the Foundation's acknowledgment (Nancy Russ to Gödel, 8 November 1956).

confirms my favorable judgment".[v] Gödel's final remark is revealing about what attracted him in Günther's project:

> In view of the great interest which a satisfactory logical theory of "total reflexion" would have and in view of the depth of the philosophical problems involved three years, in my opinion, are not an excessively long time for studying these questions.

In the event the grant was extended for two years.[w] On 28 December Günther wrote asking for a recommendation for a professorship in Hamburg, which Gödel did write.[x]

On 26 February 1957, when Günther was in California for surgery, he wrote developing the idea that the concept of reflection really originates with Leibniz, although as a logical term "reflection" derives from Hegel. He then goes on to develop some interpretation of Leibniz, using some of his own conceptual apparatus. When Gödel next wrote, on 4 April 1957 (letter 10), he reacted skeptically, finding the ideas interesting but doubting their basis in Leibniz's text. Following up their discussion in Princeton in 1956 of the manuscript of *Günther 1958*, he says that its basic ideas need clearer explanation and elucidation by examples. In its present form it is "hardly intelligible". But Gödel expresses interest in an idea Günther had expressed earlier, probably in the manuscript that he sent in 1953, that "total reflection" would be something going beyond all ascent of types. (We will return to this subject in §3.) This pleased Günther enough to overcome the effect of the critical tone of the rest, as his reply of 7 April shows. He then sent Gödel some of his papers. Gödel drafted a short note, dated 2 May, acknowledging their receipt.[y]

It seems likely that Gödel's interest in Günther's ideas had begun to decline. In the remainder of 1957, Günther wrote only one substantive

---

[v]This brief remark replaces the crossed-out statement that the paper "gives a remarkably clear exposition of Dr. Günther's leading ideas within the field of logic and also develops new means for building a system of formal logic on their basis." Gödel wrote much less favorably about the paper to Günther on 4 April 1957. John Dawson has remarked to me that Gödel avoided writing unfavorable letters of recommendation. For that reason what he says in such letters cannot always be taken at face value.

[w]Günther informed Gödel of this in his letter of 8 December 1956.

[x]Gödel to Helmut Schelsky, 8 January 1957, marked "abgeschickt". In his letter of 15 January 1957, Günther acknowledges receipt of a letter from Gödel dated 10 January, now lost. He immediately thanks Gödel for the recommendation and makes no other comment. Gödel's letter may have done no more than inform Günther that the recommendation had been sent.

[y]Although no such letter is in Günther's papers, it seems likely that it was sent; Günther may well have thought it not worth saving.

letter (22 November, with copies of *Günther 1957* and *1957a*). Gödel replied on 23 December (letter 13) with rather brief comments on some of the papers Günther had sent. About *Günther 1957*, he says it has some overlap with a manuscript Günther had sent him previously, probably the one sent in 1953.[z] He says he has read some of "your new work", apparently *1957a*,[aa] and praises its clarity, particularly in contrast to the version of *1958* that he had read.

During 1958 Günther wrote Gödel twice, sending new papers. Gödel did not reply; when Günther wrote on 1 January 1959 expressing concern, Gödel wrote an interesting reply (letter 14, 7 January), again referring to Günther's earlier idea. But that was evidently his last reply. Günther continued to write and send papers through 1959 and 1960 but got no response from Gödel. In 1961 Günther obtained the appointment at the University of Illinois. The last item in the correspondence is an undated notice of the change of his address from Richmond to Urbana, postmarked 27 July 1961. On it Günther wrote, "I have accepted a research professorship at the University of Illinois and I am looking forward to it. Herzlichst, Ihr G. G." He was perhaps letting Gödel know that he too no longer needed to continue the exchange.

3. Gödel's reaction to Günther's logical ideas

Both in the correspondence and in published writings, Günther presses his constructions in many-valued logic, taking as his point of departure the views about metaphysics and logic discussed above. Gödel's remarks about this are rather brief. In letter 4 he writes:

> Unfortunately I can't go into your 3-valued logic more in this letter. What I miss so far is a more detailed explanation of the connection in content of your truth tables with your philosophical ideas.

---

[z]Gödel does not comment now or later on the issues about minds and machines that are discussed in *Günther 1952* and *1957a* and are mentioned in Günther's letter, although he had already written on the subject in the Gibbs Lecture *1951*. (Günther had included *1952* in the packet of writings he sent in April; see his letter of 7 April.)

[aa]It is not clear whether what is being praised for clarity is *Günther 1957a* or *1957*. The fact that Gödel does not more explicitly indicate that he is talking about the same writing as in the previous sentence speaks for *1957a*, as does the contrast with *1958*, since the latter two writings set forth Günther's ideas on three-valued logic, while *1957* does not.

He doesn't reject out of hand the idea of using many-valued logic; in letter 7 he writes that it is suggested by reflection on the paradoxes[ab] and by the fact that "necessary" and "possible" can be thought of as truth-values. But Gödel is not able here or elsewhere to make very much sense of Günther's effort.

Regarding the meaning of the truth-values, Günther had attempted an explanation in his letters and in at least one publication before that time, *Günther 1953*. But in the same letter Gödel wrote that the manner in which Günther wanted to introduce the third truth-value was "not wholly intelligible" to him and that he missed any explanation of what the three truth-values really mean. Gödel may not have seen enough by August 1955 to have even the limited grasp of Günther's intention expressed in §1. Probably he had by the time he read the manuscript of *Günther 1958*, but this did not lead to a change in his expressed attitude, and his view of that paper was not very favorable. It is easy to guess what the obstacle was: Günther wasn't able to give a motivation for modifying basic logic independent of his particular metaphysical interpretation of classical logic, and although he explains the three values in terms of his ideas about consciousness, he not only does not make clear how they are *truth*-values but seems to want to cut loose from the concept of truth, more so even than intuitionistic logic does. One might well ask whether what resulted should still be called *logic*.

Günther also seems never to have tried to write down axioms and rules of inference for his many-valued calculi. Probably that would have been necessary to persuade Gödel that this was intelligible logical work. But one can see why not only technical limitations prevented Günther from undertaking the task. If the "truth" values express levels of reflection, what would be the significance of being a *theorem* of the calculus? Such calculi are formulated in order to characterize conceptions of valid logical inference or logical truth. It's not clear that Günther's scheme has any place for these notions. That the concept of truth is not at center stage for him is indicated also by the fact that when he writes about intuitionism, although it is clear to him that in some sense the conception of existence is different from that in classical mathematics, he never remarks on the fact that there is a more underlying difference about truth. Gödel, however, does not make a remark that would smoke Günther out on this issue.

[ab]Such an idea has been explored by a number of later writers on semantical paradoxes, and it was also observed that in some three-valued logical settings universal comprehension can be preserved. For a survey with its own development of these ideas see *Feferman 1984*. In these developments truth and falsity have their usual meanings and the third value has the sense "undefined". That is evidently quite different from what Günther had in mind, as he seems to say explicitly in a parenthetical remark in letter 8, p. 2.

I have left for the last what is probably the most interesting of Gödel's reactions, on the matter of the relation of levels of reflection and type theories. As I have said, it is something Günther said about this that seems to have piqued Gödel's interest at the outset and made him tolerant of Günther's evident technical limitations. It is a theme that Gödel comes back to a couple of times without being prompted by Günther.

We have to reconstruct what Günther's idea may have been, since I have not been able to locate the manuscript Günther sent in 1953. Gödel found some overlap between that manuscript and *Günther 1957*, and as we saw above, Günther does formulate an idea of "total reflection" and claim that the third R-level at which that is achieved does contain within itself all iterations of reflection and so the hierarchy of types. In *Günther 1957* no connection with a type-free logic is made, but it seems that in the manuscript of 1953 such a connection was made. It doesn't appear, however, that any concrete suggestions for such a logic were offered, and Gödel had no success in prodding Günther to work out the idea, even though it was pressing this idea that most pleased Günther in Gödel's letter of 4 April 1957, as he wrote on 7 April. Very likely Günther's invocation of many-valued logic at this point disappointed Gödel.

Evidently Gödel entertained hopes that the kind of analysis of the structure of self-consciousness that Günther undertakes and finds precedent for in the classical idealists would yield not only philosophical illumination but even a basis for constructing a type-free logic that would be philosophically better motivated than what was then available. In his last letter, letter 14, he even says that the correct axioms for such a logic should follow "*with necessity* from philosophical insights about the essence of reflection." Much earlier he had made remarks that suggest that the most satisfactory theory of what he calls "concepts" would be a type-free logic of some kind.[ac] It is not known that Gödel did any substantial work toward the construction of such a logic. If he had something in hand early in his exchange with Günther, it is imaginable that he would have brought it up. But it seems to me more likely that he was very uncertain of how one might proceed and therefore perhaps disposed to entertain hopes that light might come from somewhat unlikely sources. Gödel seems never to have resolved this uncertainty, as is indicated by his remarks in his conversations with Wang about the theory of concepts and the unsolved nature of the "intensional paradoxes".[ad]

[ac] *1944*, pp. 132, 140; cf. pp. 109–110 of the introductory note in these *Works*, vol. II.

[ad] *Wang 1996*, §§8.5–8.6.

4. Conclusions

What does this exchange tell us about Gödel's thought during this period, beyond what one can learn from remarks he makes in his letters? What he did not say is revealing. He nowhere says that in order to attain a philosophical understanding of thought and its relation to objects, one might well begin with logic rather than taking "subject" and "object" or "thought" and "being" as fundamental categories. Such an outlook might, indeed, lead us to hold out greater hopes for understanding self-consciousness with the help of the logical analysis of semantic reflection, as it is carried out in theories of truth, than for the reverse procedure of expecting to motivate the axioms of a logical theory from an analysis of self-consciousness.

Still less does Gödel invoke the linguistic turn, for example what Michael Dummett regards as an axiom of analytical philosophy, "first, that a philosophical account of thought can be attained through a philosophical account of language, and, secondly, that a comprehensive account can only be so attained."[ae] One shouldn't read too much into the absence of something from correspondence, but Günther's ideas seem so obviously to call forth a response of some such kind that we might discern at least an indication that the linguistic turn, at least as Dummett understands it, was not central to Gödel's practice as a philosopher.[af]

Given Günther's obvious technical limitations and the difficulty Gödel had in making sense of his proposals, we might find it surprising that Gödel maintained his interest in Günther for as long as he did and even up to early 1957 wrote supportive letters for him (see §2 above). There seems to have been some mutual personal sympathy; for example both write about their health problems and respond sympathetically to the other's. Both were outsiders to the American philosophical world and shared the project of dissociating modern logic from positivistic philosophy. But a deeper reason is very likely that Gödel was at the time occupied with philosophical problems about concepts (it was the time during which he worked on *1953/9*) and was very unsure about how to approach them.[ag] He was evidently prepared to entertain the

---

[ae]*Dummett 1993*, p. 4.

[af]From Wang's reports of their conversations, one would conclude that in the 1970s Gödel was actively opposed to the linguistic turn as Dummett and many others would understand it. See *Wang 1996*, remarks 5.5.7–5.5.9, pp. 180–181, and the sharp distinction between the semantic and intensional paradoxes made in §8.5.

[ag]That this uncertainty was a source of his interest in Günther was suggested by Warren Goldfarb.

possibility that post-Kantian idealism, to which he had apparently not
had a lot of exposure, would be a source of illumination. He found
Günther a clear expositor of ideas from that tradition.[ah] But he does
not seem to have been disposed to work out himself a line of thought
in which self-consciousness is a central concept, and when Günther did
not pursue what Gödel thought the most promising direction, he lost
interest. Not long after his last letter he began his study of Husserl,
whose version of idealism he seems to have found much more satisfactory.

Charles Parsons[ai]

[ah]Letter to Schelsky (see note x above).

[ai]I am indebted to Klaus Oehler, Claus Baldus and Markus Stepanians for information about Günther, to Øystein Linnebo and Thomas Teufel for assistance in locating writings of Günther and for discussion of his ideas, and to John Dawson, Solomon Feferman, Warren Goldfarb, Sally Sedgwick and Wilfried Sieg for helpful comments.

1. Günther to Gödel

Gotthard Günther
101 Oronoco Ave. (Apt. 2)
Richmond 22, Va.
April 29. 1954

Sehr geehrter Herr Professor Gödel:

Ich arbeite gerade an einer philosophischen Analyse des Satzes vom
ausgeschlossenen Dritten und bin dabei auf eine Bemerkung Karl
Mengers über Sie gestossen, die mir nicht klar ist. Menger schreibt (Krise
und Neubau in den exakten Wissenschaften, Die neue Logik, Leipzig.
Wien 1933,[a] S. 11): "Nun hat ... Gödel kürzlich gefunden, dass nicht
nur die intuitionistische Mathematik ein Teil der klassischen ist, sondern
auch der gesamte klassische Aussagenkalkül und die gesamte klassische
Zahlentheorie samt dem Satz vom ausgeschlossenen ⟨Dritten⟩ als Teil des
Intuitionismus aufgefasst werden können, indem man durch ein einfaches
Wörterbuch jeden klassischen Satz in einen intuitionistischen übersetzen
kann. . . . Die Ablehnung des Satzes vom ausgeschlossenen Dritten hat
also (da die Intuitionisten Unmöglichkeiten von Allaussagen zulassen) in

[a]*Menger 1933.*

A complete calendar of the correspondence with Günther appears on pp. 563–564 of this volume. The editors are indebted to Mr. Lothar Busch for locating Gödel's letters in Günther's papers while they were being catalogued, to Professor Tilo Brandis, Director of the Handschriftenabteilung of the Staatsbibliothek zu Berlin-Preussischer Kulturbesitz, for making copies of these letters available early in our work, as well as to Delia Graff and Øystein Linnebo for preparing typescripts of handwritten letters. The translation is by Thomas Teufel and Charles Parsons, revised using suggestions of John W. Dawson, Jr.

1. Günther to Gödel

Gotthard Günther
101 Oronoco Ave. (Apt. 2)
Richmond 22, Va.
April 29, 1954

Dear Professor Gödel,

I am currently working on a philosophical analysis of the law of the excluded middle and in the process encountered a remark about you by Karl Menger that is unclear to me. Menger writes (Krise und Neubau in den exakten Wissenschaften, Die neue Logik, Leipzig. Wien 1933,[a] p. 11): "Now... Gödel has recently discovered not only that intuitionistic mathematics is a part of classical mathematics, but also that the entire classical propositional calculus and the entire classical number theory, as well as the law of the excluded middle, can be understood as part of intuitionism, by translating every classical proposition into an intuitionistic one using a simple dictionary.... The rejection of the law of the excluded middle, therefore (since the intuitionists allow impossibilities of universal quantifications), in

Wahrheit gar keine Einschränkung, sondern *bloss eine Umbenennung* der klassischen Sätze zu Folge." Menger weisst dabei auf < "Ergebnisse eines mathematischen Kolloquiums", 4, Leipzig 1933>[b] hin, das mir hier aber leider nicht zugänglich ist, weshalb ich mich | an Sie direkt mit der Bitte um freundliche Auskunft wende.

Die intuitionistische Mathematik ist ein Teil der klassischen. Ich verstehe das. Wenn es dann aber weiter heisst, dass der klassische (also zweiwertige) Aussagenkalkül als *Teil* des Intuitionismus aufgefasst werden kann, so müssen doch wohl *zwei* verschiedene Konzeptionen des Tertium non datur im Spiel sein.

1) wenn die intuitionistische Math. Teil der Klassischen ist, dann steht sie *unter* dem *generellen* klassischen Tertium non datur
2) wenn das Tertium non datur Teil des Intuitionismus ist, muss hier eine *speziellere* Interpretation des Drittensatzes in Frage kommen.

Nun lässt sich der Drittensatz in der Tat auf verschiedene Weise formulieren so dass wir einmal eine allgemeinere (schwächere) Formulierung, das andere Mal eine stärkere (aber speziellere) Version erhalten. Als allgemeinere Formulierung möchte ich anführen:

$$[(A \to B) \cdot (\sim A \to B)] \to B \tag{1}$$

Die andere erhalten wir auf die folgende Weise. Wir gehen von dem logischen ε-Axiom

$$A(x) \to A(\varepsilon A)$$

aus. Dabei soll $\varepsilon(A)$ ein Objekt bezeichnen für das $A(x)$ wahr ist, wenn $A(x)$ überhaupt für irgend ein Objekt wahr ist. Mit "ε" lässt sich nun sowohl der Existenz- wie der Alloperator definieren: |

$$(Ex)A(x) =_{\text{Def}} A[\varepsilon(A)]$$

$$(x)A(x) =_{\text{Def}} A[\varepsilon(\sim A)]$$

Auf Grund dieser Definition liefert uns "ε" die zweite Formulierung für das Tertium non datur:

$$\sim [(x)A(x)] \to (Ex) \sim A(x) \tag{2}$$

---

[b] I.e. *Gödel 1933e*. These pointed brackets in the German text are in the letter and so do not signify an insertion. Günther fairly frequently uses pointed brackets as another style of quotation marks. For his pointed brackets we use < and >; to signify insertions we use ⟨ and ⟩. In the translation, however, the usual American conventions concerning quotation have been followed.

fact implies no limitation at all but *merely a renaming* of classical propositions." Menger refers to "Ergebnisse eines mathematischen Kolloquiums
4, Leipzig 1933"[b], which is unfortunately not accessible to me here. That
is why I am turning directly to you to ask if you would be so kind as to
give me information.

Intuitionistic mathematics is a part of classical mathematics. That
I understand. But when the passage says in addition that the classical
(that is, two-valued) propositional calculus can be conceived as *part* of
Intuitionism, this can only mean that *two* different conceptions of the
tertium non datur are at play here.

1) if intuitionist mathematics is part of classical, then it is *subject* to
the *general* classical tertium non datur;

2) if the tertium non datur is part of intuitionism, then a *more specific*
interpretation of the law of the excluded middle must be in question.

Now, the law of the excluded middle can indeed be formulated in different ways, such that we either obtain a more general (weaker) formulation, or a stronger (but more specific) version. As a more general formulation I would like to offer:

$$[(A \to B) \cdot (\sim A \to B)] \to B \tag{1}$$

The other formulation is obtained in the following manner. We start
with the logical ε-axiom:

$$A(x) \to A(\varepsilon A).$$

Here $\varepsilon(A)$ is meant to denote an object for which $A(x)$ is true, provided
$A(x)$ is true of any object at all. With "ε" the existential quantifier as
well as the universal quantifier can be defined:

$$(Ex)A(x) =_{\text{Def}} A[\varepsilon(A)]$$

$$(x)A(x) =_{\text{Def}} A[\varepsilon(\sim A)]$$

On the basis of this definition "ε" provides us with the second formulation of the tertium non datur:

$$\sim[(x)A(x)] \to (Ex) \sim A(x) \tag{2}$$

Was ich gern wissen möchte ist: ist die zitierte Bemerkung Mengers so zu
verstehen, dass die intuitionistische Mathematik ein *Teil* des Systems ist
dass durch (1) umschrieben wird und enthält sie andererseits (2) als einen
Teil ihrer selbst?

Zwischen (1) und (2) besteht ja in der Tat ein beträchtlicher ontologis-
cher Unterschied. Lassen Sie mich, bitte, diesen Unterschied auf folgende
Weise beschreiben: Paul Hofmann hat in einer Studie, die zum Besten
gehört, was über den Drittensatz geschrieben worden ist, die "unklare Fassung" des Tertium non datur gerügt.[c] Es fehle diesem logischen Axiom ein
oberster Bestimmungsgesichtspunkt unter dem sich "*A*" und "∼ *A*" kon-
tradiktorisch ausschliessen. Das ist in der Tat für Formulierung (1) der
Fall.

Dagegen ist für (2) ein solcher oberster Bestimmungsgesichtspunkt vorausgesetzt. Derselbe ist durch $\varepsilon(A)$ gegeben. D. h. es muss ein Objekt *vorhanden* sein, auf das sich die Aussage bezieht. Der oberste Bestimmungsgesichtspunkt ist also: objektiv vorhandenes Sein ... oder
4   (wenn Sie sich an | dem metaphysischen Terminus <Sein> stossen)
gegenständliche Existenz.

Der Kern der intuitionistischen Einwände scheint mir nun darin zu
liegen, dass für jede *positive* Formulierung von (2) eine allgemeinere gefunden werden kann, die die vorangehende positive Formulierung des ange-
blichen obersten Bestimmungsgesichtspunkts desavouiert. <Sein> ist kein
absolutes Datum, sondern ein Komplementärbegriff zu dem jeweiligen logischen Interpretationssystem, das wir benutzen. Deshalb ist (2) immer nur
relativ gültig für *aufgewiesene Existenz* für die der Intuitionismus den Drittensatz ja in der Tat zulässt.

Die Formulierung (1) aber repräsentiert die nach Hofmann "unklare Formulierung" des Tertium non datur. Der oberste Bestimmungsgesichtspunkt
für die totale Disjunktion fehlt. Nur ist das m. E. *kein Mangel*. Denn dieses
Fehlen indiziert dass (1) die Formulierung des Drittensatzes für eine unendliche Hierarchie auf einander folgender *möglicher* Bestimmungsgesichtspunkte ist. In diesem Sinn würde (1) den Intuitionismus einschliessen.—

Habe ich Unrecht, so wäre ich für entsprechende Bericht[ig]ung durch
Sie herzlich dankbar. Bitte vergessen Sie nicht, ich bin meiner Ausbildung
5   nach Historiker und Metaphysiker. Ich versuche das, was heute in | Logik
und Mathematik geschieht philosophisch zu interpretieren, aber nicht es
besser zu machen.

Vorläufig versuche ich allerdings meine Kollegen in der Metaphysik ohne
jeden Erfolg zu der Ansicht zu bekehren, dass man heutzutage nicht

---

[c] *Hofmann 1931*. Günther may refer to pp. 13–14 (= pp. 93–95 of *Hofmann 1931a*).

What I would like to know is: should we understand Menger's quoted comment as ⟦saying⟧ that intuitionistic mathematics is a *part* of the system characterized by (1), and does it, on the other hand, contain (2) as part of itself?

Between (1) and (2) there is indeed a significant ontological difference. Please let me describe this difference in the following way: in a study that belongs to the best things that have been written about the law of the excluded middle Paul Hofmann criticizes the "unclear formulation" of the tertium non datur.[c] ⟦He points out that⟧ this logical axiom lacks a highest determining perspective under which "*A*" and "∼ *A*" exclude each other as contradictories. This is indeed the case for formulation (1).

In contrast, for (2) such a highest determining perspective is presupposed. It is given by $\varepsilon(A)$. I.e., an object must be *given* to which the proposition refers. The highest determining perspective is therefore: objectively existing Being... or (if you take exception to the metaphysical term "Being") existence as an object.

The core of the intuitionistic objections now appears to me to be, that for any *positive* formulation of (2) a more general one can be found which disavows the preceding positive formulation of the allegedly highest standpoint of determination. "Being" is not an absolute datum but a notion complementary to the particular logical system of interpretation we use. Because of that, (2) is always only relatively valid for *exhibited existence*, for which intuitionism indeed admits the law of the excluded middle.

Formulation (1), however, represents the "unclear formulation" of the tertium non datur according to Hofmann. The highest determining perspective for the total disjunction is missing. But in my view this is *not a shortcoming*. For this lack indicates that (1) is the formulation of the law of the excluded middle for an infinite hierarchy of successive *possible* determining perspectives. In this sense (1) would include intuitionism.—

If I am mistaken, I would be most grateful for relevant correction on your part. Please do not forget that I am a historian and metaphysician by training. I try to interpret philosophically what is happening in contemporary logic and mathematics, not to improve upon it.

For the time being, though, I am trying without any success to convert my colleagues in metaphysics to the view that these days one can

mehr Metaphysik treiben kann, ohne die Arbeit der letzten hundert Jahre in symbolischer Logik und Mathematik zugrunde zu legen.

Jedenfalls bemühe ich mich für meine Person das zu tun und wäre Ihnen deshalb für Ihre Hilfe herzlich dankbar.

Mit freundlichen Grüssen und besten Dank im Voraus

   Ihr

            Gotthard Günther

2. Gödel to Günther

                                        Princeton, 15./V. 1954

      Sehr geehrter Herr Dr Günther:

Um zu erklären, in welchem Sinn die von Ihnen zitierte Mengersche Äußerung zu verstehen ist, möchte ich zunächst feststellen, daß, wörtlich genommen, weder die intuit. Math. ein Teil der klassischen ist, noch umgekehrt. Denn die intuit. Begriffe sind verschieden von den gleichbezeichneten klassischen u. können auch nicht aus diesen definiert werden (noch umgekehrt). Wie Sie richtig bemerken, hat diese Verschiedenheit ihren Grund darin, daß verschiedene Begriffe des Seins verwendet werden. Das hindert aber nicht, daß es gewisse Begriffe in der klass. Math. gibt, die *formal* denselben Gesetzen gehorchen wie die int. Begriffe[1], ~~{u. umgekehrt}~~ ~~Das erstere~~ Begriffe in der int. Math., für die formal die Grundsätze der klass. Math. gelten. Das erstere | gilt z. B. (innerhalb von Aussagen u. Funktionenkalkül 1. Stufe) für die gleichbezeichneten Begriffe. Das letztere gilt (innerhalb der Zahlentheorie), wenn man den ~~intuit.~~ klass. Begriffen:

$$\sim p, \; p \cdot q, \; p \vee q, \; p \supset q, \; (x)F(x), \; (\exists x)F(x),$$

die folgenden intuit. entsprechen läßt:

$$\sim p, \; p \cdot q, \; \sim(\sim p \cdot \sim q), \; \sim(p \cdot \sim q), \; (x)F(x), \; \sim(x) \sim F(x).$$

Für das so definierte "oder" gilt offenbar der Drittsatz in der intuit. Math. u. dasselbe gilt für die andern logischen Grundsätze[2] u. daher auch für alle Theoreme. Die klass. Zahlentheorie hat also ein vollständiges formales Bild innerhalb der int. Zahlentheorie, u. da es in der Math.

---

[1] d. h. genauer, die allen Gesetzen der intuit. Math. gehorchen ohne Rücksicht darauf, ob sie vielleicht außerdem noch andern gehorchen

[2] was nicht so trivial ist wie im Fall des Drittsatzes

no longer pursue metaphysics without taking as the basis the work of the last hundred years in symbolic logic and mathematics.

In any case, I for my part strive to do this and would therefore be most grateful for your help.

With cordial greetings and many thanks in advance,

Yours,

Gotthard Günther

2. Gödel to Günther

Princeton, 15 May 1954

Dear Dr. Günther,

In order to explain in which sense one should understand the remark of Menger that you quote, I would first like to state that, taken literally, intuitionism is neither part of classical mathematics, nor vice versa. For intuitionistic concepts are different from the classical concepts of the same name and cannot be defined in terms of them either (nor vice versa). As you correctly observe, the basis for this difference is that different concepts of Being are employed. That does not, however, prevent its being the case that there are certain concepts in classical mathematics which obey *formally* the same laws as the intuitionistic concepts,[1] and conversely concepts in intuitionistic mathematics for which formally the principles of classical mathematics hold. The former holds, for example, for concepts of the same name (within the propositional and first- order functional calculus). The latter holds (within number theory), if one correlates to the classical concepts

$$\sim p, \ p \cdot q, \ p \vee q, \ p \supset q, \ (x)F(x), \ (\exists x)F(x)$$

the following intuitionistic concepts:

$$\sim p, \ p \cdot q, \ \sim(\sim p \cdot \sim q), \ \sim(p \cdot \sim q), \ (x)F(x), \ \sim(x) \sim F(x).$$

For the "or" thus defined the law of the excluded middle evidently holds in intuitionistic mathematics, and the same is true for the other basic logical principles[2] and therefore also for all theorems. Classical number theory thus has a complete formal image within intuitionistic number

---

[1] I.e., more precisely, which obey all the laws of intuitionist mathematics without regard to whether they perhaps also obey others.

[2] which is not as trivial as in the case of the law of the excluded middle.

in erster Linie auf die Form (u. nicht den Inhalt) ankommt, so heißt das
für die Math. praktisch dasselbe, als wenn die klass. Zahlentheorie Teil
3   der intuit. wäre. | Sie haben Recht daß dabei zwei verschiedene Konzep-
tionen des Drittensatzes im Spiel sind. Die zweite, die in der intuit. Math
gilt, lautet: "p und $\neg p$ können nicht beide falsch sein". Was Sie als
zweite (speziellere) Konzeption anführen, nämlich:

$$\sim (x)A(x) . \supset . (\exists x) \sim A(x) \qquad\qquad\qquad (2)$$

gilt nicht in der intuit. Math u. ist (zumindest innerhalb der Zahlentheorie) sogar mit $p \vee \sim p$, sowie auch mit der von Ihnen angeführten 1.
Konzeption äquivalent. Für *entscheidbare* Eigenschaften A (d. h. solche
für die $(x)[A(x) \vee \sim A(x)]$ beweisbar ist), wobei x eine Variable für natürliche Zahlen ist, gilt (2) für *gewisse formalisierte Teilsysteme* $\langle S \rangle$ des
Int.[3] in dem Sinn, daß, wenn eine Ableitung eines Widerspruchs *in S* aus
4   $(x)A(x)$ vorliegt, eine Zahl n berechnet werden | kann, für die $\sim A(n)$
gilt.

Es ist mir nicht klar, was Sie unter Ihrer "unendlichen Hierarchie möglicher Bestimmungsgesichtspunkte für den Drittensatz" verstehen. Wenn
Sie damit meinen, daß es zwischen aufgewiesener u. objektiver Existenz
eine unendliche Folge von in verschiedenem Grade aufweisbarer Existenz gibt, so würde ich dem in einem gewissen Sinne zustimmen. Aber
der Kern der int. Einwände liegt wohl darin, daß gezeigt wird, daß für
aufweisbare Existenz gewisse Sätze der klass. Math. unbewiesen u. andere
sogar nachweislich falsch werden (z. B. gilt ja für gewisse Satzklassen K
im Int.: $\sim (p)(p \,\varepsilon\, K . \supset . p \vee \sim p)$)

Ich bin ganz Ihrer Meinung, daß die Anwendung der Methoden u. Resultate der Math. nicht auf die positivistische Phil. beschränkt bleiben sollte.
Ich habe mich sehr gefreut, daß Sie das Stipendium der Bollingen Foundation bekommen haben u. wünsche Ihnen besten Erfolg für Ihre Arbeit.

Mit besten Grüssen

Ihr

Kurt Gödel

---

[3]z.B. für das Heytingsche System der int. Zahlentheorie

theory, and, since mathematics is primarily concerned with form (and not content), this means for mathematics practically the same as if classical number theory were part of intuitionistic number theory. You are right that two different conceptions of the law of the excluded middle are in play here. The second, which holds for intuitionistic mathematics, says: "p und $\neg p$ cannot both be false". What you adduce as the second (more specific) conception, namely:

$$\sim (x)A(x) . \supset . (\exists x) \sim A(x) \tag{2}$$

does not hold in intuitionistic mathematics and is (at least within number theory) even equivalent with $p \vee \sim p$, as well as with the first conception you mention. For *decidable* properties A (i.e., for those for which $(x)[A(x) \vee \sim A(x)]$ can be proved), where x is a variable for natural numbers, (2) holds for *certain formalized subsystems S* of intuitionism,[3] in the sense that, if a contradiction can be derived *in S* from $(x)A(x)$, a number n can be computed for which $\sim A(n)$ holds.

It is not clear to me what you mean by your "infinite hierarchy of possible determining perspectives for the law of the excluded middle". If you mean that between exhibited and objective existence there is an infinite sequence of different degrees of exhibitable existence, I would agree with you in a certain sense. But the core of the intuitionistic objections is, rather, that it is shown that for exhibitable existence certain theorems of classical mathematics become unproved and others even demonstrably false (for example, for certain classes of propositions K, $\sim (p)(p \, \varepsilon K . \supset . p \vee \sim p))$ holds in intuitionism).

I fully agree with your view that the application of the methods and results of mathematics should not remain confined to positivistic philosophy. I was very pleased that you have received the stipend from the Bollingen Foundation and wish you the best of success with your work.

With cordial greetings,

Yours, Kurt Gödel

---

[3] e.g., for Heyting's system of intuitionistic number theory.

3. Günther to Gödel

Gotthard Günther
101 Oronoco Ave. (Apt. 2)
Richmond 22
May 23, 1954

Sehr geehrter Herr Professor Gödel:

Vor einigen Tagen erhielt ich Ihren eingehenden Brief vom 15. V., und ich möchte Ihnen hiermit recht herzlich für die darin enthaltene Auskunft danken. Ihr Schreiben hat mir in der Tat einige Punkte, die mir in der Literatur unklar geblieben waren, aufgehellt. Speziell die Mengersche Äusserung, hinter der ich etwas ganz Anderes vermutet hatte.—Ich könnte diesen Brief damit schliessen. Wenn ich dies aber nicht tue und Ihnen hiermit einige weitere Gedankengänge unterbreite, bitte ich mir zu glauben, dass ich dies mit Rücksicht auf Ihre kostbare Zeit nur äusserst zögernd tue. Aber ich befinde mich in einer gewissen Zwangslage. Ich bin—abgesehen von den Neo-Thomisten—so ziemlich der einzige Metaphysiker, der davon überzeugt ist, dass man heute nicht Metaphysik treiben kann, ohne die Ergebnisse der symbolischen mathematischen Logik vorauszusetzen. Und die symbolische Logik im Neo-Thomismus (Ivo Thomas z. B.) ist meiner Ansicht nach auf einem Irrweg. Es wird dort nämlich nicht zugegeben, dass der logische Positivismus überzeugend
2   demonstriert hat, dass die *klassische ontologische* Meta-|physik wissenschaftlich unhaltbar ist. Anstatt die Resultate der Logistik für eine *neue* Metaphysik zu verwenden, versucht man dort immer noch die mittelalterliche Kirchenmetaphysik (die Fundamental-ontologie) mit mathematischer Logik zu beweisen. Auf der anderen Seite steht Heidegger, der erst kürzlich wieder die Logistik eine "Ausartung", die sich mit einem "Schein der Produktivität" umgibt, genannt hat.[a] —Von diesen Leuten her, können meine Gedanken also keine Kontrolle erfahren. Ich muss mich also schon an mathematische Logiker wenden.

Im Folgenden möchte ich Ihnen einige grundsätzliche Gedanken unterbreiten, in denen mich Ihr Brief noch bestärkt hat. Ich bitte Sie dieselben nur zur Kenntnis zu nehmen, und Sie brauchen diesen Brief nicht zu beantworten. Es sei denn, Sie entdecken in meinen philosophischen Theoremen etwas das positiv falsch auf der Basis Ihrer eigenen Untersuchungen und Ergebnisse sein muss. In diesem Fall wäre ich für einen entsprechenden Hinweis äusserst dankbar:

---

[a] Günther probably refers to the postscript of 1943 to *Heidegger 1929*. There Heidegger raises the question whether thought stands "in the law of truth" when it "follows that thought that 'logic' grasps in its forms and rules". He mentions "die 'Logik', als deren folgerichtige Ausartung die Logistik gelten darf" (" 'logic', as whose consistent

3. Günther to Gödel

Gotthard Günther
101 Oronoco Ave. (Apt. 2)
Richmond 22
May 23, 1954

Dear Professor Gödel,

Several days ago I received your detailed letter of 15 May, and I would hereby like to thank you most warmly for the information contained in it. Your letter has indeed illuminated for me several points that had remained unclear to me in the literature. In particular Menger's remark, behind which I had conjectured something entirely different.—I could close this letter at this point. If, however, I don't do so and submit to you here several further lines of thought, I ask you to believe me that I do this only extremely hesitantly, in consideration of your valuable time. But I find myself in a certain bind. I am—apart from the neo-Thomists— pretty well the only metaphysician who is convinced that one cannot do metaphysics today without presupposing the results of symbolic mathematical logic. And symbolic logic in neo-Thomism (Ivo Thomas for example) is in my view on the wrong track. For there it is not admitted that logical positivism has convincingly demonstrated that *classical ontological* metaphysics is scientifically untenable. Instead of applying the results of symbolic logic for a *new* metaphysics, one still tries to prove medieval church metaphysics (fundamental ontology) by means of mathematical logic. On the other side stands Heidegger, who just recently again called symbolic logic a [manifestation of] "degeneracy", which surrounds itself with an "illusion of productivity".[a] —Thus from these people my thoughts cannot receive any supervision. I must therefore turn to mathematical logicians.

In the following I would like to present to you several fundamental thoughts, in which your letter has encouraged me. I only ask you to take note of them, and you do not need to answer this letter. Unless, that is, you find something in my philosophical theorems that, on the basis of your own investigations and results, must be positively false. In that case I would be very grateful for a corresponding suggestion:

---

degeneration symbolic logic may count", *1967*, p. 104). Heidegger enters into a discussion of calculation, and it is "calculating thought" that is said to be given a "Schein der Produktivität" (illusion of productivity) by the nature of calculation (ibid.). Friedman (*2000*, note 207) considers that the larger passage is a reply to the famous criticism of *Heidegger 1929* in *Carnap 1932*.

Die oberste Formel der klassischen Philosophie seit Plato/Aristoteles
lautet ὄντως ὄν D. h. Sein des Seienden. Wir haben also ein zweistu-
figes Wissen. Empirisches Wissen vom *Seienden* (Math. & Physik) und
apriorisches Wissen vom *Sein* (Logik & Metaphysik).[b] Dem zweistufigen
Wissen entspricht eine zweistufige Objektwelt. Wir haben erstens: den
Objektraum als die Vielheit der empirischen Dinge. Dahinter aber steht
3   als zweites, totales Objekt: das absolute Sein. |

Kant hat zuerst, *transzendental* gezeigt, dass das absolute Sein kein
wissenschaftliches Objekt sein kann. Seine Demonstration aber war nicht
überzeugend, weil sich die Kantische Transzendentallogik nicht formali-
sieren lässt. In der Logistik ist dann dasselbe Resultat erreicht worden,
mit dem Hinweis darauf, dass Prädikatsfunktionen eine Variable enthalten und dass der Wert der Variablen nur *empirisch* aufgenommen werden
kann. "Sein" ist der faktische Argumentwert einer Variablen. Quine: "To
be is to be the value of a variable."

Damit aber tauchte eine neue, bisher nicht dagewesene Schwierigkeit
auf. Ich will sie so kurz als möglich beschreiben. In der klassischen Tradition zählt das Subjekt des Denkens und der Reflexionsprozess über-   en
haupt nicht. Das Ziel des Denkens ist den Sinn des *absolut objektiven*
Seins zu fassen. Und Wahrheit bedeutet absolute Übereinstimmung des
Denkens mit dem absolut objektiven Gegenstand. D. h. alle Kategorien
der Logik müssen, wenn sie wahr sein sollen, absolut objektiv definierbar
sein. Alles "Subjektive" ist schlechthin zu eliminieren.

Nun kam aber erst die Kritik der reinen Vernunft und erklärte:
Dinge an sich sind grundsätzlich keine Objekte des Bewusstseins. Dann
kam der logische Positivismus und bestätigte: absolute Objektivität ist
eine blosse Fiktion. Überdies hat ca. 1930 Heisenberg in einer bedeutsamen Schrift über die Grundlagen der Quantenmechanik erklärt: "...
4   der absolut isolierte Gegenstand hat prinzipiell keine beschreibbaren |
Eigenschaften mehr."[c] D. h. die Experimentalsituation des Beobachters
muss in die Beschreibung des Objekts mit hineindefiniert werden. Schön,
wenn das aber so ist dass das absolut objektive *Sein* nur eine Fiktion ist
und wir nur *relativ* objektiven Seienden begegnen, dann ist die klassische
Logik im Irrtum, wenn sie annimmt, dass Objektivität (oder Sein) das
einzige rationale Thema des Denkens ist. Das logische Denken hat dann
zwei fundamentale Themata: 1.) das empirische (relative) Objekt und 2.)
den subjektiven Reflexionsprozess; der sich selbst nur teilweise in pseudo-
objektive Kategorien auflösen lässt.

---

[b]The line "Empirisches Wissen.. apriorisches" is marked by lines in the margin. It
is probable that all such markings are by Gödel.

The highest formula of classical philosophy since Plato/Aristotle is ὄντωζ ὄν. i.e., the Being of existing things. We thus have two-tiered knowledge. Empirical knowledge of *existing things* (mathematics & physics) and a priori knowledge of *Being* (logic & metaphysics).[b] To this two- tiered knowledge there corresponds a two-tiered world of objects. We have, first: the space of objects as the multiplicity of empirical things. Behind this, however, there stands as second, total object: absolute Being.

Kant was the first to show *transcendentally* that this absolute Being cannot be an object of science. But his demonstration was not convincing, because Kantian transcendental logic cannot be formalized. In symbolic logic the same result was then attained, by noting that predicate functions contain a variable and that the value of the variables can only be determined *empirically*. "Being" is the factual value of a variable. Quine: "To be is to be the value of a variable".

But with this a new, unprecedented difficulty emerged. I want to describe it as briefly as possible. In the classical tradition the thinking subject and the process of reflection do not count at all. The goal of thought is to grasp the sense of *absolutely objective* Being. And truth means absolute agreement of thought with the absolutely objective object. That is, all categories of logic must, if they are to be true, be absolutely objectively definable. Everything "subjective" is simply to be eliminated.

But now, first the Critique of Pure Reason came along and declared: Things in themselves are in principle not objects of consciousness. Then came logical positivism and affirmed: absolute objectivity is a mere fiction. Moreover, around 1930 Heisenberg explained in an important paper on the foundations of quantum mechanics: "...the absolutely isolated object does not, in principle, any longer have any describable properties".[c] That is, the experimental situation of the observer has to be defined into the description of the object.

Fine, but if it is the case that absolutely objective *Being* is only a fiction and we only encounter existing things that are *relatively* objective, then classical logic is in error when it assumes that objectivity (or Being) is the *only* rational topic of thought. Logical thought then has *two* fundamental topics: 1.) the empirical (relative) object and 2.) the subjective process of reflection; which itself can only be partially resolved into pseudo-objective categories.

---

[c] *Heisenberg 1931*, p. 182.

Diese beiden Themata aber werden in der gegenwärtigen Logik in einer höchst unzulässigen Weise mit einander vermischt. In Ihrem Aufsatz <Russell's Mathematical Logic" (Evanston & Chicago 1944)[d] bemerken Sie sehr richtig dass unser Ziel (aim) ist "... to set up a consistent theory of classes and concepts as objectively existing entities." (S. 152) Andererseits aber bemerken Sie in Ihrem Brief an mich, dass die Differenz zwischen klassischen und intuitionistischen Prinzipien ihren Grund darin hat "dass verschiedene Begriffe des Seins verwendet werden".

Damit aber entsteht die Frage: welcher der verschiedenen Begriffe des Seins soll den "objectively existing entities" zugrunde gelegt werden?

Die Frage ist heute nirgends zureichend beantwortet, weil man sich 5 nicht genügend Rechenschaft über | die Differenz verschiedener Seinsbegriffe gegeben hat. Um diesen Brief nicht zu lang werden zu lassen, möchte ich das Nächste in Form von Thesen (ohne längere Begründungen) konstatieren:

1) Es gibt zwei, und nur zwei, fundamentale Seinsbegriffe:

    a) Sein, das rein objektiv-thematisch definiert werden kann, d. h. ohne das denkende Subjekt in den Objektbegriff mit hinein zu definieren.

    b) Sein, doppel-thematisch interpretiert. D. h. der "subjektive" Reflexionsprozess der "Sein" denkt, muss in den Objektbegriff hineindefiniert werden.

Die Differenz dieser beiden Seinsbegriffe scheint mir dem Unterschied von klassisch:

$$\sim p, \ p \cdot q, \ p \vee q, \ p \supset q, \ (x)F(x), \ (Ex)F(x)$$

und intuitionistisch:

$$\sim p, \ p \cdot q, \ \sim(\sim p \cdot \sim q), \ p \supset q, \ (x)F(x), \sim (x) \sim F(x)$$

zu grunde zu liegen. Wenn wir nämlich die klassische Disjunktion durch "$\sim (\sim p \cdot \sim q)$" vertreten lassen, so haben wir damit ausgedrückt, dass der Drittensatz klassisch $(p \vee q)$ nur für absolut objektives Sein gilt. Für den zweiten Seinsbegriff gilt er nur abgeschwächt, insofern als wir den durch "$\sim \ldots$" ausgedrückten Reflexionsprozess in unsere Konzeption des Objekts hineinnehmen müssen.

Zwischen "$\sim \ldots$" in der Tafel

p	$\sim p$	(I)
W	F
F	W

In contemporary logic, however, these two topics are being mixed together in a highly inadmissible fashion. In your essay "Russell's mathematical logic" (Evanston & Chicago 1944)[d] you very correctly observe that our aim is "...to set up a consistent theory of classes and concepts as objectively existing entities" (p. 152). But on the other hand, you remark in your letter to me that the difference between classical and intuitionistic principles has its basis in the fact "that different notions of Being are being used".

But then the question arises: on which of the different notions of Being should the "objectively existing entities" be based?

Today this question is nowhere sufficiently answered, because the distinction between different notions of Being has not been satisfactorily accounted for. In order not to let this letter go on for too long, I would like to state the following in the form of theses (without longer justifications):

1) There are two, and only two, fundamental concepts of Being:
   a) Being that can be defined purely objective-thematically, i.e., without defining the thinking subject into the concept of the object.
   b) Being that is double-thematically interpreted. That is, the "subjective" process of reflection that thinks "Being" has to be defined into our concept of the object.

The difference between these two concepts of Being seems to me to lie at the foundation of the difference between classical:

$$\sim p, \ p \cdot q, \ p \vee q, \ p \supset q, \ (x)F(x), \ (Ex)F(x)$$

and intuitionistic:

$$\sim p, \ p \cdot q, \ \sim(\sim p \cdot \sim q), \ p \supset q, \ (x)F(x), \sim (x) \sim F(x).$$

For if we represent classical disjunction as "$\sim (\sim p . \sim q)$", we have thereby expressed [the thesis] that the law of the excluded middle in its classical form ($p \vee q$) is only valid for absolutely objective Being. For the second concept of Being it holds only in a weakened form, insofar as we have to take into our conception of the object the process of reflection expressed by "$\sim \ldots$".

Indeed, between "$\sim \ldots$" in the table:

p	$\sim p$	(I)
W	F	
F	W	

6 | und demselben Zeichen "$\sim\ldots$" in der Formel "$\sim(\sim p \cdot \sim q)$" existiert
nämlich ein Unterschied, der mir analog dem Unterschied von freier und
gebundener Variablen im Prädikatenkalkül zu sein scheint. In Tafel (I)
ist "$\sim\ldots$" sozusagen ungebunden, während in der Formel $\sim(\sim p \cdot \sim q)$
es ontologisch durch "— • —" <gebunden> ist. D. h. seine Bedeutung
Verneinung überhaupt zu sein, ist auf die Verneinung von "— • —" hin
präzisiert worden. Dasselbe gibt für die Ersetzung des Existenzoperators
durch $\sim(x) \sim F(x)$.

Der Parallelismus zwischen klass. und intuit. Begriffen scheint also
anzudeuten, dass wir zwei Kategorien von logischen Entitäten haben:
Radikal objektiv thematisiert im Sinn von a) Sein; und doppel-thema-
tisch (objektiv/subjektiv) orientiert im Sinne von b) Sein.

Dem entspricht die Zweiteilung des Prädikatenkalküls, resp. die Teilung in Quantifikationstheorie und Klassentheorie. Ihren eigenen Entdeckungen zufolge ist das Axiomensystem des engeren Funktionenkalküls
vollständig (Monatshefte für Math. v. 37 (1930)),[e] aber es existiert
Nichtentscheidbarkeit. Für den weiteren Kalkül gilt, gemäss einer zweiten
Entdeckung von Ihnen, weder Vollständigkeit noch Entscheidbarkeit.

Es scheint mir nun (wenn ich Ihre Ergebnisse und die einiger Ihrer
Kollegen, z. B. Church, richtig interpretiere [)], dass der Drittensatz in
der Quantifikationstheorie eine andere Rolle spielt als in der Klassentheo-
7 rie. Seine unbedingte, rigorose | Geltung im platonischen-aristotelischen
Sinn ist in beiden Fällen verneint. Es scheint mir aber der folgende Unterschied zu bestehen: Für die Quantifikationstheorie ist das Tertium non
datur *vorläufig* suspendiert. Für die Klassentheorie[1] ist es *schlechthin*
aufgehoben.

Unter <vorläufig suspendiert> verstehe ich: die Quantifikationstheorie
ist ein System, in dem der Drittensatz im klassischen Sinn gelten *würde*,
wenn man in der Lage wäre den ⟨inhaltlich⟩ unendlichen Umfang dieses
Systems erschöpfend darzustellen. Das ist empirisch nicht möglich. Folglich ist seine Geltung *im* System beschränkt. In anderen Worten: seine
Geltung ist nur eingeschränkt für Teilsysteme der Quantifikationstheorie.
Er gilt aber klassisch unbedingt für das (nicht-realisierbare) Gesamtsystem.

In der Klassentheorie aber scheinen mir die Dinge anders zu liegen.
Hier bedeutet seine Einschränkung nicht, dass seine klassisch-rigorose
Geltung vorläufig suspendiert ist, sondern dass sie *endgültig* aufgehoben

---

[1] Oder wenigstens Teile derselben

---

[e] *Gödel 1930.*

and the same sign "$\sim \ldots$" in the formula "$\sim (\sim p \cdot \sim q)$", there exists a difference that seems to me to be analogous to the difference between free and bound variables in the predicate calculus. In table (I) "$\sim \ldots$" is so to speak not bound, while in the formula "$\sim (\sim p \cdot \sim q)$" it is "bound" ontologically by "$— \bullet —$". That is, its meaning as negation as such has been made more precise as negation of "$— \bullet —$". The same holds for the replacement of the existential quantifier by $\sim (x) \sim F(x)$.

The parallelism between classical and intuitionistic concepts thus seems to indicate that we have two categories of logical entities: radically objectively thematized, in the sense of Being a); and double-thematically (objective/subjective) oriented in the sense of Being b).

To this corresponds the division of the predicate calculus, more precisely, the division into the theory of quantification and class theory. According to your own discoveries, the axiomatic system of the narrower functional calculus is complete (Monatshefte für Math. v. 37 (1930)),[e] yet undecidability exists. For the rest of the calculus, in accordance with a second discovery of yours, neither completeness nor decidability obtain.

Now, it seems to me (provided I interpret correctly your results and those of several of your colleagues, e.g. Church) that the law of the excluded middle plays a different role in quantification theory than in class theory. Its unconditioned and rigorous validity in the Platonic-Aristotelian sense is denied in both cases. It seems to me, however, that the following difference obtains: With respect to quantification theory the tertium non datur is *temporarily* suspended. With respect to class theory[1] it is *absolutely* abolished.

By "temporarily suspended" I understand: quantification theory is a system in which the law of the excluded middle *would* hold in the classical sense, if one were in a position to represent exhaustively the contentually infinite scope of this system. That is empirically impossible. Consequently its validity *within* the system is restricted. In other words: its validity is only limited with respect to subsystems of quantification theory. But, classically, it holds unconditionally for the (unrealizable) entire system.

In class theory, however, it seems to me that things are otherwise. Here the limitation does not mean that its classical-rigorous validity is temporarily suspended, but that it is *definitively* abolished. That is, even

---

[1] Or at least parts of it

ist. D. h. selbst wenn man die Klassentheorie vollständig, als geschlossenes System, darstellen könnte, würde man entdecken, dass das Tertium
non datur auch dann nicht gilt.

Für die Quantifikationstheorie ist das Tertium non datur wenigstens
noch ein unendliches (wenn auch nicht erreichbares) formales Ziel des
Denkens. Für die Klassentheorie existiert ein solches Ziel nicht. In phi
8   losophischer Terminologie: für die Quantifikationstheorie hat | der Drittensatz noch thematische Bedeutung. In der Klassentheorie existiert er
nur noch athematisch. D. h. wenn man die Klassentheorie vollendet und
mit generellen Entscheidungsverfahren darstellen könnte, würde man feststellen, dass das Tertium non datur auch für das Gesamtsystem nicht
gilt.

Die hier skizzierte Unterscheidung wird heute in der mathematischen
Logik noch nicht gemacht (wenigstens habe ich sie nirgends gefunden),
weil man sich, wie ich glaube nicht über den Unterschied von zweiwertiger und dreiwertiger Logik klar ist. Wird der Drittensatz vorläufig suspendiert, *aber nicht thematisch* aufgehoben, so bleibt man auf dem Boden
der zweiwertigen Logik. D. h. die logische Analyse bewegt sich auf der
Basis der Hamletschen Alternativ <Sein oder Nichtsein>. Die Korrespondierenden Werte sind: "wahr" und "falsch".

Wird der Drittensatz aber nicht nur empirisch suspendiert, sondern
endgültig für den gesammten logischen Formalismus aufgehoben, so findet
ein Übergang zum (mindestens) dreiwertigen System statt. D. h. das
Tertium non datur ist auch für die "Wahrheits"funktionen der Aussagenlogik endgültig beseitigt. Die *endgültige* Aufhebung des Tertium non
datur impliziert einen radikalen logischen Themawechsel. Im dreiwertigen System ist das Thema <Sein> aufgegeben. Das bedeutet aber, dass
die seinsthematischen Werte "wahr" und "falsch" nicht mehr zuständig
9 | sind.—Ich habe gerade Rosser und Turquette's Buch "Many valued
logics"[f] gelesen und ich finde es bezeichnend, wie diejenigen Logiker,
die sich mit dem Problem mehrwertiger Logiken befassen, immer noch
verzweifelt an der klassischen Wertdichotomie festhalten. Sie wird auf die
mehrwertigen Logiken dadurch übertragen, dass man dichotomisch zwischen "designierten" und "nicht-designierten" Werten unterscheidet. Das
geschieht, weil man sich nicht von dem Gedanken losreissen kann, dass
das einzige und alleinige Thema des Denkens objektives "Sein" ist.

Es ist aber falsch anzunehmen, dass alle Eigenschaften von Klassen
sich als Eigenschaften eines objektiven Seins darstellen lassen.[g] Erlauben
Sie mir an den Heisenbergschen Satz aus der Quantenphysik zu erinnern:

---

[f] *Rosser and Turquette 1952.*

if one could represent class theory completely as a closed system, one would find that even then the tertium non datur does not hold within it.

For quantification theory the tertium non datur is still at least an infinite (even if unattainable) formal goal of thought. For class theory such a goal does not exist. In philosophical terminology: for quantification theory the law of the excluded middle still has thematic meaning. In class theory it only exists non-thematically. That is, if one could represent class theory [as] completed and with general decision procedures, one would establish that the tertium non datur does not hold, even for the entire system.—

The distinction sketched here is not yet made today in mathematical logic (at least I have not found it anywhere), because, I believe, one is not clear about the difference between two-valued and three-valued logic. If the law of the excluded middle is temporarily suspended, *but not thematically abolished*, then one remains on the ground of two-valued logic. That is, logical analysis operates on the basis of Hamlet's alternative "To be or not to be". The corresponding values are: "true" and "false".

Yet, if the law of the excluded middle is not only empirically suspended, but *definitively* abolished for the entire logical formalism, a transition to an (at least) three-valued system takes place. I.e., the tertium non datur is also definitively eliminated for the "truth" functions of propositional logic. The *definitive* abolition of the tertium non datur implies a radical logical change of topic. In the three-valued system the topic "Being" is given up. But this means that the values "true" and "false", which thematize Being, are no longer relevant.—I just read Rosser and Turquette's book "Many valued Logics"[f] and I find it telling how these logicians, who work on the problem of many-valued logics, still hold on desperately to the classical dichotomy of values. It is transferred to many-valued logic by dichotomically distinguishing between "designated" and "non- designated" values. This happens because one cannot break free from the idea that the one and only topic of thought is objective "Being".

But it is wrong to assume that all properties of classes can be represented as properties of an objective Being.[g] Permit me to remind you of

---

[g]This sentence is marked in the margin.

der "absolut isolierte Gegenstand hat prinzipiell keine beschreibbaren Eigenschaften mehr."[h] Der physikalische Sachverhalt ist ein Zusammenhang von Subjekt *und* Objekt.

Eine Logik aber, die einer solchen Physik parallel läuft, hat eine neue Aufgabe. Sie hat sich zu fragen, was ist die objektive (irreflexive) und was ist die subjektive (reflexive) Komponente in einem theoretischen Begriff? Wenn ich eine logische Klasse als ein 100% objektives "Ding" auffasse, hat sie auch keine beschreibbaren Eigenschaften mehr, weil ihr dann die kontradiktorischen Eigenschaften "p" und "$\sim p$" gleichzeitig

10   zukommen.[i] Man ist dieser Verlegenheit | vermittels der Typentheorie oder der Quineschen Regel R3′ (in New Foundations[j]) oder mit anderen Mitteln, die aber alle auf dasselbe hinauskommen, zuleibe gegangen; aber man hat sich nie gefragt ob nicht die Klassentheorie—abgesehen von der seinsthematischen Wahrheit und Falschheit eines Begriffs—noch eine andere logische Fragestellung suggeriert, nämlich die welche Komponenten eines Begriffs *irreflexiv* gelten und welche *reflexive* Bedeutung haben.

Ich verstehe dabei als "irreflexiv", das was unabhängig vom denkenden Subjekt gilt (der klassische Begriff der Objektivität!), und als "reflexiv" dasjenige logische Motiv, in das der Denkprozess (das Subjekt nach Heisenberg) mit hineindefiniert werden muss.

Nun ist aber der Gegensatz von "irreflexiv" und "reflexiv" selbst der Gegenstand einer (iterierten) Reflexion. Wir erhalten somit einen dritten Wert, der den Denkprozess designiert, der "irreflexiv" und "reflexiv" unterscheidet. Ich nenne ihn—nach transzendentalem Vorbild—doppelt-reflexiv. In meinem Essay für den 11. Phil. Kongress in Brüssel[l] habe ich ausgeführt, dass der Unterschied von "reflexiv" und "doppelt-reflexiv" auf die Differenz von "Ich" und "Du" im Denken hinauskommt.[k] D. h. Reflexion ist grundsätzlich zweiwertig. Sie kann als *objektiver* Vorgang in der Aussenwelt interpretiert werden: dann vollzieht sie sich im *anderen*

11   Ich, d. h. im Du. Sie kann aber auch als subjektiver Erlebnis-|prozess, in dem logischen Bedeutungen innerlich erfahren werden, aufgefasst werden. Dann vollzieht sie sich im Ich.

Klassisch ist das ganz gleichgültig, weil alle denkenden Subjekte metaphysisch (angeblich) zusammenfallen. Die Differenz zwischen Reflexion im Subjekt (Ich) und Reflexion im Objekt (Du) wird aber sofort relevant, sobald man anerkennt, dass die Kommunikationstheorie (C. Shannon) ein Teil der Logik ist. Die mitteilbare (objektive) Bedeutung eines Begriffs unterscheidet sich von der denkbaren (subjektiven) Bedeutung. Und beide unterscheiden sich von der Differenz zwischen Mitteilbarkeit und Denkbarkeit.

---

[h] *Heisenberg 1931*, p. 182.

[i] This sentence is marked in the margin.

Heisenberg's dictum from quantum physics: the "absolutely isolated object does not, in principle, any longer have any describable properties".[h] The physical state of affairs is a connection of subject *and* object.

However, a logic that runs parallel to such a physics has a new task. It has to ask itself, what is the objective (irreflexive) and what is the subjective (reflexive) component in a theoretical concept? If I conceive a logical class as a 100% objective "thing", it too no longer has any describable properties, because then the contradictory properties "p" and "$\sim p$" are attached to it at the same time.[i] One has tried to counter this embarrassment by means of type theory, or Quine's rule R3$'$ (in New Foundations[j]), or by other means, all of which amount to the same thing; but one never asked whether class theory—besides the Being-thematizing truth and falsity of a concept—does not suggest yet another logical question, namely, which components of a concept hold *irreflexively* and which have *reflexive* meaning.

I understand as "irreflexive" that which holds independently of the thinking subject (the classical concept of objectivity!), and as "reflexive" the logical motif, in the definition of which the thought process (the subject according to Heisenberg) has to be included.

But the opposition between "irreflexive" and "reflexive" is itself the object of an (iterated) reflection. We thus obtain a third value, which designates the thought process that distinguishes between "reflexive" and "reflexive". I call it—following the transcendental example—double-reflexive. In my essay for the 11th philosophical congress in Brussels I explained that the distinction between "reflexive" and "double-reflexive" amounts to the difference in thought between "I" and "Thou".[k] I.e., reflection is fundamentally two-valued. It can be interpreted as an *objective* event in the external world: in this case it takes place in the *other* I, i.e., in the Thou. But it can also be understood as a process of subjective experience in which logical meanings are experienced internally. In that case it takes place in the I.

Classically, this is a matter of indifference, because all thinking subjects (allegedly) coincide metaphysically. However, the difference between reflection in the subject ("I") and reflection in the object ("Thou") becomes relevant immediately as soon as one acknowledges that communication theory (C. Shannon) is a part of logic. The communicable (objective) meaning of a concept is distinguished from the thinkable (subjective) meaning. And both are distinguished from the difference between communicability and thinkability.

---

[j] R3$'$ is the well-known schema of stratified comprehension of *Quine 1937*; cf. *Quine 1980*, p. 92.

[k] *Günther 1953*, especially pp. 47–49. See §1 of the introductory note.

Diese drei Bedeutungen werden durch meine Werte "irreflexiv" (1), "reflexiv" (2) und "doppelt-reflexiv" (3) auseinandergehalten. Die Klassentheorie verwickelt uns in Paradoxe weil wir die *objektive* Bedeutung des Klassenbegriffs als Zeichen auf dem Papier und die *subjektive* als Denkprozess im Bewusstsein ignorieren und über dies vergessen, dass die Differenz zwischen der "papiernen" Bedeutung und der in der Reflexion ebenfalls in die Logik eingeschlossen werden muss.[1]

* * *

[In the omitted portion, Günther presents and discusses some three-valued truth tables.]

15 | ...

D. h. das Thema dieser Logik ist nicht mehr "ontologisch" Sein, sondern die Differenz zwischen Sein (Objektivität) und Reflexion, die selber als logisches Reflexionsproblem begriffen wird. Das Kriterium, das zwischen einer Seinslogik und einer Reflexionslogik scheidet, ist der Drittensatz. Was nämlich im Tertium non datur ausgeschlossen wird, ist nämlich die logische Möglichkeit auf das Verhältnis von objektiv Gedachtem und Denkprozess (math: Konstruktion) noch einmal *ausserhalb* der (zweiwertigen) Logik, die beides vereint, zu reflektieren.[m]

* * *

[In two omitted paragraphs, Günther makes further remarks about the law of excluded middle and the object language–metalanguage distinction.]

16 | ... Ich bin davon ausgegangen, dass sowohl Kant wie der moderne Positivismus demonstrieren, dass der Begriff eines transzendenten, absoluten Seins wiederspruchsvoll ist. Damit fällt die ontologische Metaphysik fort.

Solange es ein transzendentes, absolutes Sein gibt, kann es auch nur *eine* Logik geben. Die Auflösung der absoluten Substanz aber hat zur Folge dass die klassische Logik sich nicht auf das reale Sein richtet (wie
17   Russell annimmt). Wir wissen ja nicht ob es das | überhaupt gibt; sondern sie richtet sich auf—und definiert den *Sinn* von Sein, d. h. sie definiert die logische Bedeutung in der identitätstheoretische Motive im Denken verwandt werden.

---

[1]From the beginning of this paragraph to 'Bewusstsein' is marked in the margin.

These three meanings are kept apart by my values "irreflexive" (1), "reflexive" (2) and "double-reflexive" (3). Class theory entangles us in paradoxes because we ignore the *objective* meaning of the concept of class as a sign on paper, and the *subjective* one as thought process in consciousness and in so doing forget that the difference between the "paper" meaning and that in reflection needs to be included in logic as well.[1]

* * *

[In the omitted portion, Günther presents and discusses some three-valued truth-tables.]

That is, the topic of this logic is no longer Being conceived "ontologically", but the difference between Being (objectivity) and reflection, where this difference is itself conceived as a logical problem for reflection. The criterion that distinguishes a logic of Being from a logic of Reflection is the law of the excluded middle. For what is excluded in the tertium non datur is the logical possibility of reflecting once again, *outside* (two-valued) logic which combines the two, upon the relation between what is objectively thought and the process of thought (mathematically: construction).[m]

* * *

[In two omitted paragraphs, Günther makes further remarks about the law of excluded middle and the object language–metalanguage distinction.]

My point of departure was that both Kant and modern positivism demonstrate that the concept of a transcendental, absolute Being, is contradictory. With that ontological metaphysics comes to an end.

As long as there is a transcendental, absolute Being, there can only be *one* logic. The dissolution of absolute substance, however, has the consequence that classical logic does not direct itself at real Being (as Russell assumes). After all, we do not know if that exists at all; rather it directs itself at—and defines the *meaning* of Being, i.e., it defines the logical meaning in which identity-theoretic motifs are employed in thought.

---

[m]This sentence is marked in the margin.

Das Thema <Sein selbst> kann durch nichts anderes ersetzt werden. Daher nur eine Logik. Dem reduzierten Thema <Sinn des Seins> aber kann als zweites Thema <Sinn der Reflexion> gegenübergestellt werden. So gelangen wir zum Konzept einer zweiten Logik.

Die klassische Logik setzt die metaphysische Identität von Denken und Sein voraus. Folglich kann das Denken kein anderes Thema als objektives Sein haben. Geben wir diese Metaphysik auf (und es ist höchste Zeit dafür), so kann das Denken als zweites ebenbürtiges Thema neben dem Sein (Objekt) das Subjekt (d. h. den Reflexionsprozess) selbst haben.—

Haben Sie herzlich Dank, dass Sie mir solange zugehört haben!

  Mit herzlichen Grüssen

                              Ihr

                      Gotthard Günther

P. S. The aim "to set up a consistent theory of classes and concepts as objektively existing entities" setzt eine Logik voraus in der formal zwischen "objektiv" und "subjektiv" unterschieden werden kann. Das ist aber nur in einer dreiwertigen Logik möglich.

4. Gödel to Günther

                                        Princeton, 30./VI. 1954

    Sehr geehrter Herr Dr Günther!

Daß irgendwelche Ihrer phil. Behauptungen meinen Resultaten widersprechen kann ich, beim gegenwärtigen Stand der Dinge, wohl verneinen, obwohl meine Resultate gewisse Formen einer subjektivistischen Interpret. der Math. unmöglich machen u. überhaupt stark gegen jede solche Interpretation sprechen. Was Ihre Deutung der Resultate bez. Unentscheidbarkeit betrifft, so möchte ich folgendes sagen: Es ist richtig, daß gewisse Theoreme von Church, Turing u.a. dahin interpretiert werden können (u. von Brouwer auch dahin interpretiert werden), daß der Drittensatz für gewisse Satzklassen A im Sinne der Formel $\sim(p)[p\,\varepsilon A. \supset .$ $p \vee \sim p]$ zu negieren ist. D. h. man kann zeigen: Die Annahme, man hätte bewiesen, daß der Drittensatz für alle p aus A gilt, führt auf einen Widerspruch. Es stimmt auch, daß nichtsdestoweniger der Drittensatz

The topic "Being itself" cannot be replaced by anything else. Hence only *one* logic. But the reduced topic "meaning of Being" can be confronted with "meaning of reflection" as second topic. In this way we reach the conception of a second logic.

Classical logic presupposes the metaphysical identity of thought and Being. Consequently thought can have no topic other than objective Being. If we give up this metaphysics (and it is high time we do so), then thought can have, as a second, equally important topic besides Being (object), the subject (i.e., the process of reflection) itself.

Thank you very much for listening to me for so long!

With warm greetings,

Yours

Gotthard Günther

P.S. The aim "to set up a consistent theory of classes and concepts as objectively existing entities" presupposes a logic which can *formally* distinguish between "objective" and "subjective". But this is only possible within a three-valued logic.

4. Gödel to Günther

Princeton, 30 June 1954

Dear Dr. Günther,

That any of your philosophical claims contradicts my results I can answer in the negative, given the way things stand at present, although my results make certain forms of a subjectivist interpretation of mathematics impossible, and generally count strongly against any such interpretation. With regard to your interpretation of the results concerning undecidability, I would like to say the following: it is correct that certain theorems of Church, Turing, et al., can be interpreted in such a way (and are indeed interpreted in this way by Brouwer) that the law of the excluded middle, in the sense of the formula $\sim(p)[p\,\varepsilon\,A. \supset .p \vee \sim p\,]$, has to be denied for certain classes of propositions A. That is, one can demonstrate: the assumption that one has proved the validity of the law of the excluded middle for all p in A leads to a contradiction. It is furthermore correct

2   als unerreichbares Ziel des Denkens bestehen bleiben könnte, | da trotz-
dem für jedes einzelne p aus A $p \vee \sim p$ beweisbar sein kann (Fall I.). Die
letztere Behauptung hat aber intuit. keinen Sinn, sondern setzt den Begriff der objektiven Existenz voraus. Die entsprechende int. formulier-
bare Behauptung aber, nämlich: $\sim(\exists p)[p \, \varepsilon A \, . \, \bullet \, . \, \sim(p \vee \sim p)]$ gilt für jedes
A. D. h. es kann innerhalb der int. Math. von keinem Satz bewiesen wer-
den, daß er unentscheidbar ist. Nichtsdestoweniger könnte das von gewissen Sätzen der int. Math. objektiv richtig u. in der Existential-Math.
sogar beweisbar sein. (Fall II). Aber die Unterscheidung der Fälle I u. II.
hat, wie gesagt, intuit. keinen Sinn,[1] ausser daß eventuell einer der beiden Fälle auf Grund gewisser Tatsachen ⟨empirisch⟩ wahrscheinlich werden könnte. Ich glaube auch nicht, daß der "Themawechsel" des Denkens
etwas mit den Fällen I, II zu tun hat; denn dieser tritt ja schon in dem
Augenblick ein, wo der Begriff der aufweisbaren Existenz eingeführt wird,
u. würde auch dann bestehen bleiben, wenn man für diese den Dritten-
3   satz beweisen könnte. Ob | die Grenze zwischen I u. II mit der zwischen
engerem u. höherem Funktionenkalkül zusammenfällt, darüber ist nichts
bekannt. Mein Resultat über die Vollständigkeit des eng. Funkt. Kalk.
hat, int. interpretiert, nichts mit dem Drittensatz zu tun, sondern betrifft
den Satz der doppelten Negation $(\sim \sim p \supset p)$. Es ist auch nicht bekannt,
ob für die ganze int. Math. der Fall II eintritt. Denn diese liegt ja nicht
in formalisierter Gestalt vor u. von Brouwer wird sogar die Möglichkeit
einer vollständigen Formalisierung bestritten.

Sie haben durchaus recht, wenn Sie sagen, daß in der int. Math. eine
Selbstreflexion des Subjekts vorliegt u. daß der Begriff der Absurdität
eine geb. Variable enthält. Er bedeutet ja: "Es gibt (im Sinne der Auf-
weisbarkeit) eine Widerlegung". Die in der idealist. Phil. behandelte
Reflexion auf das Subjekt (d. h. Ihr II Thema d. Denkens), die Unterscheidung von Reflexionsstufen etc. scheint mir sehr interessant u. wich-
4   tig. Ich halte es sogar für durchaus | möglich, daß dies "der" Weg zur
richtigen Metaphysik ist. Die damit verbundene (in Wahrheit aber davon
ganz unabhängige) Ablehnung der objektiven Bedeutung des Denkens
kann ich aber nicht mitmachen. Ich glaube nicht, daß irgend ein Kant-
sches oder positivistisches Argument oder die Antinomien d. Mengenl.,
oder die Quantenmechanik bewiesen hat, daß der Begriff des objektiven
Seins (gleichgültig ob für Dinge oder abstrakte Wesenheiten) sinnlos oder
widerspruchsvoll ist.[2] Wenn ich sage, daß man eine Theorie der Klas-

---

[1]Ich habe bloß bewiesen: "Jede Formel ist entweder widerlegbar oder es gibt eine
Realisierung", nicht aber: "oder die Existenz einer Realisierung ist beweisbar". Das
letztere ist für jedes formale System falsch.

[2]Damit will ich natürlich nicht behaupten, daß schon das naïve Denken das objektive
Sein in allen Punkten richtig erfaßt, wie die ontol. Metaphysik vielfach anzunehmen
scheint.

that the law of the excluded middle could nonetheless remain in place as an unattainable goal of thought, since $p \vee \sim p$ can be provable all the same for every individual p in A (case I.). But this latter claim has no intuitionistic sense but rather presupposes the concept of objective existence. The corresponding claim that can be formulated intuitionistically, however, namely $\sim(\exists p)[p \, \varepsilon A \, . \, \bullet \, . \, \sim(p \vee \sim p)]$ is valid for every A. That is, for no proposition within intuitionistic mathematics can one prove that it is undecidable. Nonetheless, this could be objectively true of certain propositions in intuitionistic mathematics and even be provable in existential mathematics (case II.). But the distinction between cases I. and II., as I said, has no intuitionistic sense,[1] except that one of the two cases could possibly become empirically probable on the basis of certain facts. Moreover, I do not believe that the "shift of topic" of thought has anything to do with cases I. and II.; for it already occurs at the moment when the concept of exhibitable existence is introduced, and would remain in place, even if one could prove the law of the excluded middle for this concept. Nothing is known about whether the boundary between I. and II. coincides with that between the lower and higher functional calculus. My result about the completeness of the lower functional calculus has, in its intuitionistic interpretation, nothing to do with the law of the excluded middle, but rather concerns the law of double negation ($\sim \sim p \supset p$). Moreover it is not known whether case II. occurs in all of intuitionistic mathematics. For, after all, it does not exist in a formalized version, and Brouwer even contests the possibility of a complete formalization.

You are absolutely right when you say that in intuitionistic mathematics a self-reflection of the subject is present and that the concept of absurdity contains a bound variable. It means: "There is (in the sense of exhibitable existence) a refutation". The reflection on the subject treated in idealistic philosophy (that is, your second topic of thought), the distinction of levels of reflection, etc., seem to me very interesting and important. I even consider it entirely possible that this is "the" way to the correct metaphysics. However, I cannot go along with the denial of the objective meaning of thought that is connected with it, [although] it is really entirely independent of it. I do not believe that any Kantian or positivistic argument, or the antinomies of set theory, or quantum mechanics have proved that the concept of objective being (no matter whether for things or abstract entities) is senseless or contradictory.[2] When I say that one can (or should) develop a theory of classes as ob-

---

[1] I have only proved, "Every formula is either refutable or there is a realization [of it], not, however, "or the existence of a realization is provable". The latter is false for every formal system.

[2] Of course I don't wish by that to claim that naive thought already grasps objective being correctly on all points, as ontological metaphysics often seems to suppose.

sen als objektiv existierender Gegenstände entwickeln kann (oder soll),
so meine ich damit durchaus Existenz im Sinne der ontol. Metaphysik,
womit ich aber nicht sagen will, daß die abstrakten Wesenheiten in der
Natur vorhanden sind. Sie scheinen vielmehr eine zweite Ebene der Realität zu bilden, die uns aber ebenso objektiv. u. von unserem Denken unabhängig gegenübersteht wie die Natur. Ich kann in diesem Brief leider
nicht mehr auf Ihre 3-wertige Logik eingehen. Was ich bisher vermisse,
ist eine nähere Erklärung des inhaltlichen Zusammenhangs der Wahrheit-
stabellen mit Ihren philosophischen Ideen.

Zum Schluß möchte ich noch sagen, daß ich Ihnen jederzeit gerne mit
weiteren Auskünften zur Verfügung stehe.

Mit besten Grüssen

Ihr Kurt Gödel

5. Günther to Gödel

Gotthard Günther
3407 Montrose Ave.
Richmond 22, Va.
October 2, 1954

Sehr geehrter Herr Professor Gödel!

Zuerst möchte ich mich entschuldigen dass ich Ihren freundlichen Brief
vom 30. Juni erst heute beantworte. Das hatte jedoch seine Gründe.
Ich wollte bevor ich erwiderte erst eine Publikation von Ihnen sorgfältig
studieren. Und zwar Ihren Aufsatz: Russell's Mathematical Logic (Libr.
of Liv. Philos. Evanston, Chicago 1944, S. 123-153.)[a] Und weiterhin
wollte ich den Druck eines Aufsatzes von mir: "Achilles And The Tortoise"[b] abwarten, der, wie ich glaube, den Gedanken, den ich in diesem
Brief hier zum Ausdruck bringen möchte, in seiner Weise erläutert. Die
drei Hefte von "Astounding", die die drei Teile meines "Achilles"-Auf-
satzes enthalten, gehen heute mit gesonderter Post als Drucksache an Sie
ab.—

Nun zum Thema. Sie schreiben in Ihrem letzten Brief: "Wenn ich
sage, dass man eine Theorie der Klassen als objektiv existierender Gegenstände entwickeln kann (oder soll), so meine ich damit durchaus Existenz

---

[a] *Gödel 1944.*

jectively existing entities, I do indeed mean by that existence in the sense of ontological metaphysics, by which, however, I do not want to say that abstract entities are present in nature. They seem rather to form a second plane of reality, which confronts us just as objectively and independently of our thinking as nature. Unfortunately I can't go into your 3-valued logic more in this letter. What I miss so far is a more detailed explanation of the connection in content of your truth tables with your philosophical ideas.

In conclusion, I would like to say that I will be glad to be available with further information at any time.

With cordial greetings,

Yours, Kurt Gödel

5. Günther to Gödel

Gotthard Günther
3407 Montrose Ave.
Richmond 22, Va.
October 2, 1954

Dear Professor Gödel,

First I would like to apologize for answering your kind letter of June 30 only today. But there were reasons for that. Before answering I first wanted to study carefully a publication of yours. Namely your article: Russell's Mathematical Logic (Library of Living Philosophers, Evanston, Chicago 1944, p. 123–153.)[a] In addition I wanted to wait for the publication of an article of mine: "Achilles and the Tortoise",[b] which, I believe, elucidates in its own way the thought I would like to express here in this letter. The three issues of "Astounding" which contain the three parts of my "Achilles" article, will be sent to you as printed matter today.

Now to the issues. You write in your last letter: "When I say that one can (or should) develop a theory of classes as objectively existing entities, I do indeed mean by that existence in the sense of ontological meta-

---

[b] *Günther 1954.*

im Sinne der ontol. Metaphysik; womit ich aber nicht sagen will, dass
2   die abstrakten Wesenheiten in der | Natur vorhanden sind. Sie scheinen
vielmehr eine zweite Ebene der Realität zu bilden, die uns aber ebenso
objektiv und vor unserem Denken unabhängig gegenübersteht, wie die
Natur."

Diese Feststellung in Ihrem Briefe geht Hand in Hand mit der folgenden Bemerkung in Ihrem Russell-Aufsatz: "Classes and concepts may,
however, ... be conceived as real objects; namely classes as <pluralities
of things> or as structures consisting of a plurality of things and concepts as the properties and relations of things existing independently of
our definitions and constructions. It seems to me that the assumption of
such objects is quite as legitimate as the assumption of physical bodies
and there is quite as much reason to believe in their existence." (S. 134)

Um *ein* Missverständnis von vornherein auszuschliessen: *so weit* Ihre
Sätze gehen stimme ich mit Ihnen überein—aber es scheint mir, dass sie
eine Problematik offen lassen und es ist gerade diese Problematik, in der
sich meine eigenen Bemühungen von allem, was momentan auf der Gebiet der Logik ereignet, unterscheidet.

Die bisherige Theorie des Denkens von Aristoteles bis zur Gegenwart
kennt nur *einen* Begriff des *logischen* (denkunabhängigen) *Objekts!* Und
alle Objekte werden—qua Objektivität—logisch gleich behandelt!! Nun
3   zeigt die Geschichte der Logik, dass man mit dieser | Thesis (die auf dem
Prinzip der metaph. Identität des Seins mit sich selbst beruht) von vornherein in logische Schwierigkeiten gekommen ist. In der mittelalterlichen
Logik hat man sich seit Scotus Eriugena dadurch zu helfen gesucht, dass
man innerhalb des Seins Grade der Realität angenommen hat. Nach Eriugena hat der Mensch "mehr" Sein als ein Stein und ein Engel "mehr"
Sein als der Mensch. Die logische Unhaltbarkeit dieser Auffassung ist
zum ersten Mal von Kant in seiner transzendentalen Dialektik aufgedeckt
worden.

In neuerer Zeit hat man es mit der Unterscheidung von "Ding" und
"Name" versucht. Quine's Beispiel:

Boston hat 800 000 Einwohner.

"Boston" hat sechs Buchstaben.

Um nicht missverstanden zu werden, möchte ich ausdrücklich darauf hinweisen, dass ich die sekundäre Legitimität der Unterscheidung
zwischen einem Gegenstand und seinem Namen nicht bestreite. Ich behaupte aber sie geht nicht tief genug! Beide sind logische "Objekte" und
als solche denkunabhängig. Aber auch die moderne Logik gibt uns keine
Antwort darauf, worin sich ihre Denkunabhängigkeit unterscheidet. Und
mangels einer solchen Unterscheidung wird stillschweigend angenommen,
dass diese Denkunabhängigkeit dieselbe (Identität) ist. Ich habe vor vielen Jahren Quine einmal gefragt: "What is the difference when you say

physics, by which, however, I do not want to say that abstract entities are present in nature. They seem rather to form a second plane of reality, which confronts us just as objectively and independently of our thinking as nature."

This statement in your letter goes hand in hand with the following remark in your Russell article: "Classes and concepts may, however,... be conceived as real objects; namely classes as 'pluralities of things' or as structures consisting of a plurality of things and concepts as the properties and relations of things existing independently of our definitions and constructions. It seems to me that the assumption of such objects is quite as legitimate as the assumption of physical bodies and there is quite as much reason to believe in their existence." (p. 134)

In order to avoid *one* misunderstanding right from the start: *as far as* your propositions go I agree with you—but it seems to me that there is one difficulty that you leave open, and it is just this difficulty in which my own efforts differ from everything that is currently happening in the field of logic.

Up to now the theory of thought from Aristotle to the present knows only *one* concept of *logical* (thought-independent) *object*! And all objects are—qua objectivity—treated logically in the same way!! Now, the history of logic shows that right from the start one has run into logical difficulties with this thesis (which rests on the principle of the metaphysical identity of Being with itself). In medieval logic since Scotus Eriugena one sought to help matters by assuming degrees of reality within Being. According to Eriugena human beings have "more" Being than a stone, and an angel has "more" Being than human beings. The logical untenability of this view was first revealed by Kant in his transcendental dialectic.

In more recent times one tried [to do] it by the distinction between "thing" and "name". Quine's example:

Boston has 800,000 inhabitants.

"Boston" has six letters.

In order not to be misunderstood I would like to note explicitly that I do not dispute the secondary legitimacy of the distinction between an object and its name. I do claim, however, that it does not go deep enough! Both are logical "objects" and as such are independent of thought. But even modern logic does not provide an answer as to how their thought-independence differs. And in the absence of such a distinction it is tacitly assumed that this thought-independence is the same (identity). Once many years ago I asked Quine: "What is the difference when you say

4 <A stone exists independent of my act of thinking it and a name exists equally independently?[>]" Er wusste mir keine | Antwort zu geben.

Ich behaupte nun dass es zwei grundsätzlich unterschiedene Formen der Denkunabhängigkeit gibt. Und folglich zwei prinzipiell vorschiedene Identitäten des Objektes mit sich selbst. Ich nenne die erste irreflexive Identität und die zweite Reflexionsidentität. Der Begriff eines Steines kann logisch nur als irreflexive Identität interpretiert werden. Eine Klasse hingegen hat Reflexionsidentität.

Der Unterschied ist bisher nicht gemacht worden weil er auf den Boden der zweiwertigen, klassischen Logik überhaupt nicht festgestellt werden kann. Für zweiwertiges Denken fällt irreflexive Identität und Reflexionsidentität zusammen. Das ist *eine* Interpretation von "$p = \sim \sim p$". Übrigens kommt Hegel diesem Gedanken einmal ganz nahe, wenn er entrüstet darauf hinweist dass die bisherige Philosophie keinen Unterschied zwischen der Identität eines Steins mit sich selbst und der Identität Gottes mit sich selbst mache. Für "Gott" können wir hier ruhig "Reflexion" setzen. Hegels Lösungsversuch ist selbstverständlich auch wertlos, da er zu seiner Zeit nicht wissen konnte, dass es auf dem Boden der zweiwertigen Logik eine Lösung für den Unterschied von Ding als Denkobjekt und Reflexion als Denkobjekt schlechterdings nicht gibt.

Und doch wissen wir, dass es einen solchen Unterschied gibt. Wenn wir von der Denkunabhängigkeit des Objektes <Stein> sprechen, mei-

5 nen | wir dass dieses Objekt deshalb unabhängig ist, weil es—primitiv gesprochen—von "aussen her" in das Denken tritt. Umgekehrt ist die Klasse deshalb ein denkunabhängiges Objekt weil sie "von innen kommt" und sich vom Denkprozess abgelöst hat. Das sind zwei vollkommen verschiedene Begriffe der Denkunabhängigkeit. Und meiner Ansicht nach kommen die grossen Schwierigkeiten, die wir mit der Klassentheorie haben, daher, dass dieser Unterschied logisch ignoriert wird. Es sind hier zwei vollkommen verschiedene Begriffe von "Existenz" im Spiel. Die zweiwertige ontologische Metaphysik aber kann das, da sie einfache Identitätsmetaphysik ist ("absolute" Identität von Denken und Sein) nicht zugeben.

Die Unterscheidung der beiden Formen der Existenz kann aber sofort gemacht werden, wenn wir vom zweiwertigen zum dreiwertigen System übergehen. Klassisch haben wir nur die einfache Alternative "$p \lor \sim p$", d. h. zwischen Denken-überhaupt und Gedachtem überhaupt. Die dreiwertige Logik aber fordert die Einführung einer zweiten Negation. D. h. wir haben jetzt zwischen "p" "$\sim p$" und "$\sim' p$" zu unterscheiden. D. h. zwischen Denken überhaupt und *zwei* Formen von Gedachtem (oder Existenz).

Es scheint mir nun, das gegenwärtig praktisch (aber nicht prinzipiell) die Grenze zwischen den beiden Begriffen von Denkunabhängigkeit un-

6 gefähr mit dem Unterschied von engerem und weiterem | Funktionenkal-

'A stone exists independent of my act of thinking it' and 'A name exists equally independently'?" He could not give me an answer.

Now I claim that there are two fundamentally different forms of thought-independence. And accordingly two in principle different identities of the object with itself. I call the first irreflexive identity and the second reflection-identity. The concept of a stone can, logically speaking, only be interpreted as irreflexive identity. A class, by contrast, has reflection-identity.

This distinction has not been made so far because it cannot be established at all on the basis of classical two-valued logic. For two-valued thought irreflexive identity and reflection-identity coincide. That is *one* interpretation of "$p = \sim \sim p$". Hegel, incidentally, comes very close to this thought at one point, when he indignantly observes that previous philosophy does not make a distinction between the identity of a stone with itself and the identity of God with himself. For "God" here we may well substitute "reflection". Hegel's attempted solution is of course worthless as well, since in his time he could not have known that, on the basis of two-valued logic, there is absolutely no solution to the distinction between a thing as object of thought and reflection as object of thought.

And yet we know that there is such a distinction. If we speak of the thought-independence of the object "stone" we mean that this object is independent because it—crudely put—enters into thought "from without". Conversely a class is a thought-independent object because it "comes from within" and has separated itself from the process of thought. Those are two entirely different concepts of thought-independence. And in my view the great difficulties we have with the theory of classes come from the fact that, logically, this distinction is ignored. There are two entirely different notions of "existence" in play here. But two-valued ontological metaphysics cannot admit this, since it is a simple metaphysics of identity ("absolute" identity between thought and Being).

The distinction of the two forms of existence can, however, be made right away if we pass from the two-valued to the three-valued system. Classically we have only the simple alternative "$p \vee \sim p$", that is between thinking in general and what is thought in general. Three-valued logic, however, requires the introduction of a second negation. That is, we now have to distinguish between "p", "$\sim p$", and "$\sim' p$". That is, between thinking in general and *two* forms of what is thought (or existence).

Now, it seems to me that currently the boundary between the two notions of thought independence roughly coincides practically (but not in principle) with the distinction between narrower and wider functional

kül zusammenfällt. Oder mit dem Gegensatz von Existenz und Klasse. Die Unterscheidung ist nicht präzis, weil gegenwärtig alles in das Prokrustesbett einer zweiwertigen Logik gezwängt wird. Es scheint mir dass der erweiterte Funktionenkalkül, in seiner gegenwärtigen Gestalt eine Mischform von zweiwertiger und dreiwertiger Problematik ist, wobei aber die dreiwertige Problematik *als solche* nicht anerkannt wird und in der Pseudoform einer zweiwertigen Logik erscheint.

Bei der Interpretation des Klassenbegriffes gehen zwei grundverschiedene Existenzkonzeptionen durcheinander, die getrennt behandelt werden müssten—aber nicht werden. Es scheint mir übrigens, dass meine Idee des doppelten (irreflexiven und reflexiven) Denkobjektes durch eines Ihrer Resultate gestüzt wird. Lassen Sie mich die Fussnote aus Ihrem letzten Brief zitieren: "Ich habe bloss bewiesen: <Jede Formel ist entweder widerlegbar oder es gibt eine Realisierung>, nicht aber: <oder die Existenz einer Realisierung ist beweisbar>. Das letztere ist für jedes formale System falsch." Ich würde sagen, es ist deshalb für jedes formale System falsch, weil jene Beweisbarkeit die präzise logische Eindeutigkeit des Existenzbegriffes voraussetzt. In anderen Worten: die Beweisbarkeit setzt voraus dass es nur *einen* general durchführbaren Begriff des Denkobjekts gibt. D. h. dass irreflexives Faktum und Reflexionsprozess vollständig ineinander auflösbar | seien. Gerade das aber ist nicht der Fall.—

Ich habe mir erlaubt Ihnen den Achilles-Aufsatz zu schicken, weil er das Problem zweier logisch prinzipiell verschiedener thematischer Objekte von der physikalischen Seite her aufrollt. Nach Plato und der ihm bis zur Gegenwart folgenden Tradition gibt es nur *ein* absolut generelles Objekt des Denkens, Sein-überhaupt. Folglich ist ihm im "Timaios" der leere Raum, der von *allen* Objekten entleert ist, das absolute Nichts. Es kommt ihm nicht die Idee dass die blosse Leere (die in der Tat kein Objekt im Sinne eines seienden Dinges ist) das Beispiel einer *zweiten* Objekt- und Existenzkategorie für das reflektierende Bewusstsein sein könne. Erst bei Leibniz findet sich eine vage Andeutung des Gedankens, dass der Raum die existierende Klasse aller individuellen Objekte ist. Das setzt aber voraus, das[s] der "leere" Raum in einem *anderen* Sinne "existiert" als, sagen wir, ein Stein oder ein Planet.

In dem Achilles-Aufsatz mache ich mir den Gedanken in folgender Weise zunutze: Wenn es nur eine logisch begreifbare Form der Existenz gibt, dann ist die Abwesenheit alles "Physischen" eben, wie Plato meint, das blosse Nichts. Wenn wir aber zwei Formen von Objektivität annehmen, dann ist der "leere" Raum auch ein echtes "gegenständliches" Denk-objekt. Und was echt objektiv gedacht (erfahren) werden | kann, das kann auch ⟨(technisch)⟩ behandelt werden. Folglich ist der Raum genau so *manifestierbar* wie die Körperwelt. Noch für Kant ist das unmöglich, da in der Kritik der reinen Vernunft der Raum blosse "Anschau-

calculus. Or with the contrast of existence and class. The distinction is not precise, because currently everything is forced into the Procrustean bed of a two-valued logic. It seems to me that the extended functional calculus, in its present form, is a hybrid of two- and three-valued problem areas, where, however, the three-valued problem area is not recognized *as such* and appears in the pseudo-form of a two-valued logic.

In the interpretation of the concept of class, two fundamentally different conceptions of existence are mixed up which should be treated separately—but are not. Incidentally, it seems to me that my idea of double (irreflexive and reflexive) thought-object, is supported by one of your results. Allow me to quote the footnote from your last letter: "I have only proved, 'Every formula is either refutable or there is a realization [of it]', not, however, 'or the existence of a realization is provable'. The latter is false for every formal system." I would say it is false for every formal system, because that provability presupposes precise logical uniqueness of the concept of existence. In other words: the provability presupposes that there is only *one* generally realizable concept of thought-object. I.e., that an irreflexive fact and the process of reflection can be completely resolved into each other. Just that, however, is not the case.—

I took the liberty of sending you the Achilles essay because it addresses the problem of two logically fundamentally different thematic objects from the physical side. According to Plato and the tradition that follows him up to the present day, there only is *one* absolutely general object of thought, Being in general. Consequently, for Plato in the "Timaeus" empty space, emptied of *all* objects, is absolute nothingness. He does not entertain the idea that the mere void (which indeed is not an object in the sense of a thing that is) might be an example of a *second* category of object and existence for reflecting consciousness. With Leibniz we first find a vague indication of the thought that space is the existing class of all individual objects. But that presupposes that "empty" space "exists" in *another* sense than, lets say, a stone or a planet.

In the Achilles essay I make use of this thought in the following manner: If there is only one logically comprehensible form of existence, then the absence of everything "physical" is indeed, as Plato thinks, mere nothingness. But if we assume two forms of objectivity, then "empty" space is also a genuinely "object-like" object of thought. And what can be thought (experienced) as genuinely objective, can also be (technically) treated. Consequently space is *manifest* exactly as is the world of bodies. For Kant this is still impossible, since, in the Critique of Pure Reason,

ungsform" ist. Das ist eine höfliche Umschreibung des platonischen Glaubens dass der Raum *objektiv* betrachtet Nichts ist.

Zweiwertig kann man gar nicht anders denken: Entweder die Dinge sind Etwas (Physik) oder die Abwesenheit der Dinge ist Etwas (Metaphysik). Das ganze ist ein reines *Umtauschverhältnis*, wie "rechts" und "links". Die prinzipielle Vertauschbarkeit der Parameter "Raum", "Zeit", "Materie" und "Prozess", von der ich im Achilles rede, setzt aber eine dreiwertige Logik voraus und *zwei* Konzeptionen von objektiver Existenz, die in diesem Spezialfall als "quantized" und "not quantized" charakterisiert werden.

Der Aufsatz behandelt ein generelles naturphilosophisches Problem. Als Konzession an den Leserkreis des Magazins habe ich es allerdings an dem speziellen Problem des "interstellar space-travel" durchgeführt.

In der Hoffnung, dass Sie angenehme Sommerferien gehabt haben bin ich mit warmen Grüssen

Ihr

Gotthard Günther

6. Günther to Gödel

3407 Montrose Ave.
Richmond 22, Va.
19. Juni 1955

Sehr verehrter Herr Professor Gödel:

Es scheint mir ziemlich lange her seit wir miteinander korrespondiert haben. Inzwischen hat sich manches ereignet. Die Universität Hamburg hat mich eingeladen im kommenden Wintersemester als Gastprofessor dort in der Philosophischen Fakultät zu lesen. Im Oktober werde ich hinüber fahren. Als Hauptkolleg habe ich "Metaphysik der Geschichte" gewählt. Daneben werde ich ein Seminar für Anfänger über moderne geschichtsphilosophische Theorien halten, und ein Seminar für Fortgeschrittene über transzendentale Logik.

Das aber nur nebenbei. Der Zweck meines Briefes ist ein anderer. Ich habe seit unserer letzten Korrespondenz einen Gedankengang ausgearbeitet, den ich Ihnen gern zur Kritik unterbreiten möchte.

Sie stimmten mir in Ihrem letzten Briefe bei, dass die modernen Bemühungen der mathematischen Logik im Wesentlichen eine Reflexion *auf das Denken* selbst darstellen. Wenn wir darin übereinstimmen, so sollten wir uns auch über das Folgende verständigen können.

space is a mere "form of intuition". That is a polite way of expressing the platonic belief that space, considered *objectively*, is nothing.

In two-valued terms one cannot think otherwise at all: Either things are Something (physics), or the absence of things is Something (Metaphysics). All this is a pure *relation of exchange*, like "right" and "left". But the interchangeability in principle of the parameters "space", "time", "matter" and "process", of which I speak in the Achilles, presupposes a three-valued logic and *two* conceptions of objective existence, which in this specific case are characterized as "quantized" and "not quantized".

The essay treats of a general problem of natural philosophy. As a concession to the readership of the magazine I have, to be sure, discussed it in terms of the specific problem of "interstellar space-travel".

In the hope that you had a pleasant summer vacation I remain, with warm greetings,

Yours,

Gotthard Günther

6. Günther to Gödel

3407 Montrose Ave.
Richmond 22, Va.
19 June 1955

Dear Professor Gödel:

It seems to me a rather long time since we last corresponded with one another. In the meantime many things have happened. The University of Hamburg has invited me to lecture there next winter semester as Visiting Professor in the Faculty of Philosophy. In October I will go over there. As principal lecture course I chose "Metaphysics of History". In addition I will hold a seminar for beginners about modern theories in the philosophy of history and a seminar for advanced students on transcendental logic.

But this only as an aside. The purpose of my letter is something else. Since our last correspondence I worked out a line of thought which I would like to present to you for criticism.

You agreed in your last letter that modern efforts in mathematical logic are in essence a reflection *on thought* itself. If we agree on that, we should also be able to reach agreement on the following.

2   Die obige Einsicht zwingt uns | zwischen zwei inversen Typen von Re-
flexion zu unterscheiden: 1) die Reflexion auf den bona fide Gegenstand,
der als etwas a limine vom Denken Unabhängiges gedacht wird, und 2)
die Reflexion auf das Reflektieren in 1).

Es ist unvermeidlich, dass diesen beiden Reflexionstypen zwei grundsätzlich verschiedene Konzeptionen des Begriffes <logischer Gegenstand>
entsprechen müssen. Diese Unterscheidung wird aber in der heutigen
Logik noch nicht durchgeführt.

Ich will zeigen, was ich meine. Die beiden logischen Gegenstände müssen verschiedenen Identitätscharakter haben. In der Terminologie der
älteren logischen Tradition:[1] der Gegenstand der Reflexion 1) hat Seinsidentität; der Gegenstand der Reflexion 2) aber hat Reflexionsidentität.
Der Unterschied findet sich, meines Wissens nach, zum ersten Mal bei
Hegel formuliert. Im Anschluss daran haben Sigwart und Benno Erdmann dann analysiert, was Seinsidentität logisch eigentlich bedeutet. Ihr
Resultat: Mit sich selbstidentisch sein heisst für den Gegenstand, dass
derselbe durch den Denkakt, der sich mit ihm beschäftigt, *nicht verändert
wird*. Seine (logischen) Eigenschaften bleiben dieselben, gleichgültig, ob
er gedacht oder nicht gedacht wird.—

3   Gerade das aber ist, so sage ich | nun, ein Charakteristikum, das die
Reflexion, die man zum Gegenstand der Reflexion 2) macht, unmöglich
haben kann. Die Reflexion ist ebenfalls mit sich selbst identisch. D. h.
sie hat (oder: ist?) Reflexionsidentität.

Ich frage nun, was geht vor, wenn wir die Reflexion selbst denken?
Blicken wir noch einmal zurück auf die Situation, in der sich die Reflexion 1) befindet. Dort ist der Gegenstand "da", mit *vor*-denklich gegebenen Eigenschaften, die seine Identität ausmachen.

Grundsätzlich anders aber ist die Situation für die Reflexion 2). Hier
ist kein unabhängig vom Denken gegebener Gegenstand *da*, an den das
Reflektieren jetzt herantritt. Es ist ja—per definitionem—der *subjektive* Denkakt selbst (der Reflexions*prozess*), der jetzt gedacht werden
soll. Derselbe muss also erst zum Gegenstand *gemacht* werden. D. h.,
er muss als logisches *Objekt* "gesetzt"[2] werden. Indem ich die Reflexion selbst denke, mache ich sie aus einem *subjekthaften* Prozess in einen
objekthaften Sachverhalt. Resultat: Reflexionsidentität bedeutet, dass
der logische Gegenstand, dadurch, dass er gedacht wird, sich in seinen
logischen Eigenschaften <u>ändert</u>. Die denkende Reflexion ist nicht die
gedachte Reflexion—und wenn die denkende Reflexion selbst gedacht
wird, werden ihre ursprünglich "subjektiven" Eigenschaften in inverse

---

[1]die in *diesem* Punkt auch heute noch nicht überholt ist

[2]"Gesetzt" ist ein transzendentallogischer Terminus der Fichteschen Wissenschaftslehre.

The above insight forces us to distinguish between two inverse types of reflection: 1) reflection on the *bona fide* object, which is thought as something *a limine* independent of thought, and 2) reflection on the reflecting in 1).

It is inevitable that two fundamentally different conceptions of the concept "logical object" must correspond to these two types of reflection. But this distinction is not made in contemporary logic.

I will show what I mean. The two logical objects must have a different character of identity. In the terminology of the older logical tradition[1]: the object of reflection 1) has identity of Being; the object of reflection 2), however, has identity of reflection. This distinction is to my knowledge first formulated by Hegel. Following this, Sigwart and Benno Erdmann then analyzed what, logically, identity of Being actually means. Their result: To be identical with itself means for the object that it *is not altered* by the act of thought occupied with it. Its (logical) properties remain the same, no matter whether it is thought or not.—

But just that, I now claim, is a characteristic which reflection that is made into an object of reflection 2) cannot possibly have. Reflection is also identical with itself. I.e., it has (or: is?) identity of reflection.

I now ask, what happens when we think reflection itself? Let us look once again at the situation of reflection 1). Here the object is "there", with properties given *prior* to thought that make up its identity.

The situation with respect to reflection 2), however, is fundamentally different. Here no object given independently of thought is *there*, which reflection now approaches. After all, it is—by definition—the *subjective* act of thought itself (the *process* of reflection) that is now supposed to be thought. Thus, the same act of thought first has to be *made* into an object. That is, it must be "posited"[2] as logical *object*. In thinking reflection itself I turn it from a *subject-like* process into an *object-like* state of affairs. Result: identity of reflection means that in virtue of being thought the logical object <u>alters</u> itself in its logical properties. The thinking reflection is not the reflection that is thought—and when thinking reflection is itself thought, its originally "subjective" properties are turned

---

[1]which on *this* point is not outdated even today.

[2] "Posited" is a transcendental-logical term from Fichte's Wissenschaftslehre.

4   "objektive" | verwandelt. Denn qua Reflexions*prozess* ist die Reflexion 1)
    undenkbar! Wird Reflexion 1) gedacht, so wandert ihr Prozesscharakter
    nach Reflexion 2) ab!!
        Wir haben also zwei logisch grundverschiedene Gegenstandstypen des
    Denkens überhaupt: Gegenstand 1), der dadurch, dass er gedacht wird,
    keine Veränderung erleidet und Gegenstand 2), *bei dem eine solche Ver-
    änderung einkalkuliert werden muss.*

* * *

        [In the omitted portion Günther once again explains his ideas by truth-
    tables. See the introductory note.]

10 |... Ich habe versucht meinen Gedankengang hier in äusserster Kürze
    wieder zu geben. Im Text meines Manuskriptes nimmt er mehr als 100
    Seiten ein. Ich wiederhole noch einmal: mein Grundgedanke ist, dass sich
    die Reflexion auf unser (seinsthematisches) Denken nicht mehr zweiwertig
    adäquat ausdrücken lässt, weil sie den Unterschied von Seinsidentität und
    Reflexionsidentität nicht mehr kalkülmässig bewältigen kann.
        Für Ihre Reaktion darauf wäre ich äusserst dankbar.

        Mit herzlichen Grüssen

                Ihr

                Ihnen sehr ergebenen

                Gotthard Günther

7. Gödel to Günther

                                        Princeton, 10./VIII. 1955

        Sehr geehrter Herr Dr Günther:

    Ich möchte mich zunächst entschuldigen, daß ich so lange nichts von
    mir hören ließ. Der Hauptgrund dafür war mein schlechter Gesundheits-
    zustand im letzten Herbst u. Winter, der mich daran hinderte Ihren Brief
    vom 6. Okt.[a] zu beantworten. Ich habe mich sehr über Ihre Berufung als
    Gastprofessor nach Hamburg gefreut u. gratuliere Ihnen herzlich.

---

[a] This must be an error; Gödel evidently refers to Günther's letter of 2 October 1954.

into inverse "objective" ones. For as a *process* of reflection, reflection 1) is unthinkable! If reflection 1) is thought, its process-like character migrates to reflection 2)!!

Thus, we have two logically fundamentally different types of objects of thought as such: object 1), which does not suffer any changes by being thought, and object 2) *where such a change has to be reckoned with.*

* * *

[In the omitted portion Günther once again explains his ideas by truth-tables. See the introductory note.]

... I have tried to present my line of thought here with utmost brevity. In the text of my manuscript it occupies more than 100 pages. I repeat once again: my basic idea is that reflection on our thought (that thematizes Being) can no longer be expressed adequately in two-valued terms, because in its calculus it cannot come to terms with the distinction between identity of Being and identity of reflection. I would be extremely grateful for your reaction.

With warm greetings,

Yours very sincerely,

Gotthard Günther

7. Gödel to Günther

Princeton, 10 August 1955

Dear Dr. Günther:

First I would like to apologize for your not having heard from me for so long. The main reason was the poor state of my health last fall and winter, which prevented me from answering your letter of 6 October.[a] I was very happy to hear about your appointment as visiting professor in Hamburg and extend my warm congratulations.

~~Was~~ ~~n~~ Nun ~~den~~ zum Inhalt Ihrer beiden letzten Briefe. Ich bemerke zwischen ihnen zunächst den Unterschied, daß Sie im ersten von "zwei Arten von Denkunabhängigkeit" sprechen, im zweiten aber den Gegenständen der Refl. 2.) ausdrücklich die Denkunabhängigkeit absprechen. Das erste scheint mir im Bereich der Möglichkeit zu liegen, das zweite aber unbedingt unrichtig zu sein. Ich glaube auch nicht, daß die mathematische Logik (oder überhaupt die Logik) eine Reflexion auf das *Denken* ist, höchstens *auch* auf das Denken | in erster Linie aber auf gewisse im Denken erfaßte allgemeinste Gegenstände. Die intuitionistische Logik allerdings ist eine Reflexion auf das Denken u. daher sind ihre Gegenstände nicht denkunabhängig, aber trotzdem werden sie "durch den Denkakt der sich mit ihnen beschäftigt" ebensowenig geändert wie die Gegenstände der Physik. Damit will ich natürlich durchaus nicht leugnen, daß die Gegenstände der Refl. 2) von denen der Refl. 1) toto genere verschieden sind, ja so sehr verschieden, daß man vielleicht mit Recht von zwei verschiedenen "Existenzformen" sprechen kann. Das ist es ja auch, wenn ich Sie recht verstehe, was Sie in Ihrem Achilles-Aufsatz (für dessen Zusendung ich bestens danke) zum Ausdruck bringen wollen. Das⟦s⟧ eine mehrwertige Logik das adäquate Ausdrucksmittel ist, um diesen Unterschied darzustellen, liegt durchaus im Bereich der Möglichkeit. Der Gedanke hat zweifellos etwas Bestechendes an sich, insbesondere mit Rücksicht auf die Paradoxien u. auf die Tatsache, daß ja die Prädikate | "notwendig" u. "möglich", die man zur Darstellung begrifflicher Verhältnisse braucht, auch als Wahrheitswerte interpretiert werden können. Ein Beweis für die Richtigkeit dieser Auffassung könnte nur durch eine Axiomatisierung der Logik auf dieser Grundlage u. den Erfolg der aus den Axiomen entwickelten Folgerungen erbracht werden. Daß die Mehrwertigkeit der Logik etwas mit den von mir konstruierten unentscheidbaren Sätzen zu tun haben sollte, ist allerdings kaum möglich. Das könnte eventuell für "absolut" unentscheidbare Sätze der Fall sein. Aber die von mir angegebenen sind ja in einem übergeordneten (ebenfalls richtige Gedanken ausdrückenden) Formalismus immer entscheidbar u. beweisen daher nur, daß kein Formalismus ⟨(Siehe separates Blatt!)⟩ das ganze abstrakte Denken erfassen kann. An dieser Unzulänglichkeit jedes Formalismus kann auch eine mehrwertige Logik nichts ändern.—Die Art u. Weise, wie Sie den dritten Wahrheitswert einführen wollen, ist mir auf Grund Ihrer Briefe u. des Kon-|greßvortrages leider nicht ganz verständlich. Sie gehen von einem, & u. ⋁-vertauschende͟n, dualen Verhältnis zwischen Ding u. Begriff aus. Nun ist dieser Dualitätsverhältnis [Vereinigung der Merkmale entspricht Durchschnitt der Umfänge] allerdings eine sehr interessante Tatsache, deren tiefere Bedeutung vielleicht noch der Aufklärung harrt, aber ich sehe keinen Zusammenhang zwischen diesem Charakteristikum der Begriffswelt u. ihrer verschiedenen Existenzform. Auch vermisse ich irgend eine Erklärung darüber, was eigentlich Ihre 3

n

Now to the content of your last two letters. To begin with, I note this difference between the two of them. In the first you talk about "two forms of thought-independence", but in the second you explicitly deny thought-independence for the objects of reflection 2). The former seems to me to lie within the realm of possibility, the latter, however, to be absolutely incorrect. I also do not believe that mathematical logic (or logic in general) is a reflection on *thought*. At best [it is] a reflection *also* on thought, but first and foremost on certain most general objects grasped in thought. Intuitionistic logic is indeed a reflection on thought and accordingly its objects are not thought-independent, but in spite of that, they are no more altered "by the act of thought occupied with them" than the objects of physics. By this I of course do not at all want to deny that the objects of reflection 2) are *toto genere* different from those of reflection 1), indeed so very different that one can perhaps justifiably speak of two different "forms of existence". That, after all, is what you want to express in your Achilles essay, if I understand you correctly. (Thank you very much for sending it.) It lies indeed within the realm of possibility that a many-valued logic is the adequate means of expression for representing this difference. The thought without doubt has something intriguing about it, in particular with respect to the paradoxes and the fact that the predicates "necessary" and "possible", which are used to represent conceptual relations, can also be interpreted as truth-values. A proof of the correctness of this view could only be provided by an axiomatization of logic on this basis and by the success of the consequences derived from these axioms. But that the many-valuedness of logic should have something to do with the undecidable propositions I constructed is, to be sure, hardly possible. That could potentially be the case for "absolutely" undecidable propositions. But the ones given by me are, after all, always decidable in a higher formalism (likewise expressing correct thoughts), and therefore only prove that no formalism (see separate sheet!) can capture the whole of abstract thought. This inadequacy of every formalism cannot be changed by a many-valued logic either.—The manner in which you want to introduce the third truth-value is unfortunately not fully intelligible to me on the basis of your letters and the lecture to the Congress. You take as point of departure a dual relation between thing and concept that interchanges & and ∨. Now, this relation of duality [union of characteristics corresponds to intersection of the extensions] is indeed a very interesting fact, whose deeper meaning perhaps still awaits an explanation, but I do not see a connection between this characteristic of the world of concepts and its different form of existence. I also miss any explanation at all of what your three truth-values actually

Wahrheitswerte, im Gegensatz zu dem "wahr" u. "falsch" der klassischen
Logik, bedeuten. Eine genaue Definition kann man natürlich nicht verlangen, am allerwenigsten im Rahmen der zweiwertigen Logik, aber doch
eine Erklärung in demselben Sinn, in dem man auch die Grundbegriffe
der zweiwertigen Logik (trotz ihrer Undefinierbarkeit) verständlich
machen kann. Eine Analyse des Sinnes Ihrer Wahrheitswerte scheint
mir der Kardinalpunkt zu sein, an dem Sie angreifen müssten, um Ihren
Lesern verständlich ⟨zu werden⟩ u. den Aufbau einer Ihren Ideen entsprechenden Logik weiter durchzuführen.

> Mit besten Grüssen

> ### Ihr Kurt Gödel

5 | d. h. genauer: 1.) Kein Formalismus, von dem *wir* erkennen können,
daß er richtiges (u. nur richtiges) Denken ausdrückt, kann *unser* ganzes
abstraktes Denken erfassen. 2.) Kein Formalismus, in dem nur objektiv
Richtiges ableitbar ist, kann, in seinen ableitbaren Formeln, alle objektiv
bestehenden begrifflichen Verhältnisse erfassen.

8. Günther to Gödel

> Gotthard Günther
> 3407 Montrose Ave.
> Richmond 22, Va.
> Sept. 18 55

Sehr geehrter Herr Prof. Gödel:

Nun sitze ich schon mitten in den Vorbereitungen für meine Reise
nach Hamburg, aber ich möchte nicht von hier fortgehen ohne Ihnen für
Ihren freundlichen Brief vom 10. VIII. zu danken —und Einiges darauf
zu erwidern.

Zuerst: es hat mich *wirklich* bestürzt von Ihrem schlechten Gesundheitszustand zu hören. Ich betrachte es als ein wirklich grosses Unglück
für die Wissenschaft, wenn Ihre kostbare Arbeitskraft auf diese Weise
beeinträchtigt wird. Ich hoffe ernstlich dass Ihr Brief an mich ein posi-
tives Zeichen war, dass es Ihnen definitiv besser geht.

Nun zu dem sachlichen Inhalt Ihres Schreibens. Es hat mich etwas
entmutigt *auch von Ihnen* (wie von so vielen anderen) zu hören, dass ᵴie   s
nicht einsehen können, warum ich behaupte dass beim Übergang von
einem aristotelischen (2-wertigen) zu einem nicht-aristotelischen (3-wer-
tigen) Logiksystem die Werte "wahr" und "falsch" aufgegeben werden
müssen, und dass dafür ein anderer Werttypus—der eine andere onto-

mean in contradistinction to the "true" and "false" of classical logic. One can of course not demand an exact definition, certainly not within the framework of classical two-valued logic, but still [one can demand] an explanation in the same sense in which one can make the fundamental concepts of two-valued logic perspicuous (in spite of their undefinability). An analysis of the sense of your truth-values seems to me to be the cardinal point which you should tackle in order to become comprehensible to your readers and to carry out further the construction of a logic corresponding to your ideas.

With cordial greetings

Yours, Kurt Gödel

i.e., more precisely: 1.) No formalism of which we can know that it expresses correct (and only correct) thought, can capture *our* entire abstract thought. 2.) No formalism in which only objectively correct [propositions] can be derived, can, in its derivable formulae, capture all objectively obtaining conceptual relations.

8. Günther to Gödel

Gotthard Günther
3407 Montrose Ave.
Richmond 22, Va.
18 September 1955

Dear Professor Gödel:

Now I'm already in the middle of the preparations for my trip to Hamburg, but I do not want to set out from here without thanking you for your kind letter of 10 August—and saying several things in response to it.

First: I was *truly* dismayed to hear about the poor state of your health. I consider it a real misfortune for science when your precious capacity for work is diminished in this way. I sincerely hope that your letter to me was a positive sign that you are truly doing better now.

Now to the material content of your letter. It discouraged me somewhat to hear *also from you* (as from so many others) that you cannot see why I claim that in the transition from an Aristotelian (2-valued) to a non-Aristotelian (3-valued) system of logic, the values "true" and "false" have to be abandoned, and that a different type of value—which involves

2   logische Fragestellung involviert—einge-|führt werden muss. Dass ich diese anderen Werte "irreflexiv" (contingent), "einfach reflexiv" (iterativ) und "doppelt-reflexiv" (Einheit von "irrefl." und "refl." *im Bewusstsein*) nenne, ist auf dieser Stufe der Betrachtung relativ irrelevant. Meine Terminologie hat *historische* Gründe in der Geschichte der Logik. Aber wenn jemand kommen sollte und mir eine bessere Wert*triade* demonstriert, so bin ich bereit, die meinige aufzugeben.

Was aber relevant ist, ist, dass man das Dualitätsprinzip der logischen Wertigkeit aufgeben muss. Reichenbach hat ausdrücklich festgestellt, dass dasselbe auch in der Wahrscheinlichkeitslogik, die mit beliebig vielen "Mischwerten" arbeitet, nicht aufgegeben wird. Die "Mischung" enthält immer die dualen Komponenten "wahr" und "falsch"—und ist stets dichotomisch aufteilbar.—

Nach mehrfachem Lesen Ihres Briefes ist mir der Gedanke gekommen, dass meine Interpretation einer *echten* Werttriade (wahr-unbestimmt-falsch ist *keine* echte Triade!) als eines Wertsystems, in dem die *Reflexionshöhe* eines Begriffs (und *nicht* seine wahr-falsch-Übereinstimmung mit faktischen Daten) bestimmt wird, Ihnen deshalb Schwierigkeiten macht, weil Sie meine Interpretation der hermeneutischen Struktur eines dreiwertigen Logiksystems ignorieren.

* * *

[In the omitted portion, including a P.S., Günther gives the explanation again with some technical details and gives an address in Hamburg at which he can be reached.]

9. Günther to Gödel

Gotthard Günther
3407 Montrose Ave.
Richmond 22, Va.
Sept. 20, 1956

Hoch verehrter Herr Dr. Gödel,

[Günther thanks Gödel for a conversation the previous Monday, which he says will have a substantial influence on his work. He remarks on the procedure for the new Bollingen application.]

Zum Schluss gestatten Sie mir bitte noch eine zusätzliche Bemerkung zu unserem Gespräch. Ich erwähnte, dass ich mich nicht gern als Hege-

a different ontological question—has to be introduced. That I call these other values "irreflexive" (contingent), "simply reflexive" (iterative) and "double- reflexive" (unity of "irreflexive" and "reflexive" *in consciousness*) is, at this stage of the investigation, relatively irrelevant. My terminology has a *historical* basis in the history of logic. But if someone should come and demonstrate to me a better value-*triad*, I am ready to give up my own.

What is relevant, is that one has to give up the principle of the duality of logical values. Reichenbach has explicitly stated that the latter principle is not even given up in probability logic, which works with arbitrarily many "mixed values". The "mix" always contains the dual components "true" and "false"—and can always be split up as a dichotomy.—

After reading your letter several times it occurred to me that my interpretation of a *genuine* value-triad (true—undetermined—false is *not* a genuine triad!) as a system of values in which the *level of reflection* of a concept (and *not* its true-false correspondence with factual data) is determined, creates difficulties for you, because you ignore my interpretation of the hermeneutic structure of a three-valued system of logic.

* * *

⟦In the omitted portion, including a P.S., Günther gives the explanation again with some technical details and gives an address in Hamburg at which he can be reached.⟧

9. Günther to Gödel

Gotthard Günther
3407 Montrose Ave.
Richmond 22, Va.
20 September 1956

Dear Professor Gödel,

⟦Günther thanks Gödel for a conversation the previous Monday, which he thinks will have a substantial influence on his work. He remarks on the procedure for the new Bollingen application.⟧

Finally, please allow me an additional remark about our conversation. I mentioned that I do not like to be classified as a Hegelian. My model

lianer klassifizieren lasse. Mein Vorbild wie philosophiert werden sollte,
ist nicht Hegel sondern Leibniz! Leibniz hat etwas besessen, was in der
Folgezeit völlig verloren gegangen ist: formale Exaktheit verbunden mit
metaphysischer Tiefe. Er ist bisher der Einzige, der gewusst hat, dass der
reine rechnende Formalismus eine Transparenz besitzt, die uns erlaubt,
durch ihn hindurch in uns nicht unmittelbar gegebene Realitätsdimensio-
nen zu blicken.

Ihr Ihnen dankbar verbundener

Gotthard Günther

10. Gödel to Günther

Princeton, 4./IV. 1957.

Sehr geehrter Herr Dr Günther!

Besten Dank für Ihre beiden Briefe.[a] Ich habe bedauert, daß Ihre Operation nicht ganz glatt verlaufen ist. Ich hoffe, Sie fühlen sich jetzt wieder vollkommen wohl. Was die Separata Ihrer Arbeiten betrifft, so besitze ich nur "Achilles u. die Schildkröte" u. Ihren Vortrag auf dem Phi-
losophenkongreß in [Brüssel][b].[1] Ich bin natürlich weniger an populären
Schriften interessiert als an Abhandlungen, die in philosophischen oder
wissenschaftlichen Zeitschriften erschienen sind. Von Ihrem freundlichen Angebot, mir einige Ihrer Arbeiten leihweise zu überlassen, werde
ich vielleicht später einmal Gebrauch machen. Gegenwärtig möchte ich
bloß Separata von Ihnen haben, so weit sie noch verfügbar sind. Was Ihr
neues Buch betrifft, so werde ich mich freuen, wenn ich im Herbst die
Korrekturbogen des ersten Bandes zu sehen bekomme.

2 | Nun zum Inhalt Ihrer Briefe. Was Sie über Leibniz sagen ist interes-
sant, aber ob es sich aus seinen Schriften rechtfertigen läßt, scheint mir
sehr zweifelhaft. Sie sprechen von einem "subjektiven Raum"—wenn ich
Sie recht verstehe im Sinne von Freiheitsgraden des Denkens—aber

---

[1]Außerdem besitze ich den Bericht über den "Congrès Descartes" (1937) mit einem
Vortrag von Ihnen im 8. Bd.[c]

---

[a]Meant are surely Günther's letters of 26 February and an undated letter in all
probability written in mid-March, together with a postscript dated "Montag".

of how to philosophize is not Hegel but Leibniz! Leibniz had a quality which got completely lost in ensuing times: formal exactness combined with metaphysical depth. Up to now he is the only one who knew that the pure calculating formalism has a transparency that allows us to look through it into dimensions of reality not immediately given to us.

With gratitude,

Sincerely,

Gotthard Günther

10. Gödel to Günther

Princeton, 4 April 1957

Dear Dr. Günther,

Many thanks for your two letters.[a] I am sorry that your operation did not go entirely smoothly. I hope you now feel completely well again. As regards the offprints of your works, I only have "Achilles and the Tortoise" and your lecture at the philosophers' congress in Brussels[b].[1] I am of course less interested in popular writings than in papers that appeared in philosophical or scientific journals. I will perhaps make use later of your kind offer to lend me some of your works. For now, I would like only to have offprints from you, to the extent they are still available. As regards your new book, I will be delighted to receive the proofs of the first volume in the fall.

Now to the content of your letters. What you say about Leibniz is interesting, but whether it can be justified on the grounds of his writings seems very doubtful to me. You speak of a "subjective space"—in the sense of degrees of freedom of thought, if I understand you correctly—but

---

[1] Moreover I have the report of the "Congrès Descartes" (1937) with a lecture of yours in volume 8.[c]

---

[b] *Günther 1954* and *1953.*
[c] *Günther 1937.*

ich wüßte nicht wo Leibniz davon spricht.[2] Daß das abstrakt begriffliche
Denken nur durch die Zentralmonade in die Einzelmonaden kommt, ist
zwar ein recht Leibnizscher Gedanke, aber ob Leibniz das abstrakte Denken als eine 2te Reflexionsstufe im Sinn von Hegel interpretieren wollte,
ist eine andere Frage. Es wäre interessant, wenn man Stellen aus seinen
Schriften anführen könnte, die darauf hindeuten. Schließlich imputieren
Sie Leibniz die Lehre von der Idealität der Zeit in der Form: Ein Vorgang
ist die Reflexionssituation eines Gegenstandes. Diese Formulierung ist
sehr interessant u. sollte näher ausgeführt werden, aber es würde mich
sehr wundern, wenn man darüber eine Andeutung bei Leibniz finden
3   könnte. Die Behauptung, daß Realität | ein *Verhältnis* ist (nämlich zwi-
schen Subjekt u. Objekt) ist mir als eine der idealistischen Grundpositio-
nen verständlich, aber daß sie ein *Umtausch*verhältnis ist, bedarf näherer
Erklärung. Eine solche wäre insbesondere im Zusammenhang mit Ihrer
Theorie der Negation vonnöten. Was die Beziehungen betrifft, die Sie
zwischen den Reflexionsstufen u. gewissen physikalischen Theorien her-
stellen wollen, so bin ich auf Grund Ihrer Andeutungen leider nicht im-
stande, ihnen einen verständlichen Sinn zu geben. Insbesondere sieht
man zunächst gar nicht, was die Art der Materie, aus denen die Gehirne
bestehen, mit der Möglichkeit einer Verständigung (im geistigen Sinn) zu
tun haben kann.

Was Ihr Manuscript über die nicht-Aristotelische Logik betrifft,[d] so
habe ich Ihnen ja meine Ansicht in unserem Gespräch in Princeton auseinandergesetzt. Ich glaube, daß der *Sinn* Ihrer Grundbegriffe durch
4   *Beispiele* u., soweit möglich, durch präzise Erklärungen | erläutert werden
sollte u. daß die Arbeit in ihrer gegenwärtigen Form kaum verständlich
sein wird. Ich möchte übrigens bemerken, daß ich vor einigen Jahren
ein Manuskript von Ihnen über die Grundlagen der Logik gelesen habe,[e]
das einen sehr interessanten Gedanken enthielt, den ich in Ihren neueren
Arbeiten über den Gegenstand vermisse. Sie haben nämlich damals die
Totalreflexion als etwas über alle Typenbildung hinausgehendes[3] interpretiert. Es ist plausibel, daß die Durchführung *dieser* Idee zu einer nicht
Aristotelischen Logik führen muß, da man ja auf diese Weise sofort in die
Antinomien der Mengenlehre hineinkommt. Jetzt scheinen Sie eher der
Ansicht zuzuneigen, die doppelte Reflexion mit dem zweiten logischen

---

[2]Eine "res extensa" gibt es wohl bei Leibniz überhaupt nicht, aber das ist in diesem
Zusammenhang ohne Bedeutung.

[3]was also gewissermaßen alle möglichen Typen zusammenfaßt

---

[d]Evidently a draft of *Günther 1958*.

I would not know where Leibniz talks about this[2]. That abstract conceptual thought enters individual monads only through the central monad is
a truly Leibnizian thought, but whether Leibniz meant to interpret abstract thought as a 2nd level of reflection in Hegel's sense is another question. It would be interesting if one could cite passages from his works
that point to it. Finally, you impute to Leibniz the theory of the ideality
of time in the form: A process is the situation of reflection of an object.
This formulation is very interesting and should be elaborated further, but
I would be very surprised, if one could find a suggestion of it in Leibniz.
The claim that reality is a *relation* (namely between subject and object)
is comprehensible to me as one of the fundamental positions of idealism,
but that reality is a relation of *exchange* requires further explanation.
Such explanation would be particularly necessary concerning your theory
of negation. As regards the relations you want to establish between levels of reflection and certain physical theories, I am unfortunately unable
to make sense out of them on the basis of your suggestions. In particular, initially one cannot see at all what the kind of matter that makes
up brains can have to do with the possibility of communication (in the
mental sense).

As regards your manuscript on non-Aristotelian logic,[d] I already explained my views to you in our conversation at Princeton. I believe that
the *sense* of your basic concepts should be elucidated through *examples*
and, so far as possible, through precise explanations, and that the work
in its current form, will hardly be comprehensible. By the way, I would
like to remark that a few years ago I read a manuscript of yours about
the foundations of logic[e] that contained a very interesting thought I now
miss in your more recent works on the topic. For, back then, you interpreted total reflection as something that goes beyond all formation of
types.[3] It is plausible that the implementation of *this* idea must lead to a
non-Aristotelian logic, since in this way one immediately runs up against
the antinomies of set theory. Now you seem rather to lean towards the
view of identifying double reflection with the second logical type and

---

[2] There probably is no "res extensa" in Leibniz at all, but that is unimportant in this
context.

[3] that, as it were, unifies all possible types

---

[e] Probably this was the manuscript "Der metaphysische Hintergrund der Logik und
die absolute Rationalität," which Günther sent with his letter of 2 August 1953.

Typus zu identifizieren u. die Aristotelische Logik bereits für den erweiterten Funktionenkalkül aufzugeben, was ich für unberechtigt halte.

Mit besten Grüssen u. Wünschen für eine gute Gesundheit

Ihr Kurt Gödel

11. Günther to Gödel

3407 Montrose Ave.
Richmond 22, Va.
April 7, 1957

Hochverehrter Herr Prof. Gödel,

⟦Günther makes remarks about his writings and comparative lack of publication and about three items he is about to send to Gödel.⟧

2 |...

Nun zu dem sachlichen Inhalt Ihres letzten Briefes und warum er mir eine solche besondere Freude gewesen ist! Sie schreiben, dass Sie früher eine Arbeit von mir gelesen haben, in der ich "die Totalreflexion als etwas über alle Typenbildung hinausgehendes interpretiert" habe. Und Sie fahren dann fort: "Es ist plausibel, dass die Durchführung *dieser* Idee zu einer nicht-Aristotelischen Logik führen muss, da man ja auf diese Weise sofort in die Antinomien der Mengenlehre hineinkommt. Jetzt scheinen sie der Ansicht zuzuneigen, die doppelte Reflexion mit dem zweiten logischen Typus zu identifizieren ...."

Sie wissen garnicht, wie glücklich mich diese Bestätigung gemacht hat. *Ich bin immer noch der Ansicht, dass die Totalreflexion eine solche ist, die alle überhaupt möglichen Typen zusammenfasst.* Darauf allein beruht ihr Totalitätsanspruch. Und erst wenn man die Technik einer solchen Zuammenfassung besitzt, hat man wirklich eine fertige Nicht-Aristotelik!!! Aber ich bin in den letzten Jahren schüchtern geworden, diesen Gedanken auszusprechen. Ich habe darüber vor vielen Jahren eine Anzahl Gespräche mit Quine gehabt. Er hält diese Idee für falsch und hat sie mir damals fürchterlich ver⟦r⟧issen. Er hat mich *nicht* überzeugt, dass sie falsch ist, wohl aber hat ⟨er⟩ mich damals ganz zwingend belehrt, das meine bisherige Weise das Problem anzufassen, völlig unzureichend war und nie hätte zum Ziele führen können. Inzwischen habe ich einiges gelernt, aber immer noch nicht genug, als dass ich es wagen würde die These so in den Vordergrund zu stellen, wie ich glaube, dass sie es verdient. Ich gehe jetzt vorsichtig vorwärts und formuliere meinen (vorläu-

giving up Aristotelian logic for the extended functional calculus, which I regard as unjustified.

With cordial greetings and best wishes for your health,

Yours, Kurt Gödel

11. Günther to Gödel

3407 Montrose Ave.
Richmond 22, Va.
April 7, 1957

Dear Professor Gödel:

⟦Günther makes remarks about his writings and comparative lack of publication and about three items he is about to send to Gödel.⟧

. . .

Now to the material content of your last letter, and why it was such a special pleasure for me! You write that previously you had read a work of mine, in which I "interpreted total reflection as something that goes beyond all formation of types". And you continue: "It is a plausible view that the implementation of *this* idea must lead to a non-Aristotelian logic, since in this way one immediately runs up against the antinomies of set theory. Now you seem rather to lean towards the view of identifying double reflection with the second logical type. . . ."

You don't know how happy this confirmation made me. *I am still of the opinion that total reflection is one which unites all the types that are at all possible.* That alone grounds its claim to totality. And only when one possesses the technique for such unification does one truly have a complete non-Aristotelianism!!! But in recent years I have become timid about articulating this thought. Many years ago I had a number of conversations with Quine about this. He believes this idea is false and back then tore it to pieces in a terrible way. He did *not* convince me that it is false, but he taught me very compellingly that my previous way of conceiving the problem was completely inadequate and could never have achieved its end. In the meantime I have learned a thing or two, but still not enough to dare to put the thesis as much to the fore as I believe it deserves. I now proceed cautiously and formulate my (provisional) posi-

figen) Standpunkt etwa so: Reflexion-in-Anderes (Theorie der Gegenständlichkeit) ist klassisch zweiwertig. Die Reflexion-in-sich (gleichgültig
welcher Art) beginnt mit den mehrwertigen Stellenwertsystemen. Der
unterste Fall ist die dreiwertige Logik. Die Totalreflexion aber ist nicht
dreiwertig sondern unbestimmt n-wertig, wobei immer gilt $n > 2$. Der
limes ist eine unendlich wertige Logik. ...

⟦Günther goes on to claim that three-valued logic is a "pure system of consciousness" and says that with four or more values the object, dissolved by
idealism, can be restored. He goes on to comment on the antinomies and to
defend his interpretation of Leibniz as a reconstruction.⟧

12. Günther to Gödel

3407 Montrose Ave.
Richmond 22, Va.
November 22, 1957

Hochverehrter Herr Prof. Gödel,

⟦Günther says he is enclosing two publications, *Günther 1957* and *1957a*.
Brief remarks about the former and about cybernetics.⟧

Aus diesem Grunde kann auch—wie ich in meiner zweiten Beilage
dem "Bewusstsein der Maschinen"[a] ausgeführt habe—die Kybernetik
nicht mehr zureichend mit den zweiwertigen Mitteln der Aristotelischen |
Logik interpretiert werden. Denn wenn der Mensch heute allmä⟦h⟧lich
beginnt seine Bewusstseinsfunktionen in einem electronic brain abzu-
bilden, dann ist als Grundkategorie wieder das Phänomen der Reflexion
involviert durch die jetzt entstehende Frage: wie unterscheidet sich eine
"Bewusstseinsanalogie" im Mechanismus von dem Selbstbewusstsein,
dass sie produziert. In diesem Sinn ist die Kybernetik die erste konse-
quente nicht-aristotelische Technik und dementsprechend nur mehrwertig
analysierbar.—

⟦Says that the revised first volume of his book (i.e. *Günther 1959*) is being
reviewed by the Deutsche Forschungs-Gemeinschaft.⟧

[a] *Günther 1957a.*

tion roughly like this: Reflection-in-other (theory of objecthood) is classically two-valued. Reflection-in-itself (no matter what kind) begins with many-valued place-value systems. The lowest case is three-valued logic. Yet, total reflection is not three-valued, but indeterminately n-valued, where $n > 2$. The limit is an infinite-valued logic....

〚Günther goes on to claim that three-valued logic is a "pure system of consciousness" and says that with four or more values the object, dissolved by idealism, can be restored. He goes on to comment on the antinomies and to defend his interpretation of Leibniz as a reconstruction.〛

12. Günther to Gödel

3407 Montrose Ave.
Richmond 22, Va.
November 22, 1957

Dear Professor Gödel:

〚Günther says he is enclosing two publications, *Günther 1957* and *1957a*. Brief remarks about the former and about cybernetics.〛

. . .

For this reason too—as I explained in my second supplement to the "Bewusstsein der Maschinen"[a]—cybernetics cannot any longer be interpreted adequately with the two-valued means of Aristotelian logic. For, if today man begins gradually to map the functions of his consciousness in an electronic brain, then the phenomenon of reflection is again involved as a fundamental category through the question that now arises: How is the "analogy of consciousness" in a mechanism distinguished from the self-consciousness that produces it? In this sense cybernetics is the first consistent non-Aristotelian technique and accordingly only analyzable in many-valued 〚logic〛.—

〚Says that the revised first volume of his book (i.e. *Günther 1959*) is being reviewed by the Deutsche Forschungs-Gemeinschaft.〛

13. Gödel to Günther

Princeton, 23./XII. 1957.

Sehr geehrter Herr Dr Günther!

Besten Dank für die Zusendung des Sonderdrucks u. des Buches, sowie auch Ihre freundlichen Neujahrswünsche, die ich auf das herzlichste erwidere. Ihre Arbeit: Metaphysik, Logik u. Theorie d. Reflexion[a] hatte ich bisher nicht gesehen, außer insoweit als sie mit einem alten Manuskript von Ihnen über die Metaphysik der klassischen Logik übereinstimmt.[b] Ich habe einiges aus Ihrer neuen Arbeit gelesen u. finde, daßsie sich durch besondere Klarheit auszeichnet, insbesondere auch in Vergleich mit dem Manuskript über die arist. Logik des Seins u. die nicht arist. Logik der Reflexion,[c] das ich gelesen habe. Beiliegend sende ich Ihnen mit bestem Dank das Manuskript über die Metaphysik des Todes zurück.[d] Ich habe es mit Interesse gelesen. Über die Arbeit "Logistik u. Transzendentallogik"[e] wollte ich Ihnen | einige kritische Bemerkungen schreiben, bevor ich das Separatum retourniere. Ich wurde aber leider im letzten Sommer durch einige unaufschiebbare Angelegenheiten u. durch Krankheit daran verhindert. Doch hoffe ich dies, sowie auch dasselbe bez. Ihrer neuesten Arbeit, demnächst noch nachholen zu können. Heute kann ich Ihnen leider nicht ausführlicher schreiben.

Ich hoffe, daß es Ihnen gesundheitlich und auch sonst gut geht, u. verbleibe mit besten Grüßen

Ihr Kurt Gödel

---

[a] *Günther 1957*. This is no doubt the reprint referred to in the previous sentence; the book would be *Günther 1957a*.

[b] Probably again "Der metaphysische Hintergrund der Logik und die absolute Rationalität"; see note d to letter 10 above.

14. Gödel to Günther

Princeton, Jan. 7, 1959.

Sehr geehrter Herr Dr. Gunther:

Besten Dank für Ihren Brief sowie den Sonderdruck. Ich habe mich sehr gefreut von den Fortschritten zu hoeren, die Ihre Arbeit gemacht hat, insbesondere darueber dass der erste Band des neuen Buches bereits im Druck ist. Die Frage der geschichtsphilosophischen Implikationen der

13. Gödel to Günther

Princeton, 23 December 1957

Dear Dr. Günther!

Many thanks for sending the offprint and the book, as well as your
kind wishes for the new year, which I warmly return. Your paper: Metaphysik, Logik, und Theorie der Reflexion[a] I had not seen before, except
insofar as it agrees with an old manuscript of yours about the metaphysics of classical logic.[b] I read parts of your new work and think that
it is distinguished by special clarity, in particular compared to the manuscript about the Aristotelian logic of Being and the non-Aristotelian logic
of reflection,[c] which I have read. I am returning enclosed your manuscript about the metaphysics of death.[d] I have read it with interest. I
wanted to write you several critical remarks on the paper "Logistik und
Transzendentallogik"[e] before I return the offprint. But unfortunately last
summer I was prevented from doing so by several urgent matters and by
illness. Yet I hope soon to be able to make up for this after all, as well as
for the same concerning your latest work. Today, unfortunately, I cannot
write to you at greater length.

I hope that things are going well for you with respect to health and
otherwise, and I remain with cordial greetings,

Yours, Kurt Gödel

---

[c] *Günther 1958.*
[d] Surely a version of *Günther 1957b.*
[e] *Günther 1940.*

14. Gödel to Günther

Princeton, 7 January 1959

Dear Dr. Günther,

Thank you for your letter as well as the offprint. I was very pleased
to hear of the progress that your work has made, in particular that the
first volume of the new book is already in press. The question of the implications of cybernetics for the philosophy of history seems to me very

Kybernetik scheint mir sehr interessant und aussichtsreich zu sein und ich begruesse Ihren Entschluss, sich damit zu beschaeftigen.

Was den logischen Teil Ihrer Arbeit betrifft, so scheint mir der interessanteste und aussichtsreichste Gesicht⟨s⟩punkt der zu sein, den Sie in einer Ihrer früheren Arbeiten einnahmen.[a] Sie identifizierten damals die iterierte Reflexion mit der Typentheorie und die totale Reflexion mit einer typenlosen (d. h. alle Typen in eins zusammenfassenden) Logik. Man sollte glauben, dass aus philosophischen Einsichten über das Wesen der Reflexion sich *mit Notwendigkeit* die richtigen Axiome einer typenlosen Logik ergeben muessten, was ein ungeheurer Fortschritt gegenueber dem heute angewendeten Verfahren des "trial and error" waere.

Es tut mir sehr leid, dass Sie Schwierigkeiten mit Ihren Augen haben, und ich wuensche Ihnen vom Herzen gute Besserung. Sehr gefreut hat es mich, zu hoeren, ⟨dass die Bollingen Foundation⟩ Ihnen so wohlgesinnt ist.

Ich wuensche Ihnen das beste für das kommende Jahr und verbleibe mit besten Grüssen.

Ihr

Kurt Gödel

P.S. Bitte entschuldigen Sie, dass ich so lange nichts von mir hoeren liess.

2 | P.S. Soeben fand ich Ihren Brief vom 1. Jan. im Institut vor. Ich danke bestens für die freundlichen Neujahrswünsche. Mein Gesundheitszustand ist nicht schlechter als sonst. Aber ich hatte, abgesehen von der Beschäftigung mit den mich interessierenden Fragen, allerhand zeitraubende Verpflichtungen u. sonstige Ablenkungen, was dann in Anbetracht meiner verminderten Arbeitskraft etwas zu viel wird. Der geschichtliche Teil Ihres Buches würde mich sehr interessieren, aber aus den angegebenen Gründen würde es mir kaum möglich sein, ihn im Laufe der nächsten Monate zu lesen.

Nochmals beste Grüsse

Kurt Gödel

---

[a]Gödel appears to have in mind the manuscript about the metaphysics of classical logic mentioned in letter 10; see note e thereto and the introductory note.

interesting and promising, and I welcome your decision to occupy yourself with it.

As for the logical part of your work, it seems to me that the most interesting and promising point of view is the one that you took in your earlier work.[a] At that time you identified iterated reflection with type theory and total reflection with a type-free logic, that is, one comprehending all types into one. One should think that from philosophical insights about the nature of reflection the correct axioms of a type-free logic would have to result *with necessity*, which would be an enormous advance compared to the procedure of "trial and error" applied today.

I'm very sorry that you have troubles with your eyes, and I sincerely wish you a good recovery. It has pleased me very much to hear that the Bollingen Foundation is so well disposed towards you.

I wish you the best for the coming year and remain with cordial greetings,

Yours, Kurt Gödel

P.S. Please excuse the fact that you did not hear from me for so long.

P.S. I just found your letter of 1 January in the Institute. Thank you very much for the kind New Year's greetings. My state of health is no worse than otherwise. But I had, apart from being occupied with the questions that interested me, all sorts of time-consuming duties and other distractions, which becomes somewhat too much considering my diminished capacity for work. The historical part of your book would interest me very much, but for the reasons [I have] given it would scarcely be possible for me to read it in the course of the next few months.

Once again cordial greetings

Kurt Gödel

Calendars of correspondence

Where an item was undated, an educated guess has been made on the approximate date if possible. Where it has proved impossible to make a reasonable surmise beyond the year, the date is indicated by question marks, and the letter follows all others for that year. "Form" describes the general character of the letter; incidental characteristics are footnoted. When two sources are given, the second entry in the form column or the identification column corresponds to the second source.

In the calendars, the following abbreviations are used.

For the form of the letter:

AL: Autograph letter, unsigned	SH: Shorthand
ALR: Autograph letter, retained copy	TG: Telegram
ALS: Autograph letter, signed	TL: Typed letter, unsigned
APS: Autograph postcard, signed	TLR: Typed letter, retained copy
D: Draft	TLS: Typed letter, signed

For the source:

APS: American Philosophical Society
BgN: *Briefe grosser Naturforscher und Ärzte in Handscriften*[a]
BN: Behmann *Nachlaß*
CAH: Center for American History
ETH: Eidgenössische Technische Hochschule Bibliothek
HA: Heyting Archief
IAS: Institute for Advanced Study
IIT: Menger *Nachlaß*, Illinois Institute of Technology
LC: Library of Congress, von Neumann papers
NN: Nagel *Nachlaß*
PS: Private source
SIU: SIU Carbondale, Schilpp papers
SBPK: Staatsbibliothek zu Berlin—Preussischer Kulturbesitz, Nachl. 196,[b] Box 46, Gödel
U Pitt: Carnap collection, University of Pittsburgh
UFB: Universitätsbibliothek Freiburg im Breisgau
WSB: Wiener Stadt- und Landesbibliothek, Handscriftenabteilung
YU: Yale University
ZBPW: Zentralbibliothek für Physik in Wien

---

[a] *Wiedemann 1989.*
[b] Gotthard Günther

Correspondence included in these volumes

Date	Correspondent	Letter number	To/From Gödel	Form	Source	Identifier
09/14/1928	Feigl	1	From	ALS	BgN	
11/20/1930	von Neumann	1	To	ALS	IAS	013029
11/29/1930	von Neumann	2	To	ALS	IAS	013029.5
12/24/1930	Bernays	1	To	ALS	IAS	010015.44
01/12/1931	von Neumann	3	To	ALS	IAS	013030
01/18/1931	Bernays	2	To	ALS	IAS	010015.45
01/20/1931	Tarski	1	From	ALS	IAS	012760
02/10/1931	Behmann	1	To	TLS	IAS	010015.42
02/22/1931	Behmann	2	From	ALS	BN	
02/25/1931	Behmann	3	To	TLS	IAS	010015.421
03/18/1931	Behmann	4	From	ALS	BN	
03/25/1931	Behmann	5	To	TLS	IAS	010015.422
03/30/1931	Menger	2	From	ALS	IIT	
04/02/1931	Bernays	3	From[a]	ALS	IAS	
04/07/1931	Herbrand	1	To	TLS	IAS	10837.5
04/20/1931	Bernays	4	To	ALS	IAS	010015.46
04/22/1931	Behmann	6	From	ALS	BN	
05/03/1931	Bernays	5	To	ALS	IAS	10015.47
05/18/1931	Behmann	7	To	TLS	IAS	010015.425
07/25/1931	Herbrand	2	From	ALS	IAS	010837.6
08/22/1931	Heyting	1	To	TLS	HA; IAS	No number; 010839
08/??/1931	Heyting	2	From	AL	IAS	010839.5
09/03/1931	Heyting	3	From	ALS	HA	
09/21/1931	Zermelo	1	To	ALS	IAS	013270
09/24/1931	Heyting	4	To	ALS	IAS	010840
10/12/1931	Zermelo	2	From	ALS	UFB	
10/29/1931	Zermelo	3	To	TL	UFB; IAS	013271
11/01/1931	Skolem	1	From	ALS	PS	
??/??/1931	Menger	1	To	TLS	IAS	011491
02/23/1932	Carnap	1	To	TLS	IAS	010280.86
04/11/1932	Carnap	2	From	ALS	U Pitt	

[a] Acquired after cataloging.

Date	Correspondent	Letter number	To/From Gödel	Form	Source	Identifier
06/02/1932	Menger	3	To	ALS	IAS	011492
06/11/1932	Heyting	5	To	ALS	IAS	010842
06/17/1932	Church	1	From	ALS	IAS	010329.09
07/01/1932	Heyting	6	From	ALS	HA	
07/17/1932	Heyting	7	To	ALS	IAS	010843
07/20/1932	Heyting	8	From	ALS	HA	
07/26/1932	Heyting	9	To	ALS	IAS	010844
07/27/1932	Church	2	To	ALS	IAS	010329.1
08/04/1932	Heyting	10	From	ALS	HA	
08/04/1932	Menger	5	From	ALS	IIT	
08/15/1932	Heyting	11	To	TLS	HA; IAS	No number; 010845
08/27/1932	Heyting	12	To	ALS	IAS	010847
09/11/1932	Carnap	3	From	ALS	U Pitt	
09/25/1932	Carnap	4	To	TLS	IAS	010280.87
09/27/1932	Carnap	5	To	TLS	IAS	010280.88
10/29/1932	Heyting	13	To	ALS	IAS	010848
11/15/1932	Heyting	14	From	ALS	HA	
11/24/1932	Heyting	15	To	AL	IAS	010850
11/26/1932	Heyting	15	To	ALS	IAS	010851
11/28/1932	Carnap	6	From	ALS	U Pitt	
??/??/1932	Menger	4	To	ALS	IAS	011493
??/??/1932?	Menger	6	To	ALS	IAS	011495
02/14/1933	von Neumann	4	To	ALS	IAS	013031
03/11/1933	Finsler	1	To	TLS	IAS; ETH	010632; Hs648:17
03/14/1933	von Neumann	5	From	ALS	IAS	013032
03/25/1933	Finsler	2	From	AL	IAS	010632.5
03/25/1933	Finsler	3	From	AL	IAS	010632.51
04/03/1933	Menger	7	From	ALS	IIT	
04/06/1933	Menger	8	From	ALS	IIT	
04/15/1933	Heyting	16	To	ALS	IAS	010852
05/07/1933	Heyting	17	To	APS	IAS	010852.5
05/16/1933	Heyting	18	From	ALS	HA; IAS	No number; 010853
06/13/1933	von Neumann	6	To	ALS	IAS	013033
06/19/1933	Finsler	4	To	TLS	IAS; ETH	010633; Hs648:20
08/24/1933	Heyting	19	To	ALS	IAS	010854

Date	Correspondent	Letter number	To/From Gödel	Form	Source	Identifier
09/30/1933	Heyting	20	To	ALS	IAS	010855
08/12/1934	von Neumann	7	To	ALS	IAS	013035
09/18/1934	von Neumann	8	To	ALS	IAS	013036
11/07/1934	Carnap	7	To	TLS	IAS	010280.90
05/29/1936	von Neumann	9	To	ALS	IAS	013037
05/22/1937	Menger	9	To	TLS	IAS	011497
07/03/1937	Menger	10	From	ALS	IIT	
07/13/1937	von Neumann	10	To	ALS	IAS	013038
09/12/1937	Menger	11	To	TG	IAS	011498
09/14/1937	von Neumann	11	To	ALS	IAS	013039
11/03/1937	Menger	12	To	TLS	IAS	011499
11/11/1937	von Neumann	12	To	ALS	IAS	013043
12/15/1937	Menger	13	From	ALS	IIT	
01/13/1938	von Neumann	13	To	ALS	IAS	013041
05/20/1938	Menger	14	To	TLS	IAS	011503
06/25/1938	Menger	15	From	ALS	IIT	
09/12/1938	von Neumann	14	From	ALS	LC	Container 8
10/19/1938	Menger	16	To	ALS	IAS	011505
10/??/1938	Menger	17	From	ALS	IIT	
10/29/1938	Post	1	To	APS	IAS	011717.3
10/30/1938	Post	2	To	ALS	IAS	011717.4
11/11/1938	Menger	18	From	ALS	IIT	
12/??/1938	Menger	19	To	ALS	IAS	011506
01/16/1939	Ulam	1	To	ALS	IAS	012877
02/28/1939?	von Neumann	15	To	ALS	IAS	013043
03/12/1939	Post	3	To	ALS	IAS	011717.5
03/20/1939	Post	4	From	ALS	Phyllis P. Goodman	
03/20/1939	von Neumann	16	From	ALS	LC	Container 4
04/22/1939	von Neumann	17	To	ALS	IAS	013044
05/13/1939?	von Neumann	18	To	ALS	IAS	013045
06/19/1939	Bernays	6	From	ALS	ETH	Hs.975:1692
06/21/1939	Bernays	7	To	ALS	IAS	010015.49
07/19/1939?	von Neumann	19	To	TLS	IAS	013046
07/20/1939	Bernays	8	From	ALS	ETH	Hs.975.1673
08/17/1939	von Neumann	20	From	ALS	IAS	013047
08/30/1939	Menger	20	From	ALS	IAS	011510
09/28/1939	Bernays	9	To	TLS	IAS; ETH	010015.6; Hs.975:1674

Date	Correspondent	Letter number	To/From Gödel	Form	Source	Identifier
12/12/1939	Bernays	10	To	APS	IAS	010015.7
12/29/1939	Bernays	11	From	ALS	ETH	Hs.975:1695
01/16/1942	Bernays	12	From	ALS	ETH	Hs.975:1696
06/22/1942	Menger	21	From	ALS	IIT	
08/14/1942	Ulam	2	To	TLS	IAS	012879
09/07/1942	Bernays	13	To	TLS	IAS; ETH	010016; Hs.975:1697
11/18/1942	Schilpp	1	To	TLS; TLR	IAS; SIU	012109; 20/7/4
11/30/1942	Schilpp	2	From	ALS	SIU; IAS	20/7/4; 012111
??/??/1942	Bernays	14	From	SH D	IAS	010017
??/??/1942?	Ulam	3	From	ALR	IAS	012881
03/27/1943	Schilpp	3	From	ALS	SIU; IAS	20/7/4; 012114
04/18/1943	Schilpp	4	From	ALS	SIU	20/7/4
05/18/1943	Schilpp	5	From	ALS	SIU; IAS[b]	20/7/4; 012116
05/22/1943	Schilpp	6	To	TLS; TLR	IAS; SIU	012117; 20/7/4
05/26/1943	Schilpp	7	From	ALS	SIU	20/7/4
05/31/1943	Schilpp	8	To	TLS; TLR	IAS; SIU	012118; 20/7/4
06/07/1943	Schilpp	9	From	ALS	SIU; IAS	20/7/4; 012119
06/27/1943	Schilpp	10	To	TLS; TLR	IAS; SIU	012120; 20/7/4
07/14/1943	Schilpp	11	To	TLS; TLR	IAS; SIU	012123; 20/7/4
07/28/1943	Schilpp	12	To	TLS; TLR	IAS; SIU	012126; 20/7/4
08/25/1943	Schilpp	13	To	TLS; TLR	IAS; SIU	012131; 20/7/4
09/13/1943	Schilpp	14	From	ALS; ALS[c]	SIU; IAS	20/7/4; 012132
09/20/1943	Schilpp	15	From	ALS	SIU	20/7/4
09/22/1943	Schilpp	16	To	TLS; TLR	IAS; SIU	012136; 20/7/4
09/27/1943	Schilpp	17	From	ALS	SIU	20/7/4

[b]5/17/43.
[c]Draft.

Date	Correspondent	Letter number	To/From Gödel	Form	Source	Identifier
09/28/1943	Russell	1	From	ALR	IAS	012137
10/07/1943	Schilpp	18	To	TLS; TLR	IAS; SIU	012138; 20/7/4
04/27/1944	Tarski	2	To	APS	IAS	012774
05/12/1944	Tarski	3	From	ALS	IAS	012775
07/10/1946	Schilpp	19	To	TLS; TLR	IAS; SIU	012170; 20/15/8
07/25/1946	Schilpp	20	From	ALS	SIU	20/15/8
07/30/1946	Schilpp	21	To	TLS; TLR	IAS; SIU	012171; 20/15/8
08/12/1946	Schilpp	22	From	ALS	SIU	20/15/8
08/17/1946	Schilpp	23	To	TLS; TLR	IAS; SIU	012172; 20/15/8
12/10/1946	Tarski	4	To	ALS	IAS	012780
07/15/1947	Schilpp	24	From	ALS	SIU	20/15/8
12/06/1947	Ulam	4	To	ALS	IAS	012882
03/09/1949	Schilpp	25	To	TLS; TLR	IAS; SIU	012178; 20/15/8
12/21/1949	Schilpp	26	From	ALS	SIU	20/15/8
02/27/1950	Gödel, Marianne	1	From	ALS	WSB	
05/15/1953	Schilpp	27	To	TLS; TLR	IAS; SIU	012186; 20/21/4
07/02/1953	Schilpp	28	From	ALS	SIU	20/21/4
07/06/1953	Schilpp	29	To	TLS; TLR	IAS; SIU	012188; 20/21/4
09/14/1953	Menger	22	From	ALS	IIT	
03/28/1954	Schilpp	30	From	ALS	SIU	20/21/4
04/29/1954	Günther	1	To	ALS	IAS	010754
05/15/1954	Günther	2	From	ALS	SBPK	Sheets 1–2
05/23/1954	Günther	3	To	ALS	IAS	010756
06/30/1954	Günther	4	From	ALS	SBPK	Sheets 5–7
10/02/1954	Günther	5	To	ALS	IAS	010757
06/19/1955	Günther	6	To	ALS	IAS	010758
08/10/1955	Günther	7	From	ALS	SBPK	Sheets 8–10
09/07/1955	Seelig	1	From	TLS	ETH	Hs.304:648
09/18/1955	Günther	8	To	ALS	IAS	010759
11/02/1955	Schilpp	31	To	TPS	IAS	012199
11/14/1955	Schilpp	32	From	ALS	SIU	20/21/4
11/18/1955	Seelig	2	From	TLS	ETH	Hs. 304:649
03/20/1956	von Neumann	21	From	ALS	LC	Container 5

Date	Correspondent	Letter number	To/From Gödel	Form	Source	Identifier
09/20/1956	Günther	9	To	TLS	IAS	010764
12/04/1956	Boone	1	From	TLR	IAS	010166
12/28/1956	Bernays	15	To	ALS	IAS	010018
12/31/1956	Bernays	16	To	ALS	IAS	010019
02/06/1957	Bernays	17	From	ALS; SH D	ETH; IAS	Hs.975:1698; 010020
02/25/1957	Nagel	1	From	TLS[d]	IAS	011590.05
03/09/1957	Nagel	2	To	TLS	IAS	011590.1
03/14/1957	Nagel	3	From	TLS[e]	IAS	011590.15
03/18/1957	Nagel	4	From	TLR	IAS	011590.25
03/21/1957	Nagel	5	To	TLS	IAS	011590.3
03/25/1957	Nagel	6	From	TLR	IAS	011590.35
04/04/1957	Günther	10	From	ALS	SBPK	Sheets 11–12
04/07/1957	Günther	11	To	TLS	IAS	010776
04/09/1957	Angoff	1	From	TLR	IAS	020388
04/09/1957	Nagel	7	From	TLS	NN	
04/22/1957	Angoff	2	To	TLS	IAS	020390
05/06/1957	Angoff	3	From	TLR	IAS	020391
05/28/1957	Angoff	4	To	TLS	IAS	020393
06/03/1957	Angoff	5	From	TLR	IAS	020394
06/06/1957	Angoff	6	To	TLS	IAS	020395
06/25/1957	Angoff	7	From	TLR	IAS	020396
07/11/1957	Schilpp	33	From	ALS	SIU	20/21/4
08/16/1957	Nagel	8	From	TLS	NN	
08/22/1957	Nagel	9	To	TLR	NN	
08/29/1957	Follett	1	From	TLR	IAS	020397
08/29/1957	Nagel	10	From	TLR	IAS	011590.6
09/09/1957	Schilpp	34	From	ALS[f]	IAS	012209
09/26/1957	Follett	2	To	TLS	IAS	020398
10/25/1957	Ulam	5	To	TLR	APS	
11/05/1957	Büchi	1	To	TLS	IAS	010280.62
11/08/1957	Ulam	6	From	TLS	APS	
11/20/1957	Bernays	18	From	TLS; TLR	ETH; IAS	Hs.975:1699; 010021
11/22/1957	Günther	12	To	TLS	IAS	010778

[d] Apparently unsent.

[e] Apparently unsent.

[f] Not sent.

Date	Correspondent	Letter number	To/From Gödel	Form	Source	Identifier
11/26/1957	Büchi	2	From	TL	IAS	010280.621
12/04/1957	Bernays	19	To	TLS	IAS; ETH	10022; Hs.975:1700
12/23/1957	Günther	13	From	ALS	SBPK	Sheet 13
01/28/1958	Ulam	7	From	TLR	IAS	012882.7
03/05/1958	Bernays	20	To	ALS	IAS	010026
03/14/1958	Bernays	21	From	ALS	ETH	Hs.975:1702
06/02/1958	Bernays	22	To	TLS; TL^g	IAS; ETH	010028; Hs.975:1703
09/20/1958	Pitts	1	From	TLR	IAS	011709.5
09/30/1958	Bernays	23	From	TLS; TL	ETH; IAS	Hs.975:1704; 010029
10/12/1958	Bernays	24	To	ALS	IAS	010030
10/30/1958	Bernays	25	From	ALS	ETH	Hs.975:1705
11/24/1958	Bernays	26	To	ALS; ALS^h	IAS; ETH	010031; Hs.975:1706
01/07/1959	Bernays	27	From	ALS	ETH	Hs.975:1707
01/07/1959	Günther	14	From	TLS	SBPK; IAS	Sheet 14; 010786
01/22/1959	Bernays	28	To	TLS	IAS; ETH	010033; Hs.975:1708
01/24/1959	Schilpp	35	To	TLR	SIU	20/21/4
02/03/1959	Schilpp	36	From	ALS; TLR	SIU; IAS	20/21/4; 012212
02/06/1959	Schilpp	37	To	TLR	SIU	20/21/4
10/08/1959	Bernays	29	From	ALS	ETH	Hs.975:1709
10/09/1959	Bernays	30	To	ALS	IAS	010034
05/11/1960	Bernays	31	To	ALS; SH D	IAS; ETH	010037; Hs.975:1711
12/20/1960	Bernays	32	To	ALS	IAS	010041
12/21/1960	Bernays	33	From	ALS	ETH	Hs.975:1712
03/16/1961	Bernays	34	From	ALS	ETH	Hs.975:1713
03/23/1961	Bernays	35	To	ALS; SH D	IAS; ETH	010042; Hs.975:1714
04/20/1961	Bernays	36	To	TL; TL^i	IAS; ETH	010042.5; Hs.975:1716

^{g}Initialed.

^{h}Draft.

^{i}Continued 05/05/1961.

Date	Correspondent	Letter number	To/From Gödel	Form	Source	Identifier
05/05/1961	Bernays	36	To	ALS; ALS[j]	IAS; ETH	010042.5; Hs.975:1716
05/11/1961	Bernays	37	From	ALS	ETH	Hs.975:1718
05/19/1961	Bernays	38	To	ALS	IAS	010043
07/10/1961	Bernays	39	To	TLS; TL[k]	IAS; ETH	010045; Hs.975:1719
07/23/1961	Gödel, Marianne	2	From	ALS	WSB	
08/11/1961	Bernays	40	From	ALS	ETH	Hs.975:1720
08/14/1961	Gödel, Marianne	3	From	ALS	WSB	
09/12/1961	Gödel, Marianne	4	From	ALS	WSB	
10/06/1961	Gödel, Marianne	5	From	ALS	WSB	
10/12/1961	Bernays	41	To	TLS; TL	IAS; ETH	010048; Hs.975:1721
11/08/1961	Boone	2	From	TLS	IAS	010216
12/15/1961	Bernays	42	From	ALS	ETH	Hs.975:1722
12/31/1961	Bernays	43	To	ALS	IAS	010051
??/??/1961?	Tarski	5	From	AL[l]	IAS	012237
07/30/1962	Bernays	44	From	ALS	ETH	Hs.975:1723
08/02/1962	Rappaport	1	From	TLR	IAS	011831
10/12/1962	Bernays	45	To	TLS; TL	IAS; ETH	010054; Hs.975:1724
12/31/1962	Bernays	46	To	ALS	IAS	010059
01/09/1963	Bernays	47	From	ALS	ETH	Hs.975:1725
02/23/1963	Bernays	48	To	TLS	IAS	010056
03/06/1963	Dreben	1	To	TLS	IAS	010510
03/14/1963	Bernays	49	To	ALS	IAS	010056.5
03/25/1963	van Heijenoort	1	To	TLS	IAS; CAH	012905; no number
04/23/1963	van Heijenoort	2	From	TLS	CAH; IAS	No number; 012907
06/20/1963	Cohen	1	From	TLR	IAS	010352
08/28/1963	van Heijenoort	3	From	TLS	CAH	
10/04/1963	van Heijenoort	4	From	TLS	CAH; IAS	No number; 012925
10/24/1963	van Heijenoort	5	To	ALS	IAS; CAH	012927; no number
12/18/1963	Bernays	50	From	ALS	ETH; IAS	Hs.975:1728; 010060

[j]Continuation of 04/20/1961.
[k]Initialed.
[l]Probably unsent.

Date	Correspondent	Letter number	To/From Gödel	Form	Source	Identifier
01/06/1963[m]	Ulam	8	To	TLS	IAS	012883
01/01/1964	Bernays	51	To	ALS	IAS	010061
01/22/1964	Cohen	2	From	TLR	IAS	010390.5
02/10/1964	Ulam	9	From	TLR	IAS	012885
02/16/1964	Schilpp	38	To	TLS	IAS	012213.1
02/22/1964	van Heijenoort	6	From	TLS	CAH; IAS[n]	No number; 012943
04/10/1964	Popper	1	From	TLR	IAS	011717.25, 011717.26
06/01/1964	Carnap	8	From	ALS	U Pitt	
06/24/1964	van Heijenoort	7	To	ALS	IAS; CAH	012959; no number
07/23/1964	Schilpp	39	From	TLR	IAS	012213.2
08/14/1964	van Heijenoort	8	From	TLS	CAH; IAS	No number; 012962
08/15/1964	van Heijenoort	9	From	TLS	CAH; IAS	No number; 012963
08/27/1964	van Heijenoort	10	To	TLS	IAS; CAH	012964; no number
09/18/1964	van Heijenoort	11	From	TLS	CAH; IAS	No number; 012966
11/03/1964	van Heijenoort	12	From	TLS	CAH; IAS	No number; 012973
11/14/1964	van Heijenoort	13	To	TLS	IAS; CAH	012974; no number
12/15/1964	van Heijenoort	14	From	TLS	CAH; IAS	No number; 012975
01/11/1965	Cohen	3	From	TLR	IAS	010394
02/05/1965	Bernays	52	From	ALS	ETH	Hs.975:1729
02/16/1965	Goddard	1	To	ALS	IAS	010709.3
03/02/1965	Church	3	From	TLR	IAS	010332
05/25/1965	van Heijenoort	15	To	TLS	IAS; CAH	012982; no number
07/08/1965	van Heijenoort	16	From	TLS	CAH; IAS	No number; 012987
08/09/1965	van Heijenoort	17	From	TLS	CAH	
08/13/1965	Cohen	4	From	TLR	IAS	010409

[m]Although the letter is dated 1963, internal evidence indicates that the year should be 1964.

[n]The retained copy at the I.A.S. has a postscript dated 02/21/1964, but a note by Gödel on the back says "22.II.$\overline{64}$".

Date	Correspondent	Letter number	To/From Gödel	Form	Source	Identifier
09/01/1965	Reid	1	To	TLS	IAS	011840
09/17/1965	Bernays	53	To	ALS; SH D	IAS; ETH	010071; Hs.975:1732
09/27/1965	Bernays	54	From	TLS; TL	ETH; IAS	Hs.975:1733; 010073
11/08/1965	Bernays	55	To	ALS	IAS	010074
12/02/1965	Bernays	56	From	TLS; TL	ETH; IAS	Hs.975:1734; 010075
12/10/1965	Bernays	57	To	TLS	IAS; ETH	010076; Hs.975:1735
12/13/1965	van Heijenoort	18	From	TLS	CAH	
??/??/1965?	Goddard	2	From	AL°	IAS	010709.5
01/25/1966	Bernays	58	From	TLS; TL	ETH; IAS	Hs.975:1736; 010079
02/17/1966	Ulam	10	To	TLS	IAS	012885.5
03/22/1966	Reid	2	From	TLR; TLS	IAS; ETH	011845; Hs.1001:1
04/24/1966	Bernays	59	To	ALS	IAS	010080
05/22/1966	Bernays	60	From	ALS	ETH	Hs.975:1737
05/24/1966	Dreben	2	To	TLS	IAS	010513
06/27/1966	Church	4	To	ALS	IAS	010334.2
07/19/1966	Dreben	3	From	TLR	IAS	010516
08/10/1966	Church	5	From	TLR	IAS	010334.25
09/29/1966	Church	6	From	TLR	IAS	010334.86
01/24/1967	Bernays	61	From	ALS	ETH	Hs.975:1738
04/27/1967	Cohen	5	From	TLR	IAS	010417.6
06/30/1967	Rautenberg	2	From	TLR	IAS	011834
07/07/1967	Robinson, A.	1	From	TLR	IAS	011944
07/27/1967	Wang	1	To	ALS	IAS	013069
07/31/1967	Plummer	1	From	TLR	IAS	011714, 011715
10/27/1967	Bernays	62	To	ALS	IAS	010082
12/07/1967	Wang	2	From	TLS	IAS	013073-5
12/19/1967	Wang	3	To	TLS	IAS	013076
12/20/1967	Bernays	63	From	ALS	ETH	Hs.975:1739
??/??/1967?	Rautenberg	1	To	ALS	IAS	011833
01/02/1968	Menger	23	To	TLS	IAS	011513

°Unsent.

Date	Correspondent	Letter number	To/From Gödel	Form	Source	Identifier
01/22/1968	Menger	24	From	TL	IAS	011515
03/07/1968	Wang	4	From	TLS	IAS	013085–6
04/23/1968	Wang	5	To	TLS	IAS	013087
05/16/1968	Bernays	64	From	ALS	ETH	Hs.975:1740
07/20/1968	Bernays	65	To	TLS	IAS; ETH	010084; Hs.975:1741
12/17/1968	Bernays	66	From	ALS	ETH	Hs.975:1742
12/20/1968	Perlis	1	From	TLR	IAS	011709.42
01/02/1969	Heyting	21	To	TLS	IAS	010863
01/06/1969	Bernays	67	To	ALS	IAS	010086
02/10/1969	Reid	3	To	TLS	IAS	011846
03/12/1969	Heyting	22	From	TLR[p]	IAS	010864
06/25/1969	Reid	4	From	TLS	Constance Reid	
07/??/1969	Bernays	68a	From	AL, D	IAS	010090
07/25/1969	Bernays	68b	From	ALS	ETH; IAS	Hs.975:1743; 010089
12/10/1969	Brutian	1	From	TLR	IAS	010280.5
12/30/1969	Dreben	4	To	TLS	IAS	010517
01/07/1970	Bernays	69	To	TLS; TL	IAS; ETH	010097; Hs.975:1745
04/15/1970	Dreben	5	To	TLS	IAS	010517.5
04/15/1970	Wang	6	To	TLS	IAS	013091
07/12/1970	Bernays	70	To	ALS	IAS; ETH	010097.5; Hs.975:1746
07/14/1970	Bernays	71	From	ALS	ETH	Hs.975:1747
09/12/1970	Bernays	72	To	ALS	IAS; ETH	010099; Hs.975:1748
10/02/1970	Bernays	73	From	TG; AL[q]	ETH; IAS	Hs.975:1749; 010100
12/22/1970	Bernays	74	From	ALS	ETH	Hs.975:1750
12/31/1970	Bernays	75	To	ALS	IAS	010102
??/??/1970	Balas	1	From	AL, D	IAS	010015.37
03/17/1971	Hwastecki	1	To	TLS	IAS	010897
03/??/1971	Robinson, A.	2	From	AL[r]	IAS	011956
04/14/1971	Robinson, A.	3	To	TLS; TLR	IAS; YU	011957; no number
05/25/1971	Wang	7	To	TLS	IAS	013094

[p] Unsent.

[q] Draft.

[r] Unsent draft.

Date	Correspondent	Letter number	To/From Gödel	Form	Source	Identifier
07/09/1971	Wang	8	From	TL	IAS	013097
07/20/1971	Wang	9	To	TLS	IAS	013098
07/20/1971	Wang	10	To	TLS	IAS	013100
08/04/1971	Wang	11	From	TL	IAS	013104
08/11/1971	Wang	12	To	TLS	IAS	013105
09/22/1971	Blackwell	1	To	TLS	IAS	010120.1
11/22/1971	Robinson, A.	4	From	ALS; AL, D	YU; IAS	No number; 011959
??/??/1971?	Blackwell	2	From	AL, D	IAS	010120.3
??/??/1971?	Hwastecki	2	From	D^s	IAS	010898
01/15/1972	Menger	25	To	ALS	IAS	011517
03/16/1972	Bernays	76	To	TLS; TL^t	IAS; ETH	010109; Hs.975:1751
04/10/1972	Wang	13	To	TLS	IAS	013110
04/18/1972	Henkin	1	From	TLR	IAS	010834
04/20/1972	Menger	26	From	ALS	IIT	
04/26/1972	Wang	14	To	ALS	IAS	013113
05/31/1972	Thirring	1	To	TLS	IAS	012867
06/27/1972	Honderich	1	From	TLR	IAS	013116
06/27/1972	Thirring	2	From	ALS	ZBPW	
07/19/1972	Honderich	2	From	TLR	IAS	013121
07/21/1972	Bernays	77	To	APS	IAS	010110
10/28/1972	Bernays	78	To	ALS; SH D	IAS; ETH	010111; Hs.975:1751
12/26/1972	Bernays	79	From	ALS; SH D	ETH; IAS	Hs.975:1753; 010106
12/29/1972	Robinson, A.	5	From	AL, D	IAS	011962
01/04/1973	Robinson, A.	6	To	ALS	IAS	011963
02/21/1973	Bernays	80	To	TLS	ETH	Hs.975:1754
04/23/1973	Robinson, A.	7	To	TLS; TLR	IAS; YU	011968; no number
07/02/1973	Robinson, A.	8	From	TLS; TLR	YU; IAS	No number; 011972
07/18/1973	Robinson, A.	9	From	TLR	IAS	011977
10/02/1973	Greenberg	1	From	TLS	IAS	010729.65
10/15/1973	Greenberg	2	To	TLS	IAS	010729.7

^{s}Unsent.

^{t}Initialed.

Date	Correspondent	Letter number	To/From Gödel	Form	Source	Identifier
10/16/1973	Robinson, A.	10	To	TLR	YU	
10?/??/1973?	Greenberg	3	From	AL, D	IAS	010729.75
11/22/1973	Robinson, A.	11	From	ALS	YU	
12/18/1973	Bernays	81	From	ALS	ETH; IAS	Hs.975:1755; 010107
02/01/1974	Sawyer	1	To	TLS	IAS	012108.97
03/20/1974	Robinson, A.	12	From	AL, D	IAS	011998
05/11/1974	Robinson, Renee[u]		From	AL, D	IAS	012001
05/20/1974	Mostow[v]		From	TLR	IAS	012003
07/03/1974	Bohnert	1	To	TLS	IAS	010134
07/16/1974	Grandjean	1	To	TLS	IAS	010720
07/22/1974	Suppes	1	From	TLR	IAS	012462
09/17/1974	Bohnert	2	From	TLR	IAS	010135
12/16/1974	Bernays	82	To	TLS	ETH	Hs.975:1756
12/17/1974	Bernays	83	From	ALS	ETH	Hs.975:1757
??/??/1974?	Sawyer	2	From	AL[w]	IAS	012108.98
01/12/1975	Bernays	84	From	ALS	ETH	Hs.975:1758
01/24/1975	Bernays	85	To	TLS	IAS; ETH	010114; Hs.975:1759
08/19/1975[x]	Grandjean	2	From	TLR	IAS	010728
12/02/1975	Grandjean	3	To	TLS	IAS	010729.25
12/12/1975	Wang	15	To	TLS	IAS	013166
??/??/1975?	Grandjean	4	Neither[y]		IAS	010729
??/??/1975?	Grandjean	5	From	AL[z]	IAS	010727
06/24/1976	Wang	16	To	ALS	IAS	013187
02/03/1977	Wang	17	To	TLS	IAS	013196

[u]See the end of the correspondence with Abraham Robinson.

[v]See the end of the correspondence with Abraham Robinson.

[w]Unsent.

[x]Gödel had originally dated the letter August 1975, but later changed the month to January. Presumably the year should also have been changed to 1976.

[y]Questionnaire.

[z]Unsent.

Individual calendars of correspondence

Heinrich Behmann[a]

Item	Date	To/From Gödel	Form	Source	Identifier
1	02/10/1931	To	TLS	IAS	010015.42
2	02/22/1931	From	ALS	IAS	Behmann *Nachlaß*
3	02/25/1931	To	TLS	IAS	010015.421
4	03/18/1931	From	ALS	IAS	Behmann *Nachlaß*
5	03/25/1931	To	TLS	IAS	010015.422
6	04/22/1931	From	ALS	IAS	Behmann *Nachlaß*
7	05/18/1931	To	TLS	IAS	010015.425
	09/10/1932	To	APS	IAS	010015.428

[a]Gödel's *Nachlaß* also contains copies of letters to and from Behmann to Felix Kaufmann. See Appendix A in volume V of these *Works* for one such letter.

Paul Bernays

Item	Date	To/From Gödel	Form	Source	Identifier
1	12/24/1930	To	ALS	IAS	010015.44
2	01/18/1931	To	ALS	IAS	010015.45
3	04/02/1931	From	ALS	IAS	Unnumbered[a]
4	04/20/1931	To	ALS	IAS	010015.46
5	05/03/1931	To	ALS	IAS	10015.47
	09/??/1935	From	ALS	ETH	Hs.975:1691
	09/??/1935	To	ALS	ETH	Unnumbered[b]
6	06/19/1939	From	ALS	ETH	Hs.975.1692
7	06/21/1939	To	ALS	IAS	010015.49
	07/18/1939	From	ALS[c]	IAS	010015.5
8	07/20/1939	From	ALS	ETH	Hs.975.1673
9	09/28/1939	To	TLS	IAS; ETH	010015.6; Hs.975:1674
10	12/12/1939	To	APS	IAS	010015.7
	12/28/1939	From	ALS[d]	IAS	010015.8
11	12/29/1939	From	ALS	ETH	Hs.975:1695
	01/??/1940[e]	To			
12	01/16/1942	From	ALS	ETH	Hs.975:1696
13	09/07/1942	To	TLS	IAS; ETH	010016; Hs.975:1697
14	??/??/1942	From	SH D	IAS	010017
15	12/28/1956	To	ALS	IAS	010018

[a] Acquired after cataloging.
[b] Photocopy.
[c] Copy of 07/20/1939.
[d] Copy of 12/29/39.
[e] Not extant; mentioned in Gödel's letter to Bernays of 01/16/1942.

Item	Date	To/From Gödel	Form	Source	Identifier
16	12/31/1956	To	ALS	IAS	010019?
17	02/06/1957	From	ALS	ETH; IAS	Hs.975:1698; 010020
18	11/20/1957	From	TLS	ETH; IAS	Hs.975:1699; 010021
19	12/04/1957	To	TLS	ETH; IAS	Hs.975:1700; 010022
	12/23/1957	From	AL, D	IAS	010023
	12/31/1957	To	ALS	IAS	010024
	01/01/1958	From	SH D	IAS	010025
	01/03/1958	From	ALS	ETH	Hs.975:1701
	01/06/1958	From[f]		IAS	010025.5
20	03/05/1958	To	ALS	IAS	010026
21	03/14/1958	From	ALS	ETH	Hs.975:1702
	03/15/1958	From	SH D	IAS	010027
22	06/02/1958	To	TLS; TL	IAS; ETH	010028; Hs.975:1703
23	09/30/1958	From	TLS; TL	ETH; IAS	Hs.975:1704; 010029
24	10/12/1958	To	ALS	IAS	010030
25	10/30/1958	From	ALS	ETH	Hs.975:1705
26	11/24/1958	To	ALS; D	IAS; ETH	010031; Hs.975:1706
	12/29/1958	To	ALS	IAS	010032
27	01/07/1959	From	ALS	ETH	Hs.975:1707
28	01/22/1959	To	TLS	IAS; ETH	010033; Hs.975:1708
29	10/08/1959	From	ALS	ETH	Hs.975:1709
30	10/09/1959	To	ALS	IAS	010034
	10/20/1959	From	ALS	ETH	Hs.975:1710
	11/02/1959	To	ALS	IAS	010035
31	05/11/1960	To	ALS	IAS, ETH	010037, Hs.975:1711
	08/19/1960	From	ALS[g]	IAS	010038
32	12/20/1960	To	ALS	IAS	010041

[f]Conversation notes.
[g]Unsent.

Item	Date	To/From Gödel	Form	Source	Identifier
33	12/21/1960	From	ALS	ETH	Hs.975:1712
	??/??/1960	From	SH D	IAS	010036
	??/??/1960	From	SH D	IAS	010039
	??/??/1960	To	SH D	IAS	010040
34	03/16/1961	From	ALS	ETH	Hs.975:1713
35	03/23/1961	To	ALS	IAS; ETH	010042; Hs.975:1714
	03/27/1961	To	TG	ETH	Hs.975:1715
36	04/20/1961	To	TL^h	IAS; ETH	010042.5; Hs.975:1716
36	05/05/1961	To	ALS^i	IAS; ETH	010042.5; Hs.975:1716
37	05/11/1961	From	ALS	ETH	Hs.975:1718
38	05/19/1961	To	ALS	IAS	010043
39	07/10/1961	To	TLS; TL^j	IAS; ETH	010045; Hs.975:1719
40	08/11/1961	From	ALS	ETH	Hs.975:1720
41	10/12/1961	To	TLS; TL	IAS; ETH	010048; Hs.975:1721
42	12/15/1961	From	ALS	ETH	Hs.975:1722
43	12/31/1961	To	ALS	IAS	010051
	12/??/1961	From	SH	IAS	010047
44	07/30/1962	From	ALS	ETH	Hs.975:1723
45	10/12/1962	To	TLS	IAS; ETH	010054; Hs.975:1724
46	12/31/1962	To	ALS	IAS	010059
47	01/09/1963	From	ALS	ETH	Hs.975:1725
	02/19/1963	To	TL^k	ETH	Hs.975:1726
48	02/23/1963	To	TLS	IAS	010056
49	03/14/1963	To	ALS	IAS	010056.5
50	12/18/1963	From	ALS	ETH; IAS	Hs.975:1728; 010060
	01/30/1965	To	ALS	IAS	010062

^{h}Continued 05/05/1961.
^{i}Continuation of 04/20/1961.
^{j}Initialed.
^{k}Unsent?

Item	Date	To/From Gödel	Form	Source	Identifier
51	01/01/1964	To	ALS	IAS	010061
52	02/05/1965	From	ALS	ETH	Hs.975:1729
	02/08/1965	To	APS	IAS	010064
	02/24/1965	From	ALS	ETH; IAS	Hs.975:1730; 010067
	02/25/1965	To	ALS	IAS	010068
	03/02/1965	From	ALS	ETH	Hs.975:1731
	04/16/1965	To	ALS	IAS	010070
53	09/17/1965	To	ALS; D	IAS; ETH	010071; Hs.975:1732
54	09/27/1965	From	TLS; TL	ETH; IAS	Hs.975:1733; 010073
55	11/08/1965	To	ALS	IAS	010074
56	12/02/1965	From	TLS; TL	ETH; IAS	Hs.975:1734; 010075
57	12/10/1965	To	TLS	IAS; ETH	010076; Hs.975:1735
58	01/25/1966	From	TLS; TL	ETH; IAS	Hs.975:1736; 010079
59	04/24/1966	To	ALS	IAS	010080
60	05/22/1966	From	ALS	ETH	Hs.975:1737
61	01/24/1967	From	ALS	ETH	Hs.975:1738
62	10/27/1967	To	ALS	IAS	010082
63	12/20/1967	From	ALS	ETH	Hs.975:1739
64	05/16/1968	From	ALS	ETH	Hs.975:1740
65	07/20/1968	To	TLS	IAS; ETH	010084; Hs.975:1741
	11/??/1968	From	SH	IAS	010084.5
66	12/17/1968	From	ALS	ETH	Hs.975:1742
67	01/06/1969	To	ALS	IAS	010086
68a	07/??/1969	From	ALS[1]	IAS	010090
68b	07/25/1969	From	ALS	ETH; IAS	Hs.975:1743; 010089, 010090
	12/20/1969	From	ALS	ETH	Hs.975:1744

[1]Draft of letter 68b.

Item	Date	To/From Gödel	Form	Source	Identifier
69	01/07/1970	To	TLS TL	IAS; ETH	010097; Hs.975:1745
70	07/12/1970	To	ALS	IAS; ETH	010097.5; Hs.975:1746
71	07/14/1970	From	ALS	ETH	Hs.975:1747
72	09/12/1970	To	ALS	IAS; ETH	010099; Hs.975:1748
73	10/02/1970	From	TG; D	ETH; IAS	Hs.975:1749; 010100
74	12/22/1970	From	ALS	ETH	Hs.975:1750
	12/22/1970	From	SH D	IAS	010101
75	12/31/1970	To	ALS	IAS	010102
	12/31/1971	To	ALS	IAS	010105
76	03/16/1972	To	TLS	IAS; ETH	010109; Hs.975:1751
77	07/21/1972	To	APS	IAS	010110
78	10/28/1972	To	ALS; D	IAS; ETH	010111; Hs.975:1751
79	12/26/1972	From	ALS; D	ETH; IAS	Hs.975:1753; 010106
80	02/21/1973	To	TLS	ETH	Hs.975:1754
81	12/18/1973	From	ALS	ETH; IAS	Hs.975:1755; 010107
	12/18/1973	From	SH D	IAS	010108
82	12/16/1974	To	TLS	ETH	Hs.975:1756
83	12/17/1974	From	ALS	ETH	Hs.975:1757
84	01/12/1975	From	ALS	ETH	Hs.975:1758
85	01/24/1975	From	TLS	IAS; ETH	010114; Hs.975:1759

William Boone

Item	Date	To/From Gödel	Form	Source	Identifier
	01/26/1954	From	TLS, R	IAS	010136
	04/25/1954	To	TLS	IAS	010139.5
	12/14/1954	To	TLS	IAS	010140
	03/04/1955	To	ALS	IAS	010145
	04/28/1955	To	ALS	IAS	010146
	06/15/1955	To	ALS	IAS	010147
	08/06/1955?	To	ALS	IAS	010149
	08/21/1955?	To	ALS	IAS	010150
	11/07/1955?	To	ALS	IAS	010152
	04/03/1956?	To	ALS	IAS	010157
	04/18/1956?	To	ALS	IAS	010158
	05/24/1956	To	ALS	IAS	010160
	10/26/1956	To	TLS	IAS	010162
1	12/04/1956	From	TLR	IAS	010166
	04/23/1957	To	TLS	IAS	010174
	05/01/1957	To	TLS	IAS	010175
	03/12/1958	To	ALS	IAS	010177
	04/07/1958	From	TLR	IAS	010179
	04/11/1958	To	ALS	IAS	010180
	05/08?/1958	To	TLS	IAS	010183
	06/12/1958	From	TLR	IAS	010189
	06/15/1958	To	ALS	IAS	010190
	06/19/1958	From	TLR	IAS	010192
	06/25/1958	From	TLR	IAS	010193
	06/30/1958	To	ALS	IAS	010198

Item	Date	To/From Gödel	Form	Source	Identifier
	07/24/1958	To	ALS	IAS	010202
	08/06/1958	To	ALS	IAS	010203
	01/07/1959	To	TLS	IAS	010204
	05/18/1959	To	ALS	IAS	010206
	09/09/1959	To	ALS	IAS	010207
	04/07/1960	To	TLS	IAS	010208
	10/15/1961	To	TLS	IAS	010215
2	11/08/1961	From	TLR	IAS	010216
	04/07/1962	To	TLS	IAS	010218
	08/21/1962	To	APS	IAS	010222
	11/09/1963	From	TL, D	IAS	010224
	11/19/1963	To	TLS	IAS	010226
	11/20/1963	To	TLS	IAS	010230, 010231
	02 /24/1964	To	TLS	IAS	010233
	03/18/1964	From	TLR	IAS	010234
	03/20/1964[a]	To	TLS	IAS	010235
	09/30/1965?	To	ALS	IAS	010252
	12/06/1965	To	ALS	IAS	010253
	01/27/1966?	To	ALS	IAS	010253.7
	06/01/1966	To	ALS	IAS	010254
	09/03/1966	To	ALS	IAS	010255
	09/04/1966	To	ALS	IAS	010256
	02/23/1975	To	ALS	IAS	010261
	12/23/1975	To	TLS	IAS	010262.4
	09/26/1976	To	TLS	IAS	010263
	06/29/19??[b]	To	ALS	IAS	010270

[a]In addition to the dated letters from 1964, there are four letters apparently from 1964 addressed to Gödel from Boone, with document numbers 010345–47 and 010250.

[b]There are also five other letters, document numbers 010266–69 and 010272, with no dates indicated, also from Boone to Gödel.

Rudolf Carnap

Item	Date	To/From Gödel	Form	Source	Identifier
	10/26/1930	To	TLS	IAS	010280.84
	01/01/1932	To	TLS	IAS	010280.85
1	02/23/1932	To	TLS	IAS	010280.86
2	04/11/1932	From	ALS	U Pitt	
3	09/11/1932	From	ALS	U Pitt	
4	09/25/1932	To	TLS	IAS	010280.87
5	09/27/1932	To	TLS	IAS	010280.88
6	11/28/1932	From	ALS	U Pitt	
	10/09/1933	To	TLS	IAS	010280.89
7	11/07/1934	To	TLS	IAS	010280.90
	07/18/1935	To	TLS	IAS	010280.91
	02/04/1939	To	TLS	IAS	010280.94
8	06/01/1964	From	ALS	U Pitt	

Alonzo Church

Item	Date	To/From Gödel	Form	Source	Identifier
1	06/17/1932	From	ALS	IAS	010329.09
2	07/27/1932	To	ALS	IAS	010329.1
	08/29/1946	To et al.	TLS	IAS	010329.2
	10/14/1946	To et al.	TLS	IAS	010329.3
	10/30/1946	To	TLS	IAS	010329.4
	11/29/1946	To et al.	TLS	IAS	010329.5
	12/12/1946	To	TLS	IAS	010329.6
	01/26/1965	From[a]			
	01/29/1965	To	TLS	IAS	010330
3	03/02/1965	From	TLR	IAS	010332
	03/02/1965	From	AL D	IAS	010331
	03/09/1965	To	TLS	IAS	010334
4	06/27/1966	To	ALS	IAS	010334.2
	07/16/1966	From	ALS	IAS	010334.21
	07/20/1966	To	ALS	IAS	010334.22
	08/04/1966	To	TLS	IAS	010334.23
5	08/10/1966	From	TLR	IAS	010334.25
	09/12/1966	To	TLS	IAS	010334.30
6	09/29/1966	From	TLR	IAS	010334.86
	10/03/1966	To	TL	IAS	010334.38

[a]Not extant; existence known from other letters.

Paul Cohen

Item	Date	To/From Gödel	Form	Source	Identifier
	04/24/1963	To	TLS	IAS	010347
	05/06/1963	To	ALS	IAS	010348
	05/30/1963	To	TLS	IAS	010349
	06/05/1963	From	AL, D	IAS	010350
	06/14/1963	To	ALS	IAS	010351
1	06/20/1963	From	TLR	IAS	010352
	07/02/1963	To	TLS	IAS	010354
	07/09/1963	To	ALS	IAS	010355
	07/17/1963	From	TL, D	IAS	010359[a]
	07/20/1963	To	ALS	IAS	010360
	07/24/1963	From	TLR	IAS	010361
	08/15/1963	To	TLS	IAS	010363
	08/24/1963	From	TLR	IAS	010364
	08/29/1963	To	TLS	IAS	010365
	09/04/1963	From	TLR	IAS	010368
	09/05/1963	To[b]	P	IAS	010368.5?
	09/05/1963	To	ALS	IAS	010369
	09/18/1963	From	TLR	IAS	010370
	09/25/1963	To	ALS	IAS	010371
	10/04/1963	From	TLR	IAS	010373
	10/??/1963?	To	ALS	IAS	010374
	10/14/1963	From	TLR	IAS	010375
	10/19/1963	From	ALS	IAS	010378
	10/25/1963	From	TLR	IAS	010377.6
	11/13/1963	To	TLS	IAS	010380.4

[a] Items 010358, 010358.5, 010358.6 and 010358.7 are all handwritten drafts of 010359.

[b] Transcript of a telephone message from Cohen taken by Caroline Underwood.

Item	Date	To/From Gödel	Form	Source	Identifier
	11/22/1963	From	TLR	IAS	010380.6[c]
	12/13/1963	From	TLR	IAS	010382
	01/02/1964	To	TLS	IAS	010387
2	01/22/1964	From	TLR	IAS	010390.5, 010390.6
	03/10/1964	From	TLR	IAS	010392
	11/20/1964?[d]	To	ALS	IAS	010392.5
3	01/11/1965	From	TLR	IAS	010394
	01/18/1965	To	ALS	IAS	010395
	07/07/1965	To	ALS	IAS	010402
	07/13/1965	From	TLR	IAS	010405
4	08/13/1965	From	TLR	IAS	010409
	04/01/1966	To	ALS	IAS	010417
	11/30/1966	To	TLS	IAS	010417.5
5	04/27/1967	From	TLR	IAS	010417.6
	05/21/1969	To	TLS	IAS	010430
	??/??/1969?	From	AL, D	IAS	010436
	12/09/1975	To[e]		IAS	010438.5
	??/??/????	From		IAS	010440

[c] Item 010380.5 is a handwritten draft.
[d] Dated 1965 by Cohen, internal evidence indicates the date should be 1964.
[e] Form letter to "Dear Colleague".

Gotthard Günther

Item	Date	To/From Gödel	Form	Source	Identifier
	08/02/1953	To	ALS	IAS	010742
	09/26/1953	To	TLS	IAS	010743
	12/11/1953	To	ALS	IAS	010753
1	04/29/1954	To	ALS	IAS	010754
2	05/15/1954	From	ALS	SBPK	Sheets 1–2
3	05/23/1954	To	ALS	IAS	010756
4	06/30/1954	From	ALS	SBPK	Sheets 5–7
5	10/02/1954	To	ALS	IAS	010757
6	06/19/1955	To	ALS	IAS	010758
7	08/10/1955	From	ALS	SBPK	Sheets 8–10
8	09/18/1955	To	ALS	IAS	010759
	12/20/1955	To	APS	IAS	010760
	06/22/1956	To	TLS	IAS	010762
	06/27/1956	To	APS	IAS	010763
9	09/20/1956	To	TLS	IAS	010764
	09/21/1956	To	TLS	IAS	010765
	12/08/1956	To	TLS	IAS	010769
	12/28/1956	To	ALS	IAS	010770
	01/10/1957	From	Lost[a]		
	01/15/1957	To	ALS	IAS	010772
	02/26/1957	To	ALS	IAS	010773
	03/??/1957	To	ALS	IAS	010774
	03/??/1957	To	AL[b]	IAS	010775
10	04/04/1957	From	ALS	SBPK	Sheets 11–12
11	04/07/1957	To	TLS	IAS	010776

[a] Attested by Günther's letter of 01/15/1957.
[b] Initialed "G."

Item	Date	To/From Gödel	Form	Source	Identifier
	05/02/1957	From	AL, D	IAS	010777
12	11/22/1957	To	TLS	IAS	010778
13	12/23/1957	From	ALS	SBPK	Sheet 13
	12?/??/1957?	To	AS^c	IAS	010779
	05/14/1958	To	TLS	IAS	010782
	10/22/1958	To	TLS	IAS	010783
	01/01/1959	To	TLS	IAS	010785
14	01/07/1959	From	TLS	SBPK; IAS	Sheet 14; 010786
	01/18/1959	To	TLS	IAS	010787
	01/15/1960	To	TLS	IAS	010789.5
	12/20/1960	To	AS^d	IAS	010790
	12/30/1960	To	TLS	IAS	010791
	07/27/1961	To	A^e	IAS	010792

[c] Undated greeting card.
[d] Greeting card.
[e] Initialed "G. G."

Textual notes

The individual copy-texts and the concomitant textual issues not addressed in editorial footnotes to the letters are discussed under the individual correspondents.

In these notes, the pairs of numbers on the left indicate page and line in this volume. (Line numbers do not count titles or running heads at the top of a page.) We follow our usual editorial apparatus as described in the Information for the reader.

Some of the letters originally in German appear to have been typed on an American typewriter, which explains the addition of the letter 'e' in place of the German umlauts. We have not attempted to note these.

Heinrich Behmann

Letter 1, paragraph 2, line 6, "zu erklären": These words were altered to "erklärt", but dotted lines under the strike-outs appear to indicate that Behmann wished to return to the original, as printed in the text given here.

As mentioned in footnote a to letter 1, page 2 of this letter contains a shorthand comment apparently by Gödel. It has been transcribed by Cheryl Dawson and Christian Thiel and reads as follows:

> Ferner ich möchte Sie bitten, die Fußnote 1 folgendermaßen zu ändern, denn ich glaube es ist ganz im Sinne ihres letzten Briefes, während der nähere Zeitpunkt und der Inhalt von Pr[ivat] gesprächen zwischen ~~mir~~ Ihn[en], Herrn D. und mir, den Leser kaum interessieren werden ~~und in diesem Fall sogar~~ und außerdem könnte dadurch beim Leser die irrige Meinung entstehen, als wenn ich meine ~~Ansicht seinem~~ Herrn Dub. gegenüber geäußerte Ansicht später geändert hätte, was nicht der Fall war. Die Erwähnung, daß Sie den Widerspruch Sept. 14 30 aufgeklärt haben, ist wohl deswegen ~~nicht~~ kaum nötig, weil Sie dasselbe schön früher Herrn B und Ack gegenüber getan haben müßen und ~~die~~ daß ~~Erwähnung~~ Ihre Lösung auch von mir als einwandfrei anerkannt sei, würde ich nicht ~~bedingungslos~~ ⟨ohne weiteres⟩ unterschreiben, weil ich eine einwandfreie Lösung ⟨des Antinomie Behmanns⟩ nur in der Angabe eines ⟨widerspruchsfreien⟩ Formalismus sehe, ~~von dem Ihren~~ [Here illegible text is crossed out and overwritten by the next word] ~~auch ihrer~~ ~~Berichtigung~~ k[ann?] den sie bisher nur andeutungsweise gegeben haben.

According to Christian Thiel, the typed page mentioned in letter 6 (see footnote a) and appearing at the end of the letter was originally clipped to the letter, but was apparently separated from the letter in the course of moving the Behmann *Nachlaß* to its present location. The present text was reconstructed from a half page, consisting of two parts pasted together, belonging to the item referred to in a memo attached to the pages as "Goedel an Dubislav (undatiert) 4 S. ms." and filed with the correspondence between Behmann and Dubislav. "The last of these four pages is just a further carbon copy of the final paragraph which in the original extends from page 2 bottom to page 3 top—which obviously was the reason for Goedel to cut the two parts of the passage in two separate pieces and paste them together for the version of April 22, corrected against that of March 18 by a handwritten part of a sentence. This reconstruction fits with the contents of the first three pages of the letter and of the 'correction' (which would coincide with page four [as designated above])." (Christian Thiel, communication to Charles Parsons, 1 April 1997.)

Paul Bernays

Many of Gödel's letters to Bernays bear annotations by Bernays reflecting Gödel's address changes during his years in Vienna. Those annotations have not been detailed here.

In several places, Bernays uses an apostrophe for the possessive as it would be used in English. We have retained those apostrophes without further comment.

Item		Note
108, 7	eine ~~Ding~~⟨Menge⟩	The "e" on the end of "eine" may have been added after "Ding" was altered to "Menge".
116, 8	$\{x \neq 0.$	This part of the formula was inserted by Gödel above the line.
144, 26	Es würde...	In a shorthand draft retained by Gödel, the remainder of the paragraph is replaced by the following:

Es scheint mir durchaus möglich, dass eine Erweiterung dieser Interpretation für transfinite Typen (eben bis zur ersten ϵ-Zahl) zu einem in einem gewissen Sinn konstruktiven Widerspruchsfreiheitsbeweis für Analysis führen könnte. Wie denken Sie darüber?

158, 18	Abstract	Bernays appears to be adopting the English word here. We have left his spelling intact.
284ff.	Korrekturfahnen	Inconsistent spelling has been retained.

Burton Dreben

Letter 2, p. 392, contains displayed equations that in the original copy-text had small circles either at the top or the bottom of some of the parentheses. When asked about the meaning of these, Dreben replied that they were not his and surmised that they were done by Gödel as a way of keeping track of the nesting. We have not attempted to reproduce those circles.

Gotthard Günther

In several places, Günther writes "aüsser..." instead of "äusser", as well as, for example, "vorlaüfig" instead of "vorläufig". In order to avoid cluttering up the text, we have corrected this without further notice.

Item	Note
490, 2–3 ⟨Russell's Mathematical Logic"	The use of a pointed bracket together with a closing quotation mark is Günther's.
494, 7 noch	The word is not written clearly but appears to be a "noch".
518, 3–4 Gegenständen	"Gegenstände" may be a possible reading of what Gödel writes, but "Gegenständen" is at least as likely and is grammatically correct.

References

Aanderaa, Stål
See Dreben, Burton, Peter Andrews and Stål Aanderaa.

Ackermann, Wilhelm
1924   Begründung des "tertium non datur" mittels der Hilbertschen Theorie der Widerspruchsfreiheit, *Mathematische Annalen 93*, 1–36.
1940   Zur Widerspruchsfreiheit der Zahlentheorie, *Mathematische Annalen 117*, 162–194.
1952   Widerspruchsfreier Aufbau einer typenfreien Logik. (Erweitertes System), *Mathematische Zeitschrift 55*, 364–384.
1953   Widerspruchsfreier Aufbau einer typenfreien Logik, II, *Mathematische Zeitschrift 57*, 155–166.
1954   *Solvable cases of the decision problem* (Amsterdam: North-Holland).
1956   Zur Axiomatik der Mengenlehre, *Mathematische Annalen 131*, 336–345.
See also Hilbert, David, and Wilhelm Ackermann.

Addison, John W., Jr.
1958   Separation principles in the hierarchies of classical and effective descriptive set theory, *Fundamenta mathematicae 46*, 123–135.
See also Henkin et alii.

Addison, John W., Leon Henkin and Alfred Tarski
1965   (eds.) *The theory of models. Proceedings of the 1963 International Symposium at Berkeley* (Amsterdam: North-Holland).

Albers, Donald J., Gerald L. Alexanderson and Constance Reid
1990   (eds.) *More mathematical people* (Boston, San Diego and New York: Harcourt Brace Jovanovich).

Alexanderson, Gerald L.
See Albers, Donald J., Gerald L. Alexanderson and Constance Reid.

Anderson, C. Anthony
1998   Alonzo Church's contributions to philosophy and intensional logic, *The bulletin of symbolic logic 4*, 129–171.

Andrei Sakharov Archives and Human Rights Center
2000   Alexander Esenin-Volpin, in *Faces of resistance in the USSR*
           (photo exhibit), http://www.brandeis.edu/departments/sakha-
           rov/Exhibit (accessed 3 March 2002).

Andréka, Hajnal, and István Németi
200?   Simple proof for decidability of the universal theory of cylin-
           dric set algebras of dimension 2, forthcoming; preprint, Re-
           search Group in Algebraic Logic, Institute of Mathematics,
           Hungarian Academy of Sciences; available online at
           ftp://math-inst.hu/pub/algebraic-logic/Contents.html
           (accessed 3 March 2002).

Andrews, Peter
See Dreben, Burton, Peter Andrews and Stål Aanderaa.

Asquith, Peter D., and Philip Kitcher
1985   (eds.) *PSA 1984: Proceedings of the 1984 biennial meeting of
           the Philosophy of Science Association*, vol. 2, (East Lansing,
           MI: Philosophy of Science Association).

Awodey, Steven, and André Carus
2001   Carnap, completeness, and categoricity: the *Gabelbarkeitssatz*
           of 1928, *Erkenntnis 54*, 145–172.

Ayer, Alfred Jules
1959   (ed.) *Logical positivism* (Glencoe, Ill.: Free Press).

Barbosa, Jorge Emmanuel Ferreira
1973   Sobre a consistência de *u*-sistematizações transfinitamente im-
           predicativas a consistência da matemática o metatheorema da
           escolha e a hipótese generalizada do continuum, *Boletim de
           análise e lógica matemática 5, 1*.

Barendregt, Hendrik P.
1997   The impact of the lambda calculus in logic and computer sci-
           ence, *The bulletin of symbolic logic 3*, 181–215.

Bar-Hillel, Yehoshua
1970   (ed.) *Mathematical logic and foundations of set theory* (Am-
           sterdam: North-Holland).

Bar-Hillel, Yehoshua, E. I. J. Poznanski, Michael O. Rabin and Abraham
Robinson
1961   (eds.) *Essays on the foundations of mathematics, dedicated
           to A. A. Fraenkel on his seventieth anniversary* (Jerusalem:
           Magnes Press; Amsterdam: North-Holland).

Barwise, Jon
1975   *Admissible sets and structures: an approach to definability theory* (Berlin, Heidelberg and New York: Springer).

Bauer-Mengelberg, Stefan
1965   Review of *Gödel 1962a*, *The journal of symbolic logic 30*, 359–
362.

Becker, Oskar
1930   Zur Logik der Modalitäten, *Jahrbuch für Philosophie und phä-
nomenologische Forschung 11*, 497–548.

Behmann, Heinrich
1922   Beiträge zur Algebra der Logik, insbesondere zum Entscheidungsproblem, *Mathematische Annalen 86*, 163–229.
1927   *Mathematik und Logik*, Mathematisch-Physikalische Bibliothek, vol. 71 (Leipzig and Berlin: Teubner).
1931   Zu den Widersprüchen der Logik und der Mengenlehre, *Jahresbericht der Deutschen Mathematiker-Vereinigung 40*, 37–48.
1931a   Zur Richtigstellung einer Kritik meiner Auflösung der logisch-
mengentheoretischen Widersprüche, *Erkenntnis 2*, 305–306.
1959   Der Prädikatenkalkül mit limitierten Variablen. Grundlegung
einer natürlichen exakten Logik, *The journal of symbolic logic
24*, 112–140.

Benacerraf, Paul, and Hilary Putnam
1964   (eds.) *Philosophy of mathematics: selected readings* (Engle-
wood Cliffs, NJ: Prentice-Hall; Oxford: Blackwell).
1983   Second edition of *Benacerraf and Putnam 1964* (Cambridge:
Cambridge University Press).

Bergamini, David, René Dubos, Henry Margenau and C. P. Snow
1963   (eds.) *Mathematics*, *Life* Science Library (New York: Time,
Inc.).

Bernays, Paul
1910   Das Moralprinzip bei Sidgwick und bei Kant, *Abhandlungen
der Fries'schen Schule, Neue Folge, 3. Band, 3. Heft*, 501–582
(also paginated by *Heft*, 1–82).
1913   Über den transzendentalen Idealismus, *Abhandlungen der
Fries'schen Schule, Neue Folge, 4. Band, 2. Heft*, 365–394
(also paginated by *Heft*, 1–30).
1922   Über Hilberts Gedanken zur Grundlegung der Arithmetik, *Jahresbericht der Deutschen Mathematiker-Vereinigung 31*, 10–19;
English translation by Paolo Mancosu in *Mancosu 1998*, 215–
222.

1926   Axiomatische Untersuchung des Aussagen-Kalküls der *Principia mathematica*, *Mathematische Zeitschrift 25*, 305–320.

1928   Über Nelsons Stellungnahme in der Philosophie der Mathematik, *Die Naturwissenschaften, 16. Jahrgang, Heft 9*, 142–145.

1930   Die Philosophie der Mathematik und die Hilbertsche Beweistheorie, *Blätter für deutsche Philosophie 4*, 326–367; reprinted in *Bernays 1976*, 17–61; English translation by Paolo Mancosu in *Mancosu 1998*, 234–265.

1933   Methoden des Nachweises von Widerspruchsfreiheit und ihre Grenzen, in *Saxer 1933*, 342–343.

1935   Sur le platonisme dans les mathématiques, *L'enseignement mathématique 34*, 52–69; English translation by Charles D. Parsons in *Benacerraf and Putnam 1964*, 274–286.

1937   A system of axiomatic set theory, Part I, *The journal of symbolic logic 2*, 65–77; reprinted in *Müller 1976*, 1–13.

1940   Review of *Lautman 1938*, *The journal of symbolic logic 5*, 20–22.

1940a   Review of *Gödel 1938*, *The journal of symbolic logic 5*, 117–118.

1941   A system of axiomatic set theory. Part II, *The journal of symbolic logic 6*, 1–17; reprinted in *Müller 1976*, 14–30.

1941a   Sur les questions méthodologiques actuelles de la théorie hilbertienne de la démonstration, in *Gonseth 1941*, 144–152.

1946   Review of *Gödel 1944*, *The journal of symbolic logic 11*, 75–79.

1953   Über die Fries'sche Annahme einer Wiederbeobachtung der unmittelbaren Erkenntnis, in *Specht and Eichler 1953*, 113–131.

1954a   Bemerkungen zu der Betrachtung von Alexander Wittenberg: Über adequäte Problemstellung in der Grundlagenforschung, *Dialectica 8*, 147–151.

1955   Zur Frage der Anknüpfung an die Kantische Erkenntnistheorie; eine kritische Erörterung, *Dialectica 9*, 23–65, 195–221.

1957   Von der Syntax der Sprache zur Philosophie der Wissenschaften, *Dialectica 11*, 233–246.

1959   Betrachtungen zu Ludwig Wittgensteins 'Bemerkungen über die Grundlagen der Mathematik', *Ratio 2*, 1–18, also in English as: Comments on Ludwig Wittgenstein's Remarks on the Foundations of Mathematics, in *Ratio 2*, English edition, 1–22; reprinted in *Bernays 1976*, 119–141.

1961   Zur Frage der Unendlichkeitsschemata in der axiomatischen Mengenlehre, in *Bar-Hillel et alii 1961*, 3–49; English translation by John L. Bell and M. Plänitz in *Müller 1976*, 121–172.

1961a   Zur Rolle der Sprache in erkenntnistheoretischer Hinsicht, *Synthese 18*, 185–200; reprinted in *Bernays 1976*, 155–169.

1961b   Die hohen Unendlichkeiten und die Axiomatik der Mengen-
         lehre, *Polish Academy of Sciences 1961*, 11–20.
1962   Review of *Leblanc 1962*, *The journal of symbolic logic 27*, 248–
         249.
1962a   Remarks about formalization and models, in *Nagel, Suppes
         and Tarski 1962*, 176–180.
1964   Reflections on Karl Popper's epistemology, in *Bunge 1964*, 32–
         44.
1966   Gedanken zu dem Buch "Bildung und Mathematik (Mathema-
         tik als exemplarisches Gymnasialfach)" von Alexander Witten-
         berg, *Dialectica 20*, 27–42.
1969   Bemerkungen zur Philosophie der Mathematik, in *Akten des
         XIV. Internationalen Kongresses für Philosophie, Wien, Sept.
         1968*, vol. III, 192–198; reprinted in *Bernays 1976*, 170–175.
1970   On the original Gentzen consistency proof for number theory,
         in *Myhill, Kino and Vesley 1970*, 409–417.
1971   Zum Symposium über die Grundlagen der Mathematik,
         Dialectica 25, 171–195; reprinted in *Bernays 1976*, 189–213.
1974   Concerning rationality, in *Schilpp 1974*, 597–604.
1976   *Abhandlungen zur Philosophie der Mathematik* (Darmstadt:
         Wissenschaftliche Buchgesellschaft).
1976a   See *Müller 1976*.
1976b   Kurze Biographie, in *Müller 1976*, xiv–xvi; English translation
         in *Müller 1976*, xi–xiii.
See also Hilbert, David, and Paul Bernays.

Bernays, Paul, and Abraham Adolf Fraenkel
1958   *Axiomatic set theory* (Amsterdam: North-Holland).

Bernays, Paul, and Moses Schönfinkel
1928   Zum Entscheidungsproblem der mathematischen Logik, *Ma-
         thematische Annalen 99*, 343–372.

Betsch, Christian
1926   *Fiktionen in der Mathematik* (Stuttgart: Fr. Frommanns Ver-
         lag).

Biggers, Earl Derr
1925   *The house without a key* (New York: Avenel Books).

Boolos, George
1976   On deciding the truth of certain statements involving the no-
         tion of consistency, *The journal of symbolic logic 41*, 778–781.

Boone, William Werner

1952   *Several simple unsolvable problems of group theory related to*
        the word problem (doctoral dissertation, Princeton University).

1954   Certain simple, unsolvable problems of group theory I, *Konink-*
        lijke Nederlandse Akademie van Wetenschappen, Proceedings,
        Series A: Mathematical Sciences 57, 231–237; also *Indaga-*
        tiones mathematicae 16, 231–237.

1954a   Certain simple, unsolvable problems of group theory II, *Ko-*
        ninklijke Nederlandse Akademie van Wetenschappen, Proceed-
        ings, Series A: Mathematical sciences 57, 492–497; also *Inda-*
        gationes mathematicae 16, 492–497.

1955   Certain simple unsolvable problems of group theory III, *Ko-*
        ninklijke Nederlandse Akademie van Wetenschappen, Proceed-
        ings, Series A: Mathematical sciences 58, 252–256; also *Inda-*
        gationes mathematicae 17, 252–256.

1955a   Certain simple unsolvable problems of group theory IV, *Ko-*
        ninklijke Nederlandse Akademie van Wetenschappen, Proceed-
        ings, Series A: Mathematical sciences 58, 571–577; also *Inda-*
        gationes mathematicae 17, 571–577.

1957   Certain simple unsolvable problems of group theory V, *Ko-*
        ninklijke Nederlandse Akademie van Wetenschappen, Proceed-
        ings Series A: Mathematical sciences 60, 22–27; also *Indaga-*
        tiones mathematicae 19, 22–27.

1957a   Certain simple unsolvable problems of group theory VI, *Ko-*
        ninklijke Nederlandse Akademie van Wetenschappen, Proceed-
        ings, Series A: Mathematical sciences 60, 227–232; also *Inda-*
        gationes mathematicae 19, 227–232.

1959   The word problem, *Annals of mathematics (2) 70*, 207–265.

1966   Word problems and recursively enumerable degrees of unsolv-
        ability. A first paper on Thue systems, *Annals of mathematics*
        (2) 83, 520–571.

1966a   Word problems and recursively enumerable degrees of unsolv-
        ability. A sequel on finitely presented groups, *Annals of math-*
        ematics 84, 49–84.

Booth, David, and Renatus Ziegler

1996   *Finsler set theory: platonism and circularity* (Basel, Boston
        and Berlin: Birkhäuser).

Braithwaite, Richard Bevan

1962   Introduction, in *Gödel 1962a*, 1–32.

Breger, Herbert
1992   A restoration that failed: Paul Finsler's theory of sets, in *Gillies 1992*, 249–264.

Brouwer, Luitzen Egbertus Jan
1925   Über die Bedeutung des Satzes vom ausgeschlossenen Dritten in der Mathematik, insbesondere in der Funktionentheorie, *Journal für die reine und angewandte Mathematik 154*, 1–7.

Brutian, Georg
1968   On the conception of polylogic, *Mind* (n.s.) *77*, 351–359.

Büchi, Julius Richard
1989   *Finite automata, their algebras and grammars; towards a theory of formal expressions*, edited by Dirk Siefkes (New York, Berlin and Heidelberg: Springer).
1990   *The collected works of J. Richard Büchi*, edited by Saunders Mac Lane and Dirk Siefkes (Berlin: Springer).

Büchi, Sylvia
1990   The life of J. Richard Büchi, in *Büchi 1990*, 4–6.

Buldt, Bernd
See Köhler et alii.

Bulloff, Jack J., Thomas C. Holyoke and S. W. Hahn
1969   (eds.) *Foundations of mathematics. Symposium papers commemorating the sixtieth birthday of Kurt Gödel* (New York: Springer).

Bunge, Mario
1964   *The critical approach to science and philosophy, in honor of Karl R. Popper* (Glencoe, IL: The Free Press; London: Collier-Macmillan).

Burckhardt, J. J.
1980   *Die Mathematik an der Universität Zürich 1916–1950, unter den Professoren R. Fueter, A. Speiser, P. Finsler* (Basel, Boston and Stuttgart: Birkhäuser).

Buss, Samuel R.
1986   *Bounded arithmetic*, Studies in proof theory 3 (Naples: Bibliopolis).
1995   On Gödel's theorems on lengths of proofs II: lower bounds for recognizing k symbol provability, in *Clote and Remmel 1995*, 57–90.

Butts, Robert E., and Jaakko Hintikka
  1977   (eds.) *Logic, foundations of mathematics and computability theory* (Dordrecht and Boston: Reidel).

Cantor, Georg
  1895   Beiträge zur Begründung der transfiniten Mengenlehre. I, *Mathematische Annalen 46*, 481–512.
  1932   *Gesammelte Abhandlungen mathematischen und philosophischen Inhalts. Mit erläuternden Anmerkungen sowie mit Ergänzungen aus dem Briefwechsel Cantor–Dedekind*, edited by Ernst Zermelo (Berlin: Springer); reprinted in 1962 (Hildesheim: Olms).

Carnap, Rudolf
  1930a   Bericht über Untersuchungen zur allgemeinen Axiomatik, *Erkenntnis 1*, 303–307.
  1931   Die logizistische Grundlegung der Mathematik, *Erkenntnis 2*, 91–105; English translation by Erna Putnam and Gerald J. Massey in *Benacerraf and Putnam 1964*, 31–41.
  1932   Überwindung der Metaphysik durch logische Analyse der Sprache, *Erkenntnis 2*, 219–241; English translation by Arthur Pap in *Ayer 1959*, 60–81.
  1934   Die Antinomien und die Unvollständigkeit der Mathematik, *Monatshefte für Mathematik und Physik 41*, 263–284.
  1934a   *Logische Syntax der Sprache* (Vienna: Springer); translated into English by Amethe Smeaton as *Carnap 1937*.
  1935   Ein Gültigkeitskriterium für die Sätze der klassischen Mathematik, *Monatshefte für Mathematik und Physik 42*, 163–190.
  1935c   *Philosophy and logical syntax* (London: Kegan Paul, Trench, Trubner).
  1937   *The logical syntax of language* (London: Kegan Paul, Trench, Trubner; New York: Harcourt, Brace); English translation of *Carnap 1934a*, with revisions.
  1963   Intellectual autobiography, in *Schilpp 1963*, 3–84.
  2000   *Untersuchungen zur allgemeinen Axiomatik*, edited by Thomas Bonk and Jesús Mosterín (Darmstadt: Wissenschaftliche Buchgesellschaft).
  See also Hahn et alii.

Carnes, Mark C.
  See Garraty, John A., and Mark C. Carnes.

Chang, Chen Chung
  See Henkin et alii.

Chen, Kien-Kwong
1933   Axioms for real numbers, *Tôhoku mathematical journal 37*, 94–99.

Chevalley, Claude
1934   Sur la pensée de J. Herbrand, *L'Enseignement mathématique 34*, 97–102; reprinted in *Herbrand 1968*, 17–20; translated into English by Warren Goldfarb in *Herbrand 1971*, 25–28.

Chevalley, Claude, and Albert Lautman
1931   Notice biographique sur Jacques Herbrand, *Annuaire de l'Association amicale de secours des anciens élèves de l'École normale supérieure*, 66–68; reprinted in *Herbrand 1968*, 13–15; translated into English by Warren Goldfarb in *Herbrand 1971*, 21–23.

Church, Alonzo
1932   A set of postulates for the foundation of logic, *Annals of mathematics (2) 33*, 346–366.
1933   A set of postulates for the foundation of logic (second paper), *Annals of mathematics (2) 34*, 839–864.
1934   The Richard paradox, *American mathematical monthly 41*, 356–361.
1935   A proof of freedom from contradiction, *Proceedings of the National Academy of Sciences, U.S.A. 21*, 275–281.
1965   Review of *Braithwaite 1962*, *The journal of symbolic logic 30*, 357–359.
1968   Paul J. Cohen and the continuum problem, *Proceedings of the International Congress of Mathematicians (Moscow-1966)*, 15–20.

Chwistek, Leon
1929   Neue Grundlagen der Logik und Mathematik, *Mathematische Zeitschrift 30*, 704–724.
1930   Une méthode métamathématique d'analyse, in *Sprawozdanie z 1. Kongresu Matematyków Krajów Słowiańskich (Comptes-rendus du 1. Congrès des Mathématiciens des Pays Slaves)*, *Warszawa, 1929* (Warsaw), 254–263.
1932   Neue Grundlagen der Logik und Mathematik. Zweite Mitteilung, *Mathematische Zeitschrift 34*, 527–534.
1933   Die nominalistische Grundlegung der Mathematik, *Erkenntnis 3*, 367–388.

Clote, Peter
1999   Computation models and function algebras, in *Griffor 1999*, 589–681.

Clote, Peter, and Jan Krajíček
1993   (eds.) *Arithmetic, proof theory, and computational complexity*, Oxford logic guides 23 (Oxford: Clarendon Press).

Clote, Peter, and Jeffrey B. Remmel
1995   (eds.) *Feasible mathematics II*, Progress in computer science and applied logic, vol. 13 (Boston: Birkhäuser).

Cohen, Morris Raphael, and Ernest Nagel
1934   *An introduction to logic and scientific method* (New York: Harcourt, Brace and Company).

Cohen, Paul J.
1960   On a conjecture of Littlewood and idempotent measures, *American journal of mathematics 82*, 191–212.
1963   The independence of the continuum hypothesis, *Proceedings of the National Academy of Sciences, U.S.A. 50*, 1143–1148.
1963a   A minimal model for set theory, *Bulletin of the American Mathematical Society 69*, 537–540.
1964   The independence of the continuum hypothesis, II, *Proceedings of the National Academy of Sciences, U.S.A. 51*, 105–110.
1965   Independence results in set theory, in *Addison, Henkin and Tarski 1965*, 39–54.
1966   *Set theory and the continuum hypothesis* (New York: Benjamin).

Cooper, Necia Grant
1989   (ed.) *From cardinals to chaos: reflections on the life and legacy of Stanislaw Ulam* (Cambridge: Cambridge University Press).

Craig, William
See Henkin et alii.

Dauben, Joseph Warren
1995   *Abraham Robinson: the creation of nonstandard analysis; a personal and mathematical odyssey* (Princeton: Princeton University Press).

Davis, Martin
1965   (ed.) *The undecidable: basic papers on undecidable propositions, unsolvable problems and computable functions* (Hewlett, NY: Raven Press).
1982   Why Gödel didn't have Church's thesis, *Information and control 54*, 3–24.
1998   Review of *Dawson 1997*, *Philosophia mathematica (3) 6*, 116–128.

Davis, Martin, Hilary Putnam and Julia Robinson
1961   The decision problem for exponential diophantine equations, *Annals of mathematics (2) 74*, 425–436.

Dawson, John W., Jr.
1983   The published work of Kurt Gödel: an annotated bibliography, *Notre Dame journal of formal logic 24*, 255–284; addenda and corrigenda, *ibid. 25*, 283–287.
1984   Discussion on the foundation of mathematics, *History and philosophy of logic 5*, 111–129.
1984a   Kurt Gödel in sharper focus, *The mathematical intelligencer 6 (4)*, 9–17.
1985   The reception of Gödel's incompleteness theorems, in *Asquith and Kitcher 1985*, 253–271; reprinted in *Shanker 1988a*, 74–95, and in *Drucker 1985*, 84–100.
1985a   Completing the Gödel–Zermelo correspondence, *Historia mathematica 12*, 66–70.
1993   Prelude to recursion theory: the Gödel–Herbrand correspondence, in *Wolkowski 1993*, 1–13.
1997   *Logical dilemmas: the life and work of Kurt Gödel* (Wellesley, MA: A K Peters); translated into German by Jakob Kellner as *Dawson 1999*.
1999   *Kurt Gödel: Leben und Werk*, Computerkultur XI (Vienna and New York: Springer).

Dekker, Jacob C. E.
1962   (ed.) *Recursive function theory*, Proceedings of symposia in pure mathematics, vol. 5 (Providence, RI: American Mathematical Society).

Denton, John
See Dreben, Burton, and John Denton.

DePauli-Schimanovich, Werner
See Köhler et alii.

Dilworth, Robert Palmer
1961   (ed.) *Lattice theory*, Proceedings of the Second Symposium in Pure Mathematics (Providence, RI: American Mathematical Society).

Dingler, Hugo
1931   *Philosophie der Logik und Arithmetik* (Munich: Reinhardt).

Drake, Frank R., and John K. Truss
1988   (eds.) *Logic colloquium '86* (Amsterdam: North-Holland).

Dreben, Burton
1952   On the completeness of quantification theory, *Proceedings of the National Academy of Sciences, U.S.A. 38*, 1047–1052.
1963   Corrections to Herbrand, *Notices of the American Mathematical Society 10*, 285.

Dreben, Burton, Peter Andrews and Stål Aanderaa
1963   False lemmas in Herbrand, *Bulletin of the American Mathematical Society 69*, 699–706.
1963a   Errors in Herbrand, *Notices of the American Mathematical Society 10*, 285.

Dreben, Burton, and John Denton
1966   A supplement to Herbrand, *The journal of symbolic logic 31*, 393–398.

Dreben, Burton, and Warren Goldfarb
1979   *The decision problem: solvable classes of quantificational formulas* (Reading: Addison-Wesley).

Dreben, Burton, and Jean van Heijenoort
1967   Introductory note to *Skolem 1928*, in *van Heijenoort 1967*, 508–512.

Drucker, Thomas L.
1985   *Perspectives on the history of mathematical logic* (Boston: Birkhäuser).

Dubislav, Walter
1926   *Über die Definition* (Berlin: H. Weiss).
1931   *Die Definition* (third, enlarged edition of *Dubislav 1926*), Beihefte der *Erkenntnis*, vol. I (Leipzig: Meiner).
1981   Fourth (unrevised) edition of *Dubislav 1931* with an introduction by Wilhelm K. Essler (Hamburg: Meiner).

Du Bois Reymond, Paul
1880   Der Beweis des Fundamentalsatzes der Integralrechnung: $\int_a^b F'(x)\,dx = F(b) - F(a)$, *Mathematische Annalen 16*, 115–128.

Dubos, René
See Bergamini et alii.

Dummett, Michael A. E.
1993   *Origins of analytical philosophy* (London: Duckworth; Cambridge: Harvard University Press, 1994).

Dunham, Bradford
See Wang, Hao and Bradford Dunham.

Easton, William B.
1970   Powers of regular cardinals, *Annals of mathematical logic 1*,
        139–178.

Edwards, Paul
1967   (ed.) *The encyclopedia of philosophy* (New York: Macmillan
        and The Free Press; London: Collier-Macmillan).

Ehrenfeucht, Andrzej, and Georg Kreisel
1967   Review of *Esenin-Vol'pin 1961*, *The journal of symbolic logic
        32*, 517.

Eichler, Willi
See Specht, Minna, and Willi Eichler.

Enderton, Herbert B.
1995   In memoriam, Alonzo Church, 1903–1995, *The bulletin of sym-
        bolic logic 1*, 486–488.
1998   Alonzo Church and the reviews, *The bulletin of symbolic logic
        4*, 172–180.

Erdős, Paul, and Alfred Tarski
1961   On some problems involving inaccessible cardinals, in *Bar-
        Hillel et alii 1961*, 50–82.

Esenin-Vol'pin, Alexander Sergeievich (Essenin-Volpin, Yessenin-Volpin,
Ésénine-Volpine; Есенин-Вольпин, Александр Сергеевич)
1961   Le programme ultra-intuitionniste des fondements des mathé-
        matiques, in *Polish Academy of Sciences 1961*, 201–223.
1970   The ultra-intuitionistic criticism and the antitraditional pro-
        gram for foundations of mathematics, in *Myhill, Kino and Ves-
        ley 1970*, 3–45.

Everett, C. J., and Stanisław M. Ulam
1945   Projective algebra I, *American journal of mathematics 68*, 77–
        88; reprinted in *Ulam 1974*, 231–242.

Ewald, William B.
1996   *From Kant to Hilbert: a source book in the foundations of
        mathematics*, 2 vols. (Oxford: Clarendon Press).

Feferman, Anita Burdman
1993   *Politics, logic and love: the life of Jean van Heijenoort* (Boston
        and London: Jones and Bartlett; Wellesley: AK Peters); re-
        printed as *A. B. Feferman 2001*.

1999   How the unity of science saved Alfred Tarski, in *Woleński and Köhler 1999*, 43–52.

2001   *From Trotsky to Gödel: the life of Jean van Heijenoort*, reprint of *A. B. Feferman 1993* (Natick: AK Peters).

Feferman, Solomon

1962   Transfinite recursive progressions of axiomatic theories, *The journal of symbolic logic 27*, 259–316.

1965   Some applications of the notions of forcing and generic sets, *Fundamenta mathematicae 56*, 325–345.

1969   Set-theoretical foundations of category theory (with an appendix by Georg Kreisel), in *Mac Lane 1969*, 201–247.

1982   Inductively presented systems and the formalization of metamathematics, in *van Dalen, Lascar and Smiley 1982*, 95–128.

1984   Toward useful type-free theories. I., *The journal of symbolic logic 49*, 75–111.

1984a   Kurt Gödel: conviction and caution, *Philosophia naturalis 21*, 546–562; reprinted in *Shanker 1988a*, 96–114, and in *Feferman 1998*, 150–164.

1998   *In the light of logic* (New York and Oxford: Oxford University Press).

1999   Tarski and Gödel between the lines, in *Woleński and Köhler 1999*, 53–63.

Feferman, Solomon, and Azriel Lévy

1963   Independence results in set theory by Cohen's method. II (abstract), *Notices of the American Mathematical Society 10*, 593.

Feferman, Solomon, and Clifford Spector

1962   Incompleteness along paths in progressions of theories, *The journal of symbolic logic 27*, 383–390.

Feigl, Herbert

1929   *Theorie und Erfahrung in der Physik* (Karlsruhe: G. Braun); English translation of chapter III by Gisela Lincoln (revised and edited by Robert S. Cohen) in *Feigl 1981*, 116–144.

1981   *Inquiries and provocations: selected writings 1929–1974*, edited by Robert S. Cohen, Vienna Circle collection, vol. 14 (Dordrecht: Reidel).

Fine, Arthur, and Jarrett Leplin

1988   (eds.) *PSA 1988: Proceedings of the 1988 biennial meeting of the Philosophy of Science Association* (East Lansing, MI: Philosophy of Science Association).

Finsler, Paul
  1926   Formale Beweise und die Entscheidbarkeit, *Mathematische Zeitschrift 25*, 676–682; reprinted in *Unger 1975*, 11–17; English translation by Stefan Bauer-Mengelberg in *van Heijenoort 1967*, 440–445; another English translation, by David Booth, Renatus Ziegler and David Renshaw, in *Booth and Ziegler 1996*, 50–55.
  1944   Gibt es unentscheidbare Sätze?, *Commentarii mathematici helvetici 16*, 310–320; reprinted in *Unger 1975*, 97–107; English translation by David Booth, Renatus Ziegler and David Renshaw in *Booth and Ziegler 1996*, 63–72.

Floyd, Juliet
  1995   On saying what you really want to say: Wittgenstein, Gödel, and the trisection of the angle, in *Hintikka 1995*, 373–425.

Flügge, Siegfried
  1939   Kann der Energieinhalt technisch nutzbar gemacht werden?, *Die Naturwissenschaften, 27. Jahrgang 23/24*, 402–410.

Follett, Wilson
  1966   *Modern American usage: a guide*, edited and completed by Jacques Barzun, in collaboration with Carlos Baker, Frederick W. Dupee, Dudley Fitts, James D. Hart, Phillis McGinley and Lionel Trilling (New York: Hill and Wang).
  1998   First revised edition of *Follett 1966*, revised by Erik Wensberg (New York: Hill and Wang).

Fraenkel, Abraham Adolf
  1922   Der Begriff 'definit' und die Unabhängigkeit des Auswahlaxioms, *Sitzungsberichte der Preussischen Akademie der Wissenschaften, Physikalisch-mathematische Klasse*, 253–257; English translation by Beverly Woodward in *van Heijenoort 1967*, 284–289.
See also Bernays, Paul, and Abraham Adolf Fraenkel.

Frank, Philipp
  1947   *Einstein, his life and times*, English translation of *Frank 1949* by George Rosen, edited and revised by Shuichi Kusaka (New York: Alfred A. Knopf).
  1949   *Einstein, sein Leben und seine Zeit* (Munich, Leipzig and Freiburg: P. List); reprinted as *Frank 1979*; English translation as *Frank 1947*.
  1979   Reprint of *Frank 1949*, with foreword by Albert Einstein (Braunschweig, Wiesbaden: Vieweg).

Frege, Gottlob

1879   *Begriffsschrift, eine der arithmetischen nachgebildete Formel-sprache des reinen Denkens* (Halle: Nebert); English translation by Stefan Bauer-Mengelberg in *van Heijenoort 1967*, 1–82.

1893   *Grundgesetze der Arithmetik, begriffsschriftlich abgeleitet*, vol.1 (Jena: H. Pohle); reprinted in 1962 (Hildesheim: Olms); partial English translation by Philip E. B. Jourdain and Johann Stachelroth in *Frege 1915*, *Frege 1916* and *Frege 1917*, with an introduction by Jourdain, reprinted with alterations in *Frege 1952*, 117–138; translation of excerpts also by Michael Beaney in *Frege 1997*, 194–223.

1903   *Grundgesetze der Arithmetik, begriffsschriftlich abgeleitet*, vol. 2 (Jena: Pohle); reprinted in 1962 (Hildesheim: Olms); partial English translation by Peter T. Geach and Max Black in *Frege 1952*, 159–244, partly reprinted, with additional selections translated by Michael Beaney, in *Frege 1997*, 258–289.

1915   The fundamental laws of arithmetic, *The monist 25*, 481–494; English translation of part of *Frege 1893*.

1916   The fundamental laws of arithmetic: psychological logic, *The monist 26*, 181–199; English translation of part of *Frege 1893*.

1917   Class, function, concept, relation, *The monist 27*, 114–127; English translation of part of *Frege 1893*.

1918   Der Gedanke; eine logische Untersuchung, *Beiträge zur Philosophie des deutschen Idealismus I*, 58–77; reprinted in *Frege 1967*, 342–362; English translation by Peter T. Geach and Robert H. Stoothoff in *Frege 1984*, 351–372, reprinted in *Frege 1997*, 325–345.

1918a   Die Verneinung: eine logische Untersuchung, *Beiträge zur Philosophie der deutschen Idealismus I*, 143–157; reprinted in *Frege 1967*, 362–378; English translation by Peter T. Geach in *Frege 1984*, 373–389, reprinted in *Frege 1997*, 346–361.

1952   *Translations from the philosophical writings of Gottlob Frege*, edited by Peter T. Geach and Max Black (Oxford: Basil Blackwell; New York: Philosophical Library).

1967   *Kleine Schriften*, edited by Ignacio Angelelli (Hildesheim: Georg Olms).

1976   *Wissenschaftlicher Briefwechsel*, edited by Gottfried Gabriel, Hans Hermes, Friedrich Kambartel, Christian Thiel and Albert Veraart, vol. 2 of *Nachgelassene Schriften und wissenschaftlicher Briefwechsel*, edited by Hans Hermes, Friedrich Kambartel and Friedrich Kaulbach (Hamburg: Felix Meiner, 1969–1976).

1980   *Philosophical and mathematical correspondence*, edited by
        Gottfried Gabriel, Hans Hermes, Friedrich Kambartel, Chris-
        tian Thiel and Albert Veraart, abridged from *Frege 1976* by
        Brian McGuinness, and translated by Hans Kaal (Chicago:
        University of Chicago Press; Oxford: Basil Blackwell).
1984   *Collected papers on mathematics, logic and philosophy*, edited
        by Brian McGuinness, translated by Max Black, V. H. Dud-
        man, Peter T. Geach, Hans Kaal, E.-H. W. Kluge, Brian
        McGuinness and Robert H. Stoothoff (Oxford and New York:
        Basil Blackwell).
1997   *The Frege reader*, edited by Michael Beaney (Oxford, Malden:
        Blackwell).

Friedberg, Richard
1957   Two recursively enumerable sets of incomparable degrees of
        unsolvability (solution of Post's problem, 1944), *Proceedings
        of the National Academy of Sciences, U.S.A. 43*, 236–238.

Friedman, Harvey
1975   One hundred two problems in mathematical logic, *The journal
        of symbolic logic 40*, 113–129.

Friedman, Michael
2000   *A parting of the ways. Carnap, Cassirer, and Heidegger* (Chi-
        cago and La Salle: Open Court).

Fries, Jakob Friedrich
1803   *Philosophische Rechtslehre und Kritik aller positiven Gesetzge-
        bung, mit Beleuchtung der gewöhnlichen Fehler in der Bear-
        beitung des Naturrechts* (Jena: Mauke); reprinted in *Fries
        1968-*, vol. 9 (Abteilung 2, Bd. 1, 1–206).
1924   *System der Metaphysik. Ein Handbuch für Lehrer und zum
        Selbstgebrauch.* (Heidelberg: Christian Friedrich Winter).
1968-   *Sämtliche Schriften*, edited by Gert König and Lutz Geldsetzer
        (Aalen: Scientia Verlag).

Gabelsberger, Franz Xaver
1834   *Anleitung zur deutschen Redezeichenkunst oder Stenographie*
        (Munich: Georg Franz); republished in 1908 (Wölfenbüttel:
        Heckner).

Gabriel, Pierre
1962   Des catégories abéliennes, *Bulletin de la Société Mathématique
        de France 90*, 323–448.

Garey, Michael R., and David S. Johnson
  1979   *Computers and intractability—a guide to the theory of NP-completeness* (San Francisco: W. H. Freeman).

Garraty, John A., and Mark C. Carnes
  1999   (eds.) *American national biography* (New York and Oxford: Oxford University Press).

Geiser, James
  1974   A formalization of Essenin-Volpin's proof theoretical studies by means of nonstandard analysis, *The journal of symbolic logic 39*, 81–87.

Gentzen, Gerhard
  1936   Die Widerspruchsfreiheit der reinen Zahlentheorie, *Mathematische Annalen 112*, 493–565; English translation by M. E. Szabo in *Gentzen 1969*, 132–213.
  1969   *The collected papers of Gerhard Gentzen*, edited and translated into English by M. E. Szabo (Amsterdam: North-Holland).

George, Alexander
  1994   (ed.) *Mathematics and mind* (New York and Oxford: Oxford University Press).

Geroch, Robert
  1973   Energy extraction, *Sixth Texas Symposium on Relativistic Astrophysics, Annals of the New York Academy of Sciences 224*, edited by Dennis J. Hegyi, 108–117.

Gillies, Donald
  1992   (ed.) *Revolutions in mathematics* (New York: Oxford University Press).

Girard, Jean-Yves
  See Nagel, Ernest, James R. Newman, Kurt Gödel and Jean-Yves Girard.

Gödel, Kurt
  1929   *Über die Vollständigkeit des Logikkalküls* (doctoral dissertation).
  1930   Die Vollständigkeit der Axiome des logischen Funktionenkalküls, *Monatshefte für Mathematik und Physik 37*, 349–360.
  1930a   Über die Vollständigkeit des Logikkalküls, *Die Naturwissenschaften 18*, 1068.
  1930b   Einige metamathematische Resultate über Entscheidungsdefinitheit und Widerspruchsfreiheit, *Anzeiger der Akademie der Wissenschaften in Wien 67*, 214–215.

*1930c   Vortrag über Vollständigkeit des Funktionenkalküls, in *Gödel 1995*, 16–29, with English translation by Jean van Heijenoort, John W. Dawson, Jr., William Craig and Warren Goldfarb.

1931   Über formal unentscheidbare Sätze der *Principia mathematica* und verwandter Systeme I, *Monatshefte für Mathematik und Physik 38*, 173–198; translated into English by Jean van Heijenoort as *Gödel 1967*.

1931a   Diskussion zur Grundlegung der Mathematik (Gödel's remarks in *Hahn et alii 1931*), *Erkenntnis 2*, 147–151; English translation by John W. Dawson, Jr. in *Dawson 1984*, 125–128.

1931b   Review of *Neder 1931*, *Zentralblatt für Mathematik und ihre Grenzgebiete 1*, 5–6.

1931c   Review of *Hilbert 1931*, *ibid. 1*, 260.

1931d   Review of *Betsch 1926*, *Monatshefte für Mathematik und Physik (Literaturberichte) 38*, 5.

1931e   Review of *Becker 1930*, *ibid. 38*, 5–6.

1931f   Review of *Hasse and Scholz 1928*, *ibid. 38*, 37.

1931g   Review of *von Juhos 1930*, *ibid. 38*, 39.

*1931?   [Über unentscheidbare Sätze], in *Gödel 1995*, 30–35, with English translation by Stephen C. Kleene, John W. Dawson, Jr. and William Craig.

1932   Zum intuitionistischen Aussagenkalkül, *Anzeiger der Akademie der Wissenschaften in Wien 69*, 65–66; reprinted, with additional comment, as *1933n*.

1932a   Ein Spezialfall des Entscheidungsproblems der theoretischen Logik, *Ergebnisse eines mathematischen Kolloquiums 2*, 27–28; reprinted in *Menger 1998*, 145–146.

1932b   Über Vollständigkeit und Widerspruchsfreiheit, *ibid. 3*, 12–13; reprinted in *Menger 1998*, 168–169.

1932c   Eine Eigenschaft der Realisierung des Aussagenkalküls, *ibid. 3*, 20–21; reprinted in *Menger 1998*, 176–177.

1932d   Review of *Skolem 1931*, *Zentralblatt für Mathematik und ihre Grenzgebiete 2*, 3.

1932e   Review of *Carnap 1931*, *ibid. 2*, 321.

1932f   Review of *Heyting 1931*, *ibid. 2*, 321–322.

1932g   Review of *von Neumann 1931*, *ibid. 2*, 322.

1932h   Review of *Klein 1931*, *ibid. 2*, 323.

1932i   Review of *Hoensbroech 1931*, *ibid. 3*, 289.

1932j   Review of *Klein 1932*, *ibid. 3*, 291.

1932k   Review of *Church 1932*, *ibid. 4*, 145–146.

1932l   Review of *Kalmár 1932*, *ibid. 4*, 146.

1932m   Review of *Huntington 1932*, *ibid. 4*, 146.

1932n   Review of *Skolem 1932*, *ibid. 4*, 385.

1932o   Review of *Dingler 1931*, *Monatshefte für Mathematik und Physik* (*Literaturberichte*) *39*, 3.

1933   Untitled remark following *Parry 1933*, *Ergebnisse eines mathematischen Kolloquiums 4*, 6; reprinted in *Menger 1998*, 188.

1933a   Über Unabhängigkeitsbeweise im Aussagenkalkül, *ibid. 4*, 9–10; reprinted in *Menger 1998*, 191–192.

1933b   Über die metrische Einbettbarkeit der Quadrupel des R_3 in Kugelflächen, *ibid. 4*, 16–17; reprinted in *Menger 1998*, 198–199.

1933c   Über die Waldsche Axiomatik des Zwischenbegriffes, *ibid. 4*, 17–18; reprinted in *Menger 1998*, 199–200.

1933d   Zur Axiomatik der elementargeometrischen Verknüpfungsrelationen, *ibid. 4*, 34; reprinted in *Menger 1998*, 216.

1933e   Zur intuitionistischen Arithmetik und Zahlentheorie, *ibid. 4*, 34–38; reprinted in *Menger 1998*, 216–220.

1933f   Eine Interpretation des intuitionistischen Aussagenkalküls, *ibid. 4*, 39–40; reprinted in *Menger 1998*, 221–222.

1933g   Bemerkung über projektive Abbildungen, *ibid. 5*, 1; reprinted in *Menger 1998*, 229.

1933h   (with K. Menger and A. Wald) Diskussion über koordinatenlose Differentialgeometrie, *ibid. 5*, 25–26; reprinted in *Menger 1998*, 253–254.

1933i   Zum Entscheidungsproblem des logischen Funktionenkalküls, *Monatshefte für Mathematik und Physik 40*, 433–443.

1933j   Review of *Kaczmarz 1932*, *Zentralblatt für Mathematik und ihre Grenzgebiete 5*, 146.

1933k   Review of *Lewis 1932*, *ibid. 5*, 337–338.

1933l   Review of *Kalmár 1933*, *ibid. 6*, 385–386.

1933m   Review of *Hahn 1932*, *Monatshefte für Mathematik und Physik* (*Literaturberichte*) *40*, 20–22.

1933n   Reprint of *Gödel 1932*, with additional comment, *Ergebnisse eines mathematischen Kolloquiums 4*, 40; reprinted in *Menger 1998*, 222.

*1933o   The present situation in the foundations of mathematics, in *Gödel 1995*, 45–53.

*1933?   Vereinfachter Beweis eines Steinitzchen Satzes, in *Gödel 1995*, with English translation by John W. Dawson, Jr., Israel Halperin and William Craig, 56–61.

1934   *On undecidable propositions of formal mathematical systems* (mimeographed lecture notes, taken by Stephen C. Kleene and J. Barkley Rosser); reprinted with revisions in *Davis 1965*, 39–74.

1934a   Review of *Skolem 1933*, *Zentralblatt für Mathematik und ihre Grenzgebiete 7*, 97–98.

1934b   Review of *Quine 1933, ibid. 7,* 98.
1934c   Review of *Skolem 1933a, ibid. 7,* 193–194.
1934d   Review of *Chen 1933, ibid. 7,* 385.
1934e   Review of *Church 1933, ibid. 8,* 289.
1934f   Review of *Notcutt 1934, ibid. 9,* 3.
1935   Review of *Skolem 1934, ibid. 10,* 49.
1935a   Review of *Huntington 1934, ibid. 10,* 49.
1935b   Review of *Carnap 1934, ibid. 11,* 1.
1935c   Review of *Kalmár 1934, ibid. 11,* 3–4.
1936   Untitled remark following *Wald 1936, Ergebnisse eines mathematischen Kolloquiums 7,* 6; reprinted in *Menger 1998,* 324.
1936a   Über die Länge von Beweisen, *ibid. 7,* 23–24; reprinted in *Menger 1998,* 341.
1936b   Review of *Church 1935, Zentralblatt für Mathematik und ihre Grenzgebiete 12,* 241–242.
1938   The consistency of the axiom of choice and of the generalized continuum hypothesis, *Proceedings of the National Academy of Sciences, U.S.A. 24,* 556–557.
*1938a   Vortrag bei Zilsel, in *Gödel 1995,* with English translation by Charles Parsons and Wilfried Sieg, 86–113.
1939   The consistency of the generalized continuum hypothesis, *Bulletin of the American Mathematical Society 45,* 93.
1939a   Consistency proof for the generalized continuum hypothesis, *Proceedings of the National Academy of Sciences, U.S.A. 25,* 220–224; errata in *1947,* footnote 23.
*1939b   Vortrag Göttingen, in *Gödel 1995,* with English translation by John W. Dawson, Jr. and William Craig.
*193?   ⟦Undecidable diophantine propositions⟧, in *Gödel 1995,* 164–175.
1940   *The consistency of the axiom of choice and of the generalized continuum hypothesis with the axioms of set theory* (lecture notes taken by George W. Brown), Annals of mathematics studies, vol. 3 (Princeton: Princeton University Press); reprinted with additional notes in 1951 and with further notes in 1966.
*1940a   Lecture ⟦on the⟧ consistency ⟦of the⟧ continuum hypothesis, in *Gödel 1995,* 175–185.
*1941   In what sense is intuitionistic logic constructive?, in *Gödel 1995,* 189–200.
1944   Russell's mathematical logic, in *Schilpp 1944,* 123–153; reprinted, with some alterations, as *Gödel 1964a* and as *Gödel 1972b.*

1946   Remarks before the Princeton bicentennial conference on problems in mathematics; first published in *Davis 1965*, 84–88; reprinted, with some alterations, as *Gödel 1968*.

*1946/9-B2 Some observations about the relationship between theory of relativity and Kantian philosophy, version B2, in *Gödel 1995*, 230–246.

*1946/9-C1 Some observations about the relationship between theory of relativity and Kantian philosophy, version C1, in *Gödel 1995*, 247–259.

1947   What is Cantor's continuum problem?, *American mathematical monthly 54*, 515–525; errata, *55*, 151; revised and expanded as *Gödel 1964*.

1949   An example of a new type of cosmological solutions of Einstein's field equations of gravitation, *Reviews of modern physics 21*, 447–450.

1949a   A remark about the relationship between relativity theory and idealistic philosophy, in *Schilpp 1949*, 555–562.

*1949b   Lecture on rotating universes, in *Gödel 1995*, 269–287.

*1951   Some basic theorems on the foundations of mathematics and their philosophical implications, in *Gödel 1995*, 304–323.

1952   Rotating universes in general relativity theory, *Proceedings of the International Congress of Mathematicians; Cambridge, Massachusetts, U. S. A. August 30–September 6, 1950*, I (Providence, RI: American Mathematical Society, 1952), 175–181.

*1953/9-III Is mathematics syntax of language?, version III, in *Gödel 1995*, 334–356.

*1953/9-V   Is mathematics syntax of language?, version V, in *Gödel 1995*, 356–362.

1955   Eine Bemerkung über die Beziehungen zwischen der Relativitätstheorie und der idealistischen Philosophy (German translation of *Gödel 1949a* by Hans Hartmann), in *Schilpp 1955*, 406–412.

1958   Über eine bisher noch nicht benüzte Erweiterung des finiten Standpunktes, *Dialectica 12*, 280–287; revised and expanded in English as *Gödel 1972*; translated into English by Wilfrid Hodges and Bruce Watson as *Gödel 1980*.

*1961/?   The modern development of the foundations of mathematics in the light of philosophy, in *Gödel 1995*, in German, with English translation by Eckehart Köhler, Hao Wang, John W. Dawson, Jr., Charles Parsons and William Craig, 374–387.

1962   Postscript to *Spector 1962*, 27.

1962a *On formally undecidable propositions of Principia mathematica and related systems*, translation of *Gödel 1931* by B. Melzer, with introduction by R. B. Braithwaite (Edinburgh and London: Oliver and Boyd; New York: Basic Books).

1964 Revised and expanded version of *Gödel 1947*, in *Benacerraf and Putnam 1964*, 258–273.

1964a Reprint, with some alterations, of *Gödel 1944*, in *Benacerraf and Putnam 1964*, 211–232.

1965 Expanded version of *Gödel 1934*, in *Davis 1965*, 39–74.

1967 English translation of *Gödel 1931*, in *van Heijenoort 1967*, 596–616.

1968 Reprint, with some alterations, of *Gödel 1946*, in *Klibansky 1968*, 250–253.

*1970 Ontological proof, in *Gödel 1995*, 403–404.

*1970a Some considerations leading to the probable conclusion that the true power of the continuum is $\aleph_2$, in *Gödel 1995*, 420–422.

*1970b A proof of Cantor's continuum hypothesis from a highly plausible axiom about orders of growth, in *Gödel 1995*, 422–423.

*1970c [Unsent letter to Alfred Tarski], in *Gödel 1995*, 424–425.

1972 On an extension of finitary mathematics which has not yet been used (to have appeared in *Dialectica*; first published in *Gödel 1990*, 271–280), revised and expanded English translation of *Gödel 1958*.

1972a Some remarks on the undecidability results (to have appeared in *Dialectica*; first published in *Gödel 1990*, 305–306).

1972b Reprint, with some alterations, of *Gödel 1944*, in *Pears 1972*, 192–226.

1974 Untitled remarks, in *Robinson 1974*, x.

1974a Alternate version of remark 3 of *1972a*, in *Wang 1974*, 325–326.

1980 On a hitherto unexploited extension of the finitary standpoint, English translation of *Gödel 1958*, *Journal of philosophical logic 9*, 133–142.

1986 *Collected works*, volume I: *Publications 1929–1936*, edited by Solomon Feferman, John W. Dawson, Jr., Stephen C. Kleene, Gregory H. Moore, Robert M. Solovay and Jean van Heijenoort (New York and Oxford: Oxford University Press).

1990 *Collected works*, volume II: *Publications 1938–1974*, edited by Solomon Feferman, John W. Dawson, Jr., Stephen C. Kleene, Gregory H. Moore, Robert M. Solovay and Jean van Heijenoort (New York and Oxford: Oxford University Press).

1995   *Collected works*, volume III: *Unpublished essays and lectures*, edited by Solomon Feferman, John W. Dawson, Jr., Warren Goldfarb, Charles Parsons and Robert M. Solovay (New York and Oxford: Oxford University Press).

See also Hahn et alii.

See also Nagel, Ernest, James R. Newman, Kurt Gödel and Jean-Yves Girard.

Goldfarb, Warren
1971   Review of *Skolem 1970*, *The journal of philosophy 68*, 520–530.
1979   Logic in the twenties: the nature of the quantifier, *The journal of symbolic logic 44*, 351–368.
1984   The Gödel class with identity is unsolvable, *Bulletin of the American Mathematical Society* (n.s.) *10*, 113–115.
1984a   The unsolvability of the Gödel class with identity, *The journal of symbolic logic 49*, 1237–1252.
1993   Herbrand's error and Gödel's correction, *Modern logic 3*, 103–118.
2003   In memoriam: Burton Spencer Dreben, 1927–1999, *The bulletin of symbolic logic* (to appear).

See also Dreben, Burton, and Warren Goldfarb.

Gonseth, Ferdinand
1941   (ed.) *Les entretiens de Zurich, 6–9 decémbre 1938* (Zurich: Leemann).

Grattan-Guinness, Ivor
1979   In memoriam Kurt Gödel: his 1931 correspondence with Zermelo on his incompletability theorem, *Historia mathematica 6*, 294–304.

Greenberg, Marvin Jay
1974   *Euclidean and non-Euclidean geometries: development and history* (San Francisco: Freeman).
1980   Second edition of *Greenberg 1974*.

Griffor, E. R.
1999   (ed.) *Handbook of computability theory* (Amsterdam: Elsevier).

Gross, Herbert
1971   Nachruf: Paul Finsler, *Elemente der Mathematik 26*, 19–21.

Grossi, Marie L., Montgomery Link, Katalin Makkai and Charles Parsons
1998   A bibliography of Hao Wang, *Philosophia mathematica (3) 6*, 25–38.

Günther, Gotthard

1933   *Grundzüge einer neuen Theorie des Denkens in Hegels Logik* (Leipzig: Meiner).

1937   Wahrheit, Wirklichkeit, und Zeit: die transzendentalen Bedingungen einer Metaphysik der Geschichte, *Travaux du IXe congrès international de philosophie VIII*, 105–113 (Paris: Hermann); reprinted in *Günther 1976*, 1–9.

1940   Logistik und Transzendentallogik, *Die Tatwelt 16*, 135–147; reprinted in *Günther 1976*, 11–23.

1952   Die "zweite" Maschine. Nachwort to Isaac Asimov, *Ich, der Robot* (Düsseldorf: Rauch Verlag); reprinted with some omissions in *Günther 1963*, 179–203, and in full in *Günther 1976*, 91–114.

1953   Die philosophische Idee einer nicht-aristotelischen Logik, in *Proceedings of the XIth International Congress of Philosophy, Brussels, August 20–26, 1953. Vol. V: logic, philosophical analysis, philosophy of mathematics*, 44–50 (Amsterdam: North-Holland; Louvain: Nauwelaerts); reprinted in *Günther 1976*, 24–30.

1954   Achilles and the tortoise, *Astounding science fiction 53, no. 5*, 76–88; *no. 6*, 84–97; *54, no. 1*, 80–95.

1957   Metaphysik, Logik und die Theorie der Reflexion, *Archiv für Philosophie 7/1,2*, 1–44.

1957a   *Das Bewusstsein der Maschinen: eine Metaphysik der Kybernetik* (Krefeld and Baden-Baden: Agis-Verlag).

1957b   Ideen zu einer Metaphysik des Todes. Grundsätzliche Bemerkungen zu Arnold Metzger's "Freiheit und Tod", *Archiv für Philosophie 7/3, 4*, 335–347; reprinted in *Günther 1980*, 1–13.

1958   Die aristotelische Logik des Seins und die nicht-aristotelische Logik der Reflexion, *Zeitschrift für philosophische Forschung 12*, 360–407.

1959   *Idee und Grundriss einer nicht-Aristotelischen Logik: Erster Band: die Idee und ihre philosophischen Voraussetzungen* (Hamburg: Felix Meiner).

1963   Second edition of *Günther 1957a*, with additional texts.

1975   Selbstdarstellung im Spiegel Amerikas, in *Pongratz 1975, II*, 1–76.

1976   *Beiträge zur Grundlegung einer operationsfähigen Dialektik. Band I. Metakritik der Logik. Nicht-Aristotelische Logik. Reflexion. Stellenwertheorie. Dialektik. Cybernetic Ontology. Morphogrammatik.* (Hamburg: Meiner).

1978   Second edition of *Günther 1933*, with a new preface (Hamburg: Meiner).

1978a   Second revised and expanded edition of *Günther 1959*, with a new preface and an appendix by Rudolf Kaehr (Hamburg: Meiner).

1979   *Beiträge zur Grundlegung einer operationsfähigen Dialektik. Zweiter Band: Wirklichkeit als Poly-Kontexturalität. Reflexion—Logische Paradoxie—Mehrwertige Logik—Denken—Wollen—Promielle Relation—Kenogrammatik—Dialektik der natürlichen Zahl—Dialektiker Materialismus* (Hamburg: Meiner).

1980   *Beiträge zur Grundlegung einer operationsfähigen Dialektik. Dritter Band: Philosophie der Geschichte und der Technik. Wille—Schöpfung—Arbeit. Strukturanalyse der Vermittlung. Mehrwertigkeit—Stellen- und Kontextwertlogik—Kenogrammatik.* (Hamburg: Meiner).

1991   *Idee und Grundriß einer nicht-Aristotelischen Logik: die Idee und ihre philosophischen Voraussetzungen,* third edition of *Günther 1959*, with a new preface by Claus Baldus and Bernard Mitterarer, with additional texts by Günther but without the appendix of *1978a* (Hamburg: Felix Meiner).

Haas, Elke, and Gerrit Haas
1982   Heinrich Behmann (1891–1970), *Allgemeine Zeitschrift für Philosophie 7/1*, 59–65.

Hahn, Hans
1932   *Reelle Funktionen* (Leipzig: Akademische Verlagsgesellschaft).

Hahn, Hans, Rudolf Carnap, Kurt Gödel, Arend Heyting, Kurt Reidemeister, Arnold Scholz and John von Neumann
1931   Diskussion zur Grundlegung der Mathematik, *Erkenntnis 2*, 135–151; English translation by John W. Dawson, Jr. in *Dawson 1984*, 116–128.

Hahn, S. W.
See Bulloff, Jack J., Thomas C. Holyoke and S. W. Hahn.

Hajnal, András
1956   On a consistency theorem connected with the generalized continuum problem, *Zeitschrift für mathematische Logik und Grundlagen der Mathematik 2*, 131–136.

1961   On a consistency theorem connected with the generalized continuum problem, *Acta mathematica Academiae Scientiarum Hungaricae 12*, 321–376.

Hanf, William Porter
1963   *Some fundamental problems concerning languages with infinitely long expressions* (doctoral dissertation, University of California, Berkeley).
1964   Incompactness in languages with infinitely long expressions, *Fundamenta mathematicae 53*, 309–324.

Hartmanis, Juris
1989   Gödel, von Neumann and the P=?NP problem, *Bulletin of the European Association for Computer Science 38*, 101–107.

Hasse, Helmut, and Heinrich Scholz
1928   *Die Grundlagenkrisis der griechischen Mathematik* (Charlottenburg: Metzner).

Hausdorff, Felix
1907   Untersuchungen über Ordnungstypen, *Berichte über die Verhandlungen der Königlich Sächsischen Gesellschaft der Wissenschaften zu Leipzig, Mathematische-Physische Klasse 59*, 84–159.
1909   Die Graduierung nach der Endverlauf, *Abhandlungen der Königlich Sächsischen Gesellschaft der Wissenschaften, Mathematisch-Physischen Klasse 31*, 295–334.
1914   *Grundzüge der Mengenlehre* (Leipzig: Veit); reprinted in 1949 (New York: Chelsea).
1927   *Mengenlehre*, second revised edition of *Hausdorff 1914* (Berlin: Gruyter).

Hechler, Stephen M.
1974   On the existence of certain cofinal subsets of $^{\omega}\omega$, in *Jech 1974*, 155–173.

Heidegger, Martin
1929   *Was ist Metaphysik?* (Bonn: Friedrich Cohen); reprinted in *Heidegger 1967*, 1–19.
1943   Fourth edition of *1929*, with a new postscript (Frankfurt: Vittorio Klostermann); postscript reprinted in *Heidegger 1967*, 99–108.
1949   Fifth edition of *Heidegger 1929*, with the postscript to *Heidegger 1943* and a new introduction (Frankfurt: Vittorio Klostermann); introduction reprinted in *Heidegger 1967*, 195–211.
1967   *Wegmarken* (Frankfurt: Vittorio Klostermann).
1976   Enlarged edition of *Heidegger 1967*, with marginal notes of the author and pagination of *1967* in the margins, *Gesamtausgabe I. Abteilung: Veröffentlichte Schriften 1914–1970*, Band 9 (Frankfurt: Vittorio Klostermann).

1998   *Pathmarks*, edited by William O'Neill; translation of *Heidegger 1976* with pagination of *1967* in text (Cambridge and New York: Cambridge University Press).

Heinzmann, Gerhard
1982   *Schematisierte Strukturen. Eine Untersuchung über den "Idoneïsmus" Ferdinand Gonseths auf dem Hintergrund eines konstruktivistischen Ansatzes* (Bern and Stuttgart: Verlag Paul Haupt).

Heinzmann, Gerhard and Joëlle Proust
1988   Carnap et Gödel: Échange de lettres autour de la définition de l'analyticité. Introduction, traduction et notes, *Logique et analyse 31*, 257–291.

Heisenberg, Werner
1931   Kausalgesetz und Quantenmechanik, *Erkenntnis 2*, 172–182.

Henkin, Leon
See Addison, John W., Leon Henkin and Alfred Tarski.
See also Henkin et alii.
See also Henkin, Leon, J. Donald Monk and Alfred Tarski.
See also Henkin, Leon, and Alfred Tarski.
See also Suppes et alii.

Henkin, Leon, John W. Addison, Chen Chung Chang, William Craig, Dana S. Scott and Robert L. Vaught
1974   (eds.) *Proceedings of the Tarski symposium*, Proceedings of symposia in pure mathematics, vol. 25 (Providence, RI: American Mathematical Society).

Henkin, Leon, J. Donald Monk and Alfred Tarski
1971   *Cylindric algebras*, part I (Amsterdam: North Holland).
1985   *Cylindric algebras*, part II (Amsterdam, New York and Oxford: North Holland).

Henkin, Leon, and Alfred Tarski
1961   Cylindric algebras, in *Dilworth 1961*, 83–113.

Herbrand, Jacques
1929   Sur le problème fondamental des mathématiques, *Comptes rendus hebdomadaires des séances de l'Academie des Science (Paris) 189*, 554–556, erratum, 720; reprinted in *Herbrand 1968*, 31–33; English translation by Warren Goldfarb in *Herbrand 1971*, 41–43.

| 1930 | *Recherches sur la théorie de la démonstration* (doctoral dissertation, University of Paris); also *Prace Towarzystwa Naukowego Warszawskiego, wydział III*, no. 33; reprinted in *Herbrand 1968*, 35–153; English translation by Warren Goldfarb in *Herbrand 1971*, 44–202. |

1930a   Les bases de la logique hilbertienne, *Revue de métaphysique et de morale 37*, 243–255; reprinted in *Herbrand 1968*, 155–166; English translation by Warren Goldfarb in *Herbrand 1971*, 203–214.

1931   Sur la non-contradiction de l'arithmétique, *Journal für die reine und angewandte Mathematik 166*, 1–8; reprinted in *Herbrand 1968*, 221–232; English translation by Jean van Heijenoort in *van Heijenoort 1967*, 618–628, and in *Herbrand 1971*, 282–298.

1931a   Sur le problème fondamental de la logique mathématique, *Sprawozdania z posiedzeń Towarzystwa Naukowego Warszawskiego wydział III, 24*, 12–56; reprinted in *Herbrand 1968*, 167–207; English translation by Warren Goldfarb in *Herbrand 1971*, 215–271.

1931b   Unsigned note on *Herbrand 1930*, *Annales de l'Université de Paris 6*, 186–189; reprinted in *Herbrand 1968*, 209–214; English translation by Warren Goldfarb in *Herbrand 1971*, 272–276.

1931c   Notice pour Jacques Hadamard, in *Herbrand 1968*, 215–219; English translation by Warren Goldfarb in *Herbrand 1971*, 277–281.

1968   *Écrits logiques*, edited by Jean van Heijenoort (Paris: Presses Universitaires de France).

1971   *Logical writings*, English translation of *Herbrand 1968* by Warren Goldfarb (Cambridge: Harvard University Press; Dordrecht: Reidel).

Hertz, Paul

1922   Über Axiomensysteme für beliebige Satzsysteme, *Mathematische Annalen 87*, 246–269.

1923   Über Axiomensysteme für beliebige Satzsysteme, *Mathematische Annalen 89*, 76–102.

1929   Über Axiomensysteme für beliebige Satzsysteme, *Mathematische Annalen 101*, 457–514.

Heyting, Arend

1930   Die formalen Regeln der intuitionistischen Logik, *Sitzungsberichte der Preussischen Akademie der Wissenschaften, physikalisch-mathematische Klasse*, 42–56; English translation by Paolo Mancosu in *Mancosu 1998*, 311–327.

1930a Die formalen Regeln der intuitionistischen Mathematik, *ibid.*, 57–71.
1930b Sur la logique intuitionniste, *Académie royale de Belgique, Bulletins de la classe des sciences (5) 16*, 957–963; English translation by Amy L. Rocha in *Mancosu 1998*, 306–310.
1930c Die formalen Regeln der intuitionistischen Mathematik III, *Sitzungsberichte der Preussischen Akademie der Wissenschaften, Physikalisch-mathematische Klasse*, 158–169.
1931 Die intuitionistische Grundlegung der Mathematik, *Erkenntnis 2*, 106–115; English translation by Erna Putnam and Gerald J. Massey in *Benacerraf and Putnam 1964*, 42–49.
1934 *Mathematische Grundlagenforschung. Intuitionismus. Beweistheorie*, Ergebnisse der Mathematik und ihrer Grenzgebiete 3 (Berlin: Springer).
1959 (ed.) *Constructivity in mathematics. Proceedings of the colloquium held at Amsterdam, 1957* (Amsterdam: North-Holland).
See also Hahn et alii.

Hilbert, David
1899 *Grundlagen der Geometrie. Festschrift zur Feier der Enthüllung des Gauss–Weber Denkmals in Göttingen* (Leipzig: Teubner).
1918 Axiomatisches Denken, *Mathematische Annalen 78*, 405–415; reprinted in *Hilbert 1935*, 146–156; English translation by William B. Ewald in *Ewald 1996*, vol. II, 1107–1115.
1926 Über das Unendliche, *Mathematische Annalen 95*, 161–190; English translation by Stefan Bauer-Mengelberg in *van Heijenoort 1967*, 367–392.
1928 Die Grundlagen der Mathematik, *Abhandlungen aus dem mathematischen Seminar der Hamburgischen Universität 6*, 65–85; English translation by Stefan Bauer-Mengelberg and Dagfinn Føllesdal in *van Heijenoort 1967*, 464–479.
1931 Die Grundlegung der elementaren Zahlenlehre, *Mathematische Annalen 104*, 485–494; reprinted in part in *Hilbert 1935*, 192–195; English translation by William B. Ewald in *Ewald 1996*, vol. II, 1149–1157; reprinted in *Mancosu 1998*, 266–273.
1931a Beweis des tertium non datur, *Nachrichten von der Gesellschaft der Wissenschaften zu Göttingen, mathematisch-physikalische Klasse*, 120–125.
1935 *Gesammelte Abhandlungen*, vol. 3 (Berlin: Springer).
1962 *Grundlagen der Geometrie* (Stuttgart: Teubner); ninth edition of *Hilbert 1899*, revised and expanded by Paul Bernays.

Hilbert, David, and Wilhelm Ackermann
 1928   *Grundzüge der theoretischen Logik* (Berlin: Springer).
 1938   Second, revised edition of *Hilbert and Ackermann 1928*; trans-
         lated into English by Lewis M. Hammond, George G. Leckie
         and F. Steinhardt as *Hilbert and Ackermann 1950*.
 1950   *Principles of mathematical logic*, English translation of *Hilbert
         and Ackermann 1938* (New York: Chelsea).

Hilbert, David, and Paul Bernays
 1934   *Grundlagen der Mathematik*, vol. I (Berlin: Springer).
 1939   *Grundlagen der Mathematik*, vol. II (Berlin: Springer).
 1968   Second edition of *Hilbert and Bernays 1934*.
 1970   Second edition of *Hilbert and Bernays 1939*.

Hintikka, Jaakko
 1995   (ed.) *From Dedekind to Gödel, essays on the development
         of the foundations of mathematics* (Dordrecht and Norwell:
         Kluwer Academic Publishers).
 See also Butts, Robert E., and Jaakko Hintikka.

Hoensbroech, Franz G.
 1931   Beziehungen zwischen Inhalt und Umfang von Begriffen, *Er-
         kenntnis 2*, 291–300.

Hofmann, Paul
 1931   *Das Problem des Satzes vom ausgeschlossenen Dritten* (Berlin:
         Pan-Verlagsgesellschaft); also published as *1931a* without the
         introduction.
 1931a   Das Problem des Satzes vom ausgeschlossenen Dritten, *Kant-
         Studien 36*, 83–125.

Hölder, Otto
 1924   *Die mathematische Methode. Logisch erkenntnistheoretische
         Untersuchungen im Gebiete der Mathematik, Mechanik und
         Physik* (Berlin: Julius Springer).

Holyoke, Thomas C.
 See Bulloff, Jack J., Thomas C. Holyoke and S. W. Hahn.

Huntington, Edward V.
 1932   A new set of independent postulates for the algebra of logic
         with special reference to Whitehead and Russell's *Principia
         mathematica, Proceedings of the National Academy of Sci-
         ences, U.S.A. 18*, 179–180.
 1934   Independent postulates related to C. I. Lewis's theory of strict
         implication, *Mind (n.s.) 43*, 181–198.

Janik, Allen S., and Stephen E. Toulmin
1973   *Wittgenstein's Vienna* (New York: Touchstone/Simon and Schuster).

Jech, Thomas
1974   (ed.) *Axiomatic set theory*, Proceedings of symposia in pure mathematics, vol. 13, part 2 (Providence, RI: American Mathematical Society).
1978   *Set theory* (New York: Academic Press).
1997   Second edition of *Jech 1978* (Berlin: Springer).

Jensen, Ronald Björn, and Carol Karp
1971   Primitive recursive set functions, in *Scott 1971*, 143–176.

Johnson, David S.
See Garey, Michael R., and David S. Johnson.

Johnston, William M.
1972   *The Austrian mind* (Berkeley, Los Angeles and London: University of California Press).

Joja, Athanase
See Suppes et alii.

Jørgensen, Jørgen
1931   *A treatise of formal logic. Its evolution and main branches, with its relations to mathematics and philosophy* (Copenhagen: Levin and Munksgaard; London: Oxford); republished in 1962 (New York: Russell and Russell).

Jourdain, Philip E. B.
1912   The development of the theories of mathematical logic and the principles of mathematics, *Quarterly journal of pure and applied mathematics 43*, 219–314; reprinted in *Frege 1976*, 275–301, reprinted in part in *Frege 1980*, 179–206.

Jung, Carl Gustav
1961   *Memories, dreams, reflections*, English translation of *Jung 1962* (New York: Pantheon; Toronto: Random House).
1962   *Erinnerungen, Träume, Gedanken*, recorded and edited by Aniela Jaffé (Zürich: Rascher); English translation by Richard and Clara Winston as *Jung 1961*.

Junker, Friedrich Heinrich
1919   *Höhere Analysis: I. Differentialrechnung: II. Integralrechnung*, Sammlung Göschen, vols. 87 and 88 (Leipzig and Berlin: Göschen'sche Verlagshandlung).

Kaczmarz, Stefan
1932   Axioms for arithmetic, *The journal of the London Mathematical Society* 7, 179–182.

Kalmár, László
1932   Ein Beitrag zum Entscheidungsproblem, *Acta litterarum ac scientiarum Regiae Universitatis Hungaricae Francisco-Josephinae, sectio scientiarum mathematicarum 5*, 222–236.
1933   Über die Erfüllbarkeit derjenigen Zählausdrücke, welche in der Normalform zwei benachbarte Allzeichen enthalten, *Mathematische Annalen 108*, 466–484.
1934   Über einen Löwenheimschen Satz, *Acta litterarum ac scientiarum Regiae Universitatis Hungaricae Francisco-Josephinae, sectio scientiarum mathematicarum 7*, 112–121.
1955   Über ein Problem, betreffend die Definition des Begriffes der allgemein-rekursiven Funktion, *Zeitschrift für mathematische Logik und Grundlagen der Mathematik 1*, 93–96.

Kanamori, Akihiro
1994   *The higher infinite* (Berlin, Heidelberg and New York: Springer).
1997   *The higher infinite*, corrected second printing (Heidelberg: Springer).

Kant, Immanuel
1781   *Critik der reinen Vernunft* (Riga: Hartknoch).
1787   *Kritik der reinen Vernunft*, second revised edition of *Kant 1781*.
1788   *Kritik der praktischen Vernunft* (Riga: Hartknoch); reprinted in *Kant 1902-*, vol. V; English translation by Mary J. Gregor in *Kant 1996*.
1790   *Kritik der Urteilskraft* (Berlin and Libau: Lagarde und Friederich); reprinted in *Kant 1902-*, vol. V; English translation by Paul Guyer and Eric Matthews in *Kant 2000*.
1902-   *Kants gesammelte Schriften*, edited by the Prussian Academy of Sciences, later the German Academy of Sciences (Berlin: Georg Reimer, later Walter de Gruyter).
1996   *Practical philosophy*, edited and translated by Mary J. Gregor, general introduction by Allen W. Wood, The Cambridge edition of the Works of Immanuel Kant (Cambridge and New York: Cambridge University Press).
1997   *Critique of pure reason*, translated and edited by Paul Guyer and Allen W. Wood, The Cambridge edition of the Works of Immanuel Kant (Cambridge and New York: Cambridge University Press).

2000   *Critique of the power of judgement*, edited by Paul Guyer, The Cambridge edition of the Works of Immanuel Kant (Cambridge and New York: Cambridge University Press).

Karp, Carol
See Jensen, Ronald Björn, and Carol Karp.

Kaufmann, Felix
1930   *Das Unendliche in der Mathematik und seine Ausschaltung. Eine Untersuchung über die Grundlagen der Mathematik* (Leipzig and Vienna: Franz Deuticke).

Keisler, H. Jerome, and Alfred Tarski
1964   From accessible to inaccessible cardinals: results holding for all accessible cardinal numbers and the problem of their extension to inaccessible ones, *Fundamenta mathematicae 53*, 225–308.

Kitcher, Philip
See Asquith, Peter D., and Philip Kitcher.

Kleene, Stephen C.
1935   A theory of positive integers in formal logic, *American journal of mathematics 57*, 153–173, 219–244.
1936   General recursive functions of natural numbers, *Mathematische Annalen 112*, 727–742; reprinted in *Davis 1965*, 236–252; for an erratum, a simplification and an addendum, see *Davis 1965*, 253.
1938   On notation for ordinal numbers, *The journal of symbolic logic 3*, 150–155.
1943   Recursive predicates and quantifiers, *Transactions of the American Mathematical Society 53*, 41–73; reprinted in *Davis 1965*, 254–287; for a correction and an addendum, see *Davis 1965*, 254 and 287.
1955   On the forms of predicates in the theory of constructive ordinals (second paper), *American journal of mathematics 77*, 405–428.
1955a   Arithmetical predicates and function quantifiers, *Transactions of the American Mathematical Society 79*, 312–340.
1955b   Hierarchies of number-theoretic predicates, *Bulletin of the American Mathematical Society 61*, 193–213.
1976   The work of Kurt Gödel, *The journal of symbolic logic 41*, 761–778; addendum, *ibid. 43* (1978), 613; reprinted in *Shanker 1988a*, 48–71.
1987b   Reflections on Church's thesis, *Notre Dame journal of formal logic 28*, 490–498.

Kleene, Stephen C., and J. Barkley Rosser
1935   The inconsistency of certain formal logics, *Annals of mathematics (2) 36*, 630–636.

Klein, Carsten
See Köhler et alii.

Klein, Fritz
1931   Zur Theorie der abstrakten Verknüpfungen, *Mathematische Annalen 105*, 308–323.
1932   Über einen Zerlegungssatz in der Theorie der abstrakten Verknüpfungen, *Mathematische Annalen 106*, 114–130.

Klibansky, Raymond
1968   (ed.) *Contemporary philosophy, a survey. I, Logic and foundations of mathematics* (Florence: La Nuova Italia Editrice).

Kino, Akiko
See Myhill, John, Akiko Kino and Richard E. Vesley.

Köhler, Eckehart
See Woleński, Jan and Eckehart Köhler .

Köhler, Eckehart, Bernd Buldt, Werner DePauli-Schimanovich, Carsten Klein, Michael Stöltzner and Peter Weibel
2002   (eds.) *Wahrheit und Beweisbarkeit. Leben und Werk Kurt Gödels*, Band 1: *Dokumente und historische Analysen*, Band 2: *Kompendium zu Gödels Werk* (Vienna: Hölder–Pichler–Tempsky).

Kondô, Motokiti
1938a   Sur l'uniformisation des complémentaires analytiques et les ensembles projectifs de la seconde classe, *Japanese journal of mathematics 15*, 197–230.
1958   Sur les ensembles nommables et le fondement de analyse mathématique, I, *Japanese journal of mathematics 28*, 1–116.

König, Julius (König, Gyula)
1905   Zum Kontinuum-Problem, *Mathematische Annalen 60*, 177–180, 462.
1914   *Neue Grundlagen der Logik, Arithmetik und Mengenlehre* (Leipzig: Veit).

Krajíček, Jan
See Clote, Peter, and Jan Krajíček.

Kreisel, Georg
1958a   Wittgenstein's remarks on the foundations of mathematics, *The British journal for the philosophy of science 9*, 135–158.

1959   Interpretation of analysis by means of constructive functionals of finite types, in *Heyting 1959*, 101–128.

1960   Ordinal logics and the characterization of informal concepts of proof, *Proceedings of the International Congress of Mathematicians, 14–21 August 1958* (Cambridge: Cambridge University Press), 289–299.

1962b   Review of *Davis, Putnam and Robinson 1961*, *Mathematical reviews 24A*, 573.

1965   Mathematical logic, in *Saaty 1965*, vol. 3, 95–195.

1968   A survey of proof theory, *The journal of symbolic logic 33*, 321–388.

1968a   Functions, ordinals, species, in *van Rootselaar and Staal 1968*, 143–158.

1970a   Principles of proof and ordinals implicit in given concepts, in *Myhill, Kino and Vesley 1970*, 489–516.

1971a   Observations on popular discussions of foundations, in *Scott 1971*, 189–198.

1980   Kurt Gödel, 28 April 1906–14 January 1978, *Biographical memoirs of Fellows of the Royal Society 26*, 148–224; corrections, *ibid. 27*, 697, and *28*, 718.

See also Ehrenfeucht, Andrzej, and Georg Kreisel.

Krikorian, Yervant Hovhaness
1944   (ed.) *Naturalism and the human spirit* (New York: Columbia University Press).

Kunen, Kenneth
1980   *Set theory: an introduction to independence proofs* (Amsterdam: North-Holland).

Kuratowski, Kazimierz (Kuratowski, Casimir)
1936   Les ensembles projectifs et l'induction transfinie, *Fundamenta mathematicae 27*, 269–276.

1937   Sur la géométrisation des types d'ordre dénombrable, *Fundamenta mathematicae 28*, 167–185.

1937a   Les suites transfinies d'ensembles et les ensembles projectifs, *Fundamenta mathematicae 28*, 186–196.

1937b   Sur les suites analytiques d'ensembles, *Fundamenta mathematicae 29*, 54–59.

1937c   Les types définissables et les ensembles boreliens, *Fundamenta mathematicae 29*, 97–100.

Lascar, Daniel
See van Dalen, Dirk, Daniel Lascar and Timothy J. Smiley.

Lautman, Albert
1938   *Essai sur les notions de structure et d'existence en mathématiques*, Actualités scientifiques et industrielles (Paris: Hermann et Cie), 590–591.
See also Chevalley, Claude, and Albert Lautman.

Lawvere, F. William
1964   An elementary theory of the category of sets, *Proceedings of the National Academy of Sciences, U.S.A. 52*, 1506–1511.

Leblanc, Hugues
1962   Études sur les régles d'inférence dites régles de Gentzen, *Dialogue 1*, 56–66.

Leplin, Jarrett
See Fine, Arthur, and Jarrett Leplin.

Lettvin, Jerome Y., H. R. Maturana, Warren S. McCulloch and Walter H. Pitts
1959   What the frog's eye tells the frog's brain, *Proceedings of the IRE 47*, 1940–1951; reprinted in *McCulloch 1965*, 230–255.

Lévy, Azriel
1957   Indépendance conditionnelle de $V = L$ et d'axioms qui se rattachent au système de M. Gödel, *Comptes rendus hebdomadaires des séances de l'Académie des Sciences, Paris 245*, 1582–1583.
1959   On Ackermann's set theory, *The journal of symbolic logic 24*, 154–166.
1960   Axiom schemata of strong infinity in axiomatic set theory, *Pacific journal of mathematics 10*, 223–238.
1960a   Principles of reflection in axiomatic set theory, *Fundamenta mathematicae 49*, 1–10.
1960b   A generalization of Gödel's notion of constructibility, *The journal of symbolic logic 25*, 147–155.
1963   Independence results in set theory by Cohen's method. I, III and IV (abstracts), *Notices of the American Mathematical Society 10*, 592–593.

1970   Definability in axiomatic set theory II, in *Bar-Hillel 1970*, 129–145.

See also Feferman, Solomon, and Azriel Lévy.

Lévy, Azriel, and Robert M. Solovay
1967   Measurable cardinals and the continuum hypothesis, *Israel journal of mathematics 5*, 234–248.

Lewis, Clarence I.
1918   *A survey of symbolic logic* (Berkeley: University of California Press); reprinted by Dover (New York).
1932   Alternative systems of logic, *The monist 42*, 481–507.

Link, Montgomery
See Grossi et alii.

Lipshitz, Leonard, Dirk Siefkes and P. Young
1984   J. Richard Büchi (1924–1984), *Newsletter of the Association for Symbolic Logic*, November 1984; reprinted in *Büchi 1990*, 2–3.

Lorenzen, Paul
1951b   Maß und Integral in der konstruktiven Analysis, *Mathematische Zeitschrift 54*, 275–290.
1955   *Einführung in die operative Logik und Mathematik* (Berlin: Springer).

Luzin, Nikolai (Lusin, Nicolas; Лузин, Николай Николаевич)
1930   *Leçons sur les ensembles analytiques et leurs applications* (Paris: Gauthier-Villars); reprinted with corrections, 1972 (New York: Chelsea).
1930a   Analogies entre les ensembles mesurables B et les ensembles analytiques, *Fundamenta mathematicae 16*, 48–76.

Mac Lane, Saunders
1961   Locally small categories and the foundations of set theory, in *Polish Academy of Sciences 1961*, 25–43.
1969   (ed.) *Reports of the Midwest Category Seminar III*, Lecture notes in mathematics 106 (Berlin: Springer).

Macrae, Norman
1992   *John von Neumann* (New York: Pantheon Books).

Magidor, Menachem
1976   How large is the first strongly compact cardinal? Or: a study on identity crises, *Annals of mathematical logic 10*, 33–57.

Makkai, Katalin
See Grossi et alii.

Mancosu, Paolo
1998   (ed.) *From Brouwer to Hilbert: the debate on the foundations of mathematics in the 1920s* (New York: Oxford University Press).
1999   Between Russell and Hilbert: Behmann on the foundations of mathematics, *The bulletin of symbolic logic 5*, 303–330.
1999a   Between Vienna and Berlin: the immediate reception of Gödel's incompleteness theorems, *History and philosophy of logic 20*, 33–45.
2002   On the constructivity of proofs. A debate among Behmann, Bernays, Gödel, and Kaufmann, in *Sieg, Sommer and Talcott 2002*, 349–371.

Margenau, Henry
See Bergamini et alii.

Markov, A. A. (Марков, A. A.)
1947   On the impossibility of certain algorithms in the theory of associative systems (Russian), *Doklady Akademii Nauk S.S.S.R.* (n.s.) *55*, 587–590; translated into English in *Comptes rendus (Doklady) de l'académie de l'URSS* (n.s.), *55*, 583–586.

Matiyasevich, Yuri Vladimirovich (Matijacevič; Матиясевич, Юрий Владимирович)
1970   Enumerable sets are diophantine (Russian), *Doklady Akademii Nauk S.S.S.R. 191*, 279–282; English translation, with revisions, in *Soviet mathematics Doklady 11* (1970), 354–358.

Maturana, H. R.
See Lettvin et alii.

Mauldin, R. Daniel
1981   (ed.) *The Scottish book, mathematics from the Scottish Cafe* (Boston: Birkhauser).

Mazurkiewicz, Stefan
1927   Sur une propriété des ensembles $C(A)$, *Fundamenta mathematicae 10*, 172–174; reprinted in *Mazurkiewicz 1969*, 152–154.
1969   *Travaux de topologie et ses applications*, edited by K. Borsuk, R. Engelking, B. Knaster, K. Kuratowski, J. Łoś and R. Sikorski (Warsaw: PWN).

McAloon, Kenneth
1966   *Some applications of Cohen's method* (doctoral dissertation, University of California at Berkeley).

McCulloch, Warren S.
1965   *Embodiments of mind* (Cambridge: MIT Press).
See also Lettvin et alii.

McCulloch, Warren S., and Walter Pitts
1943   A logical calculus of the ideas immanent in nervous activity,
        Bulletin of mathematical biophysics 5, 115–133; reprinted in
        McCulloch 1965, 19–39.

McKinsey, John C. C., and Alfred Tarski
1944   The algebra of topology, *Annals of mathematics (2) 45*, 141–
        191.
1946   On closed elements in closure algebras, *Annals of mathematics
        (2) 47*, 122–162.

Menger, Karl
1933   *Krise und Neuaufbau in den exakten Wissenschaften. Die
        neue Logik* (Leipzig and Vienna: Deuticke).
1953   *Calculus, a modern approach* (mimeographed notes, enlarged
        second edition) (Chicago: IIT Bookstore).
1953a   The ideas of variable and function, *Proceedings of the National
        Academy of Sciences, U.S.A. 39*, 956–961.
1955   *Calculus, a modern approach* (Boston: Ginn).
1970   Projective and related structures, part 2 of *Menger and Blu-
        menthal 1970*, 135–223.
1994   *Reminiscences of the Vienna Circle and the mathematical col-
        loquium*, edited by Louise Golland, Brian McGuinness and Abe
        Sklar (Dordrecht: Kluwer Academic).
1998   *Ergebnisse eines mathematischen Kolloquiums*, edited by E.
        Dierker and K. Sigmund (Vienna and New York: Springer).
See also *Gödel 1933h*.

Menger, Karl, and Leonard M. Blumenthal
1970   *Studies in geometry* (San Francisco: Freeman).

Menger, Karl, and Georg Nöbeling
1932   (eds.) *Kurventheorie*, Mengentheoretische Geometrie in Ein-
        zeldarstellungen II (Leipzig and Berlin: B. G. Teubner).

Menzler-Trott, Eckart
2001   *Gentzens Problem. Mathematische Logik im nationalsozialisti-
        schen Deutschland* (Basel, Boston and Berlin: Birkhäuser).

Meyerson, Émile
1931   *Du chéminement de la pensée*, 3 vols. (Paris: F. Alcan).

Mitchell, Janet
1980   (ed.) *A community of scholars* (Princeton, NJ: The Institute for Advanced Study).

Mittelstraß, Jürgen
1980   (ed. in collaboration with Gereon Wolters) *Enzyklopädie Philosophie und Wissenschaftstheorie* (Mannheim, Vienna and Zürich: Bibliographisches Institut, B. I. Wissenschaftsverlag).

Moisil, Gr. C.
See Suppes et alii.

Monk, J. Donald
See Henkin, Leon, J. Donald Monk and Alfred Tarski.

Moore, Gregory H.
1980   Beyond first-order logic: the historical interplay between mathematical logic and axiomatic set theory, *History and philosophy of logic 1*, 95–137.
1988   The origins of forcing, in *Drake and Truss 1988*, 143–173.

Mortimer, Michael
1975   On languages with two variables, *Zeitschrift für mathematische Logik und Grundlagen der Mathematik 21*, 135–140.

Mostowski, Andrzej
1939   Über die Unabhängigkeit des Wohlordnungssatzes vom Ordnungsprinzip, *Fundamenta mathematicae 32*, 201–252.
1964   Widerspruchsfreiheit und Unabhängigkeit der Kontinuumhypothese, *Elemente der Mathematik 19*, 121–125.
1965   *Thirty years of foundational studies: lectures on the development of mathematical logic and the study of the foundations of mathematics in 1930–1964* (= no. 17 of *Acta philosophica fennica*); reprinted in 1966 (New York: Barnes and Noble; Oxford: Blackwell).

Müller, Gert H.
1961   Über die unendliche Induktion, in *Polish Academy of Sciences 1961*, 75–95.
1961a   Nicht-Standard Modelle der Zahlentheorie, *Mathematische Zeitschrift 77*, 414–438.
1962   Über Rekursionsformen (Habilitationsschrift, University of Heidelberg); published in *Automatentheorie und formale Sprache: Bericht einer Tagung des mathematischen Forschungsinstituts Oberwohlfach, October 1969*.
1976   (ed.) *Sets and classes: on the work of Paul Bernays* (Amsterdam: North-Holland).

Murray, F. J.
See von Neumann, John, and F. J. Murray.

Mycielski, Jan, and Hugo Steinhaus
 1962   A mathematical axiom contradicting the axiom of choice, *Bulletin de l'Académie polonaise des sciences, série des sciences mathématiques, astronomiques et physiques 10*, 1–3.

Myhill, John
 1973   Embedding classical logic in intuitionistic logic, *Zeitschrift für mathematische Logik und Grundlagen der Mathematik 19*, 93–96.

Myhill, John, Akiko Kino and Richard E. Vesley
 1970   (eds.) *Intuitionism and proof theory* (Amsterdam: North-Holland).

Myhill, John, and Dana S. Scott
 1971   Ordinal definability, in *Scott 1971*, 271–278.

Nagel, Ernest
 1944   Logic without ontology, in *Krikorian 1944*, 210–241; reprinted in *Nagel 1956*, 55–92.
 1956   *Logic without metaphysics and other essays in the philosophy of science* (Glencoe, IL: The Free Press).
 1979   *Teleology revisited and other essays in the philosophy and history of science* (New York: Columbia University Press).
See also Cohen, Morris Raphael, and Ernest Nagel.

Nagel, Ernest, and James R. Newman
 1956   Goedel's proof, *Scientific American 194*, 6, 71–86; reprinted in *Scientific American 1968*, 221–230.
 1956a Goedel's proof, in *Newman 1956*, vol. 3, 1668–1695.
 1958   *Gödel's proof* (New York: New York University Press).
 1959   British edition of *Nagel and Newman 1958* (London: Routledge & Kegan Paul).
 2001   Revised edition of *Nagel and Newman 1958*, edited and with a new foreword by Douglas R. Hofstadter (New York: New York University Press).

Nagel, Ernest, James R. Newman, Kurt Gödel and Jean-Yves Girard
 1989   *Le théorème de Gödel*, French translations of *Nagel and Newman 1958* and *Gödel 1931*, by Jean-Baptiste Scherrer, together with an essay by Girard (Paris: Éditions du Seuil).

Nagel, Ernest, Patrick Suppes and Alfred Tarski
1962 (eds.) *Logic, methodology, and philosophy of science. Proceedings of the 1960 International Congress* (Stanford, CA: Stanford University Press).

Neder, Ludwig
1931 Über den Aufbau der Arithmetik, *Jahresbericht der Deutschen Mathematiker-Vereinigung 40*, 22–37.

Nelson, Leonard
1908 Über das sogenannte Erkenntnisproblem, *Abhandlungen aus der Fries'schen Schule, Neue Folge, 3. Band, 4. Heft*, 413–818 (also paginated by *Heft*, 1–406) (Göttingen: Vandenhoeck and Ruprecht); reprinted in *Nelson 1970–74*, vol. II, 59–393.

1914 Die kritische Ethik bei Kant, Schiller und Fries, eine Revision ihrer Prinzipen, *Abhandlungen aus der Fries'schen Schule, Neue Folge, 4. Band, 3. Heft*, 483–691; reprinted in *Nelson 1970–74*, 27–192.

1917 *Vorlesungen über die Grundlagen der Ethik*, Bd. I: *Kritik der praktischen Vernunft* (Leipzig: Veit); reprinted as volume IV of *Nelson 1970–74*.

1924 *Vorlesungen über die Grundlagen der Ethik*, Bd. III: *Rechtslehre und Politik* (Leipzig: Der neue Geist); reprinted as volume VI of *Nelson 1970–74*.

1932 *Vorlesungen über die Grundlagen der Ethik*, Bd. II: *System der philosophischen Ethik und Pädogogik* (Göttingen: Verlag Öffentliches Leben); reprinted as volume V of *Nelson 1970–74*, edited by Grete Hermann and Minna Specht; English translation of part 1 by L. H. Grunebaum as *Nelson 1956*.

1956 *System of ethics*, English translation of part 1 of *Nelson 1932* (New Haven: Yale University Press).

1962 *Fortschritte und Rückschritte der Philosophie von Hume und Kant bis Hegel und Fries*, edited by Julius Kraft (Frankfurt am Main: Verlag Öffentliches Leben); reprinted as volume VII of *Nelson 1970–74*; English translation by N. Humphrey Palmer as *Nelson 1970* and *1971*.

1964 *System der philosophischen Rechtslehre und Politik* (Frankfurt am Main: Verlag Öffentliches Leben); reprint of *Nelson 1924*.

1970 *Progress and regress in philosophy, from Hume and Kant to Hegel and Fries*, vol. I, translation of part of *Nelson 1962*, edited posthumously by Julius Kraft (Oxford: Basil Blackwell).

1970–74   *Gesammelte Schriften, in neun Bänden,* edited by Paul Bernays, Willi Eichler, Arnold Gysin, Gustav Heckmann, Grete Henry-Hermann, Fritz von Hippel, Stephan Körner, Werner Kroebel and Gerhard Weisser, 9 vols. (Hamburg: Felix Meiner).

1971   *Progress and regress in philosophy, from Hume and Kant to Hegel and Fries,* vol. II, translation of part of *Nelson 1962,* edited posthumously by Julius Kraft (Oxford: Basil Blackwell).

Németi, István
See Andréka, Hajnal, and Istvan Németi.

Neurath, Otto
1979   *Wissenschaftliche Weltauffassung, Sozialismus und logischer Empirismus,* edited by Rainer Hegselmann (Frankfurt am Main: Suhrkamp Verlag).

Newman, James R.
1956   (ed.)   *The world of mathematics* (New York: Simon and Schuster).
See also Nagel, Ernest, and James R. Newman.
See also Nagel, Ernest, James R. Newman, Kurt Gödel and Jean-Yves Girard.

Niekus, N. H., H. van Riemsdijk and Anne S. Troelstra
1981   Bibliography of A. Heyting, *Nieuw archief voor wiskunde (3) 29,* 24–35.

Nöbeling, Georg
See Menger, Karl, and Georg Nöbeling.

Notcutt, Bernard
1934   A set of axioms for the theory of deduction, *Mind* (n.s.) *43,* 63–77.

Novikov, Petr S. (Новиков, Петр С.)
1955   On the algorithmic unsolvability of the word problem in group theory (Russian), *Trudy Matematicheskogo Instituta imeni V. A. Steklova 44,* 1–143.

Ostrowski, Alexander
1920   Über Dirichletsche Reihen und algebraische Differentialgleichungen, *Mathematische Zeitschrift 8,* 241–298.

Parikh, Rohit
1971   Existence and feasibility in arithmetic, *The journal of symbolic logic 36,* 494–508.

Parry, William T.
1933   Ein Axiomensystem für eine neue Art von Implikation (ana-
        lytische Implikation), *Ergebnisse eines mathematischen Kollo-
        quiums 4*, 5–6; reprinted in *Menger 1998*, 187–188.

Parsons, Charles
1977   What is the iterative concept of set?, in *Butts and Hintikka
        1977*, 335–367; reprinted in *Parsons 1983*, 268–297.
1979   Mathematical intuition, *Proceedings of the Aristotelian Society*
        (n.s.), *80*, 145–168.
1983   *Mathematics in philosophy: selected essays* (Ithaca: Cornell
        University Press).
1995a   Platonism and mathematical intuition in Kurt Gödel's thought,
        The bulletin of symbolic logic 1, 44–74.
1996   In memoriam: Hao Wang, 1921–1995, *The bulletin of symbolic
        logic 2*, 108–111.
1998   Hao Wang as philosopher and interpreter of Gödel, *Philosophia
        mathematica (3) 6*, 3–24.
1998a   Finitism and intuitive knowledge, in *Schirn 1998*, 249–270.
See also Grossi et alii.

Pears, David F.
1972   (ed.) *Bertrand Russell: a collection of critical essays* (Garden
        City, NY: Anchor-Doubleday).

Peckhaus, Volker
1990   *Hilbertprogramm und kritische Philosophie. Das Göttinger Mo-
        dell interdisziplinärer Zusammenarbeit zwischen Mathematik
        und Philosophie* (Göttingen: Vandenhoeck & Ruprecht).

Pilet, Paul-Emile
1977   Ferdinand Gonseth—sa vie, son oeuvre, *Dialectica 31*, 23–33.

Pitcher, Everett
1988   *A history of the second fifty years, American Mathematical
        Society, 1939–1988* (Providence, RI: American Mathematical
        Society).

Pitts, Walter
See Lettvin et alii.
See also McCulloch, Warren S., and Walter Pitts.

Polish Academy of Sciences
1961   *Infinitistic methods. Proceedings of the symposium on foundations of mathematics, Warsaw, 2–9 September 1959*, International Mathematical Union and the Mathematics Institute of the Polish Academy of Sciences (Oxford and New York: Pergamon; Warsaw: PWN).

Pongratz, Ludwig J.
1975   *Philosophie in Selbstdarstellungen* II (Hamburg: Meiner).

Popper, Karl Raimund
1947   Functional logic without axioms or primitive rules of inference, *Koninklijke Nederlandse Akademie van Wetenschappen, Proceedings, Series A: Mathematical Sciences 50*, 1214–1224; also *Indagationes mathematicae 9*, Fasc. 5, 561–571.
1955   Two autonomous axiom systems for the calculus of probabilities, *The British journal for the philosophy of science 6*, 51–57.
1959   The propensity interpretation of probability, *The British journal for the philosophy of science 10*, 25–42.
1974   Intellectual autobiography, in *Schilpp 1974*, 3–181.

Post, Emil L.
1921   Introduction to a general theory of elementary propositions, *American journal of mathematics 43*, 163–185; reprinted in *van Heijenoort 1967*, 264–283 and in *Post 1994*, 21–43.
1921a   On a simple class of deductive systems, *Bulletin of the American Mathematical Society 27*, 396–397; reprinted in *Post 1994*, 545.
1936   Finite combinatory processes—formulation 1, *The journal of symbolic logic 1*, 102–105; reprinted in *Davis 1965*, 288–291.
1941   Absolutely unsolvable problems and relatively undecidable propositions: account of an anticipation, in *Davis 1965*, 338–433; reprinted in *Post 1994*, 375–441.
1943   Formal reductions of the general combinatorial decision problem, *American journal of mathematics 65*, 197–215; reprinted in *Post 1994*, 442–460.
1944   Recursively enumerable sets of positive integers and their decision problems, *Bulletin of the American Mathematical Society 50*, 284–316; reprinted in *Post 1994*, 461–494.
1947   Recursive unsolvability of a problem of Thue, *The journal of symbolic logic 12*, 1–11; reprinted in *Post 1994*, 503–513.
1965   Absolutely unsolvable problems and relatively undecidable propositions—account of an anticipation, in *Davis 1965*, 340–433; reprinted in *Post 1994*, 375–441.

1994   *Solvability, provability, definability: the collected works of Emil L. Post*, edited by Martin Davis (Boston: Birkhäuser).

Poutsma, Hendrik
1929   *A grammar of late modern English, for the use of continental, especially Dutch, students*: Part I, The sentence, second half; the composite sentence, second edition (Groningen: P. Noordhoff).

Powell, William Chambers
1972   *Set theory with predication* (doctoral thesis, State University of New York at Buffalo).

Poznanski, E. I. J.
See Bar-Hillel et alii.

Presburger, Mojżesz
1930   Über die Vollständigkeit eines gewissen Systems der Arithmetik ganzer Zahlen, in welchem die Addition als einzige Operation hervortritt, *Sprawozdanie z I Kongresu matematyków krajów słowiańskich, Warszawa 1929* (Warsaw, 1930), 92–101, 395.

Proust, Joëlle
See Heinzman, Gerhard, and Joëlle Proust.

Pudlák, Pavel
1987   Improved bounds to the length of proofs of finitistic consistency statements, in *Simpson 1987*, 309–332.
1996   On the lengths of proofs of consistency, *Collegium logicum, Annals of the Kurt Gödel Society 2*, 65–86.

Putnam, Hilary
1960a Review of *Nagel and Newman 1958*, *Philosophy of science 27*, 205–207.
See also Benacerraf, Paul, and Hilary Putnam.
See also Davis, Martin, Hilary Putnam and Julia Robinson.

Pyenson, Lewis
1999   Neugebauer, Otto Eduard, in *Garraty and Carnes 1999*, vol. 16, pp. 302–303.

Quine, Willard van Orman
1933   A theorem in the calculus of classes, *The journal of the London Mathematical Society 8*, 89–95.
1937   New foundations for mathematical logic, *American mathematical monthly 44*, 70–80; reprinted in *Quine 1953*, 80–101.

1953   *From a logical point of view. 9 logico-philosophical essays* (Cambridge: Harvard University Press).

1955a   A proof procedure for quantification theory, *The journal of symbolic logic 20*, 141–149; reprinted in *Quine 1966a*, 196–204.

1963a   *Set theory and its logic* (Cambridge: Belknap Press of Harvard University Press).

1966a   *Selected logic papers* (New York: Random House).

1969   Revised edition of *Quine 1963a* (Cambridge: Belknap Press of Harvard University Press).

1980   Third printing of *Quine 1953* (with some revisions) (Cambridge: Harvard University Press).

1995   Enlarged edition of *Quine 1966a*.

Rabin, Michael O.
See Bar-Hillel et alii.

Rautenberg, Wolfgang
1968   Die Unabhängigkeit der Kontinuumhypothese—Problematik und Diskussion, *Mathematik in der Schule 6*, 18–37.

Ravaglia, Mark
2002   *Explicating finitist reasoning* (doctoral thesis, Carnegie Mellon University).

Reid, Constance
1970   *Hilbert* (New York: Springer).
See also Albers, Donald J., Gerald Alexanderson and Constance Reid.

Reidemeister, Kurt
See Hahn et alii.

Reinhardt, William N.
1974   Remarks on reflection principles, large cardinals, and elementary embeddings, in *Jech 1974*, 189–205.

1974a   Set existence principles of Shoenfield, Ackermann, and Powell, *Fundamenta mathematicae 84*, 5–34.

Remmel, Jeffrey B.
See Clote, Peter, and Jeffrey B. Remmel.

Robinson, Abraham
1961   Non-standard analysis, *Köninklijke Nederlandse Akademie van Wetenschappen, Proceedings, Series A: Mathematical Sciences 64*, 432–440; reprinted in *Robinson 1979*, vol. 2, 3–11.

1966   *Non-standard analysis* (Amsterdam: North-Holland).

1973   Nonstandard arithmetic and generic arithmetic, in *Suppes et alii 1973*, 137–154; reprinted in *Robinson 1979*, vol. 1, 280–297.

_1974   Second edition of _Robinson 1966_.

1975   Concerning progress in the philosophy of mathematics, in _Rose and Shepherdson 1975_, 41–52; reprinted in _Robinson 1979_, vol. 2, 556–567.

1979   _Selected papers_, edited by H. Jerome Keisler, Stephan Körner, W. A. J. Luxemburg and A. D. Young; vol. 1, _Model theory and algebra_; vol. 2, _Non-standard analysis and philosophy_, edited and with introduction by W. A. J. Luxemburg and Stephan Körner; vol. 3, _Aeronautics_ (New Haven, CT: Yale University Press).

See also Bar-Hillel et alii.

Robinson, Julia
See Davis, Martin, Hilary Putnam and Julia Robinson.

Rodríguez-Consuegra, Francisco A.
1995   (ed.) _Kurt Gödel: Unpublished philosophical essays_ (Basel, Boston and Berlin: Birkhäuser).

Rose, Harvey E.
1984   _Subrecursion: functions and hierarchies_ (Oxford: Clarendon Press).

Rose, Harvey E., and John C. Shepherdson
1975   (eds.) _Logic colloquium '73_ (Amsterdam: North-Holland).

Rosser, J. Barkley
1936   Extensions of some theorems of Gödel and Church, _The journal of symbolic logic 1_, 87–91; reprinted in _Davis 1965_, 230–235.
1937   Gödel theorems for non-constructive logics, _The journal of symbolic logic 2_, 129–137.

See also Kleene, Stephen C., and J. Barkley Rosser.

Rosser, J. Barkley and Atwell R. Turquette
1952   _Many-valued logics_ (Amsterdam: North-Holland).

Russell, Bertrand
1903   _The principles of mathematics_ (London: Allen and Unwin).
1918   The philosophy of logical atomism, _The monist 28, 4_, 495–527; reprinted in _Russell 1986_, 160–180.
1919   _Introduction to mathematical philosophy_ (London: Allen and Unwin; New York: Macmillan).
1920   Second edition of _Russell 1919_.
1924   Reprint of _Russell 1920_.
1937   Second edition of _Russell 1903_, with new introduction (New York: Norton).

1940   *An inquiry into meaning and truth* (London: Allen
        win).
1945   *A history of western philosophy* (New York: Simon and Sc.
        ter).
1958   Philosophical analysis (in German), *Zeitschrift für philosophi-
        sche Forschung 12, 1*, 3–16.
1968   *The autobiography of Bertrand Russell, 1914–1944* (London:
        Allen and Unwin; Boston: Little, Brown and Co.)
1986   *The collected papers of Bertrand Russell*, vol. 8, *The philos-
        ophy of logical atomism and other essays, 1914–1919*, edited
        by John G. Slater (London, Boston and Sydney: George Allen
        and Unwin).
See also Whitehead, Alfred North, and Bertrand Russell.

Saaty, Thomas L.
1965   (ed.) *Lectures on modern mathematics* (New York: Wiley).

Savage, C. Wade
1988   Herbert Feigl: 1902–1988, in *Fine and Leplin 1988*, vol. 2, 15–
        22; available online at http://www.mcps.umn.edu/fbiofr.htm
        (accessed 15 November 2001).

Saxer, Walter
1933   (ed.) *Verhandlungen des Internationalen Mathematiker-Kon-
        gresses Zürich 1932, II. Band: Sektionsvorträge* (Zürich and
        Leipzig: Orell Füssli Verlag).

Schelling, Friedrich Wilhelm von
1809   *Philosophische Untersuchungen über das Wesen der Menschli-
        chen Freiheit und die damit zusammenhängenden Gegenstände*
        (Reutlingen: J. N. Enßlin); reprinted in *Schelling 1927*, 223–
        308.
1856–61 *Friedrich Wilhelm Joseph von Schellings Sämmtliche Werke*,
        edited by Karl Friedrich August Schelling (Stuttgart and Augs-
        burg: Cotta).
1927   *Schellings Werke, Nach der Originalausgabe in neuer Unord-
        nung herausgegeben*, edited by Manfred Schröter from *Schelling
        1856–61*, vol. 4, *Schriften zur Philosophie der Freiheit, 1804–
        1815* (München: Beck).
1964   *Philosophische Untersuchungen über das Wesen der menschli-
        chen Freiheit und die damit zusammenhängenden Gegenstände*
        (Stuttgart: Reclam).

Schilpp, Paul A.
1941   (ed.) *The philosophy of Alfred North Whitehead*, Library of living philosophers, vol. 3 (Evanston: Northwestern University); second edition (New York: Tudor, 1951).
1944   (ed.) *The philosophy of Bertrand Russell*, Library of living philosophers, vol. 5 (Evanston: Northwestern University); third edition (New York: Tudor, 1951).
1949   (ed.) *Albert Einstein, philosopher-scientist*, Library of living philosophers, vol. 7 (Evanston: Library of living philosophers); third edition (New York: Tudor, 1951).
1955   (ed.) *Albert Einstein als Philosoph und Naturforscher*, German translation by Hans Hartmann (with additions) of *Schilpp 1949* (Stuttgart: Kohlhammer).
1959   The abdication of philosophy, *Proceedings and addresses of the American Philosophical Association 32*, 19–39; reprinted with biographical notes in *The Texas Quarterly 3, no. 2*, 1–20.
1963   (ed.) *The philosophy of Rudolf Carnap*, Library of living philosophers, vol. 11 (La Salle: Open Court; London: Cambridge University Press).
1974   (ed.) *The philosophy of Karl Popper*, Library of living philosophers, vol. 14 (La Salle: Open Court).

Schirn, Matthias
1998   (ed.) *The philosophy of mathematics today* (Oxford: Clarendon Press).

Schoenflies, Arthur
1927   Die Krisis in Cantors mathematischem Schaffen, *Acta mathematica 50*, 1–23.

Scholz, Arnold
See Hahn et alii.

Scholz, Heinrich
See Hasse, Helmut, and Heinrich Scholz.

Schönfinkel, Moses
See Bernays, Paul, and Moses Schönfinkel.

Schopenhauer, Arthur
1850   "Ueber die anscheinende Absichtlicheit im Schicksale des Einzelnen", Transcendente Spekulation über die anscheinende Absichtlichkeit im Schicksale des Einzelnen, in *Schopenhauer 1972*, vol. 5, 211–237.

1972 *Parerga und Paralipomena: kleine philosophische Schriften.* Volumes 5 and 6 of *Sämtliche Werke*, edited by Arthur Hübscher, third edition (Wiesbaden: Brockhaus).

Schütte, Kurt
1951 Beweistheoretische Erfassung der unendlichen Induktion in der Zahlentheorie, *Mathematische Annalen 122*, 369–389.
1960 *Beweistheorie*, Die Grundlehren der mathematischen Wissenschaften in Einzeldarstellungen mit besonderen Berücksichtigung der Anwendungsgebiete, Bd. 103 (Berlin: Springer); revised, expanded and translated into English by J. N. Crossley as *Schütte 1977*.
1977 *Proof theory* (Berlin: Springer).

Scientific American
1968 *Mathematics in the modern world* (San Francisco and London: W. H. Freeman).

Scott, Dana S.
1961 Measurable cardinals and constructible sets, *Bulletin de l'Académie polonaise des sciences, série des sciences mathématiques, astronomiques, et physiques 9*, 521–524.
1962 A decision method for validity of sentences in two variables, *The journal of symbolic logic 27*, 477.
1967 A proof of the independence of the continuum hypothesis, *Mathematical systems theory 1*, 89–111.
1971 (ed.) *Axiomatic set theory*, Proceedings of symposia in pure mathematics, vol. 13, part 1 (Providence, RI: American Mathematical Society).
See also Myhill, John, and Dana S. Scott.

Seelig, Carl
1952 *Albert Einstein und die Schweiz* (Zürich, Stuttgart and Vienna: Europa Verlag).
1960 *Albert Einstein: Leben und Werk eines Genies unserer Zeit* (Zürich: Bertelsmann Lesering–Europa Verlag).

Shanker, Stuart G.
1987 *Wittgenstein and the turning-point in the philosophy of mathematics* (Albany: State University of New York Press).
1988 Wittgenstein's remarks on the significance of Gödel's theorem, in *Shanker 1988a*, 155–256.
1988a (ed.) *Gödel's theorem in focus* (London: Croom Helm).

Shelah, Saharon
1984   Can you take Solovay's inaccessible away?, *Israel journal of mathematics 48*, 1–47.

Shepherdson, John C.
1953   Inner models for set theory, part III, *The journal of symbolic logic 18*, 145–167.
1959   Review of *Lévy 1957*, *The journal of symbolic logic 24*, 226.
See also Rose, Harvey E., and John C. Shepherdson.

Shoenfield, Joseph R.
1971   Unramified forcing, in *Scott 1971*, 357–381.

Siefkes, Dirk
1985   The work of J. Richard Büchi, in *Drucker 1985*, 176–189; reprinted in *Büchi 1990*, 7–17.
See also Lipshitz, Leonard, Dirk Siefkes and P. Young.

Sieg, Wilfried
1988   Hilbert's program sixty years later, *The journal of symbolic logic 53*, 338–348.
1994   Mechanical procedures and mathematical experience, in *George 1994*, 71–117.
1997   Step by recursive step: Church's analysis of effective calculability, *The bulletin of symbolic logic 3*, 154–180.
1999   Hilbert's programs: 1917–1922, *The bulletin of symbolic logic 5*, 1–44.
2002   Calculations by man and machine: conceptual analysis, in *Sieg, Sommer and Talcott 2002*, 390–409.

Sieg, Wilfried, Richard Sommer and Carolyn Talcott
2002   (eds.) *Reflections on the foundations of mathematics: Essays in honor of Solomon Feferman*, Lecture notes in logic 15 (Urbana: Association for Symbolic Logic; Natick: A K Peters).

Silver, Jack H.
1975   On the singular cardinals problem, in *Proceedings of the International Congress of Mathematicians, Vancouver 1974*, vol. I, 265–268.

Simpson, Stephen G.
1987   (ed.) *Logic and combinatorics: proceedings of the AMS-IMS-SIAM joint summer research conference held August 4–10, 1985*, Contemporary mathematics 65 (Providence: American Mathematical Society).

Sinaceur, Hourya

2000   Address at the Princeton University bicentennial conference on problems of mathematics (December 17–19, 1946), by Alfred Tarski, edited with additional material and an introduction by Hourya Sinaceur, *The bulletin of symbolic logic 6*, 1–44.

Sinaceur, Mohammed Allal

1991   (ed.) *Penser avec Aristote* (Toulouse: Erès).

Skolem, Thoralf

1920   Logisch-kombinatorische Untersuchungen über die Erfüllbarkeit oder Beweisbarkeit mathematischer Sätze nebst einem Theoreme über dichte Mengen, *Skrifter utgit av Videnskapsselskapet i Kristiania, I. Matematisk-naturvidenskapelig klasse*, no. 4, 1–36; reprinted in *Skolem 1970*, 103–136; partial English translation by Stefan Bauer-Mengelberg in *van Heijenoort 1967*, 252–263.

1923a   Einige Bemerkungen zur axiomatischen Begründung der Mengenlehre, *Matematikerkongressen i Helsingfors den 4–7 Juli 1922, Den femte skandinaviska matematikerkongressen, Redogörelse* (Helsinki: Akademiska Bokhandeln), 217–232; reprinted in *Skolem 1970*, 137–152; English translation by Stefan Bauer-Mengelberg in *van Heijenoort 1967*, 290–301.

1928   Über die mathematische Logik, *Norsk matematisk tidsskrift 10*, 125–142; reprinted in *Skolem 1970*, 189–206; English translation by Stefan Bauer-Mengelberg and Dagfinn Føllesdal in *van Heijenoort 1967*, 508–524.

1929   Über einige Grundlagenfragen der Mathematik, *Skrifter utgitt av Det Norske Videnskaps-Akademi i Oslo, I. Matematisk-naturvidenskapelig klasse*, no. 4, 1–49; reprinted in *Skolem 1970*, 227–273.

1931   Über einige Satzfunktionen in der Arithmetik, *Skrifter utgitt av Det Norske Videnskaps-Akademi i Oslo, I. Matematisk-naturvidenskapelig klasse*, no. 7, 1–28; reprinted in *Skolem 1970*, 281–306.

1932   Über die symmetrisch allgemeinen Lösungen im identischen Kalkül, *Skrifter utgitt av Det Norske Videnskaps-Akademi i Oslo, I. Matematisk-naturvidenskapelig klasse*, no. 6, 1–32; also appeared in *Fundamenta mathematicae 18*, 61–76; reprinted in *Skolem 1970*, 307–336.

1933   Ein kombinatorischer Satz mit Anwendung auf ein logisches Entscheidungsproblem, *Fundamenta mathematicae 20*, 254–261; reprinted in *Skolem 1970*, 337–344.

1933a   Über die Unmöglichkeit einer vollständigen Charakterisierung
        der Zahlenreihe mittels eines endlichen Axiomensystems,
        Norsk matematisk forenings skrifter, series 2, no. 10, 73–82;
        reprinted in *Skolem 1970*, 345–354.

1934   Über die Nicht-charakterisierbarkeit der Zahlenreihe mittels
        endlich oder abzählbar unendlich vieler Aussagen mit aus-
        schließlich Zahlenvariablen, *Fundamenta mathematicae 23*,
        150–161; reprinted in *Skolem 1970*, 355–366.

1936   *Utvalgte kapitler av den matematiske logikk.* (*Efter forleesninger
        ved Universitetet i Oslo i månedene januar, februar og mail
        1936*). (*Selected chapters of mathematical logic. Based on
        lectures at the University of Oslo in the months of January,
        February and May 1936*). Christian Michelsens Institutt for
        Videnskap og Åndsfrihet, Beretninger VI, 6. (Bergen: A. S.
        John Griegs).

1937   Über die Zurückführbarkeit einiger durch Rekursionen defi-
        nierter Relationen auf "arithmetische", *Acta litterarum ac sci-
        entiarum Regiae Universitatis Hungaricae Francisco-Josephi-
        nae. Sectio scientiarum mathematicarum 8*, 73–88; reprinted
        in *Skolem 1970*, 425–440.

1940   Einfacher Beweis der Unmöglichkeit eines allgemeinen Lösungs-
        verfahrens für arithmetische Probleme, *Det Kongelige Norske
        Videnskabers Selskabs Forhandlinger BD XIII*, 1–4; reprinted
        in *Skolem 1970*, 451–454.

1962   *Abstract set theory* (Notre Dame, IN: University of Notre
        Dame).

1970   *Selected works in logic*, edited by Jens Erik Fenstad (Oslo:
        Universitetsforlaget).

Smalheiser, Neil H.
2000   Walter Pitts, *Perspectives in biology and medicine 43:2*, 217–
        226.

Smiley, Timothy J.
See van Dalen, Dirk, Daniel Lascar and Timothy J. Smiley.

Snow, C. P.
See Bergamini et alii.

Solovay, Robert M.
1965 $2^{\aleph_0}$ can be anything it ought to be, in *Addison, Henkin and
        Tarski 1965*, 435.

1965b   The measure problem (abstract), *Notices of the American
        Mathematical Society 12*, 217.

1970   A model of set theory in which every set of reals is Lebesgue measurable, *Annals of mathematics* (2) *92*, 1–56.
1971   Real-valued measurable cardinals, in *Scott 1971*, 397–428.
1974   Strongly compact cardinals and the GCH, in *Henkin et alii 1974*, 365–372.
See also Lévy, Azriel, and Robert M. Solovay.

Sommer, Richard
See Sieg, Wilfried, Richard Sommer and Carolyn Talcott.

Specht, Minna, and Willi Eichler
1953   (eds.) *Leonard Nelson zum Gedächtnis* (Frankfurt: Verlag Öffentliches Leben).

Spector, Clifford
1962   Provably recursive functionals of analysis: a consistency proof of analysis by an extension of principles formulated in current intuitionistic mathematics, in *Dekker 1962*, 1–27.
See also Feferman, Solomon, and Clifford Spector.

Speiser, Andreas
1952   *Elemente der Philosophie und der Mathematik* (Basel: Birkhäuser).

Staal, J. Frits
See van Rootselaar, Bob, and J. Frits Staal.

Steinhaus, Hugo
See Mycielski, Jan, and Hugo Steinhaus.

Stöltzner, Michael
See Köhler et alii.

Suppes, Patrick
See Nagel, Ernest, Patrick Suppes and Alfred Tarski.

Suppes, Patrick, Leon Henkin, Athanase Joja and Gr. C. Moisil
1973   (eds.) *Logic, methodology and philosophy of science IV. Proceedings of the Fourth International Congress for Logic, Methodology and Philosophy of Science, Bucharest, 1971*, Studies in logic and the foundations of mathematics 74 (Amsterdam and London: North-Holland; New York: American Elsevier).

Surányi, János
1959   *Reduktionstheorie des Entscheidungsproblems im Prädikatenkalkül der ersten Stufe* (Budapest: Verlag der ungarischen Akademie der Wissenschaften).

Tait, William W.
1961   Nested recursion, *Mathematische Annalen 143*, 236–250.
1965b   The substitution method, *The journal of symbolic logic 30*, 175–192.
1981   Finitism, *The journal of philosophy 78*, 524–546.
2001   Gödel's unpublished papers on foundations of mathematics, *Philosophia mathematica (3) 9*, 87–126.
2002   Remarks on finitism, in *Sieg, Sommer and Talcott 2002*, 410–419.

Takeuti, Gaisi
1961   Remarks on Cantor's absolute, *Journal of the Mathematical Society of Japan 13*, 197–206.
1961a   Remarks on Cantor's absolute II, *Proceedings of the Japan Academy 37*, 437–439.
1961b   Axioms of infinity of set theory, *Journal of the Mathematical Society of Japan 13*, 220–233.
1967   Consistency proofs of subsystems of classical analysis, *Annals of mathematics (2) 86*, 299-348.
1975   *Proof theory* (Amsterdam: North-Holland).
1987   Second edition of *Takeuti 1975*.

Talcott, Carolyn
See Sieg, Wilfried, Richard Sommer and Carolyn Talcott.

Tarski, Alfred
1924   Sur les principes de l'arithmétique des nombres ordinaux (transfinis), *Polskie Towarzystwo Matematyczne (Cracow), Rocznik (=Annales de la Société Polonaise de Mathématique) 3*, 148–149.
1930   Über einige fundamentale Begriffe der Metamathematik, *Sprawozdania a posiedzeń Towarzystwa Naukowego Warszawskiego, wydział III, 23*, 22–29; reprinted in *Tarski 1986*, vol. 1, 311–320; English translation by Joseph H. Woodger, with revisions, in *Tarski 1956*, 30–37.
1930a   Fundamentale Begriffe der Methodologie der deduktiven Wissenschaften I, *Monatshefte für Mathematik und Physik 37*, 361–404; reprinted in *Tarski 1986*, vol. 1, 345–390; English translation by Joseph H. Woodger in *Tarski 1956*, 60–109.
1932   Der Wahrheitsbegriff in den Sprachen der deduktiven Disziplinen, *Anzeiger der Akademie der Wissenschaften in Wien 69*, 23–25; reprinted in *Tarski 1986*, vol. 1, 613–617.

1933   Einige Betrachtungen über die Begriffe der ω-Widerspruchs-freiheit und der ω-Vollständigkeit, *Monatshefte für Mathematik und Physik 40*, 97–112; reprinted in *Tarski 1986*, vol. 1, 619–636; English translation by Joseph H. Woodger in *Tarski 1956*, 279–295.

1933a   Pojecie prawdy w jezykach nauk dedukcyjnych (The concept of truth in the languages of deductive sciences), *Prace Towarzystwa Naukowego Warszawskiego, wydział III*, no. 34; translated into German by L. Blaustein as *Tarski 1935*; English translation by Joseph H. Woodger in *Tarski 1956*, 152–278.

1935   Der Wahrheitsbegriff in den formalisierten Sprachen, *Studia philosophica* (Lemberg) *1*, 261–405; German translation of *Tarski 1933a*; reprinted in *Tarski 1986*, vol. 2, 51–198.

1938a   Der Aussagenkalkül und die Topologie, *Fundamenta mathematicae 31*, 103–134; reprinted in *Tarski 1986*, vol. 2, 473–506.

1949a   Arithmetical classes and types of Boolean algebras. Preliminary report, *Bulletin of the American Mathematical Society 55*, 64.

1956   *Logic, semantics, metamathematics: papers from 1923 to 1938*, translated into English and edited by Joseph H. Woodger (Oxford: Clarendon Press).

1962   Some problems and results relevant to the foundations of set theory, in *Nagel, Suppes and Tarski 1962*, 125–135; reprinted in *Tarski 1986*, vol. 4, 113–125.

1983   Revised second edition of *Tarski 1956*, edited by John Corcoran (Indianapolis: Hackett).

1986   *Collected papers*, edited by Steven R. Givant and Ralph N. McKenzie (Basel, Boston and Stuttgart: Birkhäuser).

1999   Letters to Kurt Gödel, 1942–1947 (translated and edited by Jan Tarski), in *Woleński and Köhler 1999*, 261–273.

See also Henkin, Leon, and Alfred Tarski.

See also Henkin, Leon, J. Donald Monk and Alfred Tarski.

See also Keisler, H. Jerome, and Alfred Tarski.

See also Nagel, Ernest, Patrick Suppes and Alfred Tarski.

Tarski, Alfred, Andrzej Mostowski and Raphael M. Robinson
1953   *Undecidable theories* (Amsterdam: North-Holland).

Thiel, Christian
1980   Behmann, Heinrich, in *Mittelstraß 1980*, vol. 1, 274–275.
2002   Gödels Anteil am Streit über Behmanns Behandlung der Antinomien, in *Köhler et alii 2002*, vol. 2, 387–394.

Toulmin, Stephen E.
    See Janik, Allen S., and Stephen E. Toulmin.

Trautman, Andrzej
    1962   Conservation laws in general relativity, in *Witten 1962*, 169–198.

Troelstra, Anne S.
    1981   Arend Heyting and his contribution to intuitionism, *Nieuw archief voor wiskunde (3) 29*, 1–23.
    See also Niekus, N. H., H. van Riemsdijk and Anne S. Troelstra.

Troelstra, Anne S., and Dirk van Dalen
    1988   *Constructivism in mathematics*, vols. I and II (Amsterdam: North-Holland).

Tseng Ting-Ho
    1938   *La philosophie mathématique et la théorie des ensembles* (doctoral dissertation, University of Paris).

Tucker, J. V.
    1963   Constructivity and grammar, *Proceedings of the Aristotelian Society 63*, 45–66.

Turing, Alan Mathison
    1937   On computable numbers, with an application to the Entscheidungsproblem, *Proceedings of the London Mathematical Society (2) 42*, 230–265; correction, *ibid. 43*, 544–546; reprinted as *Turing 1965*.
    1939   Systems of logic based on ordinals, *Proceedings of the London Mathematical Society (2) 45*, 161–228; reprinted in *Davis 1965*, 155–222.
    1950   The word problem in semi-groups with cancellation, *Annals of mathematics (2) 52*, 491–505.
    1954   Solvable and unsolvable problems, *Science news 31* (London: Penguin), 7–23; reprinted in *Turing 1992*, 99–115 and *Turing 1992a*, 187–203.
    1965   Reprint of *Turing 1937*, in *Davis 1965*, 116–154.
    1992   *Pure mathematics*, edited by J. L. Britton, *Collected works of A. M. Turing*, vol. 1 (Amsterdam, London, New York and Tokyo: North-Holland).
    1992a   *Mechanical intelligence*, edited by D. C. Ince, *Collected works of A. M. Turing*, vol. 3 (Amsterdam, London, New York and Tokyo: North-Holland).

Turquette, Atwell R.
See Rosser, J. Barkley, and Atwell R. Turquette.

Ulam, Stanisław M.
1930   Zur Masstheorie in der allgemeinen Mengenlehre, *Fundamenta mathematicae 16*, 140–150; reprinted in *Ulam 1974*, 9–19.
1958   John von Neumann, 1903–1957, *Bulletin of the American Mathematical Society 3 (2)* (May supplement), 1–49.
1960   *A collection of mathematical problems* (New York: Interscience Publishers); reprinted in *Ulam 1974*, 505–670.
1964   Combinatorial analysis in infinite sets and some physical theories, *SIAM review 6 (4), 343–355*.
1974   *Sets, numbers, and universes, selected works*, edited by William A. Beyer, Jan Mycielski and Gian-Carlo Rota (Cambridge and London: The MIT Press).
1976   *Adventures of a mathematician* (New York: Charles Scribner's).
See also Everett, C. J., and Stanisław Ulam.

Unger, Georg
1975   (ed.) *Aufsätze zur Mengenlehre* (Wissenschaftliche Buchgesellschaft: Darmstadt).

Urmson, J. O.
1956   *Philosophical analysis; its development between the two world wars* (Oxford: Clarendon Press).

van Dalen, Dirk
See Troelstra, Anne S., and Dirk van Dalen.

van Dalen, Dirk, Daniel Lascar and Timothy J. Smiley
1982   (eds.) *Logic colloquium '80* (Amsterdam: North-Holland).

van Heijenoort, Jean
1967   (ed.) *From Frege to Gödel: a source book in mathematical logic, 1879–1931* (Cambridge: Harvard University Press).
1978   *With Trotsky in exile: from Prinkipo to Coyoacán* (Cambridge and London: Harvard University Press).
1985   *Selected essays* (Naples: Bibliopolis).
See also Dreben, Burton, and Jean van Heijenoort.

van Riemsdijk, H.
See Niekus, N. H., H. van Riemsdijk and Anne S. Troelstra.

van Rootselaar, Bob, and J. Frits Staal
    1968   (eds.) *Logic, methodology and philosophy of science III. Proceedings of the Third International Congress for Logic, Methodology and Philosophy of Science, Amsterdam 1967* (Amsterdam: North-Holland).

Vaught, Robert L.
    1956   On models of some strong set theories, *Bulletin of the American Mathematical Society 62*, 601–602.

Verein Ernst Mach
    1929   (eds.) *Wissenschaftliche Weltauffassung: der Wiener Kreis* (Vienna: A. Wolf); reprinted in *Neurath 1979*, 81–101.

Vesley, Richard E.
    See also Myhill, John, Akiko Kino and Richard E. Vesley.

Vihan, Premysl
    1995   The last months of Gerhard Gentzen in Prague, *Collegium logicum, Annals of the Kurt Gödel Society 1*, 1–7.

von Hayek, Friedrich August
    1960   *The constitution of liberty* (Chicago: University of Chicago Press).

von Juhos, Béla
    1930   *Das Problem der mathematischen Wahrscheinlichkeit* (Munich: Reinhardt).

von Neumann, John
    1923   Zur Einführung der transfiniten Zahlen, *Acta litterarum ac scientiarum Regiae Universitatis Hungaricae Francisco-Josephinae. Sectio scientiarum mathematicarum 1*, 199–208; reprinted in *von Neumann 1961*, 24–33; English translation by Jean van Heijenoort in *van Heijenoort 1967*, 347–354.
    1925   Eine Axiomatisierung der Mengenlehre, *Journal für die reine und angewandte Mathematik 154*, 219–240; correction, *ibid. 155*, 128; reprinted in *von Neumann 1961*, 34–56; English translation by Stefan Bauer-Mengelberg and Dagfinn Føllesdal in *van Heijenoort 1967*, 393–413.
    1926   Zur Prüferschen Theorie der idealen Zahlen, *Acta litterarum ac scientiarum Regiae Universitatis Hungaricae Francisco-Josephinae. Sectio scientarum mathematicarum 2*, 193–227; reprinted in *von Neumann 1961*, 69–103.
    1927   Zur Hilbertschen Beweistheorie, *Mathematische Zeitschrift 26*, 1–46; reprinted in *von Neumann 1961*, 256–300.

1928   Über die Definition durch transfinite Induktion und verwandte Fragen der allgemeinen Mengenlehre, *Mathematische Annalen 99*, 373–391; reprinted in *von Neumann 1961*, 320–338.

1928a   Die Axiomatisierung der Mengenlehre, *Mathematische Zeitschrift 27*, 669–752; reprinted in *von Neumann 1961*, 339–422.

1929   Über eine Widerspruchsfreiheitsfrage in der axiomatischen Mengenlehre, *Journal für die reine und angewandte Mathematik 160*, 227–241; reprinted in *von Neumann 1961*, 494–508.

1931   Die formalistische Grundlegung der Mathematik, *Erkenntnis 2*, 116–121; reprinted in *von Neumann 1961a*, 234–239; English translation by Erna Putnam and Gerald J. Massey in *Benacerraf and Putnam 1964*, 50–54.

1936   Continuous geometry, part I, mimeographed notes by L. Roy Wilcox, The Institute for Advanced Study, 1936; reproduced in *von Neumann 1960*.

1937   Continuous geometry, parts II and III, notes by L. Roy Wilcox, planographed by Edwards Brothers, Inc., Ann Arbor, Michigan; reproduced in *von Neumann 1960*.

1958   *The computer and the brain, the 1956 Silliman Lecture*, with preface by Klara von Neumann (New Haven: Yale University Press).

1960   *Continuous geometry* (Princeton: Princeton University Press).

1961   *Collected works*, vol. I: *logic, theory of sets and quantum mechanics*, edited by A. H. Taub (New York and Oxford: Pergamon).

1961a   *Collected works*, vol. II: *operators, ergodic theory and almost periodic functions in a group*, edited by A. H. Taub (New York and Oxford: Pergamon).

2000   Second edition of *von Neumann 1958*, with a foreword by Paul M. Churchland and Patricia S. Churchland (New Haven: Yale Nota Bene).

See also Hahn et alii.

von Neumann, John, and F. J. Murray
1936   On rings of operators, *Annals of mathematics (2) 37*, 116–229.

Wajsberg, Mordchaj
1938   Untersuchungen über den Aussagenkalkül von A. Heyting, *Wiadomości matematyczne 46*, 45–101.

Wald, Abraham
1936   Über die Produktionsgleichungen der ökonomischen Wertlehre (II. Mitteilung), *Ergebnisse eines mathematischen Kolloquiums 7*, 1–6; reprinted in *Menger 1998*, 319–324.

See also *Gödel 1933h*.

Wang, Hao
1970   A survey of Skolem's work in logic, in *Skolem 1970*, 17–52.
1974   *From mathematics to philosophy* (London: Routledge and Kegan Paul; New York: Humanities Press).
1974a   Metalogic, in *Encyclopedia Britannica*, fifteenth edition, vol. 11, 1078–1086 (Chicago: Encyclopedia Britannica, Inc.); partly incorporated into chapter V of *Wang 1974*, remainder reprinted in *Wang 1990*; 325–330.
1977   Large sets, in *Butts and Hintikka 1977*, 309–333.
1978   Kurt Gödel's intellectual development, *The mathematical intelligencer 1*, 182–184.
1981   Some facts about Kurt Gödel, *The journal of symbolic logic 46*, 653–659.
1987   *Reflections on Kurt Gödel* (Cambridge: MIT Press).
1990   *Computation, logic, philosophy, a collection of essays* (Beijing: Science Press; Dordrecht: Kluwer Academic Publishers).
1991   Kurt Gödel et certaines de ses conceptions philosophiques: l'esprit, la matière, et les mathèmatiques, in *Sinaceur 1991*, 441–451.
1996   *A logical journey. From Gödel to philosophy* (Cambridge: MIT Press).

Wang, Hao, and Bradford Dunham
1973   A recipe for Chinese typewriters, IBM technical report RC4521, 5 September 1973.
1976   Chinese version of *Wang and Dunham 1973*, *Dousou bimonthly 14* (March 1976), 56–62.

Wegel, Heinrich
1956   Axiomatische Mengenlehre ohne Elemente von Mengen, *Mathematische Annalen 131*, 435–462.

Weibel, Peter
See Köhler et alii.

Weinzierl, Ulrich
1982   *Carl Seelig, Schriftsteller* (Vienna: Löcker).

Weizmann, Karl Ludwig
1915   *Lehr- und Übungsbuch der Gabelsbergerschen Stenographie*, twelfth edition (Vienna: Manzsche k.u.k. Hof-, Verlags- und Universitäts-Buchhandlung).

Wette, Eduard
1974   Contradiction within pure number theory because of a system-internal 'consistency'-deduction, *International logic review. Rassegna internazionale di logica 9*, 51–62.

Weyl, Hermann
1918   *Das Kontinuum. Kritische Untersuchungen über die Grundlagen der Analysis* (Leipzig: Veit); translated into English by Stephen Pollard and Thomas Bole as *Weyl 1987*.
1927   *Philosophie der Mathematik und Naturwissenschaft, Handbuch der Philosophie* (Munich: Oldenbourg).
1932   Second edition of *Weyl 1918*.
1949   *Philosophy of mathematics and natural science*, revised and augmented English edition of *Weyl 1927* based on a translation by Olaf Helmer (Princeton: Princeton University Press).
1987   *The continuum. A critical examination of the foundation of analysis*, English translation of *Weyl 1918* (Kirksville, MO: Thomas Jefferson University Press).

Whitehead, Alfred North, and Bertrand Russell
1910   *Principia mathematica*, vol. 1 (Cambridge: Cambridge University Press).
1925   Second edition of *Whitehead and Russell 1910*.

Wiedemann, Hans-Rudolf
1989   *Briefe großer Naturforscher und Ärzte in Handschriften* (Lübeck: Verlag Graphische Werkstätten).

Wiener, Norbert
1948   *Cybernetics, or control and communication in the animal and the machine* (Cambridge, MA: Technology Press; New York: John Wiley and Sons; Paris: Hermann et cie).
1961   Second, enlarged edition of *Wiener 1948* (Cambridge, MA: The MIT Press).

Witten, Louis
1962   (ed.) *Gravitation: an introduction to current research* (New York and London: John Wiley & Sons).

Wittenberg, Alexander Israel
1953   Über adäquate Problemstellung in der mathematischen Grundlagenforschung, *Dialectica 7*, 232–254.
1954   Über adäquate Problemstellung in der mathematischen Grundlagenforschung. Eine Antwort, *Dialectica 8*, 152–157.
1957   *Vom Denken in Begriffen, Mathematik als Experiment des reinen Denkens* (Basel and Stuttgart: Birkhäuser).
1963   *Bildung und Mathematik, Mathematik als exemplarisches Gymnasialfach* (Stuttgart: Ernst Klett).

Wittgenstein, Ludwig

1921   Logische-philosophische Abhandlung, *Annalen der Naturphilosophie 14*, 185–262; reprinted as *Logische-philosophische Abhandlung. Tractatus logico-philosophicus* in *Wittgenstein 1984*, vol. 1, 7–85; English translation by C. K. Ogden in *Wittgenstein 1922*.

1922   *Tractatus logico-philosophicus*, English translation of *Wittgenstein 1921*, with corrected German text (New York: Harcourt, Brace; London: Kegan Paul).

1956   *Bemerkungen über die Grundlagen der Mathematik: Remarks on the foundations of mathematics*, edited by G. H. von Wright, R. Rhees and G. E. M. Anscombe, with translation by G. E. M. Anscombe (Oxford: Basil Blackwell).

1984   *Werkausgabe*, Suhrkamp Taschenbuch Wissenschaft 501–508 (Frankfurt: Suhrkamp).

Woleński, Jan and Eckehart Köhler

1999   *Alfred Tarski and the Vienna circle. Austro-Polish connections in logical empiricism* (Dordrecht: Kluwer Academic Publishers).

Wolkowski, Zbigniew W.

1993   (ed.) *First international symposium on Gödel's theorems* (Singapore: World Scientific Publishing Co.).

Young, P.
See Lipshitz, Leonard, Dirk Siefkes and P. Young.

Zach, Richard

1999   Completeness before Post: Bernays, Hilbert, and the development of propositional logic, *Bulletin of symbolic logic 5*, 331–366.

2001   *Hilbert's finitism: historical, philosophical and metamathematical perspectives* (doctoral dissertation, University of California, Berkeley); available online at http://www.ucalgary.ca/~rzach/publications/html (accessed 3 March 2002).

Zermelo, Ernst

1929   Über den Begriff der Definitheit in der Axiomatik, *Fundamenta mathematicae 14*, 339–344.

1930   Über Grenzzahlen und Mengenbereiche: Neue Untersuchungen über die Grundlagen der Mengenlehre, *Fundamenta mathematicae 16*, 29–47; English translation by Michael Hallett in *Ewald 1996*, vol. II, 1219–1233.

1932   Über Stufen der Quantifikation und die Logik des Unendlichen, *Jahresbericht der Deutschen Mathematiker-Vereinigung 41*, part 2, 85–88.

Ziegler, Renatus
   See Booth, David, and Renatus Ziegler.

Addenda and corrigenda to volumes I–III
of these *Collected Works*

The following errata to volumes I and II were noted after volume III went to press. (Line numbers refer to lines of text, not titles or running heads.)

Volume I – Further corrigenda

	Text as printed	*Correction*
256, 17	ι	η
257, 8	ι	η
408, 10	Hendrick	Hendrik
413, 17/18	Ame-/rican	Amer-/ican
414, 2	hypothesis. I,	hypothesis,
451, 5	-naturvidenskabelig	-naturvidenskapelig
451, 13	kabelig	kapelig

In the translation of *Gödel 1931* published in *van Heijenoort 1967*, the symbol α in Gödel's original text was changed to κ in three places. Van Heijenoort had recommended those changes to Gödel in a letter of 25 September 1965, and in his reply of 14 October Gödel agreed that the symbols in question were "probable misprints". They were declared unequivocally to be such in footnote 46a, p. 610 of *van Heijenoort 1967*, and that judgment was accepted by the editors of volume I of these *Collected Works*. Recently, however, Susumu Hayashi has noted that the putative corrections in fact imply a contradiction; for since there are countably many types in Gödel's system, if the set κ contains an element for which type elevation is possible, it must be an infinite set, and so cannot contain a largest element. Accordingly, κ should be changed back to α on p. 178, lines 25, 26 (first occurrence) and 27 (second occurrence), and on p. 179, lines 26, 27 (first occurrence) and 28 (second occurrence). The matter may be clarified by adding the following editorial remark after the first sentence of the penultimate paragraph on those pages:

⟦The case of finitely many formulas is trivial, so it suffices to consider the case where κ is generated by type elevation from a finite set α of formulas.⟧

Volume II – Further addenda and corrigenda

	Text as printed	*Correction*
126, fn. 17	399.	399.)
234, 30	n1	h1
328, 23	Hendrick	Hendrik
334, 20/21	Ame-/rican	Amer-/ican
335, 8	hypothesis. I,	hypothesis,
340, 2	*van Dalen et alii 1982*	*van Dalen, Lascar and Smiley 1982*, 95–128
359	[Missing cross-reference]	Lascar, Daniel See van Dalen, Dirk, Daniel Lascar and Timothy J. Smiley
378, 19	-naturvidenskabelig	-naturvidenskapelig
378, 27	kabelig	kapelig
379	[Missing cross-reference]	Smiley, Timothy J. See van Dalen, Dirk, Daniel Lascar and Timothy J. Smiley
384	[Missing cross-references]	van Dalen, Dirk See Boffa, Maurice, Dirk van Dalen and Kenneth McAloon See also Troelstra, Anne S., and Dirk van Dalen

Volume III – Corrigenda

	Text as printed	*Correction*
Front of dust jacket	Robert N. Solovay	Robert M. Solovay
Back of dust jacket	[One editor's name omitted]	Add Jean van Heijenoort to list of editors for volumes I and II.
xvi, 10	implications	philosophical implications
2, 13	*Neue Stadt-bibliothek*	*Wiener Stadt- und Landes-bibliothek*
10, fn. c	*Weizman*	*Weizmann*

	Text as printed	Correction
162, 15	arithemetic	arithmetic
112, 19	]ff	ff]
290, 6	implications	philosophical implications
304, title of *1951	implications	philosophical implications
422, running head	*Gödel *1970a*	*Gödel *1970b*
423, running head	*Some considerations leading to the probable conclusion*	*A proof from a highly plausible axiom*
484, 28–29	endlich-/er	endli-/cher
496, 28/29	Vort-/räge	Vor-/träge
497, 1/2	*phenom-/enology*	*pheno-/menology*
503, 6/7	Matemati-/scheskogo	Matemati-/cheskogo
506, 19/20	Braunsch-/weig	Braun-/schweig
506, 25/26	*Re-/ichenbach*	*Rei-/chenbach*
512, 17/18	En-/glish	Eng-/lish
513, 5	Uyeda, S.	Uyeda, Seizi
514, 5/6	En-/glish	Eng-/lish
514, 13	Pergammon	Pergamon
515, 4	Weizman	Weizmann
532, 8	Weizman,	Weizmann,
532, 9	*Weizman*	*Weizmann*

Since the appearance of volume III the editors have been advised of a change in the editorship of the volume cited as *Schimanovich et alii 199?* and have received final publication information. Accordingly the following reference should be added at p. 499, line 19:

Köhler, Eckehart, Bernd Buldt, Werner DePauli-Schimanovich, Carsten Klein, Michael Stöltzner and Peter Weibel
    2002   (eds.) *Wahrheit und Beweisbarkeit. Leben und Werk Kurt Gödels, Bd. 1: Dokumente und historische Analysen, Bd. 2: Kompendium zu Gödels Werk* (Vienna: Hölder-Pichler-Tempsky).

References to other editors in the reference list to volume III should be altered as indicated, and the other editors added to the references list in the appropriate places, with corresponding corrections to the Index:

	Text as printed	*Correction*
481, 29	See Schimanovich et alii.	See Köhler et alii.
508, 16-18	[Delete reference]	
514, 37	See Schimanovich et alii.	See Köhler et alii.

Addenda and corrigenda to *1961/?*

After the publication of *1961/?* in volume III, the editors received some comments and queries from Karl Schuhmann (for the transmission of which we thank Goran Sundholm). Consultation with Wilfried Sieg and further inspection of the shorthand source for that lecture resulted in a series of corrections and alternate readings. These are listed below.

Corrections

	Text as printed	*Correction*
374, 4	nötig sich	nötig, sich
374, 16	Lehren, er-	Lehren er-
374, 28	Allem	allem
374, 30	offentsichtlich	offensichtlich
376, 12	lang	lange
376, 14	entwickelt	entwickelte
376, 24	matik sondern	matik, sondern
387, 24	auftraten und	auftraten, und
376, 25	befriedigenden sowie	befriedigenden—sowie
376, 26	selbstverständlichen Weise	selbstverständlichen—Weise
376, 27	Zeitgeist und	Zeitgeist, und
376, 35	wesentliche	Wesentliche
378, 1	A	A
378, 12	folgenden	Folgenden
378, 26	d. h., also	d. h. also

	Text as printed	Correction
378, 31	A	*A*
380, 5/6	ste-/hende	ste-/henden
380, 30	verschließen und	verschließen, und
380, 31	ebendiesem	eben diesem
380, 42	vertrauen und	vertrauen, und
382, 7	Verfahren Einsichten	Verfahren, Einsichten
382, 40	Sinnen-	Sinnes-
384, 18	neue⟦re⟧	neue
384, 22	neuerer	neuer
384, 25	neuerer und	neuer, und
384, 25	unabhängiger Axiomen	unabhängigen, Axiome
386, 1	neues	Neues
386, 2	charakteristisches	Charakteristisches

Alternative readings

	Text as printed	Alternative	Explanation
374, 13	während es	währenddessen	"Des" and "dessen" differ only in the size of a loop. Here the loop is rather small to represent "dessen".
374, 13	anderseits	andererseits	"Ander" and "anderer" differ only in the length of a slanted line. Here that line is rather short to represent a doubled "r".
374, 21	anderseits	andererseits	See note for line 13.
374, 28	anderseits	andererseits	See note for line 13.
374, 37	Rückschlag	Rückschlagen	The "en" ending here is at the discretion of the transcriber.
376, 4	muß	müsse	The shorthand supports either interpretation.
376, 36	Indem	In dem	The shorthand supports either interpretation.

Text as printed	*Alternative*	*Explanation*
378, 20 Anderseits	Andererseits	See note to 374, 13.
380, 23 anderseits	andererseits	See note to 374, 13.
380, 32 anderseits	andererseits	See note to 374, 13.
382, 25 Mächte	Motive	The shorthand sign for "-acht" has a slight resemblance to that for "-tiv". The sign for "M" is curved; thus an "o" after it is a possibility, depending on the shape of the "M".
382, 28 sollte jedenfalls	sollte [es] jedenfalls	
382, 41 anderseits	andererseits	See note to 374, 13.
384, 1 Höhen	Höhe	The "n" here is at the discretion of the transcriber.
384, 19 durch [die] früher	durch die früher	There is no separate symbol for "die" in the shorthand here, but a subtle change in the symbol for "durch" would permit deletion of the brackets.
384, 27 Kantschen	kantischen	Technically the shorthand shows no "i", but in this case, it could be inferred.
378, 20 Anderseits	Andererseits	See note to 374, 13.

Two passages on p. 378 merit further discussion. The first was brought to our attention by Michael Recktenwald. It actually begins at the bottom of p. 376 and reads: "Außerdem verlieren durch diese hypothetische Auffassung der Mathematik viele Fragen die Form: gilt der Satz *A* oder gilt er nicht? Vollkommen im Sinn dann von willkürlichen Annahmen kann ich ja nicht erwarten,...". Recktenwald suggests the following changes (indicated in italics): "Außerdem verlieren durch diese hypothetische Auffassung der Mathematik viele Fragen *der* Form: Gilt der Satz A oder gilt er nicht? *vollkommen ihren* Sinn; *denn* von willkürlichen... erwarten,...". That necessitates a change in the translation to read: "In addition, through this hypothetical conception of mathematics, many questions *of the form*

"Does the proposition *A* hold or not?" *lose their* meaning altogether; *for* from completely arbitrary assumptions I can of course not expect...".

The second passage begins on line 22 of p. 378 as "daß ein Beweis für die Richtigkeit eines solchen Satzes,..., eine sichere Begründung für diesen Satz geben muß;...". The shorthand actually reads "Daß man ein Beweis...geben muß;...". The shorthand symbol for "ein" is quite different from that for "einen"; thus we must assume an error somewhere. Either "man" is superfluous and should have been deleted (as in the published version), or we must assume "ein" should be "einen". In that case, we would alter the passage to read "daß man ein[en] Beweis für die Richtigkeit eines solchen Satzes, wie der Darstellbarkeit von allen Zahlen als Summen [von] Quadraten, [d. h.] eine sichere Begründung für diesen Satz, geben muß;...". The translation would then read "that one must give a proof for the correctness of a proposition such as the representability of every number as a sum of four squares, i.e., a secure grounding for that proposition;...".[a]

Further Corrigenda to Volume I of these *Collected Works*

Page, line	*Correction*
134, 1	S(x) should be A(x).
148, 12	'Begriffe' should be 'Begriff'
148, 19	Insert comma after 'obigen'.
186, note 55, line 3	'in der' should be 'in den'.
186, note 56 line 4	'meiner Arbeit' should be 'meine Arbeit'.

[a]We thank Wilfried Sieg for his help and comments on these matters.

Index

Note: Pages in boldface type indicate a section devoted to a particular correspondent.

656   *Index*

Printed in the United States
By Bookmasters